MELZAK—**Mathematical Ideas, Modeling and Applications** (Volume II of Companion to Concrete Mathematics)
MEYER—**Introduction to Mathematical Fluid Dynamics**
MORSE—**Variational Analysis: Critical Extramals and Sturmian Extensions**
NAYFEH—**Perturbation Methods**
NAYFEH and MOOK—**Nonlinear Oscillations**
ODEN and REDDY—**An Introduction to the Mathematical Theory of Finite Elements**
PAGE—**Topological Uniform Structures**
PASSMAN—**The Algebraic Structure of Group Rings**
PRENTER—**Splines and Variational Methods**
RIBENBOIM—**Algebraic Numbers**
RICHTMYER and MORTON—**Difference Methods for Initial-Value Problems,** 2nd Edition
RIVLIN—**The Chebyshev Polynomials**
RUDIN—**Fourier Analysis on Groups**
SAMELSON—**An Introduction to Linear Algebra**
SIEGEL—**Topics in Complex Function Theory**
    Volume 1—Elliptic Functions and Uniformization Theory
    Volume 2—Automorphic Functions and Abelian Integrals
    Volume 3—Abelian Functions and Modular Functions of Several Variables
STAKGOLD—**Green's Functions and Boundary Value Problems**
STOKER—**Differential Geometry**
STOKER—**Nonlinear Vibrations in Mechanical and Electrical System**
STOKER—**Water Waves**
WHITHAM—**Linear and Nonlinear Waves**
WOUK—**A Course of Applied Functional Analysis**

APPLIED FUNCTIONAL ANALYSIS

APPLIED FUNCTIONAL ANALYSIS

JEAN-PIERRE AUBIN
University of Paris–Dauphine

Exercises by **BERNARD CORNET** and **JEAN-MICHEL LASRY**
Translated by **CAROLE LABROUSSE**

A WILEY-INTERSCIENCE PUBLICATION
JOHN WILEY & SONS, New York · Chichester · Brisbane · Toronto

Copyright © 1979 by John Wiley & Sons, Inc.

All rights reserved. Published simultaneously in Canada.

Reproduction or translation of any part of this work beyond that permitted by Sections 107 or 108 of the 1976 United States Copyright Act without the permission of the copyright owner is unlawful. Requests for permission or further information should be addressed to the Permissions Department, John Wiley & Sons, Inc.

Library of Congress Cataloging in Publication Data

Aubin, Jean Pierre.
   Applied functional analysis.

   (Pure and applied mathematics)
   "A Wiley-Interscience publication."
   Includes index.
   1. Functional analysis. 2. Hilbert space.
I. Title.

QA320.A913   515′.7   78-20896
ISBN 0-471-02149-0

Printed in the United States of America

10 9 8 7 6 5 4 3 2 1

To My Children

HENRI-JEAN, ANNE-LAURE, and MARC AUBIN

KAM-WAH TSUI

PREFACE

The purpose of this book is to present the important results of functional analysis and to study many applications: in boundary value problems for elliptic and parabolic partial differential equations, numerical analysis, convex analysis and nonlinear analysis, optimization theory, game theory and general equilibrium in economics, and finally systems theory.

In order to consider these applications and to keep the length of the text within reasonable limits, we have decided to work entirely in the framework of *Hilbert spaces*. This allows us to prove a large number of results in the simplest fashion, at the price, of course, of generality. In particular, the weak topology is *never* used in this book, and distributions are introduced in the framework of Sobolev spaces.

By this approach we hope to convince the reader of the advantages of an abstract treatment of problems of a more concrete nature and to interest him in applied mathematics.

This book serves as the textbook for a course of 100 hours intended for fourth and fifth year students of *Mathématiques de la Décision de l'Université de Paris–Dauphine*.

A good knowledge of the topological properties of metric spaces is required in order to assimilate easily the material presented. The book *Applied Abstract Analysis*, denoted by [AAA] in the text, contains all the results used in this book with the exception of Brouwer's fixed-point theorem in Chapter 15. Nevertheless with the preliminary notions that comprise the first half of Chapter 1, this book can be considered as self-contained.

An effort has been made to deal with the various applications as soon as possible. They are indicated by an asterisk (∗), allowing the reader anxious to know the fundamental results to omit them during a first reading.

Similarly an asterisk also marks the less essential results that will be used in subsequent study or that can be considered as exercises to be worked out.

The principal results of the text are grouped at the end of the book in the hope of providing a concise resumé of what is essential.

A terminological index allows the reader to locate the definitions of terms in the text.

Two hundred exercises provide the means of applying the results that have been established and of sorting out the ones that are the most often used.

I would like to continue to express my deep gratitude to Carole Labrousse for her excellent translation of this second book.

I also thank most sincerely Bernard Cornet and Jean-Michel Lasry for their collaboration and for having prepared the collection of exercises. Chapter 15 on nonlinear analysis owes much to Bernard Cornet.

JEAN-PIERRE AUBIN

Paris, France
March 1979

CONTENTS

Introduction: A Guide to the Reader, 1

Chapter 1. The Projection Theorem   3
1. Definition of a Hilbert Space, 3
2. Review of Continuous Linear and Bilinear Operators, 9
3. Extension of Continuous Linear and Bilinear Operators by Density, 12
4. The Best Approximation Theorem, 14
5. Orthogonal Projectors, 17
6. Closed Subspaces, Quotient Spaces, and Finite Products of Hilbert Spaces, 21
7. Orthogonal Bases for a Separable Hilbert Space, 22

Chapter 2. Theorems on Extension and Separation   27
1. Extension of Continuous Linear and Bilinear Operators, 28
2. A Density Criterion, 29
3. Separation Theorems, 30
4. A Separation Theorem in Finite Dimensional Spaces, 32
5. Support Functions, 32
*6. The Duality Theorem in Convex Optimization, 35
*7. Von Neumann's Minimax Theorem, 40
*8. Characterization of Pareto Optima, 46

Chapter 3. Dual Spaces and Transposed Operators   50
1. The Dual of a Hilbert Space, 51
2. Realization of the Dual of a Hilbert Space, 55
3. Transposition of Operators, 57
4. Transposition of Injective Operators, 58

5. Duals of Finite Products, Quotient Spaces, and Closed or Dense Subspaces, 61
6. The Theorem of Lax-Milgram, 66
*7. Variational Inequalities, 67
*8. Noncooperative Equilibria in n-Person Quadratic Games, 69

Chapter 4. The Banach Theorem and the Banach-Steinhaus Theorem — 72

1. Properties of Bounded Sets of Operators, 73
2. The Mean Ergodic Theorem, 78
3. The Banach Theorem, 80
4. The Closed Range Theorem, 84
5. Characterization of Left Invertible Operators, 85
6. Characterization of Right Invertible Operators, 88
*7. Quadratic Programming with Linear Constraints, 92

Chapter 5. Construction of Hilbert Spaces — 96

1. The Initial Scalar Product, 98
2. The Final Scalar Product, 100
3. Normal Subspaces of a Pivot Space, 101
4. Minimal and Maximal Domains of a Closed Family of Operators, 102
*5. Unbounded Operators and Their Adjoints, 106
*6. Completion of a Prehilbert Space Contained in a Hilbert Space, 109
*7. Hausdorff Completion, 109
*8. The Hilbert Sum of Hilbert Spaces, 111
*9. Interpolation Spaces, 113
*10. Reproducing Kernels of a Hilbert Space of Functions, 115

Chapter 6. L^2 Spaces and Convolution Operators — 120

1. The Space $L^2(\Omega)$ of Square Integrable Functions, 121
2. The Spaces $L^2(\Omega, a)$ with Weights, 123
3. The Space $\hat{H}^s$, 125
4. The Convolution Product for Functions of $\mathscr{C}_0(\mathbb{R}^n)$ and of $L^1(\mathbb{R}^n)$, 128
5. Convolution Operators, 131
6. Approximation by Convolution, 133

*7. Example. Convolution Powers for Characteristic Functions, 135
*8. Example. Convolution Product for Polynomials: Appell Polynomials, 138

Chapter 7. Sobolev Spaces of Functions of One Variable — 144

1. The Space $H_0^m(\Omega)$ and Its Dual $H^{-m}(\Omega)$, 145
2. Definition of Distributions, 146
3. Differentiation of Distributions, 148
4. Relations Between $H_0^m(\Omega)$ and $H_0^m(\mathbb{R})$, 152
5. The Sobolev Space $H^m(\Omega)$, 153
6. Relations Between $H^m(\Omega)$ and $H^m(\mathbb{R})$, 157
*7. Characterization of the Dual of $H^m(\Omega)$, 160
8. Trace Theorems, 161
9. Convolution of Distributions, 163

Chapter 8. Some Approximation Procedures in Spaces of Functions — 166

1. Approximation by Orthogonal Polynomials, 167
2. Legendre, Laguerre, and Hermite Polynomials, 169
3. Fourier Series, 172
4. Approximation by Step Functions, 174
5. Approximation by Piecewise Polynomial Functions, 176
6. Approximation in Sobolev Spaces, 181

Chapter 9. Sobolev Spaces of Functions of Several Variables and the Fourier Transform — 185

1. The Sobolev Spaces $H_0^m(\Omega)$, $H^m(\Omega)$, and $H^{-m}(\Omega)$, 186
2. The Fourier Transform of Infinitely Differentiable and Rapidly Decreasing Functions, 188
3. The Fourier Transform of Sobolev Spaces, 194
4. The Trace Theorem for the Spaces $H^m(\mathring{\mathbb{R}}_+^n)$, 197
5. The Trace Theorem for the Spaces $H^m(\Omega)$, 204
6. The Compactness Theorem, 207

Chapter 10. Elementary Convex Analysis — 209

1. Conjugate Functions, 210
2. The Gradient, 213
3. The Subdifferential, 216
4. Extremality Conditions for a Minimization Problem, 223

*5. Hamiltonian and Lagrangian of a Minimization Problem, 228

Chapter 11. Elementary Spectral Theory — 232

1. Compact Operators, 233
2. The Theory of Riesz-Fredholm, 236
3. Characterization of Compact Operators from a Hilbert Space to Another, 239
4. The Fredholm Alternative, 241
*5. Applications: Constructions of Intermediate Spaces, 244
*6. Application: Best Approximation Processes, 247
*7. Perturbation of an Isomorphism by a Compact Operator, 251

Chapter 12. Hilbert-Schmidt Operators and Tensor Products — 255

1. The Hilbert Space of Hilbert-Schmidt Operators, 256
2. The Fundamental Isomorphism Theorem, 264
3. Hilbert Tensor Products, 265
4. The Tensor Product of Continuous Linear Operators, 269
5. The Hilbert Tensor Product by l^2, 273
6. The Hilbert Tensor Product by L^2, 274
7. The Tensor Product by the Sobolev Space H^m, 277

Chapter 13. Boundary Value Problems — 280

1. The Formal Adjoint of an Operator and Green's Formula, 282
2. Green's Formula for Bilinear Forms, 291
3. Abstract Variational Boundary Value Problems, 297
4. Examples of Boundary Value Problems, 304
5. Approximation of Solutions to Neumann Problems, 310
6. Restriction and Extension of the Formal Adjoint, 315
7. Unilateral Boundary Value Problems, 319
8. Introduction to Calculus of Variations, 322
9. Introduction to Optimal Control, 326

Chapter 14. Differential-Operational Equations and Semigroups of Operators — 331

1. Semigroups of Operators, 332

2. Characterization of Infinitesimal Generators of Semigroups, 338
3. Differential-Operational Equations, 342
4. Boundary Value Problems for Parabolic Equations, 345
5. Systems Theory: Internal and External Representations, 347

Chapter 15. Introduction to Nonlinear Analysis   **355**

1. Upper Hemicontinuous Correspondences, 356
2. Existence Theorems for a Critical Point of a Correspondence, 358
3. Fixed Point Theorems for a Correspondence, 365
4. Properties of Normal and Tangent Cones, 367
5. Variational Inequalities, 370
6. Quasi-Variational Inequalities, 372
*7. Noncooperative Equilibria in n-Person Games, 375
*8. Walras Equilibria, 377
*9. The Perron–Frobenius Theorem for Correspondences, 379

Selection of Results   **382**

1. General Properties, 382
2. Properties of Continuous Linear Operators, 383
3. Separation Theorems and Polarity, 385
4. Construction of Hilbert Spaces, 386
5. Compact Operators, 387
6. Semigroup Operators, 389
7. The Green's Formula, 390
8. Convex Analysis and Optimization, 391
9. Nonlinear Analysis, 393
10. Minimax Inequalities, 396
11. Sobolev Spaces, Convolution, and Fourier Transform, 397

Exercises   **400**

Chapter 1, 400
Chapter 2, 404
Chapter 4, 405
Chapter 5, 406
Chapter 6, 406
Chapter 7, 407

Chapter 10, 408
Chapter 11, 411
Chapter 13, 414
Chapter 14, 415
Chapter 15, 418

Index 421

TABLE OF APPLICATIONS

Optimization   §2.6, 4.7, 10.4, 10.5.

Game Theory   §2.7, 2.8, 3.8, 15.7.

Mathematical Economics   §4.7, 15.8.

Calculus of Variations   §13.8.

Optimal Control   §13.9.

Numerical Analysis   §8.4, 8.5, 8.6, 11.6, 13.5.

Systems Theory   §14.5.

Boundary Value Problems   §13.4, 13.7, 14.4.

Sobolev Spaces Chapter 7 §9.1, 9.5, 9.6, 12.7.

APPLIED FUNCTIONAL
ANALYSIS

Introduction:
A Guide to the Reader

The contents of the following chapters are summarized here. However, since most of the terms are not precisely defined in this introduction, this description serves simply as a guide to place the results in their general context.

This book can be divided into three parts. The first, consisting of Chapters 1 to 5, presents the fundamental abstract results of linear functional analysis. After recalling some basic results, we devote Chapter 1 to the theory of projectors, which is the basis of those results *specific* to Hilbert spaces. The second chapter deals with separation theorems for convex sets. We give applications immediately: the existence of a Lagrange multiplier in optimization theory, the Von Neumann minimax theorem, the characterization of Pareto optima to n-person games. Duality and transposition of continuous linear operators are treated in Chapter 3. The theorems of Lax-Milgram and of Lions-Stampacchia on the existence of variational equations and inequalities are proved. The fundamental properties of continuous linear operators are studied in Chapter 4. Finally Chapter 5 is devoted to methods of construction of Hilbert spaces. In Chapter 5, Section 4, we establish the general method of constructing Sobolev spaces.

In Chapters 6, 7, 8, and 9, which comprise the second part of the text, we study concrete examples of Hilbert spaces (spaces of square summable functions and Sobolev spaces of *functions and distributions*) and operators fundamental in analysis (differential operators, convolution operators, and Fourier transforms). Chapter 8 deals with some methods of approximation of functions.

The contents of the last part of the book are disparate. Chapter 10 introduces the reader to the fundamental results of convex analysis, essential to optimization theory. This chapter can in fact be studied after reading Chapter 3. Chapters 11 and 12 are devoted to elementary spectral theory of compact and Hilbert-Schmidt operators as well as to a brief study of

Hilbert tensor products. Chapters 13 and 14 deal with fundamental aspects of the study of boundary value problems for elliptic and parabolic partial differential equations (including unilateral boundary value problems for elliptic equations). We specify the connection that exists between optimization theory (Chapter 10, Section 5) and boundary value problems in the framework of an introduction to the calculus of variations. We also prove the Pontryagin principle for characterizing the optimal control in a control problem determined by differential equations. The treatment of operational differential equations is continued in a brief presentation of the problem of the internal representation of linear systems. A short section is devoted to a statement of approximation methods for solutions of elliptic boundary value problems and constitutes an introduction to numerical analysis.

Finally, Chapter 15 is an introduction to nonlinear analysis. We prove numerous theorems on the existence of critical points and of fixed points for correspondences (i.e., multivalued mappings) as well as fundamental theorems on the existence of solutions of variational and quasi-variational inequalities which play an important role in the recent development of applied mathematics in mechanics, economics, the modern treatment of inventory management, and, in general, impulsive control). The text concludes with a treatment of the theory of noncooperative n-person games and with the proof of the existence of a price equilibrium in Walras economic models.

CHAPTER 1

The Projection Theorem

We begin by recalling the definition of a Hilbert space; we then show that the finite dimensional spaces (which are exactly those spaces in which the unit ball is compact) are Hilbert spaces and that the space l^2 of square summable sequences is a Hilbert space. Other examples of Hilbert spaces are given in Chapter 6 (the spaces of square summable functions) and in Chapters 7 and 9 (Sobolev spaces).

We also recall in Section 2 the elementary properties of continuous linear and bilinear operators.

Section 3 is devoted to the theorem of extension by density for continuous linear and bilinear operators, which will be used very often in this book.

The specific properties of Hilbert spaces all depend on the projection theorem: We show in Section 4 the existence of a best approximation projector on a closed convex subset M of a Hilbert space, that is, a mapping t that associates to every x its best approximation $tx \in M$ by elements of M.

We study in Section 5 the properties of these projectors when M is a cone and a vector space. In particular, when M is a closed subspace, the best approximation projector is a continuous linear operator with norm equal to one, which is called the *orthogonal projector*.

We continue by showing in Section 6 that every closed subspace M of a Hilbert space, that every quotient space V/M, and that every finite product of Hilbert spaces constitute a Hilbert space.

We conclude this chapter in Section 7 by showing how to associate with every base of a Hilbert space an orthonormal base and by studying the properties of these bases.

1. DEFINITION OF A HILBERT SPACE

DEFINITION 1

Let V be a real vector space. We call a semiscalar product on $V \times V$ (or, with an abuse of the language, on V) a mapping $\{x, y\} \in V \times V \mapsto ((x, y)) \in \mathbb{R}$ that *satisfies the following conditions:*

(1) $\begin{cases} \text{i.} & \left(\left(\sum_{i=1}^{n} \lambda^i x_i, y\right)\right) = \sum_{i=1}^{n} \lambda^i ((x_i, y)) \quad \text{(linearity with respect to x).} \\ \text{ii.} & \left(\left(x, \sum_{j=1}^{m} \mu^j y_j\right)\right) = \sum_{j=1}^{m} \mu^j ((x, y_j)) \quad \text{(linearity with respect to y).} \\ \text{iii.} & ((x, y)) = ((y, x)) \quad \text{(symmetry).} \\ \text{iv.} & ((x, x)) \geq 0 \quad \text{for all} \quad x \in V \quad \text{(positivity).} \end{cases}$

We call a nonseparated prehilbert space *the pair* $\{V, ((\,,\,))\}$ *formed by a vector space and a semiscalar product. A* scalar product *is a symmetric bilinear form for which*

(2) $\qquad \forall x \neq 0, \quad ((x, x)) > 0 \quad \text{(positive definite),}$

and we call a prehilbert space the pair $\{V, ((\,,\,))\}$ *where* $((\,,\,))$ *is a scalar product.* ▲

Remark 1

Condition (1)ii is redundant since it is a consequence of conditions (1)i and iii. Condition (2) clearly implies condition (1)iv. ∎

A scalar product defines a norm and consequently a distance on the space V. To verify this we need the Cauchy-Schwarz inequality.

PROPOSITION 1
If $((x, y))$ *is a semiscalar product, then*

(3) $\qquad |((x, y))| \leq \sqrt{((x, x))} \cdot \sqrt{((y, y))} \quad \forall x, y \in V.$ ▲

Proof. First, let us suppose that $((y, y)) > 0$. Developing $((y, y))((x + \lambda y, x + \lambda y)) \geq 0$, we obtain

$$((y, y))((x, x)) + \lambda^2 ((y, y))^2 + 2\lambda ((y, y))((x, y)) \geq 0.$$

Replacing λ by $-((x, y))/((y, y))$, we obtain the desired inequality. The same reasoning applies if $((x, x)) > 0$. Now if $((x, x)) = ((y, y)) = 0$, the development of $((x + \lambda y, x + \lambda y))$ leads to $2\lambda ((x, y)) \geq 0$. Taking $\lambda = \pm \frac{1}{2}$, we deduce that $((x, y)) = 0$. ∎

The first consequence of this inequality is that $\sqrt{((x, x))}$ is a seminorm.

PROPOSITION 2

If $((x,y))$ is a semiscalar product, then $\|x\| = \sqrt{((x,x))}$ is a seminorm. It is a norm if $((x,y))$ is a scalar product. ▲

Proof. The proof depends on showing that $\|x+y\| \leq \|x\| + \|y\|$. However, according to the Cauchy-Schwarz inequality,

$$\|x+y\|^2 = ((x+y, x+y)) = \|x\|^2 + 2((x,y)) + \|y\|^2$$
$$\leq \|x\|^2 + 2\|x\|\|y\| + \|y\|^2 = (\|x\| + \|y\|)^2. \quad \blacksquare$$

Consequently, a prehilbert space is a normed space and hence a metric space for the distance $d(x,y) = \|x - y\|$. ■

DEFINITION 2

We say that a prehilbert space is a Hilbert space if it is complete under the associated distance. Therefore every Hilbert space is a Banach space (normed and complete). ▲

Remark 2

The subspace $M = \{x \in V \text{ such that } \|x\| = 0\}$ is equal to the subspace of elements $x \in V$ such that $((x,y)) = 0$ for all $y \in V$. Indeed the Cauchy-Schwarz inequality implies that if $\|x\| = 0$, $((x,y)) = 0$ for all $y \in V$. In particular, a semiscalar product is a scalar product if and only if it is *not degenerate* (degenerate means that $((x,y)) = 0$ for all $y \in V$ implies that $x = 0$). ■

*Remark 3. Complex Vector Spaces

To simplify our work, we have chosen to restrict ourselves to the study of real vector spaces. The use of complex vector spaces is nevertheless convenient in the study of Fourier series (see Chapter 8, Section 3) and the Fourier transform (see Chapter 9, Section 2).

If V is a complex vector space, we call a semiscalar product a mapping from $V \times V$ to $\mathbb{C}$ such that

(4) $\begin{cases} \text{i.} & \left(\left(\sum_{i=1}^{n} \lambda^i x_i, y\right)\right) = \sum_{i=1}^{n} \lambda^i ((x_i, y)) \quad \text{(linearity with respect to } x\text{).} \\ \text{ii.} & \left(\left(x, \sum_{j=1}^{m} \mu^j y_j\right)\right) = \sum_{j=1}^{m} \overline{\mu^j}((x, y_j)) \quad \text{(antilinearity with respect to } y\text{).} \\ \text{iii.} & ((x,y)) = \overline{((y,x))} \quad \text{(Hermitian symmetry).} \\ \text{iv.} & ((x,x)) \geq 0 \quad \text{(positivity).} \end{cases}$

where $\bar{z}$ denotes the complex conjugate of a complex number z. This is a scalar product if in addition:

(5) $\quad ((x, x)) > 0 \quad$ for all $\quad x \neq 0 \quad$ (positive definite).

The Cauchy-Schwarz inequality remains true, implying that $\|x\| = \sqrt{((x,x))}$ is a seminorm or a norm according to whether $((x, y))$ is a semiscalar product or a scalar product. The extension of the results that follow to the case of complex vector spaces is left as an exercise. ∎

Example 1. Scalar Products on $\mathbb{R}^n$

If $V = \mathbb{R}^n$, the bilinear form

(6) $$(x, y) = \sum_{i=1}^{n} x_i y_i$$

is the *Euclidean scalar product*, which is the simplest of the scalar products.

If $A = (a^{ij})_{i,j=1,\ldots,n}$ is a matrix from $\mathbb{R}^n$ to $\mathbb{R}^n$, the form

(7) $$((x, y)) = \sum_{i,j=1}^{n} a^{ij} x_i y_j$$

is bilinear. It is symmetric if and only if A is symmetric, positive (respectively, positive definite) if and only if A is positive (respectively, positive definite) in the sense that

(8) $$\forall x \neq 0, \quad \sum_{i,j=1}^{m} a^{ij} x_i x_j \geq 0 \quad \text{(respectively, } > 0\text{)}$$

We can, therefore, associate to every positive symmetric matrix a semiscalar product and to every positive definite symmetric matrix a scalar product. Conversely, every semiscalar product on $\mathbb{R}^n$ can be obtained as follows: Given $((x, y))$ a bilinear form on $\mathbb{R}^n \times \mathbb{R}^n$ and $\{e^i\}_{i=1,\ldots,n}$ the canonical base for $\mathbb{R}^n$, set $a^{ij} = ((e^i, e^j))$. The bilinearity implies that

$$((x, y)) = \left(\left(\sum_{i=1}^{n} x_i e^i, \sum_{j=1}^{n} y_j e^j\right)\right) = \sum_{i,j=1}^{n} a^{ij} x_i y_j$$

$$= \sum_{j=1}^{n} \left(\sum_{i=1}^{n} a^{ij} x_i\right) y_j = (Ax, y)$$

where $A = (a^{ij})_{i,j=1,\ldots,n}$ is the matrix of the a^{ij}'s. [In fact, we shall see that this property extends to general Hilbert spaces (see Chapter 3).] ∎

We know that $\mathbb{R}^n$ with the norm $\|x\|_\infty = \max_{i=1,\ldots,n} |x_i|$ is a *complete* space. Moreover, the Euclidean norm $|x| = \sqrt{(x, x)} = (\sum_{i=1}^{n} |x_i|)^{1/2}$ is equivalent to the preceding norm, since $\|x\|_\infty \leq |x| \leq \sqrt{n}\|x\|_\infty$ for all $x \in \mathbb{R}^n$. Therefore $\mathbb{R}^n$ is a *Hilbert space*.

We also know that the unit ball in $\mathbb{R}^n$ is compact. The finite dimensional spaces are the *only* Hilbert spaces that possess this property.

THEOREM 1 (RIESZ)
If the unit ball of a Hilbert space V is compact, the space is finite dimensional. ▲

Proof. Let B be the unit ball that we suppose to be compact. It can be covered by a finite number of balls B_i of radius $\frac{1}{2}$ and with center $x_i (i = 1, \ldots, n)$. Consider the subspace F generated by the points x_i. We shall show that $F = V$. If this were not the case, there would exist $x_0 \in V$ that does not belong to F. Since F is closed (because it is complete), $\alpha = d(x_0, F) = \inf_{y \in F} \|x_0 - y\| > 0$. Hence we deduce the existence of $y_0 \in F$ such that

$$(9) \qquad \alpha = d(x_0, F) \leq \|x_0 - y_0\| \leq d(x_0, F) + \frac{\alpha}{2} = \frac{3\alpha}{2}.$$

Consider, then,

$$z_0 = \frac{x_0 - y_0}{\|x_0 - y_0\|} \in B.$$

It belongs to one of the balls B_i; therefore

$$(10) \qquad \|z_0 - x_i\| = \frac{1}{\|x_0 - y_0\|} \|x_0 - y_0 - x_i\|x_0 - y_0\|\| \leq \tfrac{1}{2}.$$

We also introduce the element $y = x_0 - \|x_0 - y_0\|(z_0 - x_i) = y_0 + \|x_0 - y_0\| x_i$, which belongs to F, since y_0 and x_i belong to F. We therefore deduce from this that

$$(11) \qquad \alpha \leq \|x_0 - y\| = \|x_0 - y_0\|\|z_0 - x_i\| \leq \tfrac{1}{2}\|x_0 - y_0\| \leq \frac{3\alpha}{4},$$

which is impossible. ∎

Example 2

The l^2 Spaces of Square Summable Sequences. We denote by $l^2 = l^2(\mathbb{N})$ the space of square summable sequences, that is, the space of those sequences $x = \{x_k\}_k$ of elements $x_k \in \mathbb{R}$ such that

$$(12) \qquad \|x\|_2 = \left(\sum_{k=0}^{\infty} |x_k|^2\right)^{1/2} < +\infty.$$

Let us associate with every sequence $x \in l^2$ the sequence $x^{(n)}$ defined by

(13) $$x^{(n)} = \{x_0, x_1, \ldots, x_n, 0, 0, \ldots\}.$$

Then we can define the increasing sequence of positive real numbers $a_n = \sum_{k=0}^{n} |x_k y_k|$. For this sequence to converge, it is sufficient that it be bounded above. But this is indeed the case, since, applying the Cauchy-Schwarz inequality for the Euclidean scalar product, we obtain

$$a_n \leq \left(\sum_{k=0}^{n} |x_k|\right)^{1/2} \left(\sum_{k=0}^{n} |y_k|\right)^{1/2} = \|x^{(n)}\|_2 \|y^{(n)}\|_2 \leq \|x\|_2 \|y\|_2 < +\infty$$

when x and y belong to l^2. Therefore the sequence a_n converges to $a = \sum_{k=0}^{\infty} |x_k y_k|$. This implies that the series $\sum_{k=0}^{\infty} x_k y_k$ converges and that

(14) $$((x, y)) = \sum_{k=0}^{\infty} x_k y_k \quad \text{is a scalar product.} \qquad \blacksquare$$

THEOREM 2
The space l^2 with the scalar product $((x, y))$ defined by (14) is a Hilbert space.
▲

Proof. Consider a Cauchy sequence of elements x^m of l^2. For every $\varepsilon > 0$, there exists $N(\varepsilon)$ such that

(15) $$\sum_{k=0}^{\infty} |x_k^m - x_k^p|^2 = \|x^m - x^p\|_2^2 \leq \varepsilon^2 \quad \text{when} \quad m, p \geq N(\varepsilon).$$

This inequality implies that for all $k \in \mathbb{N}, |x_k^m - x_k^p| \leq \varepsilon$ when $m, p \geq N(\varepsilon)$. The sequence $\{x_k^m\}_m$ is therefore a Cauchy sequence of real numbers and hence converges to x_k. Let us denote by $x = \{x_0, x_1, \ldots, x_k, \ldots\}$ the sequence so constructed.

The increasing sequence of positive numbers $a_n = \sum_{k=0}^{n} |x_k - x_k^m|^2$ is bounded above by

(16) $$\tfrac{1}{2} a_n \leq \sum_{k=0}^{n} |x_k - x_k^p|^2 + \sum_{k=0}^{n} |x_k^p - x_k^m|^2 \leq \sum_{k=0}^{n} |x_k - x_k^p|^2 + \varepsilon^2$$

when m and $p \geq N(\varepsilon)$. Letting p approach infinity, we deduce that

(17) $$\tfrac{1}{2} a_n \leq \varepsilon^2.$$

This implies that the sequence of the a_n's converges to a number a and that

(18) $$\tfrac{1}{2} a = \sum_{k=0}^{\infty} |x_k - x_k^m|^2 \leq \varepsilon^2 \quad \text{when} \quad m \geq N(\varepsilon).$$

Consequently, x belongs to l^2 and $\|x - x^m\|_2 \leq \varepsilon\sqrt{2}$ when $m \geq N(\varepsilon)$; that is, the Cauchy sequence of elements $x^m \in l^2$ converges to $x \in l^2$. $\blacksquare$

2. REVIEW OF CONTINUOUS LINEAR AND BILINEAR OPERATORS

We recall here the characterization of continuous bilinear mappings.

PROPOSITION 1
Let V_1, V_2, and F be three Hilbert spaces and A a bilinear mapping from $V_1 \times V_2$ to F. The following conditions are equivalent:

(1) $\begin{cases} \text{i.} & A \text{ is continuous.} \\ \text{ii.} & A \text{ is continuous at the point } \{0,0\}. \\ \text{iii.} & \exists M \quad \text{such that} \quad \|A(x_1, x_2)\|_F \leq M \|x_1\|_1 \|x_2\|_2. \end{cases}$ ▲

Proof. **a.** It is clear that (1)i implies (1)ii. Let us show that this condition implies (1)iii. Since A is continuous at the origin, there exists η such that the inequalities $\|y_i\|_i \leq \eta (i = 1, 2)$ imply $\|A(y_1, y_2)\|_F = \|A(y_1, y_2) - A(0, 0)\|_F \leq 1$. If $x = \{x_1, x_2\}$ is an arbitrary element of $V_1 \times V_2$, then

$$y_i = \frac{\eta x_i}{\|x_i\|_i} \quad (i = 1, 2)$$

satisfies $\|y_i\|_i \leq \eta$ and, consequently,

$$\|A(y_1, y_2)\|_F = \frac{\eta^2}{\|x_1\|_1 \|x_2\|_2} \|A(x_1, x_2)\|_F \leq 1.$$

Hence we have established (1)iii with $M = 1/\eta^2$.

b. Let us show that (1)iii implies (1)i. To verify the continuity of A at the point $x = \{x_1, x_2\}$, we write that

$$\|A(x_1, x_2) - A(y_1, y_2)\|_F \leq \|A(x_1 - y_1, x_2)\|_F + \|A(y_1, x_2 - y_2)\|_F$$
$$\leq M \|x_2\|_2 \|x_1 - y_1\|_1 + M \|y_1\|_1 \|x_2 - y_2\|_2.$$

Thus if $\|x_i - y_i\|_i \leq \eta$, then $\|y_i\|_i \leq \eta + \|x_i\|_i \leq \eta + \|x\|$ and $\|A(x_1, x_2) - A(y_1, y_2)\|_F \leq 2(\eta + \|x\|)\eta$. Then for a given ε, choosing η such that $2(\eta + \|x\|)\eta \leq \varepsilon$, which is always possible, we deduce that $\|A(x_1, x_2) - A(y_1, y_2)\|_F \leq \varepsilon$ when $\|x_i - y_i\|_i \leq \eta (i = 1, 2)$. ■

Continuous bilinear mappings are not uniformly continuous. However, Proposition 1 implies the following.

PROPOSITION 2
Let A be a continuous mapping from a Hilbert space V to a Hilbert space F. Then A is uniformly continuous if and only if it is continuous at the origin.

This is equivalent to saying that

$$\|A\| = \sup_{\substack{x \in V \\ x \neq 0}} \frac{\|Ax\|_F}{\|x\|_V} < +\infty. \tag{2}$$

Proof. (Left as an exercise.)

We denote by $\mathscr{L}(V, F)$ the vector space of continuous linear mappings (or operators) from V to F. We recall that $\|A\|$ defined in (2) is a norm. ∎

*PROPOSITION 3
If F is a Hilbert space and V is a prehilbert space, the space $\mathscr{L}(V, F)$ is a Banach space.

Proof. Consider a Cauchy sequence of elements $A_n \in \mathscr{L}(V, F)$; then $\|A_n - A_p\| \leq \varepsilon$ when $n, p \geq N(\varepsilon)$. For every $x \in V$ the inequality $\|A_n x - A_p x\| \leq \|A_n - A_p\| \|x\| \leq \varepsilon \|x\|$ for $n, p \geq N(\varepsilon)$ implies that the sequence of elements $A_n x$ is a Cauchy sequence in F that converges to an element Ax since F is complete. Since the equalities $A_n(\lambda x + \mu y) = \lambda A_n x + \mu A_n y$ imply, by taking the limit, the equalities $A(\lambda x + \mu y) = \lambda Ax + \mu Ay$, the mapping $x \mapsto Ax$ thereby defined is linear. It is also continuous, for, since the sequence of norms $\|A_n\|$ is a Cauchy sequence, it is bounded by a constant $c > 0$; therefore the inequalities $\|A_n x\| \leq \|A_n\| \|x\| \leq c \|x\|$ imply, by taking the limit, that $\|Ax\| \leq c \|x\|$—that is, that A is continuous.

Finally by writing that for $n, p \geq N(\varepsilon)$, $\|A_n x - Ax\| \leq \|A_n x - A_p x\| + \|A_p x - Ax\| \leq \varepsilon \|x\| + \|A_p x - Ax\|$ and letting p approach infinity, we conclude that $\|A_n x - Ax\| \leq \varepsilon \|x\|$ if $n \geq N(\varepsilon)$; that is, $\|A_n - A\| \leq \varepsilon$ if $n \geq N(\varepsilon)$. Therefore the sequence of the A_n's $\in \mathscr{L}(V, F)$ does indeed converge to $A \in \mathscr{L}(V, F)$. ∎

DEFINITION 1
We call the topological dual space of a prehilbert space V the space $V^* = \mathscr{L}(V, \mathbb{R})$ of continuous linear forms on V.

PROPOSITION 4
The dual $V^* = \mathscr{L}(V, \mathbb{R})$ of a prehilbert space V is a Banach space for the norm

$$\|f\|_* = \sup_{x \in V} \frac{|f(x)|}{\|x\|}. \tag{3}$$

Proof. Indeed, $F = \mathbb{R}$ is a Hilbert space. ∎

Remark 1

We see in Section 1 of Chapter 3 that the dual V^* of a Hilbert space is in fact a *Hilbert* space.

If V and F are infinite dimensional spaces, the space $\mathscr{L}(V,F)$ is not a Hilbert space. We construct in Section 1 of Chapter 12 subspaces of $\mathscr{L}(V,F)$ that can be given a Hilbert space structure (the space of Hilbert-Schmidt operators). ∎

*****Remark 2**

Similarly, *the space $\mathscr{L}(V_1, V_2; F)$ of continuous bilinear mappings from $V_1 \times V_2$ to F is a normed vector space for the norm*

$$(4) \qquad \|A\| = \sup_{\substack{x_1 \in V_1 \\ x_2 \in V_2}} \frac{\|A(x_1, x_2)\|_F}{\|x_1\|_1 \|x_2\|_2}$$

which is complete when F is complete. (The proof is left as an exercise.) It is in fact a space isometric to a space of continuous linear mappings, namely, $\mathscr{L}(V_1, \mathscr{L}(V_2, F))$ and $\mathscr{L}(V_2, \mathscr{L}(V_1, F))$. It will be verified in an exercise that the isomorphism from $\mathscr{L}(V_1, V_2; F)$ onto $\mathscr{L}(V_1, \mathscr{L}(V_2, F))$ is the operator J, which associates to every $A \in \mathscr{L}(V_1, V_2; F)$ the linear operator $JA \in \mathscr{L}(V_1, \mathscr{L}(V_2, F))$ defined by

$$(5) \qquad JA(x_1) : x_2 \in V_2 \mapsto JA(x_1)(x_2) = A(x_1, x_2). \qquad \blacksquare$$

Let $A \in \mathscr{L}(V, F)$ be a continuous linear *bijective* mapping from V onto F. We know that its inverse A^{-1} is linear. [We shall show that A^{-1} is continuous (Theorem 4.3.1 of Banach).]

We say that an operator $A \in \mathscr{L}(V, F)$ that is bijective and *bicontinuous* (i.e., for which A and A^{-1} are continuous) is an *isomorphism*. We say that A is an *isometry* if $\|Ax\|_F = \|x\|_V$ for all $x \in V$. It is important to observe that an isometry, which is always *injective*, is not assumed to be *surjective*. If an isometry A is surjective, its inverse is clearly an isometry. It is evident that an isometry A maps every complete subset onto a complete subset and, consequently, that the *image by an isometry A is a complete subspace of F, and therefore closed*.

An operator $A \in \mathscr{L}(V, F)$ is a *Hilbert isometry* if in addition $((Ax, Ay))_F = ((x, y))_V$.

Let us call attention to the fact that for infinite dimensional spaces, the *image* $\operatorname{Im} A = A(V)$ by an operator $A \in \mathscr{L}(V, F)$ is *not necessarily closed*.

However, the kernel $\operatorname{Ker} A = A^{-1}\{0\}$ of a continuous linear operator

A is a *closed vector subspace*. This property characterizes continuous linear forms.

*PROPOSITION 5
A linear form p is continuous if and only if its kernel is closed. ▲

Proof. Let us suppose, therefore, that $H = \text{Ker } p$ is closed and let us choose $\varepsilon > 0$. We must show that there exists a ball of radius $\eta > 0$ such that $|p(x)| \leq \varepsilon$ for all $x \in B(\eta)$.

Let x_0 be such that $p(x_0) = \varepsilon$. Since H is closed, $H + x_0$ is also closed and $0 \notin H + x_0$. Hence there exists a ball $B(\eta)$ such that $B(\eta) \cap (H + x_0) = \emptyset$. Thus $|p(x)| \leq \varepsilon$ when $x \in B(\eta)$. If this were not the case, there would exist $x \in B(\eta)$ such that $|p(x)| > \varepsilon$. Set $y = \varepsilon x / |p(x)|$. Then $y \in B(\eta)$ since

$$\|y\| = \frac{\varepsilon \|x\|}{|p(x)|} \leq \|x\| \leq \eta;$$

moreover, $y \in H + x_0$, since

$$p(y - x_0) = \varepsilon \frac{p(x)}{p(x)} - p(x_0) = \varepsilon - \varepsilon = 0.$$

Therefore $y \in B(\eta) \cap (H + x_0)$, which is impossible. ∎

In particular, this proposition implies that *every linear form p on a finite dimensional space is continuous*, since its kernel, which is a finite dimensional space, is complete and therefore closed.

Consequently, *every linear mapping from a finite dimensional space to another is continuous*, since it can be expressed as a product of linear forms that are continuous. However, for *infinite dimensional spaces linear mappings are not necessarily continuous*.

3. EXTENSION OF CONTINUOUS LINEAR AND BILINEAR OPERATORS BY DENSITY

The following theorem plays a crucial role in what follows and is very frequently used.

THEOREM 1 (EXTENSION BY DENSITY)
Let V_1, V_2, and F be three Hilbert spaces, D_1 and D_2 two subspaces that are dense in V_1 and V_2, respectively, and $A \in \mathscr{L}(D_1, D_2; F)$ a continuous bilinear mapping from $D_1 \times D_2$ to F. Then there exists a unique continuous bilinear mapping $\bar{A} \in \mathscr{L}(V_1, V_2; F)$ that is an extension of A and such that $\|\bar{A}\| = \|A\|$. ▲

x^n of $D_1 \times D_2$. Then

$$\|\bar{A}(x) - \bar{\bar{A}}(x)\|_F \leq \|\bar{A}(x) - A(x_n)\|_F + \|A(x_n) - \bar{\bar{A}}(x)\|_F.$$

The right-hand side of this inequality tends to zero, and we conclude that $\bar{A}(x) = \bar{\bar{A}}(x)$. ∎

We can, therefore, express this theorem in the following form.

THEOREM 2
Let V_1, V_2, and F be three Hilbert spaces and let D_1 and D_2 be two subspaces that are dense in V_1 and V_2, respectively. Then the spaces $\mathscr{L}(D_1, D_2; F)$ and $\mathscr{L}(V_1, V_2; F)$ are isometric. ▲

Proof. Indeed, the preceding mapping $A \mapsto \bar{A}$ is clearly the desired isometry (whose inverse is the restriction to $D_1 \times D_2$ of an operator defined on $V_1 \times V_2$). ∎

In what follows, therefore, we identify the spaces $\mathscr{L}(D_1, D_2; F)$ and $\mathscr{L}(V_1, V_2; F)$ by *identifying each continuous bilinear operator A on $D_1 \times D_2$ with its unique extension $\bar{A}$, which we shall henceforth denote by A.*

With linear operators in particular this theorem becomes the following.

THEOREM 3
Let V and F be two Hilbert spaces and D a dense subspace of V. Then $\mathscr{L}(D, F) = \mathscr{L}(V, F)$ in the sense that every continuous linear operator on D has a unique extension to a continuous linear operator on V. ▲

Remark 1

We have seen (Theorem 3.5.1 of [AAA], p. 98) that Theorem 3 can be generalized to the case where A is a *uniformly continuous* (nonlinear) mapping. ∎

Remark 2

The proof of Theorem 1 shows that this theorem remains true when V_1, V_2, and F are *Banach* spaces. ∎

4. THE BEST APPROXIMATION THEOREM

Most special properties of Hilbert spaces result from the theorem concerning the best approximation of a point by points of a closed convex set. We denote by $((\,,\,))$ the scalar product in the Hilbert space V.

Proof. Let $x = \{x_1, x_2\} \in V_1 \times V_2$. There exists a sequence of elements $x^n = \{x_1^n, x_2^n\} \in D_1 \times D_2$ that converges to x. This implies that the sequences $\|x_i^n\|_i$ are bounded and that the sequences x_i^n are Cauchy sequences. There exist, therefore, a constant M and integers $N(\varepsilon)$ such that

(1) $\quad \begin{cases} \text{i.} & \|x_i^n\|_i \leq M. \\ \text{ii.} & \|x_i^n - x_i^p\|_i \leq \varepsilon \quad \text{when} \quad n, p \geq N(\varepsilon). \end{cases}$

for $i = 1, 2$. From this we deduce that the sequence $A(x_1^n, x_2^n)$ is a Cauchy sequence. Indeed,

$$\|A(x_1^n, x_2^n) - A(x_1^p, x_2^p)\|_F \leq \|A\| [\|x_2^n\|_2 \|x_1^n - x_1^p\|_1 + \|x_1^p\|_1 \|x_2^n - x_2^p\|_2]$$
$$\leq 2\|A\| M\varepsilon \quad \text{when} \quad n, p \geq N(\varepsilon).$$

Since F is complete, we find that the sequence $A(x_1^n, x_2^n)$ converges to an element $f \in F$. This element f does not depend on the choice of the sequence of elements $x^n = \{x_1^n, x_2^n\}$, which converges to $x = \{x_1, x_2\}$; indeed, if the sequence of elements $y^n = \{y_1^n, y_2^n\}$ also converges to x, we verify that $A(y_1^n, y_2^n)$ converges to f, since

$$\|f - A(y_1^n, y_2^n)\|_F \leq \|f - A(x_1^n, x_2^n)\|_F + \|A\| M(\|x_1^n - y_1^n\|_1 + \|x_2^n - y_2^n\|_2)$$

and since the right-hand side of this inequality tends to zero.

Hence we can associate to every $x = \{x_1, x_2\}$ an element $f = \bar{A}(x_1, x_2)$ that depends only on x. The mapping $\bar{A}$ from $V_1 \times V_2$ to F, which is thereby defined, is bilinear, since we deduce from

$$A\left(\sum_{i=1}^k \lambda^i x_i^n, \sum_{j=1}^l \mu^j y_j^n\right) = \sum_{i=1}^k \sum_{j=1}^l \lambda^i \mu^j A(x_i^n, y_j^n)$$

by taking the limit that

(2) $$\bar{A}\left(\sum_{i=1}^k \lambda^i x_i, \sum_{j=1}^l \mu^j x_j\right) = \sum_{i=1}^k \sum_{j=1}^l \lambda^i \mu^j \bar{A}(x_i, y_j).$$

The mapping $\bar{A}$ is continuous, since the inequalities

$$\|A(x_1^n, x_2^n)\|_F \leq \|A\| \, \|x_1^n\|_1 \, \|x_2^n\|_2$$

imply, by taking the limit, that

(3) $$\|\bar{A}(x_1, x_2)\|_F \leq \|A\| \, \|x_1\|_1 \, \|x_2\|_2$$

Thus $\|\bar{A}\| \leq \|A\|$. It is clear that the mapping $\bar{A}$ that we have constructed is an *extension* of A and, consequently, that $\|A\| \leq \|\bar{A}\|$.

It remains for us to establish that the extension is *unique*. Let $\bar{A}$ and $\bar{\bar{A}}$ be two extensions and $x \in V_1 \times V_2$ be the limit of a sequence of elements

THEOREM 1

Let V be a prehilbert space, M a complete convex subset of V, and $x \in V$. Then the following properties are equivalent:

(1)
$\quad\begin{cases} \text{i.} & tx \in M \text{ satisfies } \|x - tx\| = \min_{y \in M} \|x - y\|. \\ \text{ii.} & tx \in M \text{ satisfies } ((tx - x, tx - y)) \leq 0 \quad \text{for all} \quad y \in M. \end{cases}$

Moreover, we can associate to every $x \in V$ a unique element $tx \in M$ satisfying either one of these properties. ▲

Remark 1

If V is a *Hilbert* space, it is sufficient to suppose that M is a *closed convex* subset. ■

Proof. Let us show that (1)i *implies* (1)ii. If $y \in M$ and if $\theta \in]0, 1[$, then $(1 - \theta)tx + \theta y = tx + \theta(y - tx) \in M$. Therefore

$$\frac{\|x - tx + \theta(tx - y)\|^2 - \|x - tx\|^2}{\theta} \geq 0 \quad \text{for all} \quad y \in M.$$

Developing this expression and letting θ approach zero, we obtain (1)ii.

Now let us show that (1)ii *implies* (1)i. If $y \in M$, we deduce from the inequality (1)ii that

$$((tx - x, tx - y)) = -((x - tx, tx - x + x - y))$$
$$= \|x - tx\|^2 - ((x - tx, x - y)) \leq 0.$$

Then we use the Cauchy-Schwarz inequality

$$\|x - tx\|^2 \leq ((x - tx, x - y)) \leq \|x - tx\| \, \|x - y\|.$$

Therefore $\|x - tx\| \leq \|x - y\|$ for all $y \in M$.

There exists at most one element $tx \in M$ satisfying (1)ii. Indeed, if $sx \in M$ satisfies (1)ii, we obtain

$$\|tx - sx\|^2 = ((tx - sx, tx - sx))$$
$$= ((tx - x, tx - sx)) + ((sx - x, sx - tx)) \leq 0,$$

which implies that $sx = tx$.

There exists $tx \in M$ *satisfying* (1)i. Set $\alpha = \inf_{y \in M} \|x - y\|$ and consider a sequence of elements $y_n \in M$ satisfying

(2)
$$\alpha^2 \leq \|x - y_n\|^2 \leq \alpha^2 + \frac{1}{n}$$

(called a *minimizing sequence*). This is a *Cauchy sequence*. Indeed, we can write

$$\|y_n - y_m\|^2 = \|(y_n - x) - (y_m - x)\|^2 = \|y_n - x\|^2 + \|y_m - x\|^2 \\ + 2((y_n - x, y_m - x))$$

and

$$4\left\|\frac{y_n + y_m}{2} - x\right\|^2 = \|(y_n - x) + (y_m - x)\|^2$$
$$= \|y_n - x\|^2 + \|y_m - x\|^2 + 2((y_n - x, y_m - x)).$$

Adding these two equalities, we obtain

(3) $\quad \|y_n - y_m\|^2 = 2\|y_n - x\|^2 + 2\|y_n - x\|^2 - 4\left\|\frac{y_n + y_m}{2} - x\right\|^2.$

Since $\|y_n - x\|^2 \leq \alpha^2 + 1/n$ and $\|y_m - x\|^2 \leq \alpha^2 + 1/m$ by construction and since

$$\left\|\frac{y_n + y_m}{2} - x\right\|^2 \geq \alpha^2 \quad \text{for} \quad \frac{y_n + y_m}{2} \in M$$

from the hypothesis that M is convex, equality (3) implies the inequality

(4) $\quad \|y_n - y_m\|^2 \leq 2\left(\alpha^2 + \frac{1}{n}\right) + 2\left(\alpha^2 + \frac{1}{m}\right) - 4\alpha^2 = 2\left(\frac{1}{n} + \frac{1}{m}\right),$

which implies that the sequence of elements $y_n \in M$ is a Cauchy sequence. Since M was assumed to be complete, this sequence converges to an element $tx \in M$. The inequalities (2) show that $\|x - tx\|^2 = \alpha^2$. Hence we have established the existence of tx. ∎

DEFINITION 1
The mapping t that associates to each element $x \in V$ its best approximation $tx \in M$ defined by (1) *is called the* best approximation projector *from V onto M.* ▲

We give some other elementary properties of best approximation projectors.

PROPOSITION 1
Let M be a closed convex subset of a Hilbert space V. The best approximation projector has the following properties:

(5) $\quad \begin{cases} \text{i.} & t^2 = t \quad (t \text{ is idempotent}). \\ \text{ii.} & \|tx - ty\| \leq \|x - y\| \quad (t \text{ is a contraction}). \\ \text{iii.} & ((tx - ty, x - y)) \geq 0 \quad (t \text{ is monotone}) \end{cases}$ ▲

Proof. **a.** Property i is obvious, since if $x \in M$, $tx = x$.
b. Since $((tx - x, tx - ty)) \leq 0$ and $((ty - y, ty - tx)) \leq 0$, we obtain the inequality

(6) $$((x - y - (tx - ty), tx - ty)) \geq 0.$$

This inequality together with the Cauchy-Schwarz inequality implies

$$\| tx - ty \|^2 \leq ((x - y, tx - ty)) \leq \| x - y \| \, \| tx - ty \|,$$

that is, (5)ii.
c. We obtain property iii from inequality (6):

$$0 \leq \| tx - ty \|^2 \leq ((x - y, tx - ty)). \blacksquare$$

5. ORTHOGONAL PROJECTORS

We are now going to study the supplementary properties of best approximation projectors when the sets M are cones or closed vector subspaces.

DEFINITION 1
A subset M of a vector space V is a cone (with vertex zero) if

(1) $$\forall \lambda \geq 0, \quad \forall x \in M, \quad \lambda x \in M.$$

If V is a Hilbert space and M a subset of V, we say that the subset

(2) $\quad M^\ominus = \{y \in V \quad \text{such that} \quad ((y, z)) \leq 0 \quad \text{for all} \quad x \in M\}$

is the **Hilbertian negative polar cone** *of M and that*

(3) $\quad M^\oplus = \{y \in V \quad \text{such that} \quad ((y, z)) = 0 \quad \text{for all} \quad z \in M\}$

is the **Hilbertian orthogonal complement** *of M.* ▲

PROPOSITION 1
If $M \subset V$, then $M^\ominus$ is a closed convex cone and $M^\oplus$ is a closed vector subspace. ▲

Proof. It is obvious that $M^\ominus$ is a *cone*, since if $y \in M^\ominus$ and $\lambda \geq 0$, then $\lambda y \in M^\ominus$ because for all $z \in M, ((\lambda y, z)) = \lambda((y, z)) \leq 0$. It is also a *convex* set, since if x and $y \in M^\ominus$ and if $\theta \in \,]0, 1[$, then for every fixed $z \in M$,

$$(((1 - \theta)x + \theta y, z)) = (1 - \theta)((x, z)) + \theta((y, z)) \leq 0.$$

It is a *closed* set: If x is the limit of a sequence $\{x_n\}_n$ of elements $x_n \in M^\ominus$, we deduce that for each fixed $z \in M$,

$$((x, z)) = \lim_{n \to \infty} ((x_n, z)) \leq 0,$$

since the scalar product is a continuous bilinear form and since $((x_n, z)) \leq 0$ for all n and for all $z \in M$. Finally it is clear that

$$M^\oplus = M^\ominus \cap -M^\ominus$$

and, consequently, that $M^\oplus$ is a closed vector subspace. ∎

THEOREM 1
Let M be a closed convex cone of a Hilbert space V. The best approximation projector t from V onto M satisfies

(4) $\begin{cases} \text{i.} & t(\lambda x) = \lambda t(x) \quad \text{for all} \quad \lambda \geq 0 \\ & (t \text{ is positively homogeneous}). \\ \text{ii.} & \|x\|^2 = \|tx\|^2 + \|(1-t)x\|^2 \quad (\text{Pythagorean relation}). \\ \text{iii.} & \|t(x)\| \leq \|x\| \quad \text{and} \quad \|(1-t)(x)\| \leq \|x\|. \end{cases}$

Moreover,

(5) $\quad s = 1 - t$ is the best approximation projector onto $M^\ominus$.

The cones M and $M^\ominus$ satisfy the conditions

(6) $\begin{cases} \text{i.} & M^\ominus = \{x \in V \text{ such that } t(x) = 0\}. \\ \text{ii.} & M = \{y \in V \text{ such that } s(y) = y - ty = 0\}. \end{cases}$

Finally, the following properties are equivalent:

(7) $\begin{cases} \text{i.} & \text{every element } x \in V \text{ has a unique expression as} \\ & x = y + z \text{ where } y \in M, z \in M^\ominus, \text{ and } ((y, z)) = 0. \\ \text{ii.} & y = tx \quad \text{and} \quad z = sx. \end{cases}$ ▲

Remark 1

Condition (7) expresses the fact that $y = tx \in M$ and $z = sx \in M^\ominus$ are the only elements satisfying

$$x = y + z \quad \text{and} \quad ((y, z)) = 0.$$ ∎

First of all, we establish the following lemma:

LEMMA 1
When M is a cone, property (1)i in Section 4 characterizing the best approximation projectors is equivalent to

(8) $\begin{cases} \text{i.} & ((x - tx, tx)) = 0. \\ \text{ii.} & ((x - tx, z)) \leq 0 \quad \text{for all} \quad z \in M. \end{cases}$ ▲

Proof. Indeed, formula (1)i in Section 4 is obtained by subtracting (8)i from (8)ii. Conversely, taking successively $z = 0 \in M$ and $z = 2tx \in M$ (which is *possible because M is a cone*), we obtain from (1)i in Section 4 that $((x - tx, tx)) \leq 0$ and $((x - tx, tx)) \geq 0$; that is $((x - tx, tx)) = 0$. Consequently, (1)i in Section 4 implies (8)ii. ∎

Proof of Theorem 1. **a.** Since for all $\lambda > 0$,

$$((\lambda x - t(\lambda x), t(\lambda x))) = \lambda^2 \left(\left(x - \frac{1}{\lambda} t(\lambda x), \frac{1}{\lambda} t(\lambda x) \right) \right) = 0$$

and since

$$((\lambda x - t(\lambda x), z)) \leq 0 \quad \text{implies} \quad \left(\left(x - \frac{1}{\lambda} t(\lambda x), z \right) \right) \leq 0,$$

we deduce from Lemma 1 and from the uniqueness of the best approximation of x that

$$t(x) = \frac{1}{\lambda} t(\lambda x)$$

Moreover, the Pythagorean relation (8)ii follows from (8)i, since

$$\|x\|^2 = ((x - tx + tx, x - tx + tx))$$
$$= \|tx\|^2 + \|(1-t)x\|^2 + 2((x - tx, x)) = \|tx\|^2 + \|(1-t)x\|^2$$

From this we immediately derive the inequalities (8)iii.
b. Lemma 1 implies that the mapping $s = 1 - t$ is the best approximation projector onto $M^\ominus$, since $sx \in M^\ominus$ (because $((sx, y)) = ((x - tx, y)) \leq 0$ for all $y \in M$) and since

$$\begin{cases} \text{i.} & ((x - sx, sx)) = ((tx, (1-t)x)) = 0. \\ \text{ii.} & ((x - sx, y)) = ((tx, y)) \leq 0 \quad \text{for all} \quad y \in M^\ominus. \end{cases}$$

c. If $tx = 0$, then $x = sx + tx = sx \in M^\ominus$. Conversely, if $x \in M^\ominus$, then $\|tx\|^2 = ((tx, tx)) = ((tx, x)) \leq 0$ and, consequently, $tx = 0$.
d. Similarly, it is clear that $sx = x - tx = 0$ if and only if $x \in M$.
e. Every element $x \in V$ can be expressed $x = tx + sx$ where $tx \in M$, $sx \in M^\ominus$ and $((tx, sx)) = 0$.
Conversely, if $x = y + z$ where $y \in M$, $z \in M^\ominus$, and $((y, z)) = 0$, we obtain

$$\begin{cases} \text{i.} & ((x - y, y)) = ((z, y)) = 0. \\ \text{ii.} & ((x - y, m)) = ((z, m)) \leq 0 \quad \text{for all} \quad m \in M. \end{cases}$$

This shows that $y = t(x)$ is the unique best approximation of x and that, therefore, $z = x - tx = s(x)$. ∎

DEFINITION 2
We say that the best approximation projector t onto a closed convex cone M is the orthogonal projector *onto M.* ▲

When M is a closed vector subspace of V, we obtain the following.

THEOREM 2
Let M be a closed vector subspace of a Hilbert space V. The orthogonal projector t from V onto M is a continuous linear operator (with norm 1) that satisfies

(9) $\begin{cases} \text{i.} & \|x\|^2 = \|tx\|^2 + \|sx\|^2. \\ \text{ii.} & \|tx\| \le \|x\|, \quad \|sx\| \le \|x\| \quad \text{for all} \quad x \in V. \\ \text{iii.} & M = \text{Im } t = \text{Ker } s; \quad M^\oplus = \text{Ker } t = \text{Im } s. \\ \text{iv.} & ((tx, y)) = ((x, ty)). \end{cases}$

Moreover, tx and sx are the unique elements $y \in M$ and $z \in M^\oplus$ such that $x = y + z$ and $((y, z)) = 0$. ▲

Proof. It is sufficient to establish (9)iv. We have the following equality:
$$((tx, y)) = ((tx, ty + sy)) = ((tx, ty)) = ((x - sx, ty)) = ((x, ty)),$$
since $((y, z)) = 0$ if $y \in M$ and $z \in M^\oplus$. ■

*Remark 2

We can now construct all the other projectors s (i.e. the continuous linear operators s satisfying $s^2 = s$) onto M from the orthogonal projector t. ■

*PROPOSITION 2
The operators $s = a + (1 - a)t$ where $a \in \mathcal{L}(V, M)$ and where t is the orthogonal projector onto M are projectors onto M. Every projector s onto M is of this form. ▲

Proof. If s is a continuous projector from V onto M, then $s = s + (1 - s)t$, since $sx = sx + tx - stx = sx$ because $stx = tx$. Let us show that $s = a + (1 - a)t$ is a projector from V onto M. First of all, $sx \in M$. If $x \in M$, then $sx = ax + tx - atx = ax + x - ax = x$. Then $s^2 = s$ and s is a projector from V to M. To show that Im $s = M$, we verify that if $x = tx \in M$, then $x = tx + ax - atx = sx$ is the image of x by s. ■

PROPOSITION 3
Let M be a subset of a Hilbert space V. Then $M^{\ominus\ominus}$ is the closed convex cone

generated by M and $M^{\oplus\oplus}$ is the closed vector subspace generated by M. ▲

Proof. First, it is clear that $M \subset M^{\ominus\ominus}$. Since $M^{\ominus\ominus}$ is a closed convex cone, the closed convex cone $\tilde{M}$ generated by M is contained in $M^{\ominus\ominus}$.

We must show then that if $x \in M^{\ominus\ominus}$, x belongs to $\tilde{M}$. Let t be the orthogonal projector onto $\tilde{M}$. Then $s = (1 - t)$ is the orthogonal projector onto $(\tilde{M})^{\ominus}$. Moreover, it is easy to show that $(\tilde{M})^{\ominus} = M^{\ominus}$ by an argument analogous to that of Proposition 1. Consequently, $t = 1 - s$ is the orthogonal projector onto $M^{\ominus\ominus}$. Thus $tx = x$, which implies that $x \in \tilde{M}$. ∎

*PROPOSITION 4
Let N be a cone of a vector space V and $M \subset V$ be a subset. Then

$$(M + N)^{\ominus} = M^{\ominus} \cap N^{\ominus}$$

If $M + N$ is a closed convex cone, we obtain

$$M + N = (M^{\ominus} \cap N^{\ominus})^{\ominus}.$$ ▲

Proof. First, $M^{\ominus} \cap N^{\ominus} \subset (M + N)^{\ominus}$, since if $x \in M^{\ominus} \cap N^{\ominus}$, $y \in M$, and $z \in N$, then $((x, y + z)) = ((x, y)) + ((x, z)) \leq 0$.

Conversely, if $x \in (M + N)^{\ominus}$, then $((x, y + z)) \leq 0$ when $y \in M$, $z \in N$. Taking $z = 0$, we obtain that $x \in M^{\ominus}$. Let $z \in N$ and $y_0 \in M$ be fixed. Hence $((x, \lambda z)) \leq -((x, y_0))$ for all $\lambda > 0$, since $\lambda z \in N$. Hence by dividing by $\lambda > 0$ and letting λ tend to ∞, we deduce that $((x, z)) \leq 0$ for all $z \in N$. Hence $x \in N^{\ominus}$.

Finally, we deduce that $(M + N)^{\ominus\ominus} = (M^{\ominus} \cap N^{\ominus})^{\ominus}$. If $M + N$ is a closed convex cone, we have seen that $(M + N)^{\ominus\ominus} = M + N$. ∎

6. CLOSED SUBSPACES, QUOTIENT SPACES, AND FINITE PRODUCTS OF HILBERT SPACES

PROPOSITION 1
Let t be the orthogonal projector from a Hilbert space onto a closed vector subspace M of V. Then M, with the scalar product $((tx, ty))$, is a Hilbert space. The quotient space V/M with the scalar product

(1) $$((\varphi x, \varphi y))_{V/M} = ((sx, sy))$$

(where $\varphi : V \mapsto V/M$ is the canonical surjection and $s = 1 - t$) is also a Hilbert space. Moreover, the associated norm satisfies the following condition:

(2) $$\|\varphi x\|_{V/M} = \inf_{y \in M} \|x - y\|.$$ ▲

Proof. **a.** If x and y belong to M, then $((tx, ty))$ is a positive symmetric bilinear form on M, which is nondegenerate, for if $\|tx\|^2 = ((tx, tx)) = 0$, this implies that $tx = 0$. Since M is closed and V is complete, M is complete. It is, accordingly, a Hilbert space.

b Let $\varphi: V \mapsto V/M$ be the canonical surjection from V onto the quotient space V/M, which is a linear operator. Since $M = \text{Ker } s$ is the kernel of the orthogonal projector $s = 1 - t$ onto $M^\ominus$, we can write that $s = \tilde{s}\varphi$ where $\tilde{s}$ is a bijection from V/M onto $M^\oplus = \text{Im } s$. Since $M^\oplus$ is a Hilbert space for the scalar product $((sx, sy))$, we find that V/M is a Hilbert space for the scalar product $((u, v))_{V/M} = ((\tilde{s}u, \tilde{s}v))$ (where u and v belong to V/M). Let us observe that the norm $\|u\|_{V/M} = \sqrt{((u, u))}$ satisfies

$$\|\varphi x\|^2_{V/M} = \|\tilde{s}\varphi x\|^2 = \|sx\|^2 = \|(1-t)x\|^2 = \min_{y \in M} \|x - y\|^2. \quad \blacksquare$$

Remark 1

More generally, in the case of normed spaces, we give to the quotient space V/M the "quotient norm" defined by

$$\|\varphi x\|_{V/M} = \inf_{y \in M} \|x - y\|. \quad \blacksquare$$

Consider n Hilbert spaces V_i, with scalar products $((x_i, y_i))_i$. Let $V = \prod_{i=1}^{n} V_i$ be their product. It is clear that the bilinear form

(3) $$((x, y)) = \sum_{i=1}^{n} ((x_i, y_i))_i \quad \text{where} \quad x = \{x_i\}_i, \quad y = \{y_i\}_i$$

is a scalar product on $V \times V$ defining the norm

(4) $$\|x\| = \left(\sum_{i=1}^{n} \|x_i\|_i^2 \right)^{1/2}$$

PROPOSITION 2
The product $V = \prod_{i=1}^{n} V_i$ of n Hilbert spaces V_i is a Hilbert space for the scalar product $((x, y)) = \sum_{i=1}^{n} ((x_i, y_i))_i$. ▲

Proof. (Left as an exercise.)

*7. ORTHOGONAL BASES FOR A SEPARABLE HILBERT SPACE

DEFINITION 1
Let $\{e_n\}$ be a sequence of elements of a Hilbert space V. We shall say that

$\{e_n\}$ is an orthogonal sequence if

(1) $$\forall m \neq n, \quad ((e_m, e_n))_V = 0$$

and that it is orthonormal if, in addition,

(2) $$\forall m, \quad \|e_m\|_V = 1$$

We shall say that it is in orthogonal base if it generates a vector space that is dense in V. ▲

Remark 1

The preceding implies that V is "separable," in the sense that V is the closure of a subspace that is generated by a countable set. (See Section 5.2 of [AAA], p. 181.) ∎

Example 1

Consider the space of sequence $l^2(\mathbb{N})$. In this case the sequence $\{e_n\}$ where $e_n = (0, \ldots, 0, 1, 0, \ldots)$ is obviously an orthonormal base for $l^2(\mathbb{N})$. ∎

We see in Section 1, 2, and 3 of Chapter 8 examples of orthogonal bases in the L^2 spaces (orthogonal polynomials and Fourier series).

PROPOSITION 1
Let V and W be two Hilbert spaces, $A \in \mathscr{L}(V, W)$ an isometry from V to W, and $\{e_n\}$ an orthogonal sequence in V. Then $\{A(e_n)\}$ forms an orthogonal sequence in W. ▲

Proof. (Left as an exercise.)

We can always make an orthonormal sequence from an orthogonal sequence $\{e_n\}_n$ by replacing it by the sequence $\{e_n/\|e_n\|\}_n$. We can make a sequence $\{e_n\}_n$ of linearly independent vectors orthonormal as well.

PROPOSITION 2 (GRAM-SCHMIDT ORTHONORMALIZATION PROCESS)
Let V be a Hilbert space, $\{x_n\}_n$ a sequence of linearly independent vectors, V_k the closed subspace generated by the first k vectors x_n. Then the sequence $\{e_n\}_n$ of vectors

(3) $$e_n = x_n - t_{V_{n-1}}(x_n)$$

is an orthogonal sequence such that $\forall k, V_k$ is also generated by the first k vectors e_n. ▲

Proof. We take $e_1 = x_1$ and construct the sequence e_n by recursion. Suppose we are given $e_1, \ldots, e_{n-1}$, which generate V_{n-1}. Then by definition of the orthogonal projector $t_{V_{n-1}}$ on V_{n-1}, e_n is orthogonal to V_{n-1} and, therefore, to the $n-1$ vectors e_k. Furthermore, $x_n - e_n = t_{V_{n-1}}(x_n)$ belongs to V_{n-1}. Consequently, $e_1, \ldots, e_n$ generate the same subspace as the union of V_{n-1} and $\{x_n\}$, which is exactly V_n. ∎

Now let us establish the fundamental result on orthogonal bases.

THEOREM 1
Let V be a Hilbert space, $\{e_n\}_n$ an orthonormal sequence, and $U \subset V$ the closed subspace generated by the e_n's. We denote by t_U the orthogonal projector onto U. Then
a. *the series $\sum_{n=1}^{\infty} |((x, e_n))_V|^2$ is convergent and*

(4)
$$\begin{cases} \text{i.} \quad \sum_{n=1}^{\infty} |((x, e_n))_V|^2 = \|t_U x\|_V^2 \leq \|x\|^2 \\ \quad \text{(Bessel's inequality).} \\ \text{ii.} \quad \sum_{n=1}^{\infty} ((x, e_n))_V ((y, e_n))_V = ((t_U x, t_U y)); \end{cases}$$

b. *we can write*

(5)
$$t_U(x) = \sum_{n=1}^{\infty} ((x, e_n))_V e_n.$$

Conversely, if $\lambda = \{\lambda_n\}_n \in l^2$, there exists a unique element $x \in U$ such that $((x, e_n)) = \lambda_n$ for all n. ▲

Proof. We denote by l_0^2 the subset of sequences $\lambda \in l^2$ such that $\lambda_n = 0$ for all but a finite number of indices and by U_0 the vector space generated by the e_n's. We denote by g the linear operator from l_0^2 to U_0 defined by

(6)
$$g(\lambda) = \sum_{n=1}^{\infty} \lambda_n e_n \in U_0,$$

which is a proper definition because the sum is indeed finite. Since the sequence $\{e_n\}$ is orthonormal, we deduce that

(7)
$$((g(\lambda), g(\mu)))_V = \sum_{n=1}^{\infty} \lambda_n \mu_n = ((\lambda, \mu))_{l_0^2},$$

that is, g is an isometry from l_0^2 to U_0. Moreover, g is surjective from l_0^2 to U_0; if $x = \sum_{n=1}^{\infty} \lambda_n e_n$ is a finite sum, it is clear that the sequence $\lambda = j(x)$

CH. 1, SEC. 7 ORTHOGONAL BASES FOR A SEPARABLE HILBERT SPACE

defined by

$$((x, e_m))_V = \lambda_m \quad \text{for all} \quad m \tag{8}$$

satisfies $x = g(\lambda) = gj(x)$.

Since g is linear and continuous from the subspace l_0^2 (which is dense in l^2) to the complete space V, g has a unique extension to a continuous linear operator $\bar{g}$ from l^2 to V, according to Theorem 3.3; we deduce from (7) that $\bar{g}$ is an isometry from l^2 to V. Its image $\bar{g}(l^2)$ is therefore closed (because it is complete), and $\bar{g}(l_0^2)$ is dense in $\bar{g}(l^2)$. Thus $\bar{g}(l^2) = U$. The inverse $\bar{j}$ of $\bar{g}$ is the extension by density of $j \in \mathscr{L}(U_0, l^2)$ to U. Thus, by taking the limit, we find that if x and $y \in U$,

$$\sum_{n=1}^{\infty} ((x, e_n))_V ((y, e_n))_V = ((\bar{j}(x), \bar{j}(y)))_{l^2} = ((x, y))_V \tag{9}$$

and we can write

$$\forall x \in U, \quad x = \bar{g}\bar{j}(x) = \sum_{n=1}^{\infty} ((x, e_n))_V e_n \tag{10}$$

in the sense that

$$x = \lim_{k \to \infty} \sum_{n=1}^{k} ((x, e_n))_V e_n. \tag{11}$$

Now take x and $y \in V$. Since $t_U e_n = e_n$, we deduce, using Property (9)iv of Theorem 5.2, that

$$((x, e_n))_V = ((x, t_U e_n))_V = ((t_U x, e_n))_V. \tag{12}$$

Thus if x and $y \in V$, equalities (9) and (12) imply that

$$\sum_{n=1}^{\infty} ((x, e_n))_V ((y, e_n))_V = ((t_U x, t_U y))_V \tag{13}$$

and also Bessel's inequality (4)i by taking $y = x$. Relations (10) and (12) imply (5). ∎

THEOREM 2
Every separable Hilbert space has an orthonormal base and, consequently, is isometric to l^2. ▲

Proof. If V is a separable Hilbert space, there exists a (countable) sequence $\{x_n\}_n$ of linearly independent vectors that generate a vector space V_0 dense in V. According to Proposition 2, there exists an orthonormal base $\{e_n\}_n$ that generates the same space V_0. Thus $\{e_n\}_n$ is an orthonormal base. Theorem

1 then implies that the operator $\bar{j} \in \mathscr{L}_2(V, l^2)$ defined by

(14) $$\bar{j}(x)_n = ((x, e_n))_V \quad \forall n \in \mathbb{N}$$

is an isometry from V onto l^2, and, therefore, that every $x \in V$ can be written

(15) $$x = \sum_{n=1}^{\infty} ((x, e_n))_V e_n. \qquad \blacksquare$$

Remark 2

Orthonormal bases are used in Chapter 12 to construct Hilbert-Schmidt operators. Methods for constructing orthogonal bases are found in Chapter 11 and examples (orthogonal polynomials, Fourier series) are found in Chapter 8. ∎

CHAPTER 2

Theorems on Extension and Separation

We group in this chapter the first fundamental consequences of the projection theorem along with applications to optimization theory and game theory. We begin in Section 1 by establishing the extension of a continuous linear form f on a closed subspace M of the Hilbert space V to a continuous linear form $\tilde{f}$ on V, *having the same norm as* f (as well as the extension of continuous linear and bilinear mappings).

In Section 2 we deduce a fundamental criterion for recognizing that a set generates a dense vector subspace, that is, a criterion that permits us to prove an approximation theorem. This criterion is used constantly in what follows.

Section 3 is devoted to theorems on the strict separation of two disjoint sets M and N by a continuous linear form in the sense that

$$\sup_{x \in M} f(x) < \inf_{y \in N} f(y).$$

This is possible when $M - N$ is a closed convex subset (which is the case, for example, if M is closed and convex and N is compact and convex). In the case of finite dimensional spaces we can separate (with a weak inequality) two disjoint subsets M and N for which $M - N$ is convex.

This allows us in Section 5 to characterize the *closed convex* subsets M using their support functions σ_M defined on the dual V^* of V by

$$\sigma_M(f) = \sup_{x \in M} f(x)$$

We write that M is defined by a family of inequalities

$$M = \{x \in V \quad \text{such that} \quad f(x) \leqq \sigma_M(f) \quad \forall f \in V^*\}.$$

These separation theorems have many important consequences, notably in convex and nonconvex analysis. (See Chapters 10 and 15, which can be taken up after the study of the first five sections of this chapter.) However, we have chosen to use these theorems here to prove in Section 6 the duality theorem in convex optimization (the theorem on the existence of a Lagrange multiplier), in Section 7 the Von Neumann minimax theorem (the existence of an equilibrium in a zero-sum two-person game), and in Section 8 the characterization of Pareto optima in a cooperative n-person game, which allows us to replace the search for such an optimum by a minimization problem.

This chapter can be considered as a very brief introduction to the theory of games for one, two, and n persons.

Let us remark that in the case of normed spaces, and more generally for "locally convex" spaces, these extension and separation theorems remain true and are consequences of the Hahn-Banach theorem.

1. EXTENSION OF CONTINUOUS LINEAR AND BILINEAR OPERATORS

The following theorem is fundamental.

THEOREM 1
Let M be a closed vector subspace of a Hilbert space V. Let f be a continuous linear form defined on the subspace M. Then there exists a continuous linear form $\tilde{f}$ defined on V such that

(1) $$\begin{cases} \text{i.} & \forall x \in M, \quad \tilde{f}(x) = f(x) \quad (\tilde{f} \text{ extends } f). \\ \text{ii.} & \sup_{x \in V} \frac{|\tilde{f}(x)|}{\|x\|} = \sup_{x \in M} \frac{|f(x)|}{\|x\|}. \end{cases}$$

▲

Proof. Let t be the orthogonal projector from V onto M. The linear form $\tilde{f}$ defined on V by

(2) $$\tilde{f}(x) = f(tx)$$

is continuous and satisfies (1)i since $tx = x$ when $x \in M$. Since $\|tx\| \leq \|x\|$, we deduce that

$$|\tilde{f}(x)| \leq \left(\sup_{x \in M} \frac{|\tilde{f}(x)|}{\|x\|}\right) \|tx\| \leq \left(\sup_{x \in M} \frac{|f(x)|}{\|x\|}\right) \|x\|.$$

Consequently,

$$\sup_{x \in V} \frac{|\tilde{f}(x)|}{\|x\|} \leq \sup_{x \in M} \frac{|f(x)|}{\|x\|}.$$

The opposite inequality is trivial since $M \subset V$; therefore condition (1)ii is satisfied. ■

Remark 1

This extension theorem remains true when V is, more generally, a locally convex space; this is the Hahn-Banach theorem. ∎

*Remark 2

The proof of Theorem 1 uses no specific property of the space $\mathbb{R}$ and hence remains true for a continuous linear operator A from M to a vector space F. More generally, we obtain the following theorem (valid *only* for Hilbert spaces). ∎

THEOREM 2
Let M_1 and M_2 be two closed vector subspaces of the Hilbert spaces V_1 and V_2, respectively, and A a continuous bilinear mapping from $M_1 \times M_2$ to the Hilbert space F. Then there exists a continuous bilinear mapping $\tilde{A}$ from $V_1 \times V_2$ to F such that

(3) $\begin{cases} \text{i.} & \forall x_i \in M_i, \quad \tilde{A}(x_1, x_2) = A(x_1, x_2) \ (\tilde{A} \text{ is an extension of } A). \\ \text{ii.} & \|\tilde{A}\|_{\mathscr{L}(V_1, V_2; F)} = \|A\|_{\mathscr{L}(M_1, M_2; F)}. \end{cases}$ ▲

Proof. It suffices to verify that the mapping $\tilde{A}$ defined by

(4) $$\tilde{A}(x_1, x_2) = A(t_{M_1} x_1, t_{M_2} x_2)$$

satisfies the conclusions of the theorem. ∎

This theorem together with Theorem 1.3.1 on extension by density allows us to *extend a continuous bilinear operator on the product $M_1 \times M_2$ of (non-closed) vector subspaces of V_1 and V_2 to a Hilbert space F, to a continuous bilinear operator from $V_1 \times V_2$ to F having the same norm*. (We extend A to $\bar{A}$ on $\bar{M}_1 \times \bar{M}_2$ by density and $\bar{A}$ to $\tilde{\bar{A}}$ by Theorem 2.)

2. A DENSITY CRITERION

We establish here a criterion for density that will be frequently used.

THEOREM 1
Let D be a subset of a Hilbert space V. The following conditions are equivalent:

(1) $\begin{cases} \text{i.} & D \text{ generates a vector subspace that is dense in } V. \\ \text{ii.} & \text{Every continuous linear form } f \text{ on } V \text{ that vanishes on } D \text{ is identically zero on } V. \end{cases}$ ▲

Proof. Statement (1)i implies (1)ii. Indeed, since f is linear, f vanishes on the vector subspace $\tilde{D}$ generated by D; since f is continuous and since $\tilde{D}$ is dense in V, f vanishes on all of V (because for every $x \in V$, $f(x)$ is the limit of a sequence of elements $f(x_n) = 0$).

Now let us show that (1)ii implies (1)i. If the closure M of the vector subspace $\tilde{D}$ generated by D were distinct from V, there would exist $y_0 \notin M$. According to Theorem 1.5.2, if t is the orthogonal projector onto M, $x_0 = y_0 - ty_0$ is different from zero and belongs to $M^\oplus = D^\oplus$. Therefore the continuous linear form f defined by $f(x) = ((x_0, x))$ is not identically zero (since $x_0 \neq 0$), and yet it vanishes on D. This, therefore, is a contradiction to (1)ii. ∎

We rephrase this theorem in Section 3.4 (see Corollary 3.4.1) in terms of orthogonality.

3. SEPARATION THEOREMS

We are going to derive from the best approximation theorem the following theorems, which are known as *separation theorems*.

THEOREM 1
Let V be a Hilbert space. If M is a nonempty closed convex subset of V and if $x_0 \notin M$, there exists a continuous linear form f on V, not identically zero, such that

(1) $$f(x_0) > \sup_{x \in M} f(x).$$

Proof. The continuous linear form f defined by

(2) $$f(x) = \left(\left(\frac{x_0 - tx_0}{\|x_0 - tx_0\|}, x\right)\right)$$

(where t is the best approximation projector onto M) satisfies condition (1). Indeed, we deduce from the inequality

$$\|tx_0 - x_0\|^2 - ((x_0 - tx_0, x_0 - y))$$
$$= ((tx_0 - x_0, tx_0 - x_0 + x_0 - y)) \leq 0 \quad \text{for all} \quad y \in M$$

characterizing tx_0, that

$$0 < \|tx_0 - x_0\| \leq \left(\left(\frac{x_0 - tx_0}{\|x_0 - tx_0\|}, x_0 - y\right)\right) \quad \text{if} \quad y \in M$$

and, consequently, that

$$\sup_{y \in M} f(y) \leq f(x_0) - \|x_0 - tx_0\| < f(x_0)$$

since $x_0 \neq tx_0$. ∎

From this we derive the following result.

THEOREM 2
Let us suppose that M and N are two nonempty disjoint sets in a Hilbert space V. If

(3) $\qquad\qquad M - N$ *is a closed convex subset,*

then there exists a continuous linear form f on V, not identically zero, such that

(4) $\qquad\qquad \sup_{x \in M} f(x) < \inf_{x \in N} f(x).$ ▲

Proof. To say that M and N are disjoint amounts to saying that $0 \notin M - N$. (Indeed, $0 \in M - N$ if and only if there exists $m \in M$ and $n \in N$ such that $m = n$, that is to say, if $M \cap N \neq \varnothing$.) We apply Theorem 1 then in the case where $x_0 = 0$ and where M is replaced by $M - N$; there exists a continuous linear form on V, not identically zero, such that

$$0 = f(0) > \sup_{\substack{x \in M \\ y \in N}} f(x - y) = \sup_{x \in M} f(x) - \inf_{y \in N} f(y). \quad \blacksquare$$

DEFINITION 1
We say that a function f that satisfies (4) strictly separates M and N. ▲

Separation of a Closed Convex Set from a Compact Set

We give an example for which hypothesis (3) is satisfied.

THEOREM 3
Suppose that a nonempty closed convex set M and a nonempty compact convex set N are disjoint. Then there exists a continuous linear form, not identically zero, on V that strictly separates them. ▲

Proof. Indeed, since M and N are convex, $M - N$ is also a convex set (see Proposition 2.7.1 of [AAA], p. 73.) Moreover, since M is closed and N is compact, $M - N$ is closed. (See Proposition 2.6.5 of [AAA], p. 72.) Therefore $M - N$ is closed and convex and the conclusion is a consequence of Theorem 2. ∎

4. A SEPARATION THEOREM IN FINITE DIMENSIONAL SPACES

In the case of finite dimensional spaces we obtain a "weak separation" theorem under weaker hypotheses.

THEOREM 1
Let M be a nonempty convex subset of a finite dimensional space $\mathbb{R}^n$. If $x_0 \notin M$, there exists a linear form, not identically zero, $f \in \mathbb{R}^{n*}$, such that

(1) $$f(x_0) \geq \sup_{x \in M} f(x).$$ ▲

Proof. To every $x \in M$, we can associate $f \in \mathbb{R}^{n*}$, $\|f\| = 1$, such that $f(x_0) > f(x)$ since $\{x\}$ is a closed convex subset disjoint from x_0. Then the sets

$$S_x = \{f \in \mathbb{R}^{n*} \text{ such that } \|f\| = 1 \text{ and } f(x_0) \geq f(x)\}$$

are nonempty. They are clearly closed. Since they are contained in the unit sphere of $\mathbb{R}^{n*}$, *which is compact because $\mathbb{R}^{n*}$ is of finite dimension*, they are compact sets. It is clear that if $f \in \bigcap_{x \in M} S_x$, we have $f(x_0) \geq f(x)$ for every $x \in M$, and, consequently, $f(x_0) \geq \sup_{x \in M} f(x)$. Therefore the result depends on showing that $\bigcap_{x \in M} S_x \neq \emptyset$.

Since the sets S_x are compact, it suffices to show that $\bigcap_{1 \leq i \leq n} S_{x_i} \neq \emptyset$ for every finite sequence $\{x_i\}_{1 \leq i \leq n}$ of points of M. But $\operatorname{co}\{x_1, \ldots, x_n\}$ is a *closed convex set*. Since $x_0 \notin M$, then $x_0 \notin \operatorname{co}\{x_1, \ldots, x_n\}$: Theorem 3.1 implies the existence of $f \in \mathbb{R}^{n*}$, $\|f\|_* = 1$, such that

$$f(x_0) \geq \sup_{x \in \operatorname{co}\{x_1, \ldots, x_n\}} f(x) = \sup_{1 \leq i \leq n} f(x_i).$$

Therefore f indeed belongs to $\bigcap_{1 \leq i \leq n} S_{x_i}$.   ■

5. SUPPORT FUNCTIONS

We characterize the closed convex sets M of V by convex functions $f \mapsto \sigma_M(f)$ defined on the dual V^* of V.

DEFINITION 1
We say that the function $\sigma_M : V \mapsto \,]-\infty, +\infty]$ associated with the nonempty closed convex subset M defined by

(1) $$\sigma_M(f) = \sup_{x \in M} f(x)$$

is the **support function** *of the subset M of V.*   ▲

THEOREM 1
Let M be a subset of a Hilbert space V. The closed convex hull $\overline{co}(M)$ is defined by $\overline{co}(M) = \{x \in V$ such that $f(x) \leq \sigma_M(f)$ for every continuous linear form $f\}$. ▲

Proof. Set $\tilde{M} = \{x \in V$ such that $f(x) \leq \sigma_M(f) \quad \forall f \in V^*\}$. It is clear that $M \subset \tilde{M}$ and that $\tilde{M}$ is a closed convex set, since $\tilde{M}$ is the intersection of the closed half spaces $\{x \in V$ such that $f(x) \leq \sigma_M(f)\}$. Consequently, $\overline{co}(M) \subset \tilde{M}$. If $\overline{co}(M) \neq \tilde{M}$, there would exist $x_0 \in \tilde{M}$ such that $x_0 \notin \overline{co}(M)$. According to Theorem 1.3.1, there would exist $f \in V^*, f \neq 0$, such that $f(x_0) > \sup_{x \in M} f(x) = \sigma_M(f)$. This contradicts the fact that x_0 belongs to $\tilde{M}$. Therefore $\overline{co}(M) = \tilde{M}$. ■

*Supplementary Properties of Support Functions

We now give some other properties of support functions that prove to be useful.

PROPOSITION 1
Let M be a nonempty subset of V. Then σ_M is a positively homogeneous convex lower semicontinuous function. The function σ_M is nonnegative if $0 \in M$ and finite if M is bounded. Moreover, if M is a cone, we have

(2) $$\sigma_M(f) = \begin{cases} 0 & \text{if} \quad f \in M^- \\ +\infty & \text{if} \quad f \notin M^- \end{cases}$$

where

$$M^- = \{f \in V^* \text{ such that } f(x) \leq 0 \text{ for all } x \in V\}.$$

If M and N are two subsets, then

(3) $\quad \sigma_{\alpha M + \beta N}(f) = \alpha \sigma_M(f) + \beta \sigma_N(f) \quad \text{if} \quad \alpha, \beta \geq 0.$

Furthermore,

(4) $\quad \sigma_M(f) \leq \sigma_N(f) \quad \forall f \in V^* \quad$ *if and only if* $\quad \overline{co}(M) \subset \overline{co}(N).$

Finally,

(5) $\quad \sup_{i \in I} \sigma_{M_i}(f) = \sigma_M(f) \quad$ *where* $\quad M = \bigcup_{i \in I} M_i.$ ▲

Proof. The function $f \mapsto \sigma_M(f)$ is clearly positively homogeneous convex and lower semicontinuous. It is clear that $\sigma_M(f) \geq 0$ if $0 \in M$ and that

$$\sigma_M(f) = \sup_{x \in M} f(x) \leq \|f\|_* \sup_{x \in M} \|x\| \leq a \|f\|_*$$

is finite if M is contained in a ball of radius a. Suppose that M is a cone. If $f \in M^-$, then $\sigma_M(f) = \sup_{x \in M} f(x) \leq 0$ and since $0 \in M, 0 = f(0) \leq \sigma_M(f)$. Then $\sigma_M(f) = 0$. If $f \notin M^-$, there exists $x_0 \in M$ such that $f(x_0) > 0$. Since M is a cone, $\lambda x_0 \in M$ for all $\lambda > 0$. Thus

$$\sigma_M(f) \geq \sup_{\lambda \geq 0} f(\lambda x_0) = (\sup_{\lambda \geq 0} \lambda) f(x_0) = +\infty.$$

Moreover, it is clear that

$$\sigma_{\alpha M + \beta N}(f) = \sup_{\substack{x \in M \\ y \in N}} f(\alpha x + \beta y) = \sup_{\substack{x \in M \\ y \in N}} (\alpha f(x) + \beta f(y))$$

$$= \alpha \sup_{x \in M} f(x) + \beta \sup_{y \in N} f(y) = \alpha \sigma_M(f) + \beta \sigma_N(f).$$

Statement (4) is clear. Let us show (5). Since $M_i \subset M = \bigcup_{i \in I} M_i$ for all i, we deduce that $\sigma_{M_i}(f) \leq \sigma_M(f)$ for all i and, consequently, that $\sup_{i \in I} \sigma_{M_i}(f) \leq \sigma_M(f)$. Conversely, if $x \in M$, there exists $i \in I$ such that $x \in M_i$, that is, such that $f(x) \leq \sigma_{M_i}(f) \leq \sup_{i \in I} \sigma_{M_i}(f)$. Then $\sigma_M(f) \leq \sup_{i \in I} \sigma_{M_i}(f)$. ■

In particular, σ_M, being convex and positively homogeneous, is the *gauge* of the subset of V^* defined by

(6) $\qquad M^0 = \{f \in V^* \quad \text{such that} \quad \sigma_M(f) \leq 1\}.$

(The gauge of a convex set is defined in Section 3.7 of [AAA], p. 113.)

DEFINITION 2
If M is a nonempty subset of V, the subset M^0 of V^* defined by (6) is called the polar set *of M.*

The cone generated by M^0:

(7) $\qquad M_b = \{f \in V^* \quad \text{such that} \quad \sigma_M(f) < +\infty\}$

is called the barrier cone *of M.* ▲

We conclude this section by showing an inequality that relates the gauge and the support function of a set M.

PROPOSITION 2
Let M be a closed convex subset of a Hilbert space V that contains the origin. Then, if j_M is the gauge of M,

(8) $\qquad\qquad \forall f \in V, \quad \forall x \in V, \quad f(x) \leq \sigma_M(f) j_M(x).$ ▲

Proof. Inequality (8) is satisfied if $f \notin M_b$ and/or if x belongs to the cone M_a generated by M, since in this case the right-hand side is infinite. Therefore let us suppose that $f \in M_b$ and $x \in M_a$.

a. If $j_M(x) = 0$, then $f(x) \leq 0$. Otherwise $f(x)$ would be strictly positive and $\lambda f(x) = f(\lambda x)$ would approach infinity as λ does. Since $j_M(\lambda x) = \lambda j_M(x) = 0$, we would deduce from this that $\lambda x \in M$ for all λ, and, consequently, that

$$+\infty = \sup_{\lambda \in \mathbb{R}_+} \lambda f(x) = \sup_{\lambda \in \mathbb{R}_+} f(\lambda x) \leq \sup_{y \in M} f(y) = \sigma_M(f),$$

which is impossible.

b. If $\sigma_M(f) = 0$, then $f(x) \leq 0$ for all $x \in M_a$. Indeed, if $x \in M_a$, there exists $\lambda > 0$ such that $\lambda^{-1} x \in M$ and, consequently,

$$f(x) = \lambda f(\lambda^{-1} x) \leq \lambda \sigma_M(f) \leq 0.$$

c. If $\sigma_M(f) > 0$ and $j_M(x) > 0$, the inequality is satisfied since

$$\frac{f}{\sigma_M(f)} \in M^0, \qquad \frac{x}{j_M(x)} \in M,$$

and therefore since

$$\frac{1}{\sigma_M(f)} f\left(\frac{x}{j_M(x)}\right) = \frac{f(x)}{\sigma_M(f) j_M(x)} \leq 1. \qquad \blacksquare$$

*6. THE DUALITY THEOREM IN CONVEX OPTIMIZATION

We consider the following minimization problem:

(1) $$\alpha = \inf_{x \in K} f(x)$$

when K is a subset of a vector space U defined by

(2) $\quad K = \{x \in X \quad \text{such that} \quad A(x) \in -\mathbb{R}^n_+ \quad \text{and} \quad B(x) = v\}$

where

(3) $\begin{cases} \text{i.} & X \text{ is a } \textit{convex} \text{ subset of a vector space.} \\ \text{ii.} & f: X \to \mathbb{R} \text{ is a } \textit{convex} \text{ function.} \\ \text{iii.} & A \text{ is a mapping from } X \text{ to } \mathbb{R}^n \text{ whose } \textit{components } A_j \text{ are } \\ & \textit{convex functions.} \\ \text{iv.} & B \text{ is a } \textit{linear operator} \text{ from } U \text{ to } \mathbb{R}^m. \\ \text{v.} & v \text{ is a given element of } \mathbb{R}^m. \end{cases}$

The object of the duality theorem is to show that problem (1) is equivalent to the following problem:

(4) $\quad \alpha = \inf_{x \in X} L(x; \bar{p}, \bar{q}) \quad$ for some $\quad \bar{p} \in \mathbb{R}_+^{n*}, \quad \bar{q} \in \mathbb{R}^{m*},$

where the function L is defined on $X \times \mathbb{R}_+^{n*} \times \mathbb{R}^{m*}$ by

(5) $\quad L(x; p, q) = f(x) + \langle p, A(x) \rangle + \langle q, B(x) - v \rangle$

where $p \in \mathbb{R}_+^{n*}, \quad q \in \mathbb{R}^{m*}.$

DEFINITION 1
The function $L(x; p, q)$ is called the Lagrangian *of the minimization problem (1) on the subset $K \subset X$.* ▲

Interpretation

If we interpret $f(x)$ as a cost, we can interpret $\langle p, A(x) \rangle$ as the cost of violation of the constraint $A(x) \in -\mathbb{R}_+^n$ when $p \in \mathbb{R}_+^{n*}$ is considered as a price system, and $\langle q, B(x) - v \rangle$ as the cost of violation of the constraint $B(x) - v = 0$ when $q \in \mathbb{R}^{m*}$ is also considered as a price system.

The Lagrangian $L(x; p, q)$ is then the sum of these three costs. Thus we must find prices $\bar{p} \in \mathbb{R}_+^{n*}$ and $\bar{q} \in \mathbb{R}^{m*}$ such that problems (1) and (4) are equivalent. ∎

Before showing the existence of $\bar{p}$ and $\bar{q}$ such that (4) holds, we first verify that for arbitrary $p \in \mathbb{R}_+^{n*}, q \in \mathbb{R}^{m*},$

(6) $\qquad\qquad\qquad \inf_{x \in X} L(x; p, q) \leq \alpha.$

PROPOSITION 1
We always obtain

(7) $\qquad\qquad \sup_{p \in \mathbb{R}_+^{n*}, q \in \mathbb{R}^{m*}} L(x; p, q) = \begin{cases} f(x) & \text{if } x \in K \\ +\infty & \text{if } x \notin K \end{cases}$

and, consequently,

(8) $\qquad\qquad \sup_{p \in \mathbb{R}_+^{n*}, q \in \mathbb{R}^{m*}} \inf_{x \in X} L(x; p, q) \leq \alpha.$ ▲

Proof. If $x \in K$, then $A(x) \leq 0, B(x) - v = 0$. Since $\langle p, A(x) \rangle \leq 0$ when $p \in \mathbb{R}_+^{n*}$, we deduce that $L(x; p, q) \leq f(x)$ for all $p \in \mathbb{R}_+^{n*}, q \in \mathbb{R}^{m*}$. Moreover, $L(x; 0, 0) = f(x)$. Hence (7) is established when $x \in K$. If $x \notin K$, then there exist at least one element $\bar{p} \in \mathbb{R}_+^{n*}$ and one element $\bar{q} \in \mathbb{R}^{m*}$ such that $\langle \bar{p}, A(x) \rangle + \langle \bar{q}, B(x) - v \rangle = \theta > 0$. Since $\mathbb{R}_+^{n*} \times \mathbb{R}^{m*}$ is a cone, the elements $\{\lambda \bar{p}, \lambda \bar{q}\}$

CH. 2, SEC. 6   THE DUALITY THEOREM IN CONVEX OPTIMIZATION   37

belong to $\mathbb{R}^{n*}_+ \times \mathbb{R}^{m*}$ for all $\lambda > 0$. Therefore

$$\sup_{p,q} L(x;p,q) \geq \sup_{\lambda \geq 0} L(x;\lambda\bar{p},\lambda\bar{q})$$
$$= f(x) + \sup_{\lambda \geq 0} \lambda(\langle \bar{p}, A(x)\rangle + \langle \bar{q}, B(x) - v\rangle) = f(x) + \theta \sup_{\lambda \geq 0} \lambda = \infty.$$

Hence

$$\sup_{p \in \mathbb{R}^{n*}_+, q \in \mathbb{R}^{m*}} \inf_{x \in X} L(x;p,q) \leq \inf_{x \in X} \sup_{p \in \mathbb{R}^{n*}_+, q \in \mathbb{R}^{m*}} L(x;p,q) = \alpha. \quad\blacksquare$$

Now we are going to compare the solutions to the minimization problems (1) and (4).

PROPOSITION 2
Suppose that $\{\bar{p},\bar{q}\} \in \mathbb{R}^{n*}_+ \times \mathbb{R}^{m*}$ satisfies (4). Then $\bar{x} \in K$ minimizes f on K if and only if

(9) $\quad \begin{cases} \text{i.} & L(\bar{x};\bar{p},\bar{q}) = \min_{x \in X} L(x;\bar{p},\bar{q}). \\ \text{ii.} & \langle \bar{p}, A(\bar{x})\rangle = 0. \end{cases}$ $\quad\blacktriangle$

Proof. Suppose that $\bar{x}$ minimizes f on K. Since $\bar{x} \in K$, $\langle p, A(\bar{x})\rangle \leq 0$ for all $p \in \mathbb{R}^{n*}_+$ and in particular for $\bar{p}$. Moreover,

$$f(\bar{x}) = \alpha = \inf_{x \in X} L(x;\bar{p},\bar{q}) \leq L(\bar{x};\bar{p},\bar{q}) = f(\bar{x}) + \langle \bar{p}, A(\bar{x})\rangle + \langle \bar{q}, B\bar{x} - v\rangle.$$

Hence $\langle \bar{p}, A(\bar{x})\rangle$ is also positive. Consequently, $\langle \bar{p}, A(\bar{x})\rangle = 0$. We can then write

$$L(\bar{x};\bar{p},\bar{q}) = f(\bar{x}) + \langle \bar{p}, A(\bar{x})\rangle + \langle \bar{q}, B(\bar{x}) - v\rangle = f(\bar{x}) = \alpha = \inf_{x \in X} L(x;\bar{p},\bar{q}).$$

Conversely, suppose that $\bar{x}$ satisfies (9) and (4). Then

$$\alpha = L(\bar{x};\bar{p},\bar{q}) = f(\bar{x}) + \langle \bar{p}, A(\bar{x})\rangle + \langle \bar{q}, B(\bar{x}) - v\rangle = f(\bar{x}),$$

since $B(\bar{x}) = v$ and since $\langle \bar{p}, A(\bar{x})\rangle = 0$. Therefore $\bar{x}$ minimizes f on K. $\quad\blacksquare$

THEOREM 1
We suppose that the convexity hypotheses (3) are satisfied and, moreover, that

(10) $\quad \begin{cases} \text{i.} & \text{There exists } \tilde{x} \text{ such that } A(\tilde{x}) \in -\mathring{\mathbb{R}}^n_+ \text{ and } B(\tilde{x}) = v. \\ \text{ii.} & v \text{ belongs to the interior of } B(X). \end{cases}$

Then there exists $\bar{p} \in \mathbb{R}^{n*}_+$ and $\bar{q} \in \mathbb{R}^{m*}_+$ such that

$$\alpha = \inf_{x \in X} L(x;\bar{p},\bar{q}). \quad\blacktriangle$$

Proof of Theorem 1. Consider the mapping φ from X to $\mathbb{R} \times \mathbb{R}^n \times \mathbb{R}^m$

defined by

(11) $$\varphi(x) = (f(x), A(x), B(x) - v) \in \mathbb{R} \times \mathbb{R}^n \times \mathbb{R}^m.$$

We introduce the following:

(12) $\begin{cases} \text{i.} & w = (\alpha, 0, 0) \in \mathbb{R} \times \mathbb{R}^n \times \mathbb{R}^m \\ \text{ii.} & C =]0, \infty[\times \mathbb{R}_+^n \times \{0\} \subset \mathbb{R} \times \mathbb{R}^n \times \mathbb{R}^m. \end{cases}$

We are going to show successively that

(13)a. $$w \notin \varphi(X) + C,$$

deduce from the convexity hypotheses (3) and from the separation theorem, that there exist $\lambda \geq 0$, $\bar{p} \in \mathbb{R}_+^{n*}$, $\bar{q} \in \mathbb{R}^m$ such that

(13)b. $$\lambda \alpha \leq \inf_{x \in X} (\lambda f(x) + \langle \bar{p}, A(x) \rangle + \langle \bar{q}, B(x) - v \rangle)$$

and finally that the hypotheses (10) allow us to take $\lambda = 1$, that is, to show that

(13)c. $$\alpha \leq \inf_{x \in X} L(x; \bar{p}, \bar{q}).$$

(Let us remark that that we have the inverse inequality, according to Proposition 1.)

Therefore the proof of the theorem consists in establishing the following four lemmas:

LEMMA 1
w does not belong to $\varphi(X) + C$. ▲

Proof. Indeed, if w belonged to $\varphi(X) + C$, there would exist $\hat{x} \in X$ such that $\alpha > f(\hat{x})$, $A(\hat{x}) \leq 0$, $B(\hat{x}) - v = 0$. Therefore $\hat{x} \in K$ and $\alpha > f(\hat{x})$. This is impossible. ■

LEMMA 2
The convexity hypotheses (3) imply that

(14) $$\varphi(X) + C \text{ is a convex set.} \quad ▲$$

Proof. Let us take $x_i \in X (i = 1, 2)$ and $c_i = (\rho_i, u_i, 0)(i = 1, 2)$ where $\rho_i > 0, u_i \in \mathbb{R}_+^n$. Consider $\theta \in [0, 1]$.
Let us show that

(15) $\theta(\varphi(x_1) + c_1) + (1 - \theta)(\varphi(x_2) + c_2) = \varphi(x) + c$ where $x \in X$ and
$c = (\rho, u, 0) \in C.$

Indeed, it suffices to take

(16)
$$\begin{cases} \text{i.} & x = \theta x_1 + (1-\theta)x_2. \\ \text{ii.} & \rho = \theta \rho_1 + (1-\theta)\rho_2 + \theta f(x_1) + (1-\theta)f(x_2) - f(x) \geq 0 \\ & \text{since } f \text{ is convex.} \\ \text{iii.} & u = \theta u_1 + (1-\theta)u_2 + \theta A(x_1) + (1-\theta)A(x_2) - A(x) \in \mathbb{R}^n_+ \\ & \text{since the components } A_j \text{ of } A \text{ are convex.} \end{cases}$$

Then the equalities (16)ii and iii and the equality $B(x) - v = \theta(B(x_1) - v) + (1-\theta)(B(x_2) - v)$, which follows from the linearity of B, imply that (15) is satisfied. ∎

LEMMA 3
The convexity hypotheses (3) imply that (13)b is satisfied with $\lambda \geq 0, \bar{p} \in \mathbb{R}^{n*}_+$ and $\bar{q} \in \mathbb{R}^{m*}$. ▲

Proof. Since w does not belong to the convex set $\varphi(X) + C \subset \mathbb{R} \times \mathbb{R}^n \times \mathbb{R}^m$, we deduce from the separation theorem *in finite dimensional spaces,* Theorem 4.1, the existence of a linear form, *not identically zero,* $r = (\lambda, \bar{p}, \bar{q})$ that belongs to $\mathbb{R}^* \times \mathbb{R}^{n*} \times \mathbb{R}^{m*}$ and such that

(17) $$\langle r, w \rangle = \lambda \alpha \leq \inf_{x \in X} \langle r, \varphi(x) \rangle + \inf_{c \in C} \langle r, c \rangle.$$

Since, in particular,

$$\inf_{c \in C} \langle r, c \rangle = \inf_{\rho \geq 0} \lambda \rho + \inf_{u \in \mathbb{R}^n_+} \langle \bar{p}, u \rangle \text{ is finite,}$$

we deduce from formula (2) of Proposition 2.5.1 that

(18)
$$\begin{cases} \text{i.} & \lambda \geq 0, \quad \bar{p} \in \mathbb{R}^{n*}_+. \\ \text{ii.} & \inf_{c \in C} \langle r, c \rangle = 0. \end{cases}$$

It then suffices to remark that

$$\langle r, \varphi(x) \rangle = \lambda f(x) + \langle \bar{p}, A(x) \rangle + \langle \bar{q}, B(x) - v \rangle$$

in order to complete the proof of Lemma 3. ∎

LEMMA 4
Hypotheses (10) imply that we can take $\lambda = 1$ in (13)b. ▲

Proof. It suffices to show that $\lambda > 0$ and then to divide by $\lambda > 0$ and to replace $\bar{p}$ and $\bar{q}$ by $\bar{p}/\lambda$ and $\bar{q}/\lambda$, respectively. To this end, suppose that $\lambda = 0$; we deduce from this that $r = (0,0,0)$ is identically zero, which contradicts the fact that r is not identically zero.

Indeed, if $\lambda = 0$, (13)b implies that

(19) $$0 \leq \inf_{x \in X}(\langle \bar{p}, A(x)\rangle + \langle \bar{q}, B(x) - v \rangle)$$

Then we deduce from hypothesis (10)i that $\bar{p} = 0$, since for the element $\tilde{x}$ we would have

(20) $$0 \leq \langle \bar{p}, A(\tilde{x})\rangle < 0 \quad \text{if } \bar{p} \text{ were not zero}$$

because $A(\tilde{x}) \in -\overset{\circ}{\mathbb{R}}^n_+$ and $B(\tilde{x}) - v = 0$.

Finally hypothesis (10)ii implies that $\bar{q} = 0$; indeed, since v belongs to the interior of $B(X)$, we can associate a ball of radius $\eta > 0$ such that $v + B(\eta) \subset B(X)$. We deduce then from (19) with $\bar{p} = 0$ that

(21) $$0 \leq \inf_{w \in B(\eta)} \langle \bar{q}, w \rangle < 0 \quad \text{if } \bar{q} \text{ were not zero.}$$

Therefore the condition $\lambda = 0$ implies that $\bar{p} = 0$ and $\bar{q} = 0$. ∎

*7. VON NEUMANN'S MINIMAX THEOREM

Let us consider the problem of a zero-sum two-person game presented in Section 2.10 of [AAA], p. 120. We take a real-valued function $f : E \times F \mapsto \mathbb{R}$ where

(1)
- i. E is the set of strategies of the first player (Emily).
- ii. F is the set of strategies of the second player (Frank).
- iii. $f(x, y)$ is both the loss of the first player and the gain of the second player (the sum of the gains is zero).

We are going to show that appropriate convexity hypotheses imply that

(2) $\quad \beta = \alpha \quad \text{where} \quad \beta = \sup_{y \in F} \inf_{x \in E} f(x, y) \text{ and } \alpha = \inf_{x \in E} \sup_{y \in F} f(x, y)$

rather than the inequality $\beta \leq \alpha$, which always holds.

THEOREM 1
We suppose that

(3)
- i. E is a compact convex subset.
- ii. $\forall y \in F,\ x \mapsto f(x, y)$ is convex and lower semicontinuous

and that

(4) $\begin{cases} \text{i.} & F \text{ is a convex subset.} \\ \text{ii.} & \forall x \in E,\ y \mapsto f(x, y) \text{ is concave.} \end{cases}$

Then there exists $\bar{x} \in E$ such that

(5) $$\sup_{y \in F} f(\bar{x}, y) = \beta = \alpha.$$ ▲

Interpretation in Game Theory

We are going to deduce from this theorem the well-known minimax theorem proved by Von Neumann.

THEOREM 2
We suppose that

(6) *the sets E and F are compact and convex*

and that

(7) $\begin{cases} \text{i.} & \forall y \in F,\ x \mapsto f(x, y), \text{ is convex and lower semicontinuous.} \\ \text{ii.} & \forall x \in E,\ y \mapsto f(x, y), \text{ is concave and upper semicontinuous.} \end{cases}$

Then there exist $\bar{x} \in E$ and $\bar{y} \in F$ such that

(8) $\quad \forall x \in E, \quad \forall y \in F, \quad f(\bar{x}, y) \leq f(\bar{x}, \bar{y}) \leq f(x, \bar{y}).$ ▲

Proof. We apply Theorem 1 to the function f and to the function $-f$ where the roles of E and F are interchanged. Thus we obtain the existence of $\bar{x} \in E$ and $\bar{y} \in F$ such that

(9) $$\sup_{y \in F} f(\bar{x}, y) = \alpha = \beta = \inf_{x \in E} f(x, \bar{y}).$$

In particular, taking $y = \bar{y}$, we obtain $f(\bar{x}, \bar{y}) \leq \alpha = \beta$ and taking $x = \bar{x}$, we obtain $\alpha = \beta \leq f(\bar{x}, \bar{y})$. Consequently, $\alpha = \beta = f(\bar{x}, \bar{y})$. ■

DEFINITION 1
A pair $\{\bar{x}, \bar{y}\} \in E \times F$ that satisfies (8) is called a saddle point *of f.* ▲

A saddle point is a point of *equilibrium* in a two-person zero-sum game in the following sense: If the player controlling the strategy x modifies his strategy when the second player plays $\bar{y}$, he increases his loss; hence it is in his interest to play $\bar{x}$. Similarly, if the player controlling the strategy y modifies his strategy when the first player plays $\bar{x}$, he diminishes his gain; thus it is in his interest to play $\bar{y}$. This property of equilibrium of saddle points justifies their use as a (reasonable) solution in a two-person zero-sum game. ∎

Remark 1

In addition to its use in game theory, Theorem 1 proves to be a remarkable theoretic "tool" in convex analysis. ∎

As the proof of the theorem is rather long, we derive Theorem 1 from Theorem 3 and Propositions 1 and 2 below, which have an intrinsic interest themselves.

First we shall denote by

(10) $\qquad \mathscr{S} = $ the set of finite subsets $K = \{y_1, \ldots, y_n\}$ of F

and

(11) $$v = \sup_{K \in \mathscr{S}} \inf_{x \in E} \max_{y \in K} f(x, y).$$

Since each point $y \in F$ can be identified with a subset $\{y\} \in \mathscr{S}$, we deduce that

(12) $$\beta = \sup_{\{y\}} \inf_{x \in E} \max_{y \in \{y\}} f(x, y) \leq v.$$

Moreover, since $\max_{y \in K} f(x, y) \leq \sup_{y \in F} f(x, y)$, we obtain $\inf_{x \in E} \max_{y \in K} f(x, y) \leq \inf_{x \in E} \sup_{y \in F} f(x, y) = \alpha$ and, consequently, $v \leq \alpha$. Putting these results together, then, we always have

(13) $$\beta \leq v \leq \alpha.$$

We are going to show that under suitable topological hypotheses, $v = \alpha$ (Theorem 3) and that under suitable convexity hypotheses, $\beta = v$. (See Propositions 1 and 2.)

THEOREM 3
Let us suppose that

(14) $\qquad\qquad E$ *is compact*

and that

(15) $\qquad \forall x \in F, \quad x \mapsto f(x, y) \quad$ *is lower semicontinuous.*

Then there exists $\bar{x} \in E$ such that

(16) $$\sup_{y \in F} f(\bar{x}, y) = \alpha = v.$$ ▲

Proof. It is sufficient to show that there exists $\bar{x} \in E$ such that

(17) $$\sup_{y \in F} f(\bar{x}, y) \leq v.$$

(Since $\alpha \leq \sup_{y \in F} f(\bar{x}, y)$ and $v \leq \alpha$, we deduce from this that $v = \alpha$.) Set

(18) $$S_y = \{x \in E \quad \text{such that} \quad f(x, y) \leq v\}.$$

Inequality (17) means that

(19) $$\bar{x} \in \bigcap_{y \in F} S_y.$$

Thus we must show that this intersection is nonempty. To this end, since E is compact, we shall verify that the S_y's are closed sets satisfying the finite intersection property. The set S_y is closed since S_y is the lower section of the lower semicontinuous function $x \mapsto f(x, y)$. Let us show that for every finite sequence $K = \{y_1, \ldots, y_n\} \in \mathscr{S}$ of F, the intersection $\bigcap_{i=1,\ldots,n} S_{y_i} \neq \emptyset$. Since E is compact and since $x \mapsto \max_{i=1,\ldots,n} f(x, y_i) = \max_{y \in K} f(x, y)$ is lower semicontinuous, we deduce that there exists $\bar{x} \in E$ that minimizes this function. (See Theorem 3.9.2 of [AAA], p. 117.) Such an $\bar{x} \in E$ satisfies

$$\max_{y \in K} f(\bar{x}, y) = \inf_{x \in E} \max_{y \in K} f(x, y) \leq \sup_{K \in \mathscr{S}} \inf_{x \in E} \max_{y \in K} f(x, y) \leq v$$

and, consequently, $\bar{x} \in \bigcap_{i=1,\ldots,n} S_{y_i}$. Therefore the intersection of the compact sets S_y is nonempty, and there exists $\bar{x}$ satisfying (19), and thus (17). ∎

To show now that $v = \beta$, we write

(20) $$v = \sup_{K \in \mathscr{S}} \alpha_K \quad \text{where} \quad \alpha_K = \min_{x \in E} \max_{y \in K} f(x, y).$$

Proposition 1 consists in showing that

(21) for all K, $\quad \alpha_K = \beta_K \quad$ where $\quad \beta_K = \max_{\lambda \in M^n} \inf_{x \in E} \sum_{i=1}^{n} \lambda_i f(x, y_i)$

where $K = \{y_1, \ldots, y_n\}$ and $M^n = \{\lambda \in \mathbb{R}_+^n \text{ such that } \sum_{i=1}^n \lambda_i = 1\}$. Proposition 2 consists in showing that

(22) $$\text{for all } K, \quad \beta_K \leq \beta.$$

Properties (21) and (22) imply that $v = \sup_{K \in \mathscr{S}} \alpha_K = \sup_{K \in \mathscr{S}} \beta_K \leq \beta$ and, consequently, that $v = \beta$ since $\beta \leq v$.

PROPOSITION 1
Let us suppose that

(23) $\begin{cases} \text{i.} & E \text{ is a convex subset of a vector space.} \\ \text{ii.} & \forall y \in F, \quad x \mapsto f(x, y) \text{ is convex.} \end{cases}$

Then for all $K = \{y_1, \ldots, y_n\} \in \mathscr{S}$, we have $\alpha_K = \beta_K$. ▲

Proof. We are going to associate to K and to f the function $\tilde{f}$ defined on $E \times M^n$ by

(24) $$\tilde{f}(x, \lambda) = \sum_{i=1}^{n} \lambda_i f(x, y_i).$$

We remark then that

(25) $$\alpha_K = \inf_{x \in E} \sup_{\lambda \in M^n} \tilde{f}(x, \lambda) = \inf_{x \in E} \sup_{y \in K} f(x, y),$$

since for all fixed x, we always have

(26) $$\sup_{\lambda \in M^n} \sum \lambda_i f(x, y_i) = \sup_{1 \le i \le n} f(x, y_i) = \sup_{y \in K} f(x, y).$$

Consequently, we always have the inequality $\beta_K \le \alpha_K$. Now we are going to show that

(27) $$\alpha_K \le a \quad \text{when} \quad a > \beta_K.$$

Then, by letting a approach β_K, we deduce that $\beta_K \ge \alpha_K$ and therefore that $\beta_K = \alpha_K$. Indeed, we are going to show that

(28) $$\exists x_a \in E \quad \text{such that} \quad \sup_{\lambda \in M^n} \tilde{f}(x_a, \lambda) \le a \quad \text{when} \quad a > \beta_K.$$

We denote by Φ the mapping from E to $\mathbb{R}^n$ defined by

(29) $$\Phi(x) = \{f(x, y_1), \ldots, f(x, y_i), \ldots, f(x, y_n)\}.$$

We also introduce

(30) $$\begin{cases} \text{i.} & \Phi_+(E) = \Phi(E) + \mathbb{R}_+^n. \\ \text{ii.} & \theta = \{1, \ldots, 1, \ldots, 1\} \in \mathbb{R}^n. \end{cases}$$

LEMMA 1
We can write statement (28) in the form

(31) $$a\theta \in \Phi_+(E) \quad \text{when} \quad a > \beta_K.$$

Proof. Indeed, to say that $a\theta \in \Phi_+(E)$ is the same as to say that there exists $x_a \in E$ and $u = (u_1, \ldots, u_n) \in \mathbb{R}_+^n$ such that $a\theta = \Phi(x_a) + u$, that is, that $a = f(x_a, y_i) + u_i$ for all i. Since $\lambda_i \ge 0$ for all i and $\sum_{i=1}^{n} \lambda_i = 1$, this is equivalent to writing

$$a = \sum_{i=1}^{n} \lambda_i a = \sum_{i=1}^{n} \lambda_i f(x, y_i) + \sum_{i=1}^{n} \lambda_i u_i \ge \tilde{f}(x_a, \lambda) \quad \text{for all} \quad \lambda \in M^n.$$

LEMMA 2
The convexity hypotheses (23) imply that $\Phi_+(E)$ is a convex set.

Proof. The proof is analogous to that of Lemma 6.2. ∎

LEMMA 3
The convexity hypotheses (25) imply (31). ▲

Proof. Let us suppose on the contrary that $a\theta \notin \Phi_+(E)$. Since $\Phi_+(E)$ is a convex subset of $\mathbb{R}^n$, we deduce from the separation theorem in finite dimensional spaces that there exists a linear form $\lambda = (\lambda_1, \ldots, \lambda_n)$, *not identically zero*, that separates $a\theta$ from $\Phi_+(E)$, that is to say, that satisfies

$$\langle \lambda, a\theta \rangle = a \langle \lambda, \theta \rangle \leq \inf_{u \in \Phi_+(E)} \langle \lambda, u \rangle$$

(32)
$$= \inf_{x \in E} \langle \lambda, \Phi(x) \rangle + \inf_{v \in \mathbb{R}^n_+} \langle \lambda, v \rangle.$$

From this we deduce first that $\inf_{v \in \mathbb{R}^n_+} \langle \lambda, v \rangle$ is finite, which implies, by formula (2) of Proposition 2.5.1, that

(33) $\begin{cases} \textbf{i.} & \text{The components } \lambda_i \text{ are nonnegative.} \\ \textbf{ii.} & \inf_{v \in \mathbb{R}^n_+} \langle \lambda, v \rangle = 0. \end{cases}$

Dividing the inequality (32) by

$$\langle \lambda, \theta \rangle = \sum_{i=1}^n \lambda_i > 0 \quad (\text{since } \lambda \neq 0 \text{ and } \lambda \in \mathbb{R}^n_+),$$

we obtain the existence of $\bar{\lambda} = \lambda / \langle \lambda, \theta \rangle \in M^n$ satisfying

$$a \leq \inf_{x \in E} \langle \bar{\lambda}, \Phi(x) \rangle = \inf_{x \in E} \tilde{f}(x, \bar{\lambda})$$
$$\leq \sup_{\lambda \in M^n} \inf_{x \in E} f(x, \lambda) = \beta_K.$$

We have thus obtained a contradiction, since we assumed that $\beta_K < a$. ∎

It now remains to establish Proposition 2.

PROPOSITION 2
Let us suppose that

$\begin{cases} \textbf{i.} & F \text{ is a convex subset of a vector space.} \\ \textbf{ii.} & \forall x \in E, \; y \mapsto f(x, y) \text{ is concave.} \end{cases}$

Then for all $K = \{y_1, \ldots, y_n\}$, *we have* $\beta_K \leq \beta$. ▲

Proof. Let λ be a fixed element of M^n and $y_\lambda = \sum_{i=1}^n \lambda_i y_i$. Since f is concave with respect to y, we have

$$\sum_{i=1}^{n} \lambda_i f(x, y_i) \leq f(x, y_\lambda)$$

and therefore

$$\inf_{x \in E} \sum_{i=1}^{n} \lambda_i f(x, y_i) \leq \inf_{x \in E} f(x, y_\lambda) \leq \sup_{y \in F} \inf_{x \in E} f(x, y) = \beta.$$

Taking the supremum with respect to λ, we deduce that $\beta_K \leq \beta$. ∎

*8. CHARACTERIZATION OF PARETO OPTIMA

Consider an n-person game $i = 1, \ldots, n$ described when we are given

(1) $\begin{cases} \textbf{i.} & \text{A set } X \text{ of strategies.} \\ \textbf{ii.} & n \text{ loss functions } f_i : X \mapsto \mathbb{R} \text{ associating to each strategy } x \in X \\ & \text{the loss } f_i(x) \text{ of the } i\text{th player.} \end{cases}$

These loss functions define a *partial preordering* on X by

(2) $\qquad x \leq y \qquad \text{if for all} \qquad i = 1, \ldots, n, \quad f_i(x) \leq f_i(y).$

It is natural to distinguish the maximal elements for this preordering.

DEFINITION 1
We shall say that $\bar{x}$ is a weak Pareto minimum if there exists no element $y \in X$ such that

(3) $\qquad\qquad f_i(y) < f_i(\bar{x}) \qquad \text{for all} \qquad i = 1, \ldots, n.$ ▲

It is possible to select a Pareto minimum by minimizing on X a convex combination $f_\lambda(x) = \sum_{i=1}^{n} \lambda_i f_i(x)$ of loss functions.

PROPOSITION 1
Consider $\lambda = (\lambda_1, \ldots, \lambda_n) \in \mathbb{R}_+^{n*}$ such that $\sum_{i=1}^{n} \lambda_i = 1$. If $\bar{x} \in X$ minimizes $f_\lambda(x) = \sum_{i=1}^{n} \lambda_i f_i(x)$, then $\bar{x}$ is a weak Pareto minimum. ▲

Proof. If $\bar{x}$ were not a weak Pareto minimum, there would exist y such that $f_i(y) < f_i(\bar{x})$ for all i. Since the components λ_i are positive, we deduce that

$$f_\lambda(y) = \sum_{i=1}^{n} \lambda_i f_i(y) < f_\lambda(\bar{x}) = \sum_{i=1}^{n} \lambda_i f_i(\bar{x}),$$

and, consequently, arrive at a contradiction. ∎

We are going to show that appropriate convexity hypotheses imply the converse, namely, that every weak Pareto minimum can be obtained by minimizing on X a suitable loss function f_λ.

THEOREM 1
Suppose that

(4) $\quad \begin{cases} \text{i.} & X \text{ is a convex subset of a vector space } U. \\ \text{ii.} & \text{The loss functions } f_i \text{ are convex.} \end{cases}$

If $\bar{x}$ is a weak Pareto minimum, there exists $\lambda \in \mathbb{R}_+^{n*}$ satisfying $\sum_{i=1}^n \lambda_i = 1$ and

(5) $\qquad\qquad \bar{x} \text{ minimizes } f_\lambda(x) = \sum_{i=1}^n \lambda_i f_i(x) \text{ on } X.$ ▲

Proof. We denote by F the mapping from X to $\mathbb{R}^n$ defined by

(6) $\qquad\qquad F(x) = \{f_1(x), \ldots, f_n(x)\}.$

We also set

(7) $\qquad\qquad \mathring{F}_+(X) = F(X) + \mathring{\mathbb{R}}_+^n.$

We are going to show successively that

(8) $\quad F(\bar{x})$ does not belong to $\mathring{F}_+(X)$ if $\bar{x}$ is a weak Pareto minimum

and that the convexity hypotheses imply that

(9) $\qquad\qquad \mathring{F}_+(X)$ is a convex subset of $\mathbb{R}^n$,

and we shall deduce from the separation theorem the existence of $\bar{\lambda}$ satisfying the conclusion of the theorem. The proof consists in establishing the following three lemmas.   ∎

LEMMA 1
An element $\bar{x} \in X$ is a weak Pareto minimum if and only if $F(\bar{x})$ does not belong to $\mathring{F}_+(X)$.   ▲

Proof. Indeed, since $\mathring{\mathbb{R}}_+^n$ is the cone of vectors u all of whose components u_i are strictly positive, we deduce from this that $F(\bar{x}) \in \mathring{F}_+(X)$ if and only if there exists $y \in X$ such that $f_i(\bar{x}) > f_i(y)$ for all i, that is, if and only if $\bar{x}$ is not a weak Pareto minimum.   ∎

LEMMA 2
The convexity hypotheses (4) imply that $\mathring{F}_+(X)$ is convex.   ▲

Proof. The proof is analogous to that of Lemma 6.2.   ∎

LEMMA 3

There exists $\bar{\lambda}$ satisfying the conclusions of Theorem 1.   ▲

Proof. Since $F(\bar{x})$ does not belong to the convex set $\overset{\circ}{F}_+(X)$, we deduce from the separation Theorem 4.1 in finite dimensional spaces the existence of a linear form $\lambda \in \mathbb{R}^{n*}$, *not identically zero*, such that

(10) $\quad \langle \lambda, F(\bar{x}) \rangle \leq \inf_{u \in \overset{\circ}{F}_+(X)} \langle \lambda, u \rangle = \inf_{x \in X} \langle \lambda, F(x) \rangle + \inf_{v \in \mathbb{R}^n_+} \langle \lambda, v \rangle.$

Since $\inf_{v \in \mathbb{R}^n_+} \langle \lambda, v \rangle$ is finite, we obtain, by formula (2) of Proposition 2.5.1,

(11) $\quad \begin{cases} \text{i.} & \lambda \in \mathbb{R}^{n*}_+. \\ \text{ii.} & \inf_{v \in \mathbb{R}^n_+} \langle \lambda, v \rangle = 0. \end{cases}$

Since λ is not identically zero, $\sum_{i=1}^n \lambda_i > 0$. Hence dividing the two sides of inequality (10) by $\sum_{i=1}^n \lambda_i$, we deduce that

$$\langle \bar{\lambda}, F(\bar{x}) \rangle = f_{\bar{\lambda}}(\bar{x}) \leq \inf_{x \in X} \langle \bar{\lambda}, F(x) \rangle = \inf_{x \in X} f_{\bar{\lambda}}(x); \qquad \bar{\lambda} = \frac{\lambda}{\sum_{i=1}^n \lambda_i}. \qquad \blacksquare$$

Remark 1

The *cooperative* concepts of a solution in *n*-person games consist in defining selection processes for Pareto minima.   ■

Remark 2

We shall briefly describe a procedure for selecting a Pareto minimum. Let us denote by

(12) $\qquad\qquad \alpha_i = \inf_{x \in X} f_i(x)$

the minimal loss of the *i*th player (when he is the only one to play), and suppose we are given a strategy $x_0 \in X$ such that

(13) $\qquad\qquad f_i(x_0) > \alpha_i \qquad$ for all $\qquad i.$

We set

(14) $\qquad\qquad u(x) = \max_{i=1,\ldots,n} \frac{f_i(x) - \alpha_i}{f_i(x_0) - \alpha_i},$

which measures the maximum of the relative losses of the players yielded by x.   ■

PROPOSITION 2

Let $d = \inf_{x \in X} u(x)$. Then $\bar{x} \in X$ minimizes u on X if and only if

(15) $\qquad \forall i = 1, \ldots, n, \qquad f_i(\bar{x}) \leq (1-d)\alpha_i + df_i(x_0).$

It is a weak Pareto minimum, which exists if we suppose, for example, that

(16) $\qquad X$ is compact and the functions f_i are lower semicontinuous. ▲

Proof. **a.** If $u(\bar{x}) = d$, we deduce (15) from (14). Conversely, it is clear that (15) implies that $u(\bar{x}) \leq d$.
b. Suppose now that $\bar{x}$ minimizes u and is not a weak Pareto minimum: There exists $y \in X$ such that $f_i(y) < f_i(\bar{x})$ for all i. We deduce then that

$$\frac{f_i(y) - \alpha_i}{f_i(x_0) - \alpha_i} < \frac{f_i(\bar{x}) - \alpha_i}{f_i(x_0) - \alpha_i}$$

for all i and, consequently, that $u(y) < u(\bar{x})$, which is impossible.
c. If the loss functions f_i are lower semicontinuous, then so is the function u defined by (14). If X is compact, we obtain the existence of $\bar{x} \in X$ minimizing u on X, that is, *the existence of a weak Pareto minimum $\bar{x}$ satisfying* (15) *and, therefore, such that $f_i(\bar{x}) \leq f_i(x_0)$ for all i.* ∎

Remark 3

This selection procedure depends only on the initial strategy $x_0 \in X$ (which is thereby "improved"); it is *invariant* under linear transformations with positive coefficients operating on the loss functions of the players. ∎

CHAPTER 3

Dual Spaces and Transposed Operators

In this chapter we take up the study of duality, which is a remarkable tool in numerous problems. It is within the framework of Hilbert spaces that the theory of duality is simplest. This is the second advantage of restricting ourselves to Hilbert spaces, but with the drawback that many results do not generalize to Banach spaces.

We begin in Section 1 by establishing the fundamental isomorphism theorem: *If V is a Hilbert space, there exists an isometry J from V onto its dual*, which associates to every x the differential $Jx \in V^*$ of the quadratic function $x \mapsto \frac{1}{2}\|x\|^2$.

Moreover, we use this isometry to construct a completion of a prehilbert space. At this point, then, one can study a method for constructing spaces of square summable functions found in Sections 1 and 2 of Chapter 6. Our first concern here is to characterize in Section 2 the spaces F isometric to the dual V^*, to be able, when convenient, to replace $\mathscr{L}(V, \mathbb{R})$ by such an isometric space chosen explicitly or implicitly and called a *realization of the dual*. We can always take $F = V$, since it is isometric to V^*. When we choose this realization, we say that V is a *pivot space*. We show that V is *rarely* a good choice for a realization. (This will become clear after reading Section 5.) Indeed, each problem motivates the particular choice of a realization of the dual. Sections 2 and 3 of Chapter 6 show the advantage in identifying the dual of the space $L^2(\Omega, a)$ with the space $L^2(\Omega, 1/a)$ rather than with itself.

We show that a Banach space F is isometric to V^* if and only if there exists a nondegenerate bilinear form

$$\{f, x\} \in V^* \times V \to \langle f, x \rangle \in \mathbb{R}$$

satisfying

$$\|f\|_F = \sup_{x \in V} \frac{|\langle f, x \rangle|}{\|x\|}.$$

This criterion enables us to recognize easily whether a space F is isometric to V.

In Section 3 we associate to every operator $A \in \mathscr{L}(V, F)$ its transpose $A^* \in \mathscr{L}(F^*, V^*)$ defined by

$$\forall f \in F^*, \quad \forall x \in V, \quad \langle A^* f, x \rangle = \langle f, Ax \rangle.$$

We show that the transposition $A \to A^*$ is an isometry from $\mathscr{L}(V, F)$ onto $\mathscr{L}(F^*, V^*)$.

We show in Section 4 that $A \in \mathscr{L}(V, F)$ is injective if and only if $A^* \in \mathscr{L}(F^*, V^*)$ has a dense image in V^*. This justifies calling any injective operator $A \in \mathscr{L}(V, F)$ with a dense image an *embedding*, in such a way that the transpose of an embedding is an embedding. We conclude with the lemma of Farkas by which we can characterize the inverse image of a cone and the direct image of a cone when it is closed.

We define in Section 3.6 the concept of a V-elliptic operator $A \in \mathscr{L}(V, V^*)$ satisfying

$$\exists c > 0 \quad \text{such that} \quad \langle Ax, x \rangle \geq c \|x\|^2 \quad \forall x \in V.$$

The theorem of Lax-Milgram states that every V-elliptic operator A is an isomorphism from V onto its dual. We generalize this result in Section 7 by showing the existence of a solution $x \in K$ to the *variational inequalites*

$$\langle Ax, x - y \rangle \leq \langle f, x - y \rangle \quad \forall y \in K$$

when A is V-elliptic, $f \in V^*$, and K is a closed convex subset of V.

We apply the latter result to establish the existence of a noncooperative equilibrium in an n-person game where the loss functions are quadratic.

1. THE DUAL OF A HILBERT SPACE

We shall deduce from the best approximation theorem the fundamental fact that *a Hilbert space is isometric to its dual*.

Let us consider a prehilbert space V for a scalar product $((x, y))$ and its topological dual $V^* = \mathscr{L}(V, \mathbb{R})$, the space of continuous linear forms on V. We already know that V^* is a Banach space for the dual norm defined by $\|f\|_* = \sup_{x \in V} |f(x)| / \|x\|$. (See proposition 1.2.4.)

PROPOSITION 1
Let V be a prehilbert space. There exists an isometry J from V to its dual V^*: for every $x \in V$, Jx is the differential of the function $x \mapsto \frac{1}{2} \|x\|^2$. ▲

Proof. We associate to every $x \in V$ the continuous linear form Jx on V

defined by

(1) $$Jx : y \in V \mapsto Jx(y) = ((x,y)).$$

This is clearly a linear form that is continuous since

(2) $$\|Jx\| = \sup_{y \in V} \frac{|Jx(y)|}{\|y\|} = \sup_{y \in V} \frac{|((x,y))|}{\|y\|} = \frac{|((x,x))|}{\|x\|} = \|x\|$$

according to the Cauchy-Schwarz inequality.

Moreover, Jx is the differential of $x \mapsto \frac{1}{2}\|x\|^2$, since

$$\lim_{\|y\| \to 0} \frac{(\frac{1}{2}\|x+y\|^2 - \frac{1}{2}\|x\|^2 - \langle Jx, y \rangle)}{\|y\|} = \lim_{\|y\| \to 0} (\frac{1}{2}\|y\|)) = 0$$

The mapping $J : x \in V \mapsto Jx \in V^*$ is clearly linear since

$$J(\lambda_1 x_1 + \lambda_2 x_2)(y) = ((\lambda_1 x_1 + \lambda_2 x_2, y))$$
$$= \lambda_1((x_1, y)) + \lambda_2((x_2, y)) = \lambda_1(Jx_1)(y) + \lambda_2(Jx_2)(y).$$

Equality (2) shows that J is an isometry. ∎

DEFINITION 1

We say the isometry $J \in \mathscr{L}(V, V^*)$ defined by (1) is the duality operator *from V to its dual.* ▲

We are going to prove the fundamental theorem establishing that if V is a Hilbert space, this isometry is surjective.

THEOREM 1

Let V be a Hilbert space for the scalar product $((x,y))$ and $J \in \mathscr{L}(V, V^*)$ be the duality operator. Then J is a surjective isometry from V onto V^*. The dual space V^* is a Hilbert space for the scalar product

(3) $$((f,g))_* = ((J^{-1}f, J^{-1}g)) = f(J^{-1}g).$$ ▲

Proof. We must show that J is surjective. To this end, we take $f \in V^*$, $f \neq 0$, and we shall construct $x \in V$ such that $f = Jx$, that is, such that $f(y) = Jx(y) = ((x,y))$ for every $y \in V$. We introduce the kernel $H = \operatorname{Ker} f$ of F (which is a closed hyperplane), an element $x_0 \in V$ that does not belong to H (that exists because $f \neq 0$), and the orthogonal projection tx_0 of x_0 onto H (which *exists* according to Theorem 1.5.2).

The element $y_0 = (x_0 - tx_0)/[f(x_0)]$ satisfies $f(y_0) = 1$ (because $f(tx_0) = 0$ since $tx_0 \in H$). Consider the operator s which associates to every y the element

(4) $$sy = y - f(y)y_0.$$

This operator s sends V to H since
$$f(sy) = f(y) - f(y)f(y_0) = 0.$$
Then, since $((y_0, z)) = 0$ for all z in H, we obtain
(5) $$((y_0, y)) - f(y)\|y_0\|^2 = ((y_0, sy)) = 0.$$
This implies that the element $x = y_0/\|y_0\|^2$ is a solution to the equation $Jx = f$, since (5) implies

$$f(y) = \left(\left(\frac{y_0}{\|y_0\|^2}, y\right)\right) = ((x, y)) = Jx(y) \quad \text{for all} \quad y \in V.$$

Consider then the dual space V^*. It is clear that
(6) $$((f, g))_* = ((J^{-1}f, J^{-1}g))$$
is a positive symmetric bilinear form on V^*. Since J is an isometry, we deduce that $((f, f))_* = \|J^{-1}f\|^2 = \|J(J^{-1}f)\|_*^2 = \|f\|_*^2$, that is, that the dual norm $\|f\|_*^2$ on V^* also comes from the scalar product $((f, g))_*$ defined by (6). Since V^* is complete, it is therefore a Hilbert space. ∎

As this point we give some applications.

Completion of a Prehilbert Space

THEOREM 2
Let V be a prehilbert space: There exists a completion $\hat{V}$ of V, that is, an isometry j from V to the Hilbert space $\hat{V}$ such that $j(V)$ is dense in $\hat{V}$. ▲

Proof. Indeed, we know that J is an isometry from V to V^*. Take $\hat{V} = \overline{J(V)}$, the closure of $J(V)$ in V^*. This is a Banach space, since V^* is complete. Moreover, the scalar product defined on $J(V)$ by (6), being a continuous bilinear form, can be extended by density to a scalar product on $\hat{V} = \overline{J(V)}$ by applying Theorem 1.3.1. Then $\hat{V}$ is a Hilbert space in which $J(V)$ is dense, that is, a completion of V. ∎

Remark 1

Let us recall that the completions of a Hilbert space are all mutually isomorphic. (See [AAA], p. 159, Theorem 4.7.2.) We have given a rapid (although artificial) example of how a completion can be constructed. We study the possibility of choosing a completion in an overspace of V in Chapter 5, Section 6. ∎

Dual Base

DEFINITION 2
Let $\{e_n\}_n$ be an orthogonal sequence in a Hilbert space V. We say that the sequence $\{e_n^*\}_n$ in V^* defined by $e_n^* = Je_n$ is the *dual sequence* of the sequence $\{e_n\}$. ▲

In particular, we obtain

(7) $\qquad \langle e_n^*, e_m \rangle = 0 \quad \text{if} \quad m \neq n, \quad \langle e_n^*, e_n \rangle = 1,$

and if $f \in V^*$,

(8) $\qquad ((f, e_n^*))_* = \langle f, J^{-1} e_n^* \rangle = \langle f, e_n \rangle.$

Then if $\{e_n\}_n$ is an orthonormal base, we deduce from Proposition 1.7.1 that $\{e_n^*\}_n$ is also an orthonormal base and that

(9) $\qquad \forall x \in V, \quad x = \sum_{n=1}^{\infty} \langle e_n^*, x \rangle e_n; \quad \forall f \in V^*, \quad f = \sum_{n=1}^{\infty} \langle f, e_n \rangle e_n^*.$

Bidual of V

We call the *bidual* V^{**} of V the dual of V^*. Since V^* is a Hilbert space, there exists an isometry K from V^* onto the bidual. Then if V is a Hilbert space, the product KJ of the isometries $J \in \mathcal{L}(V, V^*)$ and $K \in \mathcal{L}(V^*, V^{**})$ is an isometry from V onto the bidual V^{**} of V.

Conventional Identification

We agree to identify the bidual V^{**} of a Hilbert space V with itself, that is to identify J^{-1} with the isometry from V^* onto V.

Interpretation of Duality

In *economics* Hilbert spaces U represent commodity spaces (or functions with values in commodity spaces). An element f of the dual associates to every "commodity" $x \in U$ a scalar $\langle f, x \rangle \in \mathbb{R}$. If we interpret this scalar as the *value* of the commodity x, f can then be considered as a "*price*," the role of which is to associate a value to every commodity.

In *mechanics* we associate to a given mechanical system a Hilbert space U of the velocities of the system. There is then a duality between the space U and another Hilbert space F whose elements constitute, in a general way, the *forces* that may act on the system.

The value $\langle f, x \rangle$ is the *power*. The traditional method of "virtual works" is none other than the establishing of the duality between a space of velocities or displacements and a space of forces.

2. REALIZATION OF THE DUAL OF A HILBERT SPACE

We are going to characterize those spaces isometric to the dual V^* of a Hilbert space. This is useful in analysis, since it is natural to try to represent the duals of spaces of functions by spaces isomorphic to spaces of functions (or *generalized functions*).

In other words, we choose among all the spaces isomorphic to the dual of a Hilbert space a well-determined representative, which will not necessarily be the space $\mathscr{L}(V, \mathbb{R})$ of continuous linear forms on V. In most cases the choice of such a representative in a given problem is determined by reasons of simplicity.

This is illustrated by the spaces $L^2(\Omega, a)$ of square summable functions with positive "weight" a. The reader is, therefore, advised to read, along with this section, Section 2, in Chapter 6, in which it is established that $L^2(\Omega, 1/a)$ is a realization of the dual of $L^2(\Omega, a)$. The problem of taking distribution spaces as realizations of the duals of Sobolev spaces is the subject of Chapters 7 and 9.

DEFINITION 1
Let V be a Hilbert space. We shall say that the pair $\{F, j\}$ where

(1) $\quad\begin{cases} \text{i.} & F \text{ is a Hilbert space.} \\ \text{ii.} & j \text{ is an isometry from } F \text{ onto the space } \mathscr{L}(V, \mathbb{R}). \end{cases}$

is a realization of the dual of V. We set

(2) $\qquad\qquad \langle f, x \rangle = jf(x) \qquad \forall f \in F, \quad \forall x \in V$

and say that $\langle f, x \rangle$ is a bilinear form on $F \times V$ that establishes a duality between F and V. ▲

More briefly, we also say that $\langle f, x \rangle$ is the *duality pairing* on $F \times V$.

PROPOSITION 1
Let $\{F, j\}$ be a realization of the dual V^* of a Hilbert space V. The duality pairing has the following properties:

(3) $\quad\begin{cases} \text{i.} & \langle f, x \rangle \text{ is not degenerate on } F \times V \text{ (that is, } \langle f, x \rangle = 0 \\ & \text{for all } f \text{ implies } x = 0 \text{ and } \langle f, x \rangle = 0 \text{ for all } x \text{ implies } f = 0.) \\ \text{ii.} & \|f\|_F = \displaystyle\sup_{x \in V} \frac{|\langle f, x \rangle|}{\|x\|}. \end{cases}$ ▲

Proof. (Left as an exercise.)

The fundamental fact is that the properties (3) of a bilinear form on $F \times V$ are sufficient to show that F is isomorphic to the dual V^* of V. They provide a convenient method in applications for choosing a well-determined representative of the dual of a Hilbert space.

THEOREM 1
Let V be a Hilbert space and F a Banach space. Suppose that there exists a bilinear form $\{f, x\} \mapsto \langle f, x \rangle$ on $F \times V$ satisfying the properties (3). Then there exists an isometry j from the space F onto the space V^*. ▲

Proof. Consider the mapping j from F to V^* defined by

(4) $\qquad\qquad jf(x) = \langle f, x \rangle \qquad$ for all $\quad f \in F, \quad x \in V$.

Indeed, $x \mapsto jf(x) = \langle f, x \rangle$ is linear and also continuous, since $|jf(x)| = |\langle f, x \rangle| \leq \|f\|_F \|x\|$ according to (3)ii. This mapping j is linear, since

$$j(\lambda f + \mu g)(x) = \langle \lambda f + \mu g, x \rangle = \lambda \langle f, x \rangle + \mu \langle g, x \rangle$$
$$= \lambda jf(x) + \mu jg(x) \qquad \text{for all} \quad x \in V.$$

It is an isometry, since, according to (3)ii,

$$\|jf\|_* = \sup_{x \in V} \frac{|jf(x)|}{\|x\|} = \sup_{x \in V} \frac{|\langle f, x \rangle|}{\|x\|} = \|f\|_F.$$

It remains to for us *show that j is surjective*, that is, that the image $j(F)$ is *complete and dense* in V^*.

The image is *complete*: Let $\{jf_n\}_n$ be a Cauchy sequence of elements jf_n of $j(F)$. Since j is an isometry, the sequence of f_n is a Cauchy sequence that converges to an element $f \in F$ since F is complete. Thus jf_n converges to jf since j is continuous.

The image $j(F)$ is *dense*: We use the density criterion (Theorem 2.2.1). Let $x_0 \in V = V^{**}$ be a continuous linear form on V^* that vanishes on $j(F)$. Then $\langle f, x_0 \rangle = jf(x_0) = 0$ for all $f \in V^*$. Since the bilinear form $\langle f, v \rangle$ is not degenerate, we conclude that $x_0 = 0$. Therefore, $j(F)$ is dense in V^*. ∎

We deduce from Theorem 1.1 and 2.1 that $K = j^{-1}J \in \mathscr{L}(V, F)$ *is a surjective isometry* satisfying

(5) $\qquad\qquad ((x, y)) = \langle Kx, y \rangle \qquad \forall x, y \in V,$

since $((x, y)) = Jx(y) = \langle j^{-1}Jx, y \rangle = \langle Kx, y \rangle$.

Conventional Notation

To facilitate this exposition we agree henceforth that a realization $\{F, j\}$ of the dual has been chosen once and for all (explicitly or implicitly). We agree to set

(6) $\qquad F = V^* \quad \text{and} \quad jf(x) = \langle f, x \rangle$

and we say that the isometry K from V onto $F = V^*$ is the *duality operator* associated with the scalar product on V and with the duality pairing on $V^* \times V$ by the relation (5).

The theorem establishing the isomorphism between V and its dual (Theorem 1.1) is an important example of a realization of the dual; we can state it in the following form: *The pair $\{V, J\}$ is a realization of the dual of a Hilbert space V for the duality pairing $((f, v))$ defined on $V \times V$.*

We shall see in examples that apart from exceptional cases, it is not convenient to choose (V, J) as a realization of the dual V. We emphasize this point by giving a special name to those Hilbert spaces for which such a realization is chosen.

DEFINITION 2
We say that a Hilbert space H for the scalar product (x, y) is a pivot space *if it is identified with its dual. In this case we set $H^* = H$, $J = 1$, and $\langle x, y \rangle = (x, y)$.* ▲

In other words, a pivot space is a space that is "equal to its dual", whereas a Hilbert space is only "isometric to its dual."

3. TRANSPOSITION OF OPERATORS

Let V_1 and V_2 be two Hilbert spaces, $A \in \mathscr{L}(V_1, V_2)$ a continuous linear operator from V_1 to V_2, and V_1^* and V_2^* the duals of V_1 and V_2, respectively.

PROPOSITION 1
Let $A \in \mathscr{L}(V_1, V_2)$. The linear operator A^* from V_2^* to V_1^* defined by

(1) $\qquad \langle A^* f_2, x_1 \rangle = \langle f_2, A x_1 \rangle \quad \forall f_2 \in V_2^*, \quad \forall x_1 \in V_1^*$

is continuous and satisfies

(2) $\qquad \| A^* \|_{\mathscr{L}(V_2^*, V_1^*)} = \| A \|_{\mathscr{L}(V_1, V_2)}.$ ▲

DEFINITION 1
We say that A^* is the transpose *of A.* ▲

Proof of Proposition. It is clear that the mapping A^* defined by (1) sends V_2^* to V_1^* and is linear. It is continuous since

$$\|A^*\| = \sup_{f_2 \in V_2^*} \frac{\|A^*f_2\|_*}{\|f_2\|_*} = \sup_{f_2 \in V_2^*} \sup_{x_1 \in V_1} \frac{|\langle A^*f_2, x_1 \rangle|}{\|f_2\|_* \|x_1\|}$$

$$= \sup_{x_1 \in V_1} \sup_{f_2 \in V_2^*} \frac{|\langle f_2, Ax_1 \rangle|}{\|x_1\| \|f_2\|_*} = \sup_{x_1 \in V_1} \frac{\|Ax_1\|}{\|x_1\|} = \|A\|. \quad \blacksquare$$

PROPOSITION 2
The mapping $A \mapsto A^*$ is an isometry from $\mathscr{L}(V_1, V_2)$ onto $\mathscr{L}(V_2^*, V_1^*)$. If $B \in \mathscr{L}(V_2, V_3)$, then $(BA)^* = A^*B^* \in \mathscr{L}(V_3^*, V_1^*)$. ▲

Proof. It is clear that $(\lambda A + \mu B)^* = \lambda A^* + \mu B^*$, which, together with (2), shows that $A \mapsto A^*$ is an isometry. If $B \in \mathscr{L}(V_2, V_3)$ and $A \in \mathscr{L}(V_1, V_2)$, then

$$\langle (BA)^*f_3, x_1 \rangle = \langle f_3, BAx_1 \rangle = \langle B^*f_3, Ax_1 \rangle = \langle A^*B^*f_3, x_1 \rangle$$

for all $x_1 \in V_1$, which implies that $(BA)^* = A^*B^*$. $\blacksquare$

*Remark 1. "Adjoint" Operators

Let $J_1 \in \mathscr{L}(V_1, V_1^*)$ and $J_2 \in \mathscr{L}(V_2, V_2^*)$ be duality operators. We associate to every $A \in \mathscr{L}(V_1, V_2)$ the operator $A^\circledast = J_1^{-1} A^* J_2 \in \mathscr{L}(V_2, V_1)$. Since $((x_i, y_i)) = \langle J_i x_i, y_i \rangle$ for $i = 1, 2$, this operator satisfies

(3) $\qquad ((A^\circledast x_2, x_1))_1 = ((x_2, Ax_1))_2 \qquad \forall x_1 \in V_1, \quad x_2 \in V_2.$ ▲

DEFINITION 2
We say that the operator $A^\circledast = J_1^{-1} A^* J_2 \in \mathscr{L}(V_2, V_1)$ is the adjoint *of the operator $A \in \mathscr{L}(V_1, V_2)$.* ▲

Naturally, if V_1 and V_2 are both chosen as pivot spaces, the concepts of transpose and of adjoint coincide. In this book we use only the concept of transpose.

4. TRANSPOSITION OF INJECTIVE OPERATORS

We begin to studying the transpose of an injective operator and the transpose of an operator with a dense image. First we examine the relation between the image and the kernel of an operator and its transpose.

LEMMA 1

Let M be a subset of a Hilbert space V. Then

(1) $\begin{cases} \text{i.} & J(M^{\ominus}) = \{f \in V^* \text{ such that } \langle f, x \rangle \leq 0 \text{ for all } x \in M\}. \\ \text{ii.} & J(M^{\oplus}) = \{f \in V^* \text{ such that } \langle f, x \rangle = 0 \text{ for all } x \in M\}. \end{cases}$ ▲

Proof. (Left as an exercise.)

DEFINITION 1

We set

(2) $\begin{cases} \text{i.} & M^{-} = \{f \in V^* \text{ such that } \langle f, x \rangle \leq 0 \text{ for all } x \in M\}. \\ \text{ii.} & M^{\perp} = \{f \in V^* \text{ such that } \langle f, x \rangle = 0 \text{ for all } x \in M\}. \end{cases}$

We say that M^{-} is the **negative polar cone** of M and that $M^{\perp}$ is the **orthogonal complement** (or **annihilator**) of M. In short, we call $M^{\perp}$ the **orthogonal** of M. ▲

It is convenient to use this definition to reformulate the density criterion (Theorem 2.2.1).

COROLLARY 1 (DENSITY CRITERION)

A subset D generates a vector subspace that is dense in V if and only if $D^{\perp} = \{0\}$. ▲

Proof. Indeed, to say that $D^{\perp} = \{0\}$ is the same as saying that every continuous linear form that vanishes on D also vanishes on V. ■

PROPOSITION 1

Let V_1 and V_2 be two Hilbert spaces, $A \in \mathscr{L}(V_1, V_2)$ and $A^* \in \mathscr{L}(V_2^*, V_1^*)$ its transpose. Then

(3) $\begin{cases} \text{i.} & \operatorname{Ker} A = (\operatorname{Im} A^*)^{\perp}. \\ \text{ii.} & \operatorname{Ker} A^* = (\operatorname{Im} A)^{\perp}. \end{cases}$ ▲

Proof. **a.** Indeed, $Ax_1 = 0$ if and only if $\langle f_2, Ax_1 \rangle = \langle A^* f_2, x_1 \rangle = 0$ for all $f_2 \in V_2^*$, that is, if and only if $x_1 \in A^*(V_2^*)^{\perp} = (\operatorname{Im} A^*)^{\perp}$.
b. Similarly, $A^* f_2 = 0$ if and only if $\langle A^* f_2, x_1 \rangle = \langle f_2, Ax_1 \rangle = 0$ for all $x_1 \in V_1$, that is, if and only if $f_2 \in A(V_1)^{\perp} = (\operatorname{Im} A)^{\perp}$. ■

Remark 1

Proposition 1 remains true for Banach spaces. ■

COROLLARY 2
Let V_1 and V_2 be Hilbert spaces. The closure $\overline{\operatorname{Im} A}$ of $\operatorname{Im} A$ in V_2 is equal to $(\operatorname{Ker} A^*)^\perp$. The closure $\overline{\operatorname{Im} A^*}$ of $\operatorname{Im} A^*$ in V_1^* is equal to $(\operatorname{Ker} A)^\perp$. ▲

Proof. We know that $\operatorname{Im} A = (\operatorname{Im} A)^{\oplus\oplus} = (\operatorname{Im} A)^{\perp\perp} = (\operatorname{Ker} A^*)^\perp$ according to Proposition 1.5.2, Lemma 1, and Proposition 1. The same argument yields the second statement. ■

Remark 2

The first assertion remains true in the case of Banach spaces. The second extends to the case of reflexive Banach space and is *false in the case of non-reflexive Banach spaces*. ■

THEOREM 1
Let V_1 and V_2 be Hilbert spaces. An operator $A \in \mathscr{L}(V_1, V_2)$ has a dense image if and only if $A^* \in \mathscr{L}(V_2^*, V_1^*)$ is injective. The operator A is injective if and only if $\operatorname{Im} A^*$ is dense in V_1^*. ▲

Proof. The operator A has a dense image if and only if $(\operatorname{Im} A)^\perp = \operatorname{Ker} A^* = \{0\}$ according to Corollary 1 and Proposition 1. An analogous argument implies the second statement. ■

Remark 3

The assertions of Theorem 1 extend to the case of *reflexive Banach spaces*. The assertion "A is injective implies $\operatorname{Im} A^*$ is dense in V_1^*" is false if V_1 is not reflexive. ■

This theorem motivates the following definition.

DEFINITION 2
We shall say that a continuous linear operator j from V_1 to V_2 is an embedding from V_1 into V_2 if

(4) $\quad \begin{cases} \textbf{i.} & j \text{ is injective.} \\ \textbf{ii.} & j(V_1) \text{ is dense in } V_2. \end{cases}$ ▲

From this we deduce the following result.

PROPOSITION 2
Let V_1 and V_2 be two Hilbert spaces. A continuous operator j from V_1 to V_2 is an embedding if and only it its transpose $j^* \in \mathscr{L}(V_2^*, V_1^*)$ is also an embedding. ▲

*Lemma on Bipolars

Proposition 1 extends to the case of cones in the following fashion.

PROPOSITION 3
Let $A \in \mathscr{L}(V_1, V_2)$ be a continuous linear operator from a Hilbert space V_1 to a Hilbert space V_2, and let P be a closed convex cone in V_1. Then

(5) $$A(P)^- = (A^*)^{-1}(P^-),$$

consequently,

(6) the closure $\overline{A(P)}$ of $A(P) = [(A^*)^{-1}(P^-)]^-.$ ▲

Proof. Indeed, to say that $f_2 \in [A(P)]^-$ amounts to saying that $\langle f_2, Ax_1 \rangle = \langle A^*f_2, x_1 \rangle \leq 0$ for all $x_1 \in P$, that is, that A^*f_2 belongs to P^-, or that f_2 belongs to $(A^*)^{-1}P^-$. Therefore

$$A(P)^{\ominus\ominus} = A(P)^{--} = [(A^*)^{-1}P^-]^-$$

is the closure of the cone $A(P)$ in V_2, according to Proposition 1.5.2. ■

Remark 4

We can reformulate this result in the following fashion, known as the *Farkas lemma*. ■

PROPOSITION 4
Let V_1 and V_2 be Hilbert spaces, $A \in \mathscr{L}(V_1, V_2)$, and P a closed convex cone contained in V_1. Then the property

(7) there exists $x_1 \in P$ such that $Ax_1 = x_2$, for x_2 given in V_2

implies the property

(8) $\qquad\qquad \langle f_2, x_2 \rangle \leq 0 \qquad$ when $\qquad A^*f_2 \in P^-.$

Conversely, if the cone $A(P)$ is closed, property (8) implies property (7). ▲

Proof. Indeed, (7) is equivalent to saying that $x_2 \in A(P)$ and (8) is equivalent to saying that $x_2 \in [(A^*)^{-1}(P^-)]^-.$ ■

5. DUALS OF FINITE PRODUCTS, QUOTIENT SPACES, AND CLOSED OR DENSE SUBSPACES

The Dual of a Finite Product

Let us consider, first of all, n Hilbert spaces V_i, with the scalar products $((x_i, y_i))_i$, and their product $V = \prod_{i=1}^n V_i$, a Hilbert space for the scalar product

$$((x, y)) = \sum_{i=1}^{n} ((x_i, y_i))_i \quad \text{where} \quad x = \{x_i\}_i, \quad y = \{y_i\}_i, \tag{1}$$

The bilinear form

$$\langle f, x \rangle = \sum_{i=1}^{n} \langle f_i, x_i \rangle \tag{2}$$

defined on $V^* \times V$, where $V^* = \prod_{i=1}^{n} V_i^*$, establishes a duality between V^* and V. If $J_i \in \mathscr{L}(V_i, V_i^*)$ denotes the duality operator, the relation

$$((x, y)) = \sum_{i=1}^{n} ((x_i, y_i))_i = \sum_{i=1}^{n} \langle J_i x_i, y_i \rangle \tag{3}$$

implies that the duality operator J from V onto V^* is defined by

$$Jx = \{J_i x_i\}_i \quad \text{if} \quad x = \{x_i\}_i \in V. \tag{4}$$ ∎

Duals of Closed Subspace and of Quotient Spaces

Let M be a closed subspace of a vector space V and V/M the quotient of V by M. We are going to characterize their duals.

PROPOSITION 1
The dual of a closed subspace M of a Hilbert space V is isometric to the quotient $V^*/M^\perp$ of V^* by the orthogonal complement $M^\perp$ of M.

The dual of the quotient space V/M of V by M is isometric to the closed subspace $M^\perp$ of V^*. ▲

Proof. **a.** We deduce the second assertion from the first since $V^{**} = V$ and $M^{\perp\perp} = M$. Indeed, $V/M = V^{**}/M^{\perp\perp}$ is isometric to the dual $(M^\perp)^*$ of $M^\perp$, and, consequently, $(V/M)^*$ is isometric to $M^\perp$.
b. Let us establish the first assertion: Let $\rho \in \mathscr{L}(V^*, M^*)$ be the restriction operator associating to every $f \in V^* = \mathscr{L}(V, \mathbb{R})$ its restriction $\rho f \in M^* = \mathscr{L}(M, \mathbb{R})$ to the subspace M. It is clear that $\rho \in \mathscr{L}(V^*, M^*)$ is linear and continuous from V^* to M^*. The kernel $\operatorname{Ker} \rho$ of ρ is $M^\perp$. Indeed, to say that $\rho f = 0$ amounts to saying that $f(x) = 0$ for all $x \in M$, that is, that $f \in M^\perp$.

If φ denotes the canonical surjection from V^* onto $V^*/M^\perp$ and $\tilde{\rho} \in \mathscr{L}(V^*/M^\perp, M^*)$ the operator derived from ρ by taking the quotient, we know that $\tilde{\rho}$ is a linear operator from $V^*/M^\perp$ to M^*.

Let $g \in V^*/M^\perp$ and let $f \in V^*$ be a representative of g satisfying $\varphi f = g$.

Then, since $\tilde{\rho}\varphi = \rho$,

$$\|\tilde{\rho}g\|_{M^*} = \|\rho f\|_{M^*} = \sup_{x\in M}\frac{|\langle f,x\rangle|}{\|x\|} \leq \sup_{x\in V}\frac{|\langle f,x\rangle|}{\|x\|} = \|f\|_{V^*}.$$

Consequently,

(5) $$\|\tilde{\rho}g\|_{M^*} \leq \inf_{\varphi f=\varphi g}\|f\|_{V^*} = \|g\|_{V^*/M^\perp}.$$

Moreover, if $h \in M^*$ and if t denotes the orthogonal projection from V onto M, $f = ht$ is the extension of h to V with norm equal to $\|h\|_{M^*}$. Then $g = \varphi f$ satisfies $\tilde{\rho}g = \rho f = h$ and

(6) $$\|g\|_{V^*/M^\perp} \leq \|f\|_{V^*} = \|h\|_{M^*}.$$

Inequalities (5) and (6) show that $\tilde{\rho}$ is a surjective isometry. ∎

Remark 1

We are going to give an application of Proposition 1 for finding *lower bounds* of the error in approximation theory. ∎

PROPOSITION 2
Let M be a closed subset of a Hilbert space V. Then

(7) $$\inf_{y\in M}\|x - y\| = \sup_{f\in M^\perp}\frac{|\langle f,x\rangle|}{\|f\|_*}. \quad \blacktriangle$$

Proof. Let φ be the canonical surjection from V onto V/M. Then, since V/M is isometric to the dual of $M^\perp$, we have

$$\inf_{y\in M}\|x-y\| = \|\varphi x\|_{V/M} = \|\varphi x\|_{(M^\perp)^*} = \sup_{f\in M^\perp}\frac{|\langle f,x\rangle|}{\|x\|_V}. \quad \blacksquare$$

If we know an element $f \in M^\perp$, we derive a lower estimate for the minimal error $\inf_{y\in M}\|x-y\|$ of the approximation of x by the elements $y \in M$, since (7) implies that

$$\frac{\langle f,x\rangle}{\|f\|_*} \leq \inf_{y\in M}\|x-y\|.$$

Dual of a Dense Subspace

PROPOSITION 3
Let V and H be two Hilbert spaces with scalar products $((x,y))$ and (x,y)

respectively. Suppose that $V \subset H$ and that

(8)   the canonical injection j from V to H is an embedding.

Then

(9) $j^* \in \mathscr{L}(H^*, V^*)$ is an embedding from H^* into V^*.

Suppose that we identify j^* with the canonical injection from H^* to V^*. Then, if $\langle f, x \rangle$ is the duality pairing between V^* and V and $[f, x]$ is the duality pairing between H^* and H, we obtain

(10) $[f, x] = \langle f, x \rangle$ for all $f \in H^*$, $\forall x \in V$.

If J is the duality operator from V^* onto V and K the duality operator from H onto H^*, then the scalar products are related by the formula

(11) $(x, y) = ((J^{-1}Kx, y))$ for all $x, y \in V$.   ▲

Proof. We already know that j^* is an embedding (Proposition 4.2). If we agree to identify j^* with the canonical injection from H^* to V^* (i.e., to identify H^* with a subspace of V^*), we obtain for $f \in H^*$ and $x \in V$:

$$[f, x] = [f, jx] = \langle j^*f, x \rangle = \langle f, x \rangle.$$

In other words, the *bilinear forms* $\langle f, x \rangle$ and $[f, x]$ *defined on* $V^* \times V$ *and* $H^* \times H$ *coincide on the subspace* $H^* \times V$ contained (and dense) in the spaces $V^* \times V$ and $H^* \times H$. In particular, if x and $y \in V$, we derive from (10) that

$$(x, y) = [Kx, y] = [JJ^{-1}Kx, y] = \langle JJ^{-1}Kx, y \rangle = ((J^{-1}Kx, y)),$$

since $Kx \in H^*$ and $y \in V$.   ■

Remark 2

This proposition shows that we *cannot at the same time identify H with H^* and V with V^**. Indeed, if we identity H with H^*, then we set $[f, v] = (f, v)$ and $K = 1$. If we also identified V with V^*, then $J = 1$ and (10) would imply that the scalar products $((\ ,\ ))$ and $(\ ,\ )$ coincide and, consequently, that $V = H$.   ■

Now let us consider *the case where H is taken to be a pivot space*. Since $H \times V$ is dense in $V^* \times V$ and since the duality pairings $\langle f, v \rangle = [f, v] = (f, v)$ coincide on $H \times V$, we deduce that $\langle f, v \rangle$ is the *unique extension by density* of the bilinear form (f, v) on $H \times V$.

The uniqueness of this extension leads us to set

(12) $\langle f, v \rangle = (f, v)$ for $f \in V^*$ and $v \in V$.

Then we derive from (12) that the canonical isometry from V onto V^* is equal to J. Proposition 3 becomes

PROPOSITION 4
Let us suppose that the Hilbert space $\{V,((\,,))\}$ is embedded into the pivot space $\{H,(\,,)\}$. Then we can embed H into the dual V^* of V and V in its dual V^*. In other words, the canonical injections j from V to H, j^* from H to V^*, and j^*j from V to V^* are embeddings:

(13) $$V \xrightarrow{j} H \xrightarrow{j^*} V^*$$

and the bilinear form (f,v) on $V^* \times V$ is the unique extension by continuity of the scalar product (x,y) of H restricted to $H \times V$. ▲

Proof. It only remains to show that j^*j is an embedding from V into V^*. It is clear that j^*j is injective: It has a dense image, since its transpose $(j^*j)^* = j^*j$ is injective. ∎

Remark 3

This proposition motivates the terminology of pivot space since if V is embedded into H, *we can* embed H into V^*. ∎

Remark 4

When there exist several embeddings from a space H into several spaces V^*, *we can identify at most only one of these embeddings with a canonical injection*. For example, suppose that H is a pivot space, that V is a Hilbert space, and that V_0 is a *closed subspace* of V such that

(14) $\begin{cases} \textbf{i.} & \text{the injection } j \text{ from } V \text{ to } H \text{ is an embedding.} \\ \textbf{ii.} & \text{the injection } j_0 \text{ from } V_0 \text{ to } H \text{ is \textit{also} an embedding.} \end{cases}$

Then

(15) $\begin{cases} \textbf{i.} & \text{the transpose } j^* \text{ is an embedding from } H \text{ into } V^* \\ \textbf{ii.} & \text{the transpose } j_0^* \text{ is an embedding from } H \text{ into } V_0^* \end{cases}$

and we have seen that V_0^* can be identified with $V^*/V_0^\perp$.

Consequently, it is impossible to identify at the same time the embeddings j^* and j_0^* with the canonical injections. ∎

This remark plays a fundamental role in the study of duals of Sobolev spaces. (See Section 4 of Chapter 5 and see Chapters 7 and 9.) It is the basis of the distinction between the transpose and the formal adjoint that we make in Chapter 13 and also of the Green formula relating these two concepts.

6. THE THEOREM OF LAX-MILGRAM

Let us recall that if V is of finite dimension, every positive definite linear mapping A from V to V^* is an isomorphism. Since the unit spheres in a finite dimensional space is *compact* (Theorem 1.1.1), we deduce that the constant $c = \inf_{x \in S} \langle Ax, x \rangle$ is *strictly positive*. Thus if A is positive definite, $\langle Ax, x \rangle \geq c\|x\|^2$ for all $x \in V$ where $c > 0$. We are going to show that this latter condition also implies that A is an isomorphism in the case of infinite dimensional spaces.

DEFINITION 1
Let A be a continuous linear operator from V onto V^*. We shall say that A is V-elliptic if there exists a constant $c > 0$ such that

(1) $$\forall x \in V, \quad \langle Ax, x \rangle \geq c\|x\|^2. \qquad \blacktriangle$$

THEOREM 1 (LAX-MILGRAM)
A continuous linear V-elliptic operator A is an isomorphism from a Hilbert space V onto its dual. Moreover,

(2) $$\|A^{-1}\|_{\mathscr{L}(V^*, V)} \leq \frac{1}{c}. \qquad \blacktriangle$$

Proof. First of all, A is *injective* since if $Ax = 0$, we derive $c\|x\|^2 \leq \langle Ax, x \rangle = 0$ and, thus, $x = 0$.

To show that A is *surjective*, we are going to show successively that the image $A(V)$ is *closed* (i.e., *complete*) and *dense* in V^*.

The image $A(V)$ is complete. Indeed, if $\{A(x_n)\}$ is a Cauchy sequence of elements $A(x_n)$ of $A(V)$, then $\{x_n\}$ is a Cauchy sequence of elements of V, since A being V-elliptic implies

$$\|x_n - x_m\|^2 \leq c^{-1} \langle A(x_n - x_m), x_n - x_m \rangle \leq c^{-1} \|A(x_n - x_m)\|_* \|x_n - x_m\|$$

and, consequently,

$$\|x_n - x_m\| \leq c^{-1} \|A(x_n - x_m)\|_* \to 0 \quad \text{when} \quad n, m \to \infty.$$

Since V is complete, the sequence x_n converges to an element x. Since A is continuous, $A(x_n)$ converges to $A(x) \in A(V)$.

The image $A(V)$ is dense in V^*. Using Theorem 4.1, it suffices to verify that the transpose $A^* \in \mathscr{L}(V, V^*)$ is injective, which is the case, since A^* is also V-elliptic (according to the first part of the proof).

Finally, A^{-1} *is continuous.* We are going to establish inequality (2). For every $f \in V^*$, consider $x = A^{-1}f$. The V-ellipticity condition implies that

$$\|A^{-1}f\|^2 = \|x\|^2 \leq c^{-1} \langle A(x), x \rangle = c^{-1} \langle f, A^{-1}f \rangle \leq c^{-1} \|f\|_* \|A^{-1}f\|.$$
■

We use this theorem to establish in Chapter 13 existence and uniqueness in boundary-value problems.

*7. VARIATIONAL INEQUALITIES

Let us consider now

(1) $\qquad$ a closed convex subset K of V.

If $A \in \mathscr{L}(V, V^*)$ is the duality operator from V onto V^*, the projection $x = tA^{-1}$ of $A^{-1}f$ (where $f \in V^*$) onto K is characterized by

(2) $\begin{cases} \text{i.} & x \in K. \\ \text{ii.} & \langle Ax, x - y \rangle \leq \langle f, x - y \rangle \quad \forall y \in K. \end{cases}$

More generally, we are going to show that the "variational inequalities" (2) has a solution when A is V-elliptic.

THEOREM 1 (LIONS-STAMPACCHIA)

Suppose that the operator $A \in \mathscr{L}(V, V^*)$ is V-elliptic: $\exists c > 0$ *such that*

(3) $\qquad \forall x \in V, \quad \langle Ax, x \rangle \geq c \|x\|^2,$

Then $\forall f \in V^*$, there exists a unique solution $x = G(f)$ to the variational inequalities (2) *satisfying*

(4) $\qquad \|G(f) - G(g)\| \leq c^{-1} \|f - g\|_*.$ ▲

Proof. **a.** We begin by proving the inequalities (4). Suppose that solutions $x = G(f)$ and $y = G(g)$ exist. Then we deduce from the inequalities

$$\langle Ax, x - y \rangle \leq \langle f, x - y \rangle$$
$$\langle Ay, y - x \rangle \leq \langle g, y - x \rangle$$

that

$$c \|x - y\|^2 \leq \langle A(x - y), x - y \rangle \leq \langle f - g, x - y \rangle \leq \|f - g\|_* \|x - y\|,$$

according to the V-ellipticity hypothesis. Hence (4) is established. This inequality implies *uniqueness*, since $G(f) - G(g) = 0$ when $f - g = 0$.

b. We are going to show *existence* by a constructive method. Let $((x, y)) = \langle Jx, y \rangle$ be the scalar product on V and $J \in \mathscr{L}(V, V^*)$ be the duality operator. We can rewrite variational inequalities in the form

(5) $\quad \rho((J^{-1}(Ax - f), x - y)) = ((x - x + \rho J^{-1}(Ax - f), x - y)) \leq 0$
$\quad$ for all $\quad y \in K \quad$ where $\quad \rho > 0$.

Then if t denotes the best approximation projector on K, every solution x to the variational inequalities (2) is a solution of the nonlinear equation

(6) $\quad x = t(x - \rho J^{-1}(Ax - f)), \quad \rho > 0$

and conversely.

This suggests introducing the following iterative method, which generalizes the algorithm for solving equations introduced in [AAA], pp. 130–132.

(7) $\quad x_{n+1} = t(x_n - \rho J^{-1}(Ax_n - f)), \quad x \in K$

and using the Banach-Picard theorem on contractions. (See Theorem 3.11.1 in [AAA], p. 121.)

Since the projection t satisfies $\|tx - ty\| \leq \|x - y\|$, we obtain

(8) $\quad \|x_{n+1} - x_n\| = \|t(x_n - \rho J^{-1}(Ax_n - f)) - t(x_{n-1} - \rho J^{-1} A(x_{n-1} - f))\|$
$\quad \leq \|(1 - \rho J^{-1} A)(x_n - x_{n-1})\|$

We must evaluate the norm of $(1 - \rho J^{-1} A)$. To this end we remark that

$$\|(1 - \rho J^{-1} A)x\|^2 = \|x\|^2 + \rho^2 \|J^{-1} Ax\|^2 - 2\rho((x, J^{-1} Ax)).$$

Since

$$\|J^{-1} Ax\| = \|Ax\|_* \leq \|A\| \|x\|$$

and since

$$((x, J^{-1} Ax)) = \langle Ax, x \rangle \geq c \|x\|^2,$$

we deduce that

(9) $\quad \|(1 - \rho J^{-1} A)x\|^2 \leq (1 + \rho^2 \|A\|^2 - 2\rho c) \|x\|^2.$

Consequently, if we choose ρ such that $\theta^2 = 1 + \rho^2 \|A\|^2 - 2\rho c < 1$ (which is the case when $0 < \rho < 2c/\|A\|^2$), we obtain the inequality

(10) $\quad \|x_{n+1} - x_n\| \leq \theta \|x_n - x_{n-1}\|.$

Hence $\{x_n\}$ is a Cauchy sequence that converges to some $x \in K$. Since the mapping $x \mapsto t(x - \rho J^{-1}(Ax - f))$ is continuous, we deduce that the sequence x_n converges to the solution $x = G(f)$ of the variational inequalities. ∎

Remark 1

We can replace the V-ellipticity hypothesis by the weaker "K-ellipticity" hypothesis:

(11) $\quad \forall x, y \in K, \quad \langle A(x-y), x-y \rangle \geq c \|x-y\|^2 \quad$ where $\quad c > 0.$ ∎

Remark 2

Chapter 15, Section 5, gives the generalization of this theorem to the case of nonlinear operators. ∎

*8. NONCOOPERATIVE EQUILIBRIA IN N-PERSON QUADRATIC GAMES

We suppose that we are given

(1) $\begin{cases} \text{i.} & n \text{ Hilbert spaces } V^i, \text{with scalar products} \quad ((x^i, y^i))_i. \\ \text{ii.} & \text{elements} \quad u^i \in V^i. \\ \text{iii.} & \text{closed convex subsets} \quad K^i \subset V^i. \\ \text{iv.} & \text{operators} \quad M_j^k \in \mathscr{L}(V^j, V^k) \quad \text{where} \quad M_j^j = 1. \end{cases}$

We consider the following problem: to find $x = \{x^1, \ldots, x^n\}$ satisfying

(2) $\begin{cases} \text{i.} & x^i \in K^i \quad \text{for all} \quad i = 1, \ldots, n, \\ \text{ii.} & \left\| x^i + \sum_{j \neq i} M_j^i x^j - u^i \right\|_i^2 = \min_{y^i \in K^i} \left\| y^i + \sum_{j \neq i} M_j^i x^j - u^i \right\|_i^2 \\ & \text{for all} \quad i = 1, \ldots, n. \end{cases}$

We suppose that the operators M_j^i satisfy the following consistency condition:

There exists a constant $c > 0$ such that for all

(3) $\quad x = \{x^1, \ldots, x^n\}, \quad \text{we have} \quad \sum_{j,k=1}^n ((M_j^k x^j, x^k))_k \geq c \sum_{i=1}^n \|x^i\|_i^2.$

THEOREM 1

Under hypotheses (1) *and* (3), *there exists a unique solution* x *to problem* (2). ▲

Remark 1

Problem (2) can be written

(4) $$x^i = t_i(u^i - \sum_{j \neq i} M^i_j x^j) \quad \text{for all} \quad i = 1, \ldots, n$$

where t_i is the best approximation projector on K^i. ∎

Remark 2

In game theory, the solution x to (2) is called the *noncooperative equilibrium* (or *Nash equilibrium*) in an n-person game. The set K^i denotes the set of strategies of the ith player and u^i the objective he tries to attain at best. If the players $j \neq i$ choose the strategies x^j, the objective u^i of the ith player is modified and becomes $u^i - \sum_{j \neq i} M^i_j x^j$, so that the player i is led to choose $t_i(u^i - \sum_{j \neq i} M^i_j x^j)$. Condition (2) is therefore a condition of equilibrium expressing the fact that no player finds it in his interest to modify *by himself* the solution of equilibrium. The study of noncooperative equilibria in n-person games is continued in Chapter 15, Section 7.

The proof of the theorem is accomplished in two stages. First we prove the following: ∎

LEMMA 1
We set

$$K = \prod_{i=1}^n K^i, \quad V = \prod_{i=1}^n V^i, \quad V^* = \prod_{i=1}^n V_i^*, \quad J^i \in \mathscr{L}(V^i, V^{i*})$$

the isometry from V^i onto V^{i*}, $f = \{f^1, \ldots, f^n\} \in V^*$ where $f^i = J^i u^i$ and $A \in \mathscr{L}(V, V^*)$ the operator defined by $Ax = \{A^1 x, \ldots, A^k x, \ldots, A^n x\}$ where $A^k x = \sum_{j=1}^n J^k M^k_j x^j$. Then x is a solution to (2) if and only if

(5) $\begin{cases} \text{i.} & x \in K. \\ \text{ii.} & \langle Ax - f, x - y \rangle \leq 0 \quad \text{for all} \quad y \in K. \end{cases}$ ▲

Proof of Lemma 1. To say that $x^i = t_i(u^i - \sum_{j \neq i} M^i_j x^j)$ amounts to saying that

(6) $((x^i + \sum_{j \neq i} M^i_j x^j - u^i, x^i - y^i))_i = \langle A^i x - f^i, x^i - y^i \rangle \leq 0 \quad \text{for all} \quad y^i \in K^i.$

Adding these inequalities from $i = 1$ to n, we obtain

$$\langle Ax - f, x - y \rangle = \sum_{i=1}^n \langle A^i x - f^i, x^i - y^i \rangle \leq 0 \quad \text{for all} \quad y \in K.$$

Conversely, to show that (5) implies the inequalities (6), we take $y^j = x^j$ for all $j \neq i$ and y^i an arbitrary element of K^i. Then in this case

$$\langle Ax - f, x - y \rangle = \langle A^i x - f^i, x^i - y^i \rangle \leq 0 \quad \text{for all} \quad y^i \in K^i. \blacksquare$$

Proof of Theorem 1. According to Lemma 1, it suffices to show the existence of a solution to (5). We can then apply the Lions-Stampacchia theorem, Theorem 7.1, if we first verify that the operator A is V-elliptic. But this is precisely hypothesis (3). ∎

Remark 3

Theorem 1 then gives us again the theorem of Nash in the quadratic case (see Theorem 5.7.1 of [AAA], p. 204) without our needing to use the Brouwer fixed point theorem. ∎

Remark 4

Theorem 1 furnishes an algorithm that converges to the equilibrium. At the $n+1$st step, player i chooses the strategy x^i_{n+1}, which minimizes on K^i the distance to the objective $u^i - \rho \sum_{j \neq i} M^i_j x^j_n + (1-\rho)(x^i_n - u_i)$ for a convenient scalar $\rho > 0$. ∎

CHAPTER 4

The Banach Theorem and the Banach-Steinhaus Theorem

This chapter is devoted to the study of the properties of linear operators.

We begin by characterizing in Section 1 the bounded sequences $\{A_h\}$ of continuous linear operators

$$A_h \in \mathscr{L}(V, F)$$

by showing that

$$\sup_h \|A_h\|_{\mathscr{L}(V,F)} < +\infty$$

if and only if

$$\forall x \in V, \quad \sup_h \|A_h x\| < +\infty$$

We next recall the fundamental properties of bounded sets of operators, in particular that pointwise convergence on a dense subset is equivalent to uniform convergence on the compact subsets. We obtain as a corollary the well-known Banach-Steinhaus theorem.

In Section 2 we apply this theorem together with the best approximation theorem to prove the mean ergodic theorem, that is, the convergence of the Cesaro means $1/n \sum_{k=0}^{n-1} A^k$ where $A \in \mathscr{L}(V, F)$.

We study in Section 3 the stability of a *surjective* operator $A \in \mathscr{L}(V, F)$. Since there exist solutions x of the equation

$$Ax = f \quad \text{for all} \quad f \in F,$$

we show that one can choose among them a solution x that also satisfies the condition

$$\|x\|_V \leq C \|f\|_F,$$

where C is a constant.

This is the well-known Banach theorem, which can also be expressed by saying that the image under A of every open set is open. This implies in particular that if $A \in \mathscr{L}(V,F)$ is bijective, $A^{-1} \in \mathscr{L}(F,V)$ is automatically continuous.

We use this result to show in Section 4 that the image of A is closed if and only if the image of A^* is closed.

We then characterize in Sections 6 and 7 those linear operators that have continuous and linear left or right inverses. We construct, as well, particular left and right inverses, called orthogonal left and right inverses, and study their properties.

To conclude this chapter we resolve explicitly the problem of quadratic programming with linear equality constraints.

1. PROPERTIES OF BOUNDED SETS OF OPERATORS

Let V and F be Hilbert spaces and $\mathscr{L}(V,F)$ the Banach space of continuous linear operators from V to F.

We say that a family $A_h \in \mathscr{L}(V,F)$ is *bounded* if

$$(1) \qquad \sup_h \|A_h\|_{\mathscr{L}(V,F)} = \sup_h \sup_{x \in V} \frac{\|A_h x\|_F}{\|x\|_V} < +\infty.$$

It is clear that if a set $\{A_h\}_h$ is bounded, then the set $\{A_h(x)\}_h$ is bounded in F for all $x \in V$. We conclude from Baire's theorem (see Remark 1 below) that the converse is true.

THEOREM 1 (UNIFORM BOUNDEDNESS)
For a family of continuous linear operators $A_h \in \mathscr{L}(V,F)$ to be bounded it is necessary and sufficient that

$$(2) \qquad \forall x \in V, \quad \sup_h \|A_h x\|_F < +\infty. \qquad \blacktriangle$$

Proof. Let us assume (2) is true and show that $\{A_h\}$ is bounded. We introduce the function φ defined on V by

$$(3) \qquad \varphi(x) = \sup_h \|A_h x\|_F.$$

Hypothesis (2) implies that φ is a function from V to $\mathbb{R}$. Since $x \mapsto \|A_h x\|_F$ is a continuous seminorm, it follows that φ *is a lower semicontinuous seminorm on V.*

Suppose now that we knew that φ was continuous; then there would exist a constant M such that

$$(4) \qquad \forall x, \quad \sup_h \|A_h x\|_F = \varphi(x) \leq M \|x\|_V.$$

Hence
$$\sup_h \|A_h\|_{\mathscr{L}(V,F)} \leq M$$
Therefore it remains to prove the following. ∎

LEMMA 1
Every lower semicontinuous seminorm φ on a Hilbert space is continuous. ▲

Proof. Since φ is lower semicontinuous, the subsets $F_n = \{x \in V \text{ such that } \varphi(x) \leq n\}$ are closed. Hence $V = \bigcup_{n=1}^{\infty} F_n$. The Baire theorem recalled below implies that one of the closed sets F_n has a nonempty interior; that is, it contains a ball $x_0 + B(\eta)$ with center x_0 and radius η. Hence for all $x \in V$ we have

$$\frac{\eta}{\|x\|}\varphi(x) = \varphi\left(\frac{\eta}{\|x\|}x + x_0 - x_0\right) \leq \varphi\left(x_0 + \frac{\eta x}{\|x\|}\right) + \varphi(x_0) \leq 2n.$$

Thus we obtain the inequality (4) with $M = 2n/\eta$. ∎

*Remark 1

For the convenience of the reader we state and prove Baire's theorem because of its important consequences. ∎

BAIRE'S THEOREM
Let E be a complete metric space. Then one of the following two equivalent properties holds

(a)
  i. *For every countable family of dense open sets U_n in E, $\bigcap_n U_n$ is dense in E.*
  ii. *For every family of closed sets F_n in E with empty interior, $\bigcup_n F_n$ has an empty interior.* ▲

Proof. It is clear that properties (a)i and (a)ii are equivalent by taking complements. We shall prove (a)i.

Let us consider a countable family of open sets U_n such that $\bar{U}_n = E$ and show that $U_\infty = \bigcap_n U_n$ is dense in E; that is, for all $x \in E$ and all $\varepsilon > 0$,

$$B(x, \varepsilon) \cap U_\infty \neq \emptyset$$

For this we are going to show that the limit x_* of a sequence x_n that we shall construct by recursion belongs to

$$B(x, \varepsilon) \cap U_\infty.$$

Since U_1 is dense in E, the intersection $\mathring{B}(x,\varepsilon) \cap U_1$ is nonempty. Since U_1 is open there exists a ball $B(x_1, \varepsilon_1)$ such that

$$B(x_1, \varepsilon_1) \subset \mathring{B}(x, \varepsilon) \cap U_1, \qquad \varepsilon_1 < \tfrac{1}{2}$$

Suppose that we have constructed the $(n-1)$ first balls $B(x_k, \varepsilon_k)$ such that

$$B(x_k, \varepsilon_k) \subset \mathring{B}(x_{k-1}, \varepsilon_{k-1}) \cap U_k, \qquad \varepsilon_k < \frac{1}{2^k}.$$

We then construct the ball $B(x_n, \varepsilon_n)$ in the following fashion: Since U_n is dense in E, then the open set $\mathring{B}(x_{n-1}, \varepsilon_{n-1}) \cap U_n$ is nonempty and contains a ball $B(x_n, \varepsilon_n)$ satisfying

$$B(x_n, \varepsilon_n) \subset \mathring{B}(x_{n-1}, \varepsilon_{n-1}) \cap U_n, \qquad \varepsilon_n < \frac{1}{2^n}.$$

The sequence of centers x_n of these balls is a Cauchy sequence since if $m \geq n$, then $B(x_m, \varepsilon_m) \subset B(x_n, \varepsilon_n)$ and, consequently, $d(x_n, x_m) \leq \varepsilon_n \leq 2^{-n}$. Hence this sequence converges to an element x_* of E, since E is complete. Moreover, if $m \geq n$,

$$d(x_n, x_*) \leq d(x_n, x_m) + d(x_m, x_*) \leq \varepsilon_n + d(x_m, x_*).$$

Letting m approach infinity, we deduce that $d(x_n, x_*) \leq \varepsilon_n$; that is, $x_* \in B(x_n, \varepsilon_n)$. Consequently, since $B(x_n, \varepsilon_n) \subset \mathring{B}(x, \varepsilon) \cap U_n$,

$$x_* \in \bigcap_n B(x_n, \varepsilon_n) \subset \mathring{B}(x, \varepsilon) \cap \bigcap_n U_n = \mathring{B}(x, \varepsilon) \cap U_\infty$$

which completes the proof of the theorem. ∎

Let us also mention the following properties of bounded families of operators. (See [AAA], p. 150, Theorem 4.6.1, in the case of equicontinuous sets of functions on metric spaces.)

THEOREM 2
Let $\{A_n\}_n \in \mathscr{L}(V, F)$ be a bounded sequence of continuous linear operators. If D is a dense subspace of V and if

(5) $$\forall x \in D, \quad \lim_{n \to \infty} \|A_n x - Ax\|_F = 0$$

then, for every compact set K of V,

(6) $$\lim_{n \to \infty} \sup_{x \in K} \|A_n x - Ax\|_F = 0$$

and in particular A_n converges pointwise to A. If, moreover, $\{x_p\}_p$ is a sequence of elements $x_p \in V$ which converges to x in V, then

(7) $$\lim_{n, p \to \infty} \|A_n x_p - Ax\|_F = 0. \qquad \blacktriangle$$

Proof. Suppose we are given $M = \sup_n \|A_n\|_{\mathscr{L}(V,F)}$, K a compact subset of U, and $\varepsilon > 0$ fixed. Since K is compact, there exist p points x_i such that for all $x \in K$ there exists at least one x_i satisfying

$$\|x - x_i\| \leq \frac{\varepsilon}{4M + 1}.$$

Since D is dense in V, there exist p points $y_i \in D$ such that $\|x_i - y_i\| \leq \varepsilon/(4M + 1)$ for all $i = 1, \ldots, p$. Finally, hypothesis (5) implies that there exists n_0 such that $\|A_n y_i - A y_i\| \leq \varepsilon/(4M + 1)$ for all $i = 1, \ldots, p$ and for all $n \geq n_0$. Therefore for all $x \in K$,

$$\|A_n x - Ax\|_F \leq \|A_n(x - x_i)\|_F + \|A_n(x_i - y_i)\|_F + \|A_n y_i - A y_i\|_F$$
$$+ \|A(y_i - x_i)\|_F + \|A(x_i - x)\|_F$$
$$\leq (4M + 1)\frac{\varepsilon}{4M + 1} = \varepsilon.$$

Consequently, $\sup_{x \in K} \|A_n x - Ax\|_F \leq \varepsilon$ when $n \geq n_0$, which establishes (6). To show (7), we use the fact that $\|x_p - x\|_V \leq \varepsilon/(M + 1)$ when $p \geq p_0$ and that $\|A_n x - Ax\|_F \leq \varepsilon/(M + 1)$ when $n \geq n_0$. Then

$$\|A_n x_p - Ax\|_F \leq \|A_n(x_p - x)\|_F + \|A_n x - Ax\|_F$$
$$\leq M\|x_p - x\|_V + \|A_n x - Ax\|_F \leq (M + 1)\frac{\varepsilon}{M + 1} = \varepsilon,$$

when $n \geq n_0, p \geq p_0$. ∎

Remark 2

We only used the Baire Theorem; so theorems 1 and 2 can be extended to the case of Banach and Fréchet spaces. ∎

We deduce from these results the following.

THEOREM 3 (BANACH-STEINHAUS)
Let V and F be Hilbert spaces, $\{A_n\}_n$ a countable sequence of continuous linear operators from V to F. Suppose that

(8) $\qquad \forall x \in V, \quad A_n x$ *converges to an element f of F.*

Then there exists an operator $A \in \mathscr{L}(V, F)$ such that, for every compact set K of V,

(9) $\qquad \lim_{n \to \infty} \sup_{x \in K} \|A_n x - Ax\|_F = 0.$ ▲

Proof. Let us denote by $A(x)$ the limit in F of $A_n x$. Then A is a linear operator, since the equality $A_n(\lambda x + \mu y) = \lambda A_n x + \mu A_n y$ becomes $A(\lambda x + \mu y) = \lambda A x + \mu A y$ by taking the limit.

Since the *countable sequence* $A_n x$ is convergent, we deduce that it is bounded. Consequently,

$$\forall x \in V, \quad \sup_n \|A_n x\|_F < +\infty.$$

It follows, then, from Theorem 1 that $\sup_n \|A_n\|_{\mathscr{L}(V,F)} = M$ is finite. The inequalities $\|A_n x\|_F \leq M \|x\|_V$ imply, by taking limits, that $\|A x\|_F \leq M \|x\|_V$ for all x and, therefore, that A is continuous. Finally, Theorem 2 implies property (9). ∎

We mention the following consequence, which is important in approximation theory.

*THEOREM 4
Let $U, V,$ and F be Hilbert spaces and let $\{A_n\}_n \in \mathscr{L}(V, F)$ be a countable sequence of continuous linear operators from V to F. Suppose that

(10)   the unit ball of U is relatively compact in V.

If we suppose that

(11) $\forall x \in V,\ A_n x$ converges to an element f of F,

then there exists a continuous linear operator $A \in \mathscr{L}(V, F)$ such that

(12) $$\lim_{n \to \infty} \|A_n - A\|_{\mathscr{L}(U, F)} = 0.$$ ▲

Proof. Letting K be the compact set that is the closure of the unit ball of U in V, we deduce from (9) that

$$\|A_n - A\|_{\mathscr{L}(U, F)} = \sup_{x \in K} \|A_n x - A x\|_F \quad \text{converges to zero.} \quad \blacksquare$$

The Banach-Steinhaus theorem also implies that every separately continuous bilinear mapping is continuous.

*THEOREM 5
Let $U, V,$ and F be three Hilbert spaces and $B \in \mathscr{L}(U, V; F)$ a bilinear mapping from $U \times V$ to F that is separately continuous. Then B is in fact continuous ▲

Proof. It suffices to show that if two sequences of elements $x_n \in U$ and $y_n \in V$ converge to zero, then the sequence $B(x_n, y_n)$ converges to zero in F. Let us consider the operators A_n from V to F defined by

(13) $$A_n(y) = B(x_n, y).$$

These operators are clearly linear and continuous, since B is bilinear and continuous with respect to y when x_n is fixed. The continuity of B with respect to x (when y is fixed) shows that $A_n(y)$ converges to zero for every fixed y. This implies that the elements $A_n y$ are bounded; consequently, it follows from Theorem 1 that the operators A_n are bounded in $\mathscr{L}(V, F)$. Since A_n converges pointwise to zero and since y_n converges to zero in F, the final assertion of Theorem 2 implies that $B_n(x_n, y_n) = A_n(y_n)$ converges to zero in F. ∎

2. THE MEAN ERGODIC THEOREM

The abstract version of the mean ergodic theorem is used to study the asymptotic behavior of the means $1/n \sum_{k=0}^{n-1} A^k$ where A is an operator with norm 1. We begin with the following.

PROPOSITION 1
Let V be a Hilbert space (We identify V with its dual V^*.) Let $\mathscr{A}$ be a family of continuous linear operators $A \in \mathscr{L}(V, V)$ satisfying:

$$\text{(1)} \quad \begin{cases} \text{i.} & \forall A \in \mathscr{A} \quad \|A\|_{\mathscr{L}(V,V)} \leq 1 \\ \text{ii.} & \forall A, B \in \mathscr{A} \quad AB \in \mathscr{A} \end{cases}$$

Let $F_{\mathscr{A}} = \bigcap_{A \in \mathscr{A}} \operatorname{Ker}(A - 1)$ be the closed subspace of fixed points of operators A and $G_{\mathscr{A}} = \bigcup_{A \in \mathscr{A}} \operatorname{Im}(A - 1)$. Then

$$\text{(2)} \quad F_{\mathscr{A}} = G_{\mathscr{A}}^{\oplus}$$

If t is the orthogonal projector onto $F_{\mathscr{A}}$, we have, for any $A \in \mathscr{A}$

$$\text{(3)} \quad At = tA = t \qquad \blacktriangle$$

Proof. We note that $F_{\mathscr{A}^*} = \bigcap_{A \in \mathscr{A}} \operatorname{Ker}(A^* - 1) = F_{\mathscr{A}}$. Indeed, if $x \in F_{\mathscr{A}}$, then

$$\|A^*x - x\|^2 = \|A^*x\|^2 + \|x\|^2 - 2((x, A^*x))$$
$$= \|A^*x\|^2 + \|x\|^2 - 2((Ax, x))$$
$$= \|A^*x\|^2 - \|x\|^2 \leq \|x\|^2 - \|x\|^2 \leq 0.$$

Hence $F_{\mathscr{A}} \subset F_{\mathscr{A}^*}$ and, by symmetry, $F_{\mathscr{A}^*} \subset F_{\mathscr{A}}$.

It is clear that $G_{\mathscr{A}}^{\oplus} = \bigcap_{A \in \mathscr{A}} \operatorname{Im}(A - 1)^{\oplus} = \bigcap_{A \in \mathscr{A}} \operatorname{Ker}(A^* - 1) = F_{\mathscr{A}^*} = F_{\mathscr{A}}$.

Let t be the orthogonal projector. Since tx is a fixed point of A, we have $Atx = tx$ for all $x \in V$.

Since $Ax - x$ belongs to $G_{\mathscr{A}} \subset F_{\mathscr{A}}^{\oplus}$, we deduce that $t(Ax - x) = 0$ for all $x \in V$. ∎

Remark 1

Let $\mathscr{A}x = \{Ax\}_{A\in\mathscr{A}}$ be the "orbit" of x. It is clear that $\mathscr{A}x$ is invariant by any $A\in\mathscr{A}$, that $co(\mathscr{A}x)$ is also invariant by any $A\in\mathscr{A}$ (by linearity) and that $\overline{co}(\mathscr{A}x)$ remains invariant by any $A\in\mathscr{A}$ (by continuity). Therefore, since $tA = t$, we deduce that

(4) $\qquad\qquad tx = ty \qquad$ for all $\qquad y\in\overline{co}(\mathscr{A}x)$.

In particular, if there exists a fixed point $\bar{x}$ of $\mathscr{A}$ belonging to $\overline{co}(\mathscr{A}x)$, then it is necessarily equal to $\bar{x} = t(\bar{x}) = t(x)$ and, therefore, is *unique* (in $\overline{co}(\mathscr{A}x)$). Such a fixed point does exist: we take $\bar{x}\in\overline{co}(\mathscr{A}x)$ to be the unique projection of 0 onto $\overline{co}(\mathscr{A}x)$, which exists by Theorem 1.4.1. Since $A\bar{x}\in\overline{co}(\mathscr{A}x)$ and since $\|A\bar{x}\| \leq \|\bar{x}\|$, we deduce that $\|A\bar{x}\| = \|\bar{x}\|$, and thus that $A\bar{x} = \bar{x}\in\overline{co}(\mathscr{A}\bar{x})$. We have proved the following result. ∎

PROPOSITION 2

Let $\mathscr{A}$ be a family of operators satisfying conditions (1) *and t be the orthogonal projector onto $F_\mathscr{A}$. For any $x\in V$, tx is the unique fixed point of $\mathscr{A}$ that belongs to $\overline{co}(\mathscr{A}x)$, which is the element of $\overline{co}(\mathscr{A}x)$ with minimal norm.* ▲

THEOREM 1

Let $A\in\mathscr{L}(V,V)$ be a continuous linear operator from V to itself, with norm at most equal to 1. Consider the operators

$$T_n = \frac{1}{n}\sum_{k=0}^{n-1} A^k \quad (Cesaro\ means)$$

The operators T_n converge pointwise to the orthogonal projector t onto $F = \operatorname{Ker}(A-1)$. ▲

Proof. We associate with A the family $\mathscr{A} = \{A^k\}_{k\in\mathbb{N}}$ that satisfies obviously property (1). Let t be the orthogonal projector onto $F_\mathscr{A} = \operatorname{Ker}(A-1)$. Since $T_n tx = tx$, it remains to prove that $T_n(1-t)x$ converges to 0, that is, that T_n converges pointwise to 0 on the Hilbert space $F_\mathscr{A}^\oplus$. Since the vector space $G_\mathscr{A} = \operatorname{Im}(A-1)$ is dense in the Hilbert space $F_\mathscr{A}^\oplus$ (by Proposition 1) and since $\|T_n\| \leq 1/n\sum_{k=0}^{n-1}\|A^k\| \leq 1$ for all n, Theorem 1.2 implies that we have to check that $T_n z$ converges to 0 for all $z\in G_\mathscr{A}$. Therefore, we have to verify that $T_n(Ay - y)$ converges to 0 for all $y\in V$. But

$$\|T_n(Ay - y)\| = \frac{1}{n}\|Ay - y\| \leq \frac{2}{n}\|y\| \qquad \text{converges to 0.}$$

Hence the theorem is proved. ∎

Application **The Mean Ergodic Theorem of Von Neumann**

Let Ω be an open subset of $\mathbb{R}^n$, a a continuous, strictly positive, real-valued function, and $L^2(\Omega, a)$ the space of square summable functions for the weight a (See section 6.2.)

Let φ be a mapping from Ω to Ω such that

$$\int_E a(\omega)\,d\omega = \int_{\varphi^{-1}(E)} a(\omega)\,d\omega$$

for all Borel subsets E of Ω. We associate to φ the operator A from $L^2(\Omega, a)$ to itself defined by

$$Ax(\omega) = x[\varphi(\omega)] \qquad \text{for almost all } \omega.$$

This is a continuous linear operator with norm 1. It follows that *for all $x \in L^2(\Omega, a)$, the functions*

$$y_n : \omega \to \frac{1}{n}[x(\omega) + x(\varphi(\omega)) + \ldots + x(\varphi^{(n-1)}(\omega))]$$

converge in quadratic mean to a function x_∞ satisfying $x_\infty(\omega) = x_\infty[\varphi(\omega)]$ for almost all ω.

Interpretation

If Ω is a set of "states," a discrete dynamical system is defined by a mapping φ from Ω to Ω which maps an initial state $\omega \in \Omega$ to $\varphi(\omega)$ in the first period, $\varphi^j(\omega) = \varphi(\varphi^{j-1}(\omega))$ in the jth period.

We take for the function x the characteristic function of a subset Γ of Ω. Then

$$T_n x = \frac{1}{n} \sum_{j=0}^{n-1} x[\varphi^j(\omega)]$$

is the proportion of the number of points of the trajectory $\{\varphi^j(\omega)\}_{j=0,1,\ldots,n}$ which belong to Γ, that is the average time spent in Γ for the successive states. The mean ergodic theorem shows that, in a certain sense, this average time converges.

The origin of ergodic theory, of which we have presented the simplest of the theorems, is found in statistical mechanics, where, rather than seeking the state of a dynamical system at a given instant, one tries to find the probability that the state of a system at a given instant belongs to a given subset. ∎

3. THE BANACH THEOREM

THEOREM 1 (BANACH)
Let V and F be two Hilbert spaces. Suppose that the operator $A \in \mathscr{L}(V, F)$

is surjective. Then there exists a constant $c > 0$ such that

(1) $\quad \forall f \in F, \quad \exists x \in V \quad$ satisfying $\quad Ax = f \quad$ and $\quad \|x\| \leq c^{-1}\|f\|.\quad$ ▲

In fact we are going to reformulate this theorem in the following form.

THEOREM 2
Let V and F be two Hilbert spaces. Suppose that the operator $A \in \mathcal{L}(V, F)$ is surjective. Then we can associate to every $\varepsilon > 0$ a number $\eta > 0$ such that

(2) $\qquad\qquad\qquad B_F(\eta) \subset AB_V(\varepsilon).\qquad\qquad$ ▲

Proof of the equivalence of Theorems 1 and 2. It is clear that (2) results from (1) with $\eta = c\varepsilon$. Conversely, if $f \in F$, then $\eta \dfrac{f}{\|f\|_F} \in B_F(\eta)$ and there exists y such that $Ay = \eta \dfrac{f}{\|f\|_F}$ and $\|y\| \leq \varepsilon$. Hence $x = (y\|f\|)/\eta$ satisfies (1) with $c = \eta/\varepsilon$. ■

In particular, we derive from the Banach theorem the following consequence.

THEOREM 3
Let V and F be Hilbert spaces. Let $A \in \mathcal{L}(V, F)$ be a continuous linear bijective operator from V onto F. Then A^{-1} is continuous. ▲

Proof. Indeed, if A is bijective (2) can be written as follows:

$$\forall \varepsilon > 0, \quad \exists \eta > 0 \quad \text{such that} \quad A^{-1}(B_F(\eta)) \subset B_V(\varepsilon),$$

which expresses the fact that A^{-1} is continuous. ■

We also derive the following proposition.

PROPOSITION 1
Let V be a vector space, $((x, y))_1$ and $((x, y))_2$ two scalar products satisfying

(3) $\quad\begin{cases} \text{i.} & \{V, ((\ ,\ ))_1\} \text{ is complete.} \\ \text{ii.} & \{V, ((\ ,\ ))_2\} \text{ is complete.} \\ \text{iii.} & \|x\|_1 \leq c\|x\|_2, \quad \forall x \in V, \quad \text{where} \quad c > 0. \end{cases}$

Then the norms are equivalent: $\exists k > 0$ such that

(4) $\qquad\qquad\qquad \|x\|_2 \leq k\|x\|_1 \qquad \text{for all} \quad x \in V.\qquad$ ▲

Proof. The hypotheses show that the identity is continuous from the

Hilbert space $\{V,((\ ,\ ))_2\}$ to the space $\{V,((\ ,\ ))_1\}$. Hence its inverse, which is the identity from $\{V,((\ ,\ ))_1\}$ to $\{V,((\ ,\ ))_2\}$, is continuous. ∎

The proof of Theorem 2 consists in proving Lemmas 1 and 2.

LEMMA 1
Suppose that V is a Hilbert space. If there exists $\eta > 0$ such that

(5) $$2B_F(\eta) \subset \overline{AB_V(\varepsilon)},$$

then

(6) $$B_F(\eta) \subset AB_V(\varepsilon). \qquad \blacktriangle$$

Proof. Indeed, hypothesis (5) implies that for every integer n

(7) $\quad 2B_F(\eta_n) \subset 2^{-n}\bar{K} \quad$ where $\quad K = AB_V(\varepsilon) \quad$ and $\quad \eta_n = 2^{-n}\eta$.

Let f be an element of $B_F(\eta)$. Since $2f \in 2B_F(\eta) \subset \bar{K}$, we can find $z_0 \in K$ such that

(8) $$2f - z_0 \in 2B_F(\eta_1).$$

Suppose that we have constructed a sequence $z_k (0 \leq k \leq n-1)$ such that

(9) $$z_k \in K, \quad 2f - \sum_{k=0}^{n-1} 2^{-k}z_k \in 2B_F(\eta_n).$$

Since $2B_F(\eta_n) \subset 2^{-n}\bar{K}$, we can find

(10) $\quad z_n \in K \quad$ such that $\quad 2f - \sum_{k=0}^{n-1} 2^{-k}z_k - 2^{-n}z_n \in 2B_F(\eta_{n+1}),$

according to (7).
Thus we have constructed a sequence $\{z_n\}$ of elements z_n of K such that

(11) $$\lim_{n \to \infty} \left(f - \frac{1}{2}\sum_{k=0}^{n} 2^{-k}z_k \right) = 0.$$

Since $z_k \in K = AB_V(\varepsilon)$, we can find $y_k \in B_V(\varepsilon)$ such that $A(y_k) = z_k$. Then (11) shows that f is the limit of the sequence of elements Ax_n where

(12) $$x_n = \frac{1}{2}\sum_{k=0}^{n} 2^{-k}y_k.$$

We are going to show that the sequence x_n converges. Indeed, it is a Cauchy sequence of elements of V, since

$$\|x_n - x_m\| = \left\|\frac{1}{2}\sum_{k=n+1}^{m} 2^{-k}y_k\right\| \leq \frac{1}{2}\sum_{k=n+1}^{m} 2^{-k}\|y_k\| \leq \frac{\varepsilon}{2}\sum_{k=n+1}^{m} 2^{-k}.$$

Then, since *V is complete*, this Cauchy sequence x_n converges to an element x_* of V. This limit x_* belongs to $B_V(\varepsilon)$, since this is a closed convex set and since

(13) $\begin{cases} \text{i.} \quad \dfrac{1}{1-2^{-n-1}}x_n = \dfrac{1-2^{-1}}{1-2^{-n-1}}\sum_{k=0}^{n}2^{-k}y_k \in B_V(\varepsilon). \\ \text{ii.} \quad x_* = \lim_{n\to\infty}\dfrac{1}{1-2^{-n-1}}x_n \in B_V(\varepsilon). \end{cases}$

Since $\frac{1}{2}\sum_{k=0}^{n}2^{-k}z_k = Ax_n$, it follows from (11) and from the continuity of A that $f = Ax_* \in AB_V(\varepsilon)$. ∎

For Theorem 2 to be true, it suffices that hypothesis (5) of Lemma 1 is satisfied; this is the object of Lemma 2.

LEMMA 2
Suppose that F is a Hilbert space. If A is surjective, there exists $\eta > 0$ such that

(5) $\qquad\qquad 2B_F(\eta) \subset \overline{AB_V(\varepsilon)}.$ ▲

Proof. We set $K = AB_V(\varepsilon)$ and $K_n = nAB_V(\varepsilon)$. Since $V = \bigcup_{n=1}^{\infty} nB_V(\varepsilon)$ and A is surjective, we deduce that $F \subset \bigcup_{n=1}^{\infty}\bar{K}_n$. Since F is a complete metric space, Baire's theorem (see Remark 4.1.1) implies that at least one of these $\bar{K}_n$'s has a nonempty interior. If y_n belongs to the interior of $\bar{K}_n$, $y = 1/n\, y_n$ belongs to the interior of $\bar{K}$; hence there exists a ball of radius 4η such that $B_F(4\eta) \subset \bar{K} - y \subset \bar{K} + \bar{K} = 2\bar{K}$, since $-y \in K$. ∎

Remark 1

We have only used the fact that the spaces V and F are complete metric spaces in the proof of Theorem 2. Hence this theorem extends to the case of Banach spaces and Fréchet spaces (complete metric vector spaces). ∎

*The Closed Graph Theorem

We know that the graph of a *continuous mapping* A from V to F is closed. (See Proposition 4.4.8 of [AAA], p. 145.) We are going to prove, using the Banach theorem, that the converse is true when A is linear and V and F are Hilbert spaces.

THEOREM 4 (THE CLOSED GRAPH THEOREM)
Let V and F be two Hilbert spaces. For a linear mapping A from V to F to be continuous, it is necessary and sufficient that the graph of A be closed; that is, if x_n converges to 0 in V and if Ax_n converges to f in F, then $f = 0$. ▲

Proof. We know by hypothesis that the graph

$$G = \{x, Ax\}_{x \in V}$$

is a closed vector subspace of the Hilbert space $V \times F$ and is, consequently, a Hilbert space.

The projection κ of G onto V is continuous and bijective, and is therefore an *isomorphism* according to Theorem 3.

If ω denotes the projection of G onto F, we can write that $A = \omega \kappa^{-1}$ is the product of two continuous operators and that, consequently, it is continuous. ∎

Remark 2

This theorem is true even if V and F are Banach spaces or Fréchet spaces. ∎

Remark 3

We also know another converse. (See Proposition 4.4.9 of [AAA], p. 146.) If V is a topological space and F is compact, every mapping A whose graph is closed is continuous. ∎

4. THE CLOSED RANGE THEOREM

THEOREM 1

Let V_1 and V_2 be two Hilbert spaces and $A \in \mathscr{L}(V_1, V_2)$. If $\operatorname{Im} A$ is closed in V_2, then $\operatorname{Im} A^*$ is also closed in V_1^* and

(1) $\begin{cases} \text{i.} & \operatorname{Im} A^* = (\operatorname{Ker} A)^\perp. \\ \text{ii.} & \operatorname{Im} A = (\operatorname{Ker} A^*)^\perp. \end{cases}$ ▲

Proof. We already know that $\operatorname{Im} A^* \subset (\operatorname{Ker} A)^\perp$. The opposite inclusion proves that $\operatorname{Im} A^* = (\operatorname{Ker} A)^\perp$ is a closed subspace of V_1^*. Thus we take $f_1 \in (\operatorname{Ker} A)^\perp$, and we show that there exists $f_2 \in V_2^*$ such that $f_1 = A^* f_2$.

To this end we associate to $f_1 \in (\operatorname{Ker} A)^\perp \subset V_1^*$ the linear form g_2 defined on the *closed subspace* $\operatorname{Im} A$ of V_2 by

(2) $\langle g_2, x_2 \rangle = \langle f_1, x_1 \rangle \qquad$ if x_1 is a solution of $Ax_1 = x_2$.

Since $f_1 \in (\operatorname{Ker} A)^\perp$, the form g_2 is indeed defined because it does not depend on the choice of the solution x_1 of the equation $Ax_1 = x_2$. (For if $Ax_1 = Ay_1$, $x_1 - y_1 \in \operatorname{Ker} A$ and $\langle f_1, x_1 \rangle = \langle f_1, y_1 \rangle$.) It is clear that g_2 is linear. *Let us show that g_2 is continuous.* According to the Banach theorem, Theorem 3.1, there exists a constant $c > 0$ such that we can associate to $x_2 \in V_2$ a solution x_1 of the equation $Ax_1 = x_2$ that satisfies $\|x_1\| \leq c^{-1}\|x_2\|$. Thus

(3) $\qquad |\langle g_2, x_2 \rangle| = |\langle f_1, x_1 \rangle| \leq \|f_1\|_* \|x_1\| \leq c^{-1} \|f_1\|_* \|x_2\|,$

that is, g_2 is continuous.

According to the extension theorem, Theorem 2.1.1, we can extend g_2 to a continuous linear form f_2 on V_2. Then $A^*f_2 = f_1$, since for all $x_1 \in V_1$,

(4) $\qquad \langle f_1, x_1 \rangle = \langle g_2, Ax_1 \rangle = \langle f_2, Ax_1 \rangle = \langle A^*f_2, x_1 \rangle.$ ∎

Remark 1

Since we have only used Theorem 2.1.1 and Theorem 3.1, which are both true in the case of Banach spaces, it follows that this proof is still true for Banach spaces. ∎

In particular, we obtain the following property of surjective operators.

THEOREM 2

Let V_1 and V_2 be Hilbert spaces. Then $A \in \mathscr{L}(V_1, V_2)$ is surjective, if and only if $A^* \in \mathscr{L}(V_2^*, V_1^*)$ is an isomorphism from V_2^* onto its closed image (range) $\text{Im } A^* = (\text{Ker } A)^\perp$. ▲

Proof. If A is surjective, $\text{Im } A = V_2$ is closed and hence $\text{Im } A^* = (\text{Ker } A)^\perp$ is closed in V_1^*. Moreover, since $\text{Im } A$ is dense in V_2, A^* is injective. Therefore A^* is a bijection from the Hilbert space V_2^* onto the Hilbert space $\text{Im } A^*$. According to Theorem 3.3, A^* is an isomorphism. ∎

Remark 2

Since V_1 and V_2 are Hilbert spaces, the converse of Theorem 1 clearly holds:

(5) $\qquad\qquad$ If $\text{Im } A^*$ is closed, $\text{Im } A$ is also closed.

When V_1 and V_2 are Banach spaces, the converse (5) is still true; a special proof is needed to establish this. The situation is analogous for Theorem 2. ∎

5. CHARACTERIZATION OF LEFT INVERTIBLE OPERATORS

Let $A \in \mathscr{L}(V, F)$ be a continuous linear operator from V to F.

DEFINITION 1

We say that $A \in \mathscr{L}(V, F)$ is left invertible (or, respectively, right invertible) if there exists a continuous linear operator $B \in \mathscr{L}(F, V)$ such that

(1) $\qquad BA = 1_V \qquad$ (or $\qquad AB = 1_F$).

We say that such an operator B is a continuous linear "left" inverse (or, "right"). ▲

When the context permits no confusion, we say only that B is a left (or right) inverse without mentioning that it is *also continuous and linear.*

We begin with a simple statement.

PROPOSITION 1
A linear operator $A \in \mathscr{L}(V, F)$ is left invertible if and only if its transpose $A^* \in \mathscr{L}(F^*, V^*)$ is right invertible. Moreover, B is a left inverse of A if and only if B^* is a right inverse of A^*. ▲

PROPOSITION 2
Let V and F be two Hilbert spaces. The following assertions are equivalent:

(2) $\quad \begin{cases} \textbf{i.} & A \in \mathscr{L}(V, F) \quad \text{is injective and} \quad \text{Im } A \quad \text{is closed in } F. \\ \textbf{ii.} & A \quad \text{is left invertible.} \end{cases}$ ▲

Proof. Suppose that (2)i is true. Since Im A is closed, it is a Hilbert space and the continuous bijective operator A_0 from V onto Im A has, therefore, an inverse A_0^{-1}, which is continuous by Theorem 3.3. Moreover, if t is the orthogonal projection from F onto Im A (which is closed), then $A^- = A_0^{-1} t$ is a continuous linear operator from F onto V, which is a left inverse of A:

$$A^- A x = A_0^{-1} t A x = A_0^{-1} A x = A_0^{-1} A_0 x = x.$$

Conversely, if A has a left inverse B, A is injective because $Ax = 0$ implies that $x = BAx = 0$. Im A is closed, since if Ax_n converges to an element $f \in F$, then $x_n = BAx_n$ converges to $x = Bf$ and, therefore, Ax_n converges to $Ax \in$ Im A. ■

Remark 1

We can define $A^- = A_0^{-1} t$ even when V is only a vector space. This is a left inverse that is only linear (not continuous). Since t is the orthogonal projector onto Im A, it is convenient to introduce the following definition. ■

DEFINITION 2
Suppose that F is a Hilbert space and that

A is an injective linear operator from V to F with closed image in F.

CH. 4, SEC. 5   CHARACTERIZATION OF LEFT INVERTIBLE OPERATORS   87

We shall say that the operator A^-, which associates to every $f \in F$ the unique element A^-f, which minimizes on V the function $x \mapsto \|f - Ax\|_F$, is the orthogonal left inverse of A.   ▲

We are going to give an explicit formula for A^-.

PROPOSITION 3
Let V and F be two Hilbert spaces, $K \in \mathscr{L}(F, F^*)$ the duality operator from F onto F^* and

(3) $\qquad A \in \mathscr{L}(V, F) \qquad$ *a left invertible operator.*

Then the orthogonal left inverse is defined by

(4) $\qquad A^- = (A^*KA)^{-1}A^*K.$ ▲

Proof. Since AA^-f is the orthogonal projection from f onto the closed subspace Im A, we obtain

$$((AA^-f - f, Ax))_F = 0 \qquad \text{for all} \qquad x \in V.$$

Consequently,

$$((AA^-f - f, Ax))_F = \langle KAA^-f - Kf, Ax \rangle$$
$$= \langle (A^*KA)A^-f - A^*Kf, x \rangle = 0 \qquad \text{for all} \qquad x \in V.$$

This implies that A^-f is a solution of the equation

$$(A^*KA)A^-f = A^*Kf.$$

Formula (4) then follows from the following lemma.   ■

LEMMA 1
Suppose that

(5) $\qquad A \in \mathscr{L}(V, F) \qquad$ *is left invertible.*

Then

(6) $\qquad A^*KA \qquad$ *is an isomorphism of V onto V^*.*   ▲

Proof. We use the Lax-Milgram theorem, Theorem 3.6.1; for this it is necessary to verify that A^*KA is V-elliptic. Now

$$\langle A^*KAx, x \rangle = \langle KAx, Ax \rangle = \|Ax\|_F^2.$$

But according to the Banach theorem, Theorem 3.3, there exists a constant $c > 0$ such that $\|Ax\|_F \geq \sqrt{c}\|x\|_V$. It follows, therefore, that

(7) $\qquad \langle A^*KAx, x \rangle \geq c\|x\|_V^2 \qquad \text{for all} \qquad x \in V.$ ■

Remark 2

The Method of Least Squares. The solution A^-f to the problem $Ax = f$ is, in the case of finite dimensional spaces, the one given by the method of least squares. We suppose that the vector $f \in F = \mathbb{R}^m$ represents m "data" that depend linearly on n unknown parameters represented by a vector $x \in \mathbb{R}^n (n < m)$; that is, that $f = Ax$ where A is a linear injective operator. The problem is to find a vector x. Since the solution A^-f is that which minimizes $x \mapsto \|f - Ax\|$, we say that it is obtained by the method of least squares. This elementary technique is often used in statistics. ■

*Remark 3

We can construct all the other left inverses B of A by using A^-. ■

*PROPOSITION 4
Let t be the orthogonal projection from F onto $\mathrm{Im}\ A$ where

(8) $\qquad\qquad A \in \mathscr{L}(V, F) \quad$ *is a left invertible operator.*

Then every left inverse B of A can be written

(9) $\qquad\qquad B = A^- + C(1 - t) \quad$ where $\quad C \in \mathscr{L}(F, V)$. ▲

Proof. Let $C \in \mathscr{L}(F, V)$ be an arbitrary operator from F to V. Let us show that B is a left inverse of A:

$$BAf = A^-Af + C((1 - t)Af) = A^-Af = f$$

since $A^-Af = f$ and since $(1 - t)Af = 0$ because $Af \in \mathrm{Im}\ A$ and, therefore, $tAf = Af$.

Conversely, every left inverse B can be written

$$B = A^- + B(1 - t),$$

since $A^-f = A_0^{-1}tf = Btf$. Indeed, the restriction to $\mathrm{Im}\ A = \mathrm{Im}\ t$ of every left inverse B of A is equal to A_0^{-1}. ■

6. CHARACTERIZATION OF RIGHT INVERTIBLE OPERATORS

PROPOSITION 1
Let V and F be two Hilbert spaces. The following assertions are equivalent.

(1) $\qquad \begin{cases} \text{i.} & A \in \mathscr{L}(V, F) \quad \text{is surjective.} \\ \text{ii.} & A \quad \text{is right invertible.} \end{cases}$ ▲

Proof. **a.** It is clear that (1)ii implies (1)i.

b. Suppose that A is surjective. Let $M = \operatorname{Ker} A$ and let $N = (\operatorname{Ker} A)^{\oplus}$ be the Hilbert orthogonal complement of M. We denote by t and s the orthogonal projectors onto M and N. Since $x = tx + sx$ and $Ax = Asx$, the operator $A_1 = As$ is a continuous bijection from $(\operatorname{Ker} A)^{\oplus}$ onto F. Hence $A^+ = A_1^{-1}$ is a linear operator from F to $(\operatorname{Ker} A)^{\oplus}$, which is continuous according to Theorem 3.3. It is clear that A^+ is a right inverse of A, since

$$AA^+f = AA_1^{-1}f = A_1 A_1^{-1}f = f \quad \text{for all} \quad f \in F. \qquad \blacksquare$$

Moreover, since A^+f is orthogonal to $\operatorname{Ker} A$ and since $A^+f - x \in \operatorname{Ker} A$ when x is a solution of the equation $Ax = f$, it follows that

(2) $\qquad ((A^+f, A^+f - x))_V = 0 \quad \text{when} \quad Ax = f$

and, consequently, that A^+f is the orthogonal projection of zero onto $A^{-1}(f)$.
$\blacksquare$

Remark 1

We can define A^+ even when F is only a vector space. This is a right inverse that is only linear (not continuous). $\blacksquare$

DEFINITION 1
Suppose that V is a Hilbert space and that

(3) $\qquad \begin{cases} A \text{ is a surjective linear operator from } V \text{ onto } F \\ \text{whose kernel } \operatorname{Ker} A \text{ is closed in } V. \end{cases}$

We shall say that the operator A^+, which associates to every $f \in F$ the unique element A^+f which minimizes $x \mapsto \|x\|_V$ on $A^{-1}(f)$, is the *orthogonal right inverse* of A. $\blacktriangle$

We are going to give an explicit formula for A^+.

PROPOSITION 2
Let V and F be two Hilbert spaces, $J \in \mathscr{L}(V, V^*)$ the duality operator from V onto V^*, and

(4) $\qquad A \in \mathscr{L}(V, F)$ is a right invertible operator.

Then the orthogonal right inverse A^+ of A is equal to

(5) $\qquad A^+ = J^{-1}A^*(AJ^{-1}A^*)^{-1}.$ $\blacktriangle$

Proof. Since $\bar{x} = A^+f$ is the unique projection of zero onto the (nonempty) set of solutions x of the equation $Ax = f$, it is characterized by

(6) $\quad ((\bar{x}, \bar{x} - x)) = 0 \quad$ for all x such that $\quad Ax = f = A\bar{x}$.

In particular, $y = \bar{x} - x$ runs over the kernel Ker A of A. It follows, then, that

(7) $\quad ((\bar{x}, y)) = \langle J\bar{x}, y \rangle = 0 \quad$ for all $\quad y \in \text{Ker } A$.

Hence $J\bar{x} \in (\text{Ker } A)^{\perp} = \overline{\text{Im } A^*} = \text{Im } A^*$, since the image of A^* is closed, A^* being left invertible (See Propositions 3.4.1 and 5.2.) Thus there exists a unique element $p \in F^*$ such that

(8) $\quad\quad\quad\quad\quad\quad\quad\quad \bar{x} = J^{-1} A^* p.$

This element p is, therefore, a solution of the equation

(9) $\quad\quad\quad\quad\quad\quad\quad\quad A\bar{x} = f = (AJ^{-1} A^*) p.$

Since $A^* \in \mathcal{L}(F^*, V^*)$ is left invertible, Lemma 5.1 implies that $AJ^{-1}A^*$ is invertible. Hence (8) and (9) imply that

(10) $\quad\quad\quad\quad\quad\quad \bar{x} = J^{-1} A^* p = J^{-1} A^* (AJ^{-1} A^*)^{-1} f.$ ∎

We now give some properties of left and right invertible operators.

PROPOSITION 3
Suppose that $A \in \mathcal{L}(V, F)$ is right invertible. Then

(11) $\quad \begin{cases} \text{i.} & (A^+)^* = (A^*)^- . \\ \text{ii.} & A = (A^+)^- . \end{cases}$

and

(12) $\quad\quad\quad\quad\quad\quad\quad (A^+)^* J A^+ = (AJ^{-1} A^*)^{-1}.$

If $A \in \mathcal{L}(V, F)$ is left invertible, then

(13) $\quad \begin{cases} \text{i.} & (A^-)^* = (A^*)^+ . \\ \text{ii.} & A = (A^-)^+ . \end{cases}$ ▲

Proof. We verify these properties only in the case where $A \in \mathcal{L}(V, F)$ is right invertible. We can write

$$(A^+)^* = (AJ^{-1} A^*)^{-1} AJ^{-1}$$
$$= ((A^*)^*(J^{-1})A^*)^{-1} (A^*)^* J^{-1},$$

which shows that $(A^+)^* = (A^*)^-$ is the orthogonal left inverse of $A \in \mathcal{L}(F^*, V^*)$. (We take $K = J^{-1}$, the canonical isometry from V^* to V.) In particular, it follows that

$$(A^+)^* J A^+ = (AJ^{-1} A^*)^{-1}.$$

The operator $A^+ \in \mathscr{L}(F, V)$ is left invertible. Its orthogonal left inverse is defined by

$$(A^+)^- = ((A^+)^* J A^+)^{-1} (A^+)^* J = (AJ^{-1} A^*)(A^+)^* J$$
$$= AJ^{-1} J = A$$

and is, therefore, equal to A. ∎

*Remark 2

We can characterize all the other right inverses of $A \in \mathscr{L}(V, F)$. ∎

PROPOSITION 4
Let

(14) $\qquad A \in \mathscr{L}(V, F) \quad \text{be a right invertible operator}$

and let s be the orthogonal projector of V onto Ker A.
 Then every right inverse B of A can be written

(15) $\qquad B = A^+ + sC \quad \text{where} \quad C \in \mathscr{L}(F, V).$ ▲

Proof. It is clear that $B = A^+ + sC$ is a right inverse of A, since $AB = AA^+ + AsC = AA^+$ (indeed, $As = 0$ because s is a projector onto Ker A).
 Conversely, if B is a right inverse of A, then $B = A^+ + s(B - A^+)$, since

$$Bf = sBf + (1-s)Bf = sBf + (1-s)A^+ f$$
$$= A^+ f + s(B - A^+)f$$

because $Bf - A^+ f \in \text{Ker } A$, consequently,

$$Bf - A^+ f = s(Bf - A^+ f).$$ ∎

*Pseudoinverses

We can extend the notion of minimal right and left inverses to the case of operators with a closed image.

*PROPOSITION 5
Let V and F be two Hilbert spaces and J and K the canonical isometries from V onto V^* and from F onto F^*. Suppose that

(16) $\qquad A \in \mathscr{L}(V, F) \quad \text{has a closed image.}$

Then there exists a unique continuous linear operator $B \in \mathscr{L}(F, V)$, which associates to each $f \in F$ the unique element $\bar{x} = Bf$, which minimizes $x \mapsto \|x\|_V$

on the set $K(f)$ defined by

$$K(f) = \{x \in V, \quad \text{which minimizes} \quad y \mapsto \|f - Ay\|_F\}. \qquad \blacktriangle$$

Proof. Let s be the orthogonal projection onto $(\operatorname{Ker} A)^{\oplus}$ and t the orthogonal projection onto $\operatorname{Im} A$ (which is closed by hypothesis). Then As is a continuous bijection from $(\operatorname{Ker} A)^{\oplus}$ onto $\operatorname{Im} A$, whose inverse $(As)^{-1}$ is continuous from $\operatorname{Im} A$ to $(\operatorname{Ker} A)^{\oplus}$, according to the Banach theorem, Theorem 3.3. Then $(As)^{-1}t$ is a continuous linear operator from F to V, which minimizes the norm on the set $K(f)$. $\blacksquare$

DEFINITION 2
Suppose that $A \in \mathscr{L}(V, F)$ has a closed image. We say that the operator $B = (As)^{-1}t$ defined by Proposition 4 is the pseudoinverse of A. $\blacktriangle$

*7. QUADRATIC PROGRAMMING WITH LINEAR CONSTRAINTS

The problem of *quadratic programming with linear constraints* consists in seeking an element $\bar{x} \in V$ such that

(1) $\quad \begin{cases} \text{i.} & A\bar{x} = v \quad \text{where } v \text{ is given in } V. \\ \text{ii.} & \|\bar{x} - u\| = \min_{Ax=v} \|x - u\| \quad \text{where } u \text{ is given in } V. \end{cases}$

where

(2) $\qquad A \in \mathscr{L}(V, F) \quad$ is a surjective operator from V onto F.

PROPOSITION 1
Suppose (2): there exists a unique solution $\bar{x}$ to the problem (1) defined by

(3) $\quad \begin{cases} \text{i.} & \bar{x} = u - J^{-1}A^*\bar{p}. \\ \text{ii.} & \bar{p} = (AJ^{-1}A^*)^{-1}(Au - v) \in F^*. \end{cases}$

where $J \in \mathscr{L}(V, V^*)$ is the isometry from V onto V^*. $\blacktriangle$

Proof. Making the change of variable $y = x - u$ reduces our task to finding $\bar{y} = \bar{x} - u$, which is a solution to the following problem:

(4) $\quad \begin{cases} \text{i.} & A\bar{y} = v - Au \\ \text{ii.} & \|\bar{y}\| = \min_{Ay = v - u} \|y\| \end{cases}$

However, by definition of A^+, this amounts to saying that $\bar{y} = A^+(v - Au)$

and, therefore, that

$$\bar{x} = u + \bar{y} = u - A^+Au + A^+v = u - J^{-1}A^*(AJ^{-1}A^*)^{-1}(Au - v)$$
$$= u - J^{-1}A^*\bar{p} \text{ where } \bar{p} = (AJ^{-1}A^*)^{-1}(Au - v). \qquad \blacksquare$$

In formula (3) $\bar{p}$ plays a special role for the following reason. Set

(5) $$\alpha(v) = \inf_{Ax=v} \tfrac{1}{2} \|x - u\|^2.$$

PROPOSITION 2
For each $v \in F$, the element $-\bar{p}(v) = -(AJ^{-1}A^*)^{-1}(Au - v) \in F^*$ is the differential with respect to v of the function α defined by (5):

(6) $$\lim_{\theta \to 0} \frac{\alpha(v + \theta w) - \alpha(v)}{\theta} = \langle -\bar{p}(v), w \rangle \quad \text{for all} \quad w \in F. \qquad \blacktriangle$$

Proof. Indeed,

$$\alpha(v) = \tfrac{1}{2}\|\bar{x} - u\|^2 = \tfrac{1}{2}\|\bar{y}\|^2 = \tfrac{1}{2}\|A^+(v - Au)\|^2$$

Therefore, developing $\alpha(v + \theta w) - \alpha(v)$, we obtain

$$\alpha(v + \theta w) - \alpha(v) = \theta((A^+w, A^+v - A^+Au)) + \frac{\theta^2}{2}\|A^+w\|^2.$$

It follows that

$$\lim_{\theta \to 0} \frac{\alpha(v + \theta w) - \alpha(v)}{\theta} = ((A^+w, A^+v - A^+Au)),$$

and

$$((A^+v - A^+Au, A^+w)) = \langle (A^+)^*JA^+(v - Au), w \rangle$$
$$= \langle (AJ^{-1}A^*)^{-1}(v - Au), w \rangle = -\langle \bar{p}(v), w \rangle$$

since $(A^+)^*JA^+ = AJ^{-1}A^1$ according to Proposition 6.3. $\qquad \blacksquare$

Another reason justifies the presentation of (3).

PROPOSITION 3
For every $p \in F^*$ the element $x(p) = u - J^{-1}A^*p$ minimizes the function $x \mapsto \tfrac{1}{2}\|x - u\|^2 + \langle p, Ax \rangle.$ $\qquad \blacktriangle$

Proof. (Left as an exercise.)

Remark 1

We can interpret V as the space of commodities produced, u as the demand for the produced commodities, F as the space of resources, v as the available resource, $A \in \mathscr{L}(V, F)$ as the operator that associates to every produced commodity x the resource Ax necessary for its production. Problem (1) becomes that of finding among all the commodities that can be produced with the resource v that which is the closest to the demand u. The function $-\alpha(v)$ denotes the satisfaction obtained with the resources v. Propositions 1 and 2 show that $\bar{p} = p(v)$ is the marginal increase of satisfaction due to a variation of resources; it defines the price of the resources. Propositions 1 and 3 show that knowing $\bar{p}$ and minimizing the sum of the cost $\langle \bar{p}, Ax \rangle$ (of resources necessary to produce x) and of $\frac{1}{2}\|x - u\|^2$ yields the optimal production $\bar{x}$. ∎

COROLLARY 1

Given $u \in V$ and $p \in V^*$, the solution $\bar{x}$ of

(7) $\quad \begin{cases} \text{i.} & \langle p, \bar{x} \rangle = v. \\ \text{ii.} & \|\bar{x} - u\| = \min_{\langle p, x \rangle} \|x - u\|. \end{cases}$

is equal to

(8) $$\bar{x} = u - \frac{\langle p, u \rangle - v}{\|p\|_*^2} J^{-1} p.$$

▲

Proof. (Left as an exercise.)

Remark 2

In particular, if $w \in V$ and if we set $v = \langle p, w \rangle$, then

$$\bar{x}(p) = u - \frac{\langle p, u - w \rangle}{\|p\|_*^2} J^{-1} p$$

is the optimal solution of the problem $\min_{\langle p, x - w \rangle = 0} \|x - u\|$. The mapping $p \mapsto \bar{x}(p)$ is an example of what economists call *demand functions*: If V is a commodity space, w the vector of the initial endowment of a consumer, u his consumption objective, $p \in V^*$ the prevailing price system, the budget constraint $\langle p, x - w \rangle = 0$ forces the consumer to choose a commodity $x \in V$ whose value $\langle p, x \rangle$ is equal to the value $\langle p, w \rangle$ of his initial endowment. Then $\bar{x}(p)$ is the consumption closest to the consumption objective. ∎

Remark 3

If we consider n consumers $i = 1, \ldots, n$ characterized by their consumption objectives u_i and their initial resources w_i, they associate to every price system $p \in V^*$ the demands

$$\bar{x}_i(p) = u_i - \frac{\langle p, u_i - w_i \rangle}{\|p\|^2} J^{-1} p.$$

We call a *Walras equilibrium* a price $\bar{p}$ for which the sum of the demands is equal to the sum of the resources, that is, such that $\sum_{i=1}^{n} \bar{x}_i(\bar{p}) = \sum_{i=1}^{n} w_i$. It can be easily verified that $\bar{p}$ (defined up to a scalar multiple) is equal to $\bar{p} = J^{-1}(\sum_{i=1}^{n}(u_i - w_i))$. In other words, the equilibrium price depends linearly on the excess of the demand (the sum of the consumption objectives) over the supply (the sum of the resources). ∎

Remark 4

The study of Walras equilibria in a general model is presented in Chapter 15, Section 8.

CHAPTER 5

Construction of Hilbert Spaces

We have already constructed finite products, closed subspaces, and quotient spaces of Hilbert spaces in Section 1.6 and studied their duals in Section 3.5. We have also constructed the completion of a prehilbert space in Section 3.1. We devote this chapter to other construction methods (with the exception of the Hilbert tensor product, which is studied in Chapter 12). Sections 5 to 10 may be omitted in a first reading. However, the aim of the first four sections is *to construct Hilbert spaces for which given linear mappings are continuous;* these are, therefore, important.

First of all, in Section 1 we consider the case of a finite family of linear mappings A_p from V to the *Hilbert* spaces F_p. We then give conditions under which the vector space V is a Hilbert space for the scalar product $\Sigma((A_p x, A_p y))$ (called the *initial* scalar product).

On the other hand, in Section 2 we provide a vector space F with a *final* scalar product for which a linear operator A from a Hilbert space V to F is then continuous.

We attempt in Sections 3 and 4 to formalize the construction procedures for Hilbert spaces in analysis. In particular, we use these methods to construct Sobolev spaces. (See Chapters 7 and 9.)

The reader is encouraged to study the construction of Sobolev spaces at the same time. The underlying idea is the following: We start from a (small) vector space $\mathscr{D}$ [the space $\mathscr{D}(\Omega)$ of functions that possess derivatives of all orders ("infinitely differentiable") and with compact support, for example] and from the *simplest* scalar product (φ, ψ) on $\mathscr{D}$ [$(\varphi, \psi) = \int_\Omega \varphi(\omega)\psi(\omega)\,d\omega$, for example]. We then let the completion H of $\mathscr{D}$ for (φ, ψ) play the role of a *pivot space* [for example, $L^2(\Omega)$ is considered as a pivot space].

We divide into two categories those Hilbert spaces V such that

$$\mathscr{D} \subset V \subset H,$$

calling *normal* those in which $\mathscr{D}$ is *dense*. We agree then to identify H with a dense subspace of the dual of a normal space. Details are given in Section 3.

CH. 5 INTRODUCTION

In Section 4 we associate to a finite family $\mathscr{A}$ of linear operators A_p from $\mathscr{D}$ to $\mathscr{D}$ the scalar product.

$$((\varphi, \psi))_{\mathscr{A}} = (\varphi, \psi) + \sum_p (A_p \varphi, A_p \psi).$$

The completion of $\mathscr{D}$ for *this scalar product is denoted by* $H_0(\mathscr{A})$ and is called the *minimal domain* of $\mathscr{A}$. It is contained in H when we suppose that the family $\mathscr{A}$ is closed. Then A_p has a unique extension to a *continuous* linear operator *from* $H_0(\mathscr{A})$ *to* H. Moreover, $H_0(\mathscr{A})$ is a *normal* space.

Now suppose that we can associate to $\mathscr{A}$ a family $\mathscr{A}^*$ of linear operators A_p^* from $\mathscr{D}$ to $\mathscr{D}$ such that

$$(A_p \varphi, \psi) = (\varphi, A_p^* \psi) \qquad \forall \varphi, \psi \in \mathscr{D}.$$

We show then that A_p has a unique extension to a continuous linear operator A_p from H to $H_0^*(\mathscr{A}^*)$ [which contains H, because $H_0(\mathscr{A}^*)$ is *normal*]. In other words, we show that for all p,

A_p is a linear operator from $\mathscr{D}$ to $\mathscr{D}$,

A_p is a continuous linear operator from $H_0(\mathscr{A})$ to H,

A_p is a continuous linear operator from H to $H_0^*(\mathscr{A}^*)$.

We are, therefore, led to introduce the space

$$H(\mathscr{A}) = \{x \in H \quad \text{such that} \quad A_p x \in H \quad \text{for all} \quad p\},$$

which is a Hilbert space containing $H_0(\mathscr{A})$. It is called the *maximal domain* of $\mathscr{A}$. It is generally different from $H_0(\mathscr{A})$ and in this case is not a normal space.

We introduce the concept of an unbounded operator $(D(A), A)$, and we study its properties (Section 5).

In Section 6 we give a condition that allows us to choose a realization of the completion of V contained in a Hilbert space H when V is itself a subspace of H. We then construct the *Hausdorff completion* of a nonseparated prehilbert space (Section 7).

We construct in Section 8 a Hilbert subspace of the countable product $\prod V_n$ of a sequence of Hilbert spaces, which we call the *direct sum* of the spaces V_n.

In Section 9 we present the intermediate spaces and the interpolation spaces V between two spaces U and W and establish the interpolation property.

Finally we conclude this chapter with the study of the reproducing kernels of Aronsajn-Bergmann, which characterize the Hilbert spaces of functions on a set X by means of a function

$$k : X \times X \mapsto \mathbb{R}$$

that is symmetric of positive type and is called the *reproducing kernel* of this space.

1. THE INITIAL SCALAR PRODUCT

We begin with the case of an operator A from V to F.

PROPOSITION 1

Let us consider

(1) $\begin{cases} \text{i.} & \text{a vector space } V. \\ \text{ii.} & \text{a Hilbert space } F. \\ \text{iii.} & \text{an injective linear operator } A \text{ from } V \text{ to } F \text{ whose image is closed in } F. \end{cases}$

If $((f,g))_F$ denotes the scalar product of F and K is the duality operator from F into F^*, then V is a Hilbert space for the initial scalar product

(2) $\qquad\qquad ((x, y)) = ((Ax, Ay))_F,$

and the duality operator J from V onto V^* is equal to

(3) $\qquad\qquad J = A^*KA.$

In this case, A is an isometry from V to F, and the norm of its orthogonal left inverse A^- is equal to one. ▲

Proof. It is clear that $((x, y)) = ((Ax, Ay))_F$ is a positive bilinear form on V. It is positive-definite because, A being injective, $((x, x)) = ((Ax, Ax))_F = \|Ax\|_F^2 = 0$ implies that $Ax = 0$ and, consequently, that $x = 0$. Therefore $((x, y))$ is a scalar product and A is an isometry from V onto F. Since Im A is closed, it follows that V is complete; indeed, if $\{x_n\}$ is a Cauchy sequence of elements x_n of V, then Ax_n is a Cauchy sequence in the closed subspace Im A, which is complete since F is complete. Hence Ax_n converges to an element Ax, which amounts to saying that x_n converges to x. Thus V is a Hilbert space.

Since we can write that

$$((x, y)) = ((Ax, Ay))_F = \langle KAx, Ay \rangle = \langle A^*KAx, y \rangle,$$

it follows that $J = A^*KA$ is the duality operator from V onto V^*.

The orthogonal left inverse $A^- = (A^*KA)^{-1}A^*K$ of A has a norm equal to one, since

$$\|A^-f\|_V = \|AA^-f\|_F = \|tf\|_F \le \|f\|_F,$$

where t is the orthogonal projector onto Im A. ∎

Let us consider now the case where we are given

(4) $\begin{cases} \text{i.} & \text{a vector space } V, \\ \text{ii.} & n \text{ Hilbert spaces } F_i, \\ \text{iii.} & n \text{ linear operators } A_i \text{ from } V \text{ to } F_i. \end{cases}$

Suppose that

(5)   if $A_i x = 0$ for all $i = 1, \ldots, n,$ then $x = 0.$

(This property holds if one of the A_i's is injective.) We shall say then that the family of the A_i's is *collectively injective*. Suppose, as well, that

(6) $\begin{cases} \text{if a sequence } \{x_m\}_m \text{ of elements } x_m \in V \text{ satisfies } \lim_n A_i x_n = f_i \\ \text{for all} \quad i = 1, \ldots, n, \text{ then there exists } \bar{x} \in V \quad \text{such that} \\ f_i = A_i \bar{x} \quad \text{for all} \quad i = 1, \ldots, n. \end{cases}$

Then we shall say that the *family of operators A_i is closed*.

PROPOSITION 2
Suppose that the conditions (5) and (6) are satisfied. Then V is a Hilbert space for the scalar product:

(7) $$((x, y)) = \sum_{i=1}^{n} ((A_i x, A_i y))_{F_i}.$$

The duality operator J from V onto V^* is then equal to

(8) $$J = \sum_{i=1}^{n} A_i^* K_i A_i$$

where K_i denotes the duality operator from F_i onto its dual. The dual V^* is equal to the set of continuous linear forms of the form:

(9) $$f = \sum_{i=1}^{n} A_i^* f_i \quad \text{where} \quad f_i \in F_i^*.$$ ▲

Proof. Consider the Hilbert space $F = \prod_{i=1}^{n} F_i$ and the operator A from V to F defined by

$$Ax = \{A_i x\}_{i=1,\ldots,n}.$$

Condition (5) amounts to saying that A is injective, and condition (6) means that the image of A is closed. Proposition 1 implies, then, that V is a Hilbert space for the initial scalar product:

$$((x, y)) = ((Ax, Ay))_F = \sum_{i=1}^{n} ((A_i x, A_i y))_{F_i}$$

$$= \sum_{i=1}^{n} \langle K_i A_i x, A_i y \rangle = \sum_{i=1}^{n} \langle A_i^* K_i A_i x, y \rangle$$

and that $J = \sum_{i=1}^{n} A_i^* K_i A_i$. Since $A^* = \sum_{i=1}^{n} A_i^*$ is a surjective operator from $F^* = \prod_{i=1}^{n} F_i^*$ onto V^*, it follows that every $f \in V^*$ can be written (in a nonunique fashion) in the form $f = \sum_{i=1}^{n} A_i^* f_i$ where $f_i \in F_i$ for all i. ∎

2. THE FINAL SCALAR PRODUCT

PROPOSITION 1
Let us consider

(1) $\begin{cases} \text{i.} & \text{a Hilbert space } V. \\ \text{ii.} & \text{a vector space } F. \\ \text{iii.} & \text{a surjective linear operator from } V \text{ onto } F \text{ whose kernel is closed in } V. \end{cases}$

If $((x, y))_V$ denotes the scalar product of V and A^+ is the orthogonal right inverse of A, then F is a Hilbert space for the final *scalar product*

(2) $\qquad\qquad ((f, g))_F = ((A^+ f, A^+ g))_V.$

If $J \in \mathscr{L}(V, V^*)$ is the duality operator from V onto V^*, then the duality operator K from F onto F^* is equal to

(3) $\qquad\qquad K = (AJ^{-1}A^*)^{-1}.$

The norm of A is equal to one. ▲

Proof. It is clear that the orthogonal right inverse A^+, which exists according to hypotheses (1), is injective and that its image $A^+(F) = (\text{Ker } A)^{\oplus}$ is closed. Proposition 1.1, then, implies that the space F is a Hilbert space for the scalar product defined by (2), that the duality operator from F onto F^* is defined by

$$K = (A^+)^* J A^+ = (AJ^{-1}A^*)^{-1}$$

(according to Proposition 4.6.3), and that $A = (A^+)^-$ has norm 1. ∎

Remark 1

When F has the final scalar product, A is right invertible and $A^* \in \mathscr{L}(F^*, V^*)$

is left invertible. We can, therefore, give F^* the initial scalar product defined by A^* and also the dual scalar product of the final scalar product of F. These two coincide. ∎

PROPOSITION 2
Suppose that (1) *is the case. The dual scalar product of the final scalar product on F^* is equal to*

$$((f,g))_{F^*} = ((A^*f, A^*g))_{V^*}.$$ ▲

Proof. Indeed,

$$((f,g))_{F^*} = \langle f, K^{-1}g \rangle = \langle f, AJ^{-1}A^*g \rangle = \langle A^*f, J^{-1}A^*g \rangle$$
$$= ((A^*f, A^*g))_{V^*}.$$ ∎

3. NORMAL SUBSPACES OF A PIVOT SPACE

Suppose we are given

(1) $\begin{cases} \text{i.} & \text{a vector space } \mathscr{D} \text{ whose elements are denoted by } \varphi, \psi \\ \text{ii.} & \text{a scalar product } (\varphi, \psi) \text{ on } \mathscr{D}. \end{cases}$

We denote by

(2) H = the completion of $\mathscr{D}$ for the scalar product (φ, ψ),

and *we agree to identify the space H with its dual*, that is, to suppose that

(3) H is a pivot space.

Remark 1

We would have the same situation if we were given

$\begin{cases} \text{i.} & \text{a pivot space } H \text{ for the scalar product } (\varphi, \psi). \\ \text{ii.} & \text{a vector subspace } \mathscr{D} \text{ dense in } H. \end{cases}$ ∎

Among the Hilbert spaces V between $\mathscr{D}$ and H, we are going to distinguish the *normal* spaces.

DEFINITION 1
We say that a Hilbert space V containing $\mathscr{D}$ and embedded in H is normal *if $\mathscr{D}$ is dense in V. We denote in general by*

(4) V_0 = closure of $\mathscr{D}$ in V

when V is not a normal space, and we say that V_0 is the normal space *associated with V*. ▲

Consider the canonical injections j_0 and j from V_0 and V to H, respectively. Since these injections are embeddings, their transposes j_0^* and j^* are also embeddings of H in V_0^* and V^*, respectively. (See Proposition 3.4.2.)

We agree

(5) $\begin{cases} \text{to identify the transpose } j_0^* \in \mathscr{L}(H, V_0^*) \text{ of the canonical} \\ \text{injection from a normal space } V_0 \text{ to } H \text{ with a canonical injection,} \end{cases}$

that is, *to identify H with a dense subspace of the dual of a normal space*. In this case we denote by

(6) $\qquad (f, \varphi) = $ the duality pairing on $V_0^* \times V_0$.

(See Proposition 3.5.4.)

PROPOSITION 1
The space $\mathscr{D}$ is dense in the dual V_0^* of a normal space V_0. The space V_0^* is a subspace of the algebraic dual space of $\mathscr{D}$. ▲

Proof. We use the density criterion (Theorem 2.2.1); we must show that if $\varphi \in V_0 = (V_0^*)^*$ is a continuous linear form on V_0^* that vanishes on $\mathscr{D}$—that is, satisfying $(\varphi, \psi) = 0$ for all $\psi \in \mathscr{D}$—then φ is identically zero. But since $\varphi \in H$, this follows from the density of $\mathscr{D}$ in H. Moreover, every $f \in V_0^*$ defines a linear form ρf on $\mathscr{D}$. Let us show that the mapping $\rho : f \in V_0^* \to \rho f \in \mathscr{D}^{\circledast}$ (the *algebraic* dual space of $\mathscr{D}$) is *injective*; indeed, if $\rho f = 0$, then the continuous linear form f on V_0 vanishes on the dense subspace $\mathscr{D}$ of V_0, and, therefore, is identically zero. ∎

Consequently, if we have three normal spaces $U_0, V_0,$ and W_0 such that $U_0 \subset V_0 \subset W_0$, it follows that

(7) $\begin{cases} \textbf{i.} \quad \mathscr{D} \subset U_0 \subset V_0 \subset W_0 \subset H \subset W_0^* \subset V_0^* \subset U_0^* \subset \mathscr{D}^{\circledast} \\ \textbf{ii.} \quad \text{each space is dense in the following ones (except } \mathscr{D}^{\circledast}). \end{cases}$

Hence, by analogy, we can say that the *notion of embedding is "transitive" in the family of normal spaces and their duals*. ∎

4. MINIMAL AND MAXIMAL DOMAINS OF A CLOSED FAMILY OF OPERATORS

We are going to consider a family

(1) $\qquad \mathscr{A} = \{A_p\}_{p \in P}$ of linear operators from $\mathscr{D}$ to $\mathscr{D}$

where P is a *finite family* of indices.

DEFINITION 1

We say that the family $\mathscr{A}$ is closed if it satisfies the following property:

(2) $\quad\begin{cases} \text{If } \{\varphi_n\} \text{ is a sequence of elements } \varphi_n \in \mathscr{D} \text{ such that the sequence} \\ \varphi_n \text{ converges to zero in } H \text{ and the sequences } A_p\varphi_n \text{ converge to} \\ f_p \text{ in } H, \text{ then } f_p = 0 \text{ for all } p \in P. \end{cases}$ ▲

We associate to the family $\mathscr{A}$ the scalar product defined on $\mathscr{D}$ by

(3) $\quad ((\varphi, \psi))_{\mathscr{A}} = (\varphi, \psi) + \sum_{p \in P} (A_p\varphi, A_p\psi),$

which is an initial scalar product. We are going to show that there exists a completion of $\mathscr{D}$ for this scalar product that is contained in the space H.

More generally, if $Q \subset P$ and if $\mathscr{B} = \{A_p\}_{p \in Q} \subset \mathscr{A}$, we also consider the scalar product

(4) $\quad ((\varphi, \psi))_{\mathscr{B}} = (\varphi, \psi) + \sum_{p \in Q} (A_p\varphi, A_p\psi).$

PROPOSITION 1

Let $\mathscr{A}$ be a closed family of operators A_p from $\mathscr{D}$ to $\mathscr{D}$. There exist completions $H_0(\mathscr{A})$ and $H_0(\mathscr{B})$ of $\mathscr{D}$ for the scalar products $((\varphi, \psi))_{\mathscr{A}}$ and $((\varphi, \psi))_{\mathscr{B}}$ such that $H_0(\mathscr{A}) \subset H_0(\mathscr{B}) \subset H$, where the injections are embeddings. ▲

Proof. Let $H_0(\mathscr{A})$ be a completion of $\mathscr{D}$ for the scalar product $((\varphi, \psi))_{\mathscr{A}}$, θ the canonical injection from $\mathscr{D}$ to $H_0(\mathscr{A})$, and j the canonical injection from $\mathscr{D}$ to H, which is continuous on $\mathscr{D}$ for the norm $\|\varphi\|_{\mathscr{A}}$. Then there exists a continuous mapping $\hat{j} \in \mathscr{L}(H_0(\mathscr{A}), H)$, which clearly has a dense image. We are going to show that this is an embedding, that is, that $\hat{j}$ is injective. Hence let $\varphi \in H_0(\mathscr{A})$ be such that $\hat{j}\varphi = 0$ and show that $\varphi = 0$. There exists a sequence of elements $\varphi_n \in \mathscr{D}$ such that $\theta(\varphi_n)$ converges to φ in $H_0(\mathscr{A})$. Thus the sequence $\theta(\varphi_n)$ is a Cauchy sequence in $H_0(\mathscr{A})$; consequently, since θ is an isometry, the sequence φ_n is a Cauchy sequence for the norm $\|\varphi\|_{\mathscr{A}}$. This implies that the sequence φ_n and the sequences $A_p\varphi_n$ are Cauchy sequences in H. But since H is complete, the sequence φ_n converges to $f_0 \in H$, and the sequences $A_p\varphi_n$ to the elements $f_p \in H$. Since $\varphi_n = j(\varphi_n) = \hat{j}\theta(\varphi_n)$, it follows that φ_n converges to $f_0 = \hat{j}\varphi = 0$ (because j is continuous). The hypothesis that $\mathscr{A}$ is closed implies, therefore, that the limits $f_p \in H$ vanish.

Thus $|\varphi_n|_H^2 + \sum_{p \in P} |A_p\varphi_n|^2 = \|\varphi_n\|_{\mathscr{A}}^2 = \|\theta\varphi_n\|_{H_0(\mathscr{A})}^2$ converges to zero. This means that φ_n converges to zero in $(\mathscr{D}, \|\cdot\|_{\mathscr{A}})$ and, consequently, that $\theta(\varphi_n)$ converges both to φ and to zero. Therefore $\varphi = 0$, which shows that $\hat{j}$ is injective. We can, accordingly, identify $H_0(\mathscr{A})$ with a dense subspace of $\mathscr{D}$.

Similarly, if $\mathscr{B} \subset \mathscr{A}, \mathscr{B}$ is a closed family, and we can identify $H_0(\mathscr{B})$ with a dense subspace of H. Finally the same argument shows that if k is the (continuous) injection from $(\mathscr{D}, \|\varphi\|_{\mathscr{A}})$ to $H_0(\mathscr{B})$, its extension $\hat{k}$ defined by $k = \hat{k}\theta$ is injective. Thus we can also identify $H_0(\mathscr{A})$ with a dense subspace of $H_0(\mathscr{B})$. ∎

DEFINITION 2

Let $\mathscr{A}$ be a closed family of linear operators A_p from $\mathscr{D}$ to itself. We shall say that $H_0(\mathscr{A})$ is the minimal domain *of the family $\mathscr{A}$.*

We identify the unique extension $\hat{A}_p \in \mathscr{L}(H_0(\mathscr{A}), H)$ of A_p with A_p by setting $\hat{A}_p = A_p$. ▲

The minimal domain $H_0(\mathscr{A})$ is, therefore, a normal space. ∎

If we denote by H^P the product of the $H_p = H$, it follows that the operator $\vec{A} : \mathscr{D} \to H^P$, which associates to $\varphi \in \mathscr{D}$ the sequence $\{A_p \varphi\}_{p \in P} \in H^P$, has a unique extension to a continuous linear operator from $H_0(\mathscr{A})$ to H^P, also denoted by $\vec{A}$.

Dual of the Minimal Domain

Let us consider two closed families $\mathscr{A} = \{A_p\}_{p \in P}$ and $\mathscr{A}^* = \{A_p^*\}_{p \in P}$ of linear operators A_p and A_p^* from $\mathscr{D}$ to itself satisfying

(5) $\quad\quad\quad \forall p \in P, \quad (A_p^* \varphi, \psi) = (\varphi, A_p \psi) \quad\quad \forall \varphi, \quad \psi \in \mathscr{D}.$

Consider the normal space $H_0(\mathscr{A}^*)$; the space H is then identified with a dense subspace of the dual $H_0^*(\mathscr{A}^*)$ of $H_0(\mathscr{A}^*)$. (See Proposition 3.5.4.)

PROPOSITION 2

Suppose that the closed families $\mathscr{A}$ and $\mathscr{A}^*$ satisfy hypothesis (5). Then, $\forall p \in P, (A_p^*)^* \in \mathscr{L}(H, H_0^*(\mathscr{A}^*))$ is the unique extension of the operator A_p from $\mathscr{D}$ to $\mathscr{D}$, and $(\vec{A^*})^* \in \mathscr{L}(H^P, H_0^*(\mathscr{A}^*))$ is the unique extension of the operator $\sum_{p \in P} A_p$ from $\mathscr{D}^P$ to $\mathscr{D}$. ▲

Proof. We show that for every $\varphi \in \mathscr{D}, (A_p^*)^* \varphi = A_p \varphi$ and that $(\vec{A^*})^* \vec{\varphi} = \sum_{p \in P} A_p \varphi_p$ when $\vec{\varphi} = \{\phi_p\}_{p \in P} \in \mathscr{D}^P$. Indeed, for every $\psi \in \mathscr{D}$, we deduce from hypothesis (5) that $((A_p^*)^* \varphi, \psi) = (\varphi, A_p^* \psi) = (A_p \varphi, \psi)$. Since $\mathscr{D}$ is dense in $H_0(\mathscr{A}^*)$, we deduce that $(A_p^*)^* \varphi = A_p \varphi$. Similarly, we verify that if $\vec{\varphi} = \{\varphi_p\}_{p \in P} \in \mathscr{D}^P$, then we have $(\vec{A^*})^* \vec{\varphi} = \sum_{p \in P} A_p \varphi_p$.

The fact that $\mathscr{D}$ is *dense* in H and that $\mathscr{D}^P$ is *dense* in H^P implies that the operators $(A_p^*)^*$ and $(\vec{A^*})^*$ are the unique extensions of A_p and of $\sum_{p \in A} A_p$, respectively. ∎

DEFINITION 3

If the closed families $\mathscr{A}$ and $\mathscr{A}^*$ satisfy (5), we agree to make the following identifications:

(6) $\quad \begin{cases} \textbf{i.} & A_p = (A_p^*)^* \in \mathscr{L}(H, H_0^*(\mathscr{A}^*)). \\ \textbf{ii.} & \sum_{p \in P} A_p = (\vec{A^*})^* \in \mathscr{L}(H^p, H_0^*(\mathscr{A}^*)). \end{cases}$ ▲

PROPOSITION 3

The duality operator from $H_0(\mathscr{A})$ to its dual $H_0^*(\mathscr{A})$ is, therefore, equal to

(7) $$J = 1 + \sum_{p \in P} A_p^* A_p.$$

Every continuous linear form $f \in H_0^*(\mathscr{A})$ can be written in a non unique fashion in the form

(8) $$f = f_0 + \sum_{p \in P} A_p^* f_p \qquad \text{where } f_0, f_p \in H.$$ ▲

Proof. If φ and $\psi \in H_0(\mathscr{A})$, then the elements $A_p \varphi$ belong to H and $((\varphi, \psi))_\mathscr{A} = (\varphi, \psi) + \sum_{p \in P}(A_p \varphi, A_p \psi) = (\varphi + \sum_{p \in P} A_p^* A_p \varphi, \psi) = (J\varphi, \psi)$. Then $J = 1 + \sum_{p \in P} A_p^* A_p$. Moreover, since the operator $\vec{A}$ is an isomorphism from $H_0(\mathscr{A})$ onto its closed image in H^p, $\vec{A^*} = \sum_{p \in P} A_p^*$ is a surjection from H^p onto $H_0^*(\mathscr{A})$, which implies the final assertion. ∎

Maximal Domain

Since the operators A_p are defined from H to $H_0^*(\mathscr{A}^*)$, we can define the domain of the family $\mathscr{A} = \{A_p\}_{p \in P}$ of operators A_p in the following fashion.

PROPOSITION 4

Suppose that the closed families $\mathscr{A}$ and $\mathscr{A}^*$ satisfy (5). Then the subspace

(9) $\quad H(\mathscr{A}) = \{\varphi \in H \quad \text{such that} \quad A_p \varphi \in H \quad \text{for all} \quad p \in P\}$

with the scalar product

(10) $$((\varphi, \psi))_\mathscr{A} = (\varphi, \psi) + \sum_{p \in P}(A_p \varphi, A_p \psi)$$

is a Hilbert space contained in H; moreover, $H_0(\mathscr{A})$ is the closure of $\mathscr{D}$ in $H(\mathscr{A})$. ▲

Proof. We must show that $H(\mathscr{A})$ is complete. Let $\{\varphi_n\}_n$ be a Cauchy sequence in $H(\mathscr{A})$. Then the sequences $\{\varphi_n\}$ and $\{A_p \varphi_n\}$ are Cauchy sequences in H, which converge to $f_0 \in H$ and $f_p \in H$, respectively. Since $A_p \in \mathscr{L}(H, H_0^*(\mathscr{A}^*))$, it follows that $A_p \varphi_n$ converges, as well, to $A_p f_0$ in $H_0^*(\mathscr{A}^*)$. Therefore we conclude that $A_p f_0 = f_p \in H$ for all $p \in P$ and, consequently, that $f_0 \in H(\mathscr{A})$ and that φ_n converges to f_0 in $H(\mathscr{A})$. ∎

DEFINITION 4
Let $\mathscr{A}$ and $\mathscr{A}^*$ be two closed families of operators from $\mathscr{D}$ to $\mathscr{D}$ satisfying (5). We shall say that $H(\mathscr{A})$ is the maximal domain of the family $\mathscr{A}$ of operators A_p. ▲

*5. UNBOUNDED OPERATORS AND THEIR ADJOINTS

DEFINITION 1
Consider two Hilbert spaces H and F. An unbounded operator from H to F is a pair $(D(A), A)$ where

(1) $\begin{cases} \text{i.} & D(A) \text{ is a vector subspace of } H. \\ \text{ii.} & A \text{ is a linear operator from } D(A) \text{ to } F. \end{cases}$

The space $D(A)$ is called the domain of A. The operator $(D(A), A)$ is closed if

(2) $\begin{cases} \text{for every sequence } \{x_n\}_n \text{ of elements } x_n \in D(A) \text{ converging to } x \in H \\ \text{such that } Ax_n \text{ converges to } f \text{ in } F, \text{ we have } x \in D(A) \text{ and } f = Ax \end{cases}$

and has dense domain if the domain $D(A)$ is dense in H.

We shall say that $D(A)$ has the norm of the graph when $D(A)$ is a prehilbert space for the scalar product

(3) $$((x, y)) = ((x, y))_H + ((Ax, Ay))_F.$$ ▲

We then deduce from Proposition 1.2 the following result.

PROPOSITION 1
The domain $D(A)$ of a closed unbounded operator A is a Hilbert space for the norm of the graph. ▲

Consider an unbounded operator $(D(A), A)$ from H to F, which has *dense domain*. We are going to associate to it an unbounded operator $(D(A^*), A^*)$ from F^* to H^* that will also be closed and have dense domain.

Therefore let $(D(A), A)$ be a closed unbounded operator from H to F, that is, a continuous linear operator from $D(A)$ to F when the space $D(A)$ has the norm of the graph.

Since the canonical injection j from $D(A)$ to H is continuous and has a dense image, it is an embedding. According to Proposition 3.4.2, j^* is an embedding from H^* in $D(A)^*$, which we identify with a canonical injection.

Then the transpose $A^* \in \mathscr{L}(F^*, D(A)^*)$ is a continuous linear operator. Thus we define

(4) $\qquad D(A^*) = \{f \in F^* \quad \text{such that} \quad A^*f \in H^*\}.$

DEFINITION 2
We say that the unbounded operator $(D(A^*), A^*)$ from F^* to H^* defined by (4) is the adjoint of the closed unbounded operator with dense domain $(D(A), A)$. ▲

PROPOSITION 2
The adjoint $(D(A^*), A^*)$ of a closed unbounded operator with dense domain $(D(A), A)$ is closed and has dense domain. ▲

We need the following.

LEMMA 1
The set $G(A^*) = \{f, A^*f\}_{f \in D(A^*)} \subset F^* \times H^*$ is the orthogonal complement of the subset $G(A) = \{\{Ax, -x\}\}_{x \in D(A)}$ of $F \times H$. ▲

Proof of the Lemma. It is clear that $G(A^*) \subset G(A)^\perp$. Conversely, let $\{f, g\} \in F^* \times H^*$ be orthogonal to $G(A)$; then $\langle f, Ax \rangle_{F^* \times F} - \langle g, x \rangle_{H^* \times H} = 0$ for all $x \in D(A)$. This implies that

$$\langle A^*f, x \rangle_{D(A)^* \times D(A)} - \langle g, x \rangle_{H^* \times H} = \langle A^*f - g, x \rangle_{D(A^*) \times D(A)} = 0$$

for all $x \in D(A)$. Thus $A^*f = g$. Since $g \in H^*$, this implies that $f \in D(A^*)$ and that $g = A^*f$ and, consequently, that $\{f, g\} \in G(A^*)$. ■

Proof of Proposition 2. To say that $(D(A^*), A^*)$ is a closed operator from F^* to H^* is to say that $G(A^*)$ is closed. This, then, is a consequence of Lemma 1.

To show that $D(A^*)$ is dense in F^*, we suppose that an arbitrary continuous linear form $y \in F = (F^*)^*$ on F^* vanishes on $D(A^*)$:

(5) $\qquad \langle f, y \rangle_{F^* \times F} = \langle f, y \rangle_{F^* \times F} - \langle A^*f, 0 \rangle_{H^* \times H} = 0 \qquad \forall f \in D(A^*).$

According to Theorem 2.2.1, we have to show that $y = 0$. But condition (5) expresses the fact that the pair $\{y, 0\}$ is orthogonal to the subset

(6) $\qquad G(A^*) = \{\{f, A^*f\}\}_{f \in D(A^*)} = G(A)^\perp.$

The hypothesis that $(D(A), A)$ is closed implies that $G(A)$ is a closed subset of $F \times H$. Thus $\{y, 0\} \in G(A^*)^\perp = G(A)^{\perp\perp} = G(A)$; this implies, then that $y = A0 = 0$. ■

Remark 1
If $(D(A), A)$ is an unbounded operator from H to F and if $B \in \mathscr{L}(H, F)$ is a

continuous linear operator from H to F, then $(D(A), A + B)$ is an unbounded operator that is closed when $(D(A), A)$ is closed and has dense domain when $(D(A), A)$ has dense domain. ∎

Example 1

Suppose we are given

(7) $\begin{cases} \text{i.} & \text{a pivot space } H. \\ \text{ii.} & \text{a Hilbert space } V \text{ dense in } H \text{ with a stronger topology.} \end{cases}$

We agree to identify H with a dense subspace of V^*. Consider an operator $A \in \mathscr{L}(V, V^*)$. We associate to it the unbounded operator $(D(A), A)$ from H to itself whose domain is defined by

(8) $\qquad D(A) = \{x \in V \text{ such that } Ax \text{ belongs to } H\}.$ ∎

PROPOSITION 3

If $A \in \mathscr{L}(V, V^*)$ is an isomorphism (in particular, if A is V-elliptic), the associated unbounded operator $(D(A), A)$ is closed and has dense domain. Its adjoint is the unbounded operator $(D(A^*), A^*)$ associated to the transpose $A^* \in \mathscr{L}(V, V^*)$ of A. Moreover,

(9) $\qquad D(A) \text{ and } D(A^*) \text{ are dense in } V.$ ▲

Proof. If A is an isomorphism from V onto V^*, then

$$D(A) = A^{-1}(H)$$

and

$$V = A^{-1}(V^*).$$

Since H is dense in V^*, it follows that $D(A)$ is dense in V. Hence, $D(A)$ is also dense in H since V is dense in H.

We temporarily denote by $A' \in \mathscr{L}(V, V^*)$ the transpose of $A \in \mathscr{L}(V, V^*)$. To show that the adjoint of $(D(A), A)$ is the unbounded operator $(D(A^*), A^*)$ associated to A', it suffices to show that

$$D(A^*) = D(A').$$

It is clear that $D(A') \subset D(A^*)$ because if $x \in D(A')$, then $A'x \in H = H^*$.

Conversely, let $x \in D(A^*)$; that is, let $x \in H$ such that $A^*x \in H$. We must show that $x \in D(A')$, that is, that $x \in V$. But since A' is an isomorphism, $x = A'^{-1}(A'x)$ indeed belongs to V. Therefore $(D(A'), A')$ is the adjoint of $(D(A), A)$. These unbounded operators are obviously closed. ∎

*6. COMPLETION OF A PREHILBERT SPACE CONTAINED IN A HILBERT SPACE

Let V_0 be a prehilbert space contained in a Hilbert space H. Under what conditions can we choose a realization V of the completion $\hat{V}_0$ that is contained in H?

PROPOSITION 1
Suppose that

(1) $\quad \begin{cases} \textit{if a sequence } \{y_n\} \textit{ of elements } y_n \in V_0 \textit{ that satisfies} \\ \|y_n\|_{V_0} \leq 1 \textit{ converges to } y \in V_0 \textit{ in } H, \textit{ then } \|y\|_{V_0} \leq 1. \end{cases}$

Then there exists a completion V of V_0 contained in H. ▲

Proof. Let j be the continuous injection from V_0 to H, θ the canonical embedding from V_0 to its completion $\hat{V}_0$. Then the injection j from V_0 to H has a unique extension to a continuous mapping $\hat{j}$ from $\hat{V}_0$ to H. If $\hat{j}$ is injective, then $V = \hat{j}(\hat{V}_0)$ is a Hilbert space for the scalar product $((\hat{j}x_0, \hat{j}y_0)) = ((x_0, y_0))_{\hat{V}_0}$, which is isometric to the completion $\hat{V}_0$. Thus V is a realization of the completion $\hat{V}$.

Let us show, then, that hypothesis (1) implies that $\hat{j}$ is injective, that is, that if $\hat{j}(\hat{x}_0) = 0$, then $\hat{x}_0 = 0$. Since $\theta(V_0)$ is dense in $\hat{V}_0$, there exists a sequence $x_0^n \in V_0$ such that $\theta(x_0^n)$ converges to $\hat{x}_0$. Hence we can associate to every $\varepsilon > 0$ an integer N such that

$$\|\theta(x_0^n) - \theta(x_0^m)\|_{\hat{V}_0} = \|x_0^n - x_0^m\|_{V_0} \leq \varepsilon$$

when $m, n \geq N$. Moreover, since $\hat{j}$ is continuous, $x_0^m = j(x_0^m) = \hat{j}\theta(x_0^m)$ converges to $j(\hat{x}_0) = 0$. Hence $x_0^n - x_0^m$ belongs to the ball with radius ε in V_0 and converges to x_0^n in H when $m \to \infty$. According to (1) it follows that x_0^n belongs to this ball, that is, that $\|x_0^n\|_{V_0} \leq \varepsilon$ when $n \geq N$. Thus x_0^n converges to zero; consequently, $\theta(x_0^n)$ converges to zero. This implies that $\hat{x}_0 = 0$ and that $\hat{j}$ is therefore injective. ∎

Remark 1

Clearly this proposition can be extended to the case where V is a normed space and where H is a topological vector space. ∎

*7. HAUSDORFF COMPLETION

Let us consider a nonseparated prehilbert space with the semiscalar product $((x, y))$.

We introduce the vector subspace.

$$M = \{x \in V \quad \text{such that} \quad ((x, y)) = 0 \quad \text{for all} \quad y \in V\}.$$

Let $\theta : V \mapsto V/M$ be the canonical surjection from V onto the quotient of V by M. We define on V/M the bilinear form

$$((u, v))_{V/M} = ((x, y)) \quad \text{if} \quad u = \theta(x), v = \theta(y).$$

This is possible since the expression does not depend on the choice of the representatives x and y of u and v; indeed, if $x_1 = x + x_0$ and $y_1 = y + y_0$ (where x_0 and $y_0 \in M$) are other representatives of u and v, we have

$$((x_1, y_1)) = ((x + x_0, y + y_0)) = ((x, y)) + ((x_0, y)) + ((y_0, x)) + ((x_0, y_0))$$
$$= ((x, y)) = ((u, v))_{V/M}.$$

It is clear that the bilinear form is symmetric; it is positive definite. Indeed, if $((u, u))_{V/M} = 0$, then $((x, x)) = 0$ for every representative x of u. Since $((x, y))$ is symmetric and positive, the Cauchy-Schwarz inequality implies that $|((x, y))| \leq \sqrt{((x, x))} \sqrt{((y, y))} = 0$ for all $y \in V$, and therefore, that $x \in M$ and that $u = \theta(x) = 0$.

DEFINITION 1

We say that the prehilbert space V/M is the separated *or* (Hausdorff) *prehilbert space associated to V and that its completion $\widehat{V/M}$ is the* Hausdorff completion *of V.* ▲

We are going to characterize the dual of a nonseparated prehilbert space.

PROPOSITION 1

The dual $\mathscr{L}(V, \mathbb{R})$ is isomorphic to the dual $\mathscr{L}(\widehat{V/M}, \mathbb{R})$ of its Hausdorff completion $\widehat{V/M}$. ▲

Proof. Indeed the mapping that associates to every continuous linear form $\tilde{f} \in \mathscr{L}(V/M, \mathbb{R})$ the continuous linear form $f = \tilde{f}\theta \in \mathscr{L}(V, \mathbb{R})$ is a continuous injective linear mapping from $\mathscr{L}(V/M, \mathbb{R})$ to $\mathscr{L}(V, \mathbb{R})$.

This mapping is *surjective*; indeed, let $f \in \mathscr{L}(V, \mathbb{R})$ be a continuous linear form on V. Then there exists a constant c such that, for all $x \in V$, $|f(x)| \leq c\sqrt{((x, x))}$. Consequently, if $x \in M$, $f(x)$ is equal to zero. We can, therefore, take the quotient by M: There exists a unique linear function $\tilde{f}$ from V/M to $\mathbb{R}$ such that $f = \tilde{f}\theta$. Moreover, $\tilde{f}$ is continuous, since for all $u \in V/M$ and for every representative x of u we have

$$|\tilde{f}(u)| = |\tilde{f}(\theta x)| = |f(x)| \leq c\sqrt{((x, x))} = c\sqrt{((u, u))}_{V/M}.$$

Therefore $\mathscr{L}(V/M, \mathbb{R})$ and $\mathscr{L}(V, \mathbb{R})$ are isometric. Moreover, according to Theorem 1.3.3 we know that $\mathscr{L}(V/M, \mathbb{R})$ and $\mathscr{L}(\widehat{V/M}, \mathbb{R})$ are isometric. ■

CH. 5, SEC. 8 THE HILBERT SUM OF HILBERT SPACES

*8. THE HILBERT SUM OF HILBERT SPACES

We have seen that the product of a finite number of Hilbert spaces is a Hilbert space. This is no longer the case for an infinite product $\prod_{n=1}^{\infty} V_n$ of Hilbert spaces. Nevertheless, we are going to give a Hilbert space structure to a suitable subset of this product.

DEFINITION 1
We say that the subset $V = \bigoplus_{n=1}^{\infty} V_n$ *of the product* $\prod_{n=1}^{\infty} V_n$ *of the Hilbert spaces* V_n *defined by*

(1) $\quad V = \left\{ x = \{x_n\}_n \in \prod_{n=1}^{\infty} V_n \quad \text{such that} \quad \left(\sum_{n=1}^{\infty} \|x_n\|_{V_n}^2 \right)^{1/2} < +\infty \right\}$

is the Hilbert sum *of the* V_n's. ▲

We are going to show that this is a Hilbert space.

PROPOSITION 1
The Hilbert sum $\bigoplus_{n=1}^{\infty} V_n$ *is a Hilbert space for the scalar product*

(2) $\quad\quad\quad\quad\quad ((x, y)) = \sum_{n=1}^{\infty} ((x_n, y_n))_{V_n}.$ ▲

Proof. If $x = \{x_n\}_n$ and $y = \{y_n\}_n$ belong to $\bigoplus_{n=1}^{\infty} V_n$, we set

$$\|x\| = \left(\sum_{n=1}^{\infty} \|x_n\|_{V_n}^2 \right)^{1/2} \quad \text{and} \quad \|y\| = \left(\sum_{n=1}^{\infty} \|y_n\|_{V_n}^2 \right)^{1/2}.$$

It is clear that $\|\lambda x\| = \lambda \|x\|$, so that $\lambda x \in \bigoplus_{n=1}^{\infty} V_n$. Since $\|x_n + y_n\|_{V_n}^2 \leq 2(\|x_n\|_{V_n}^2 + \|y_n\|_{V_n}^2)$, it follows that $\|x + y\| \leq \sqrt{2(\|x\|^2 + \|y\|^2)} < +\infty$, so that $x + y \in \bigoplus_{n=1}^{\infty} V_n$. Thus $\bigoplus_{n=1}^{\infty} V_n$ is a vector space. Let us show that the series $((x_n, y_n))_{V_n}$ converges. Indeed, according to the Cauchy-Schwartz inequality, we have

$$|((x_n, y_n))_{V_n}| \leq \|x_n\|_{V_n} \|y_n\|_{V_n} \leq \tfrac{1}{2}(\|x_n\|_{V_n}^2 + \|y_n\|_{V_n}^2).$$

Consequently, (2) is a meaningful definition. It is easy to verify that this is indeed a scalar product on $\bigoplus_{n=1}^{\infty} V_n$.

Finally let us show that $\bigoplus_{n=1}^{\infty} V_n$ is complete, by an argument analogous to the one we used to show that l^2 is complete (See Theorem 1.1.2.)

Let $x^{(p)} = \{x_n^{(p)}\}$ be a Cauchy sequence in $\bigoplus_{n=1}^{\infty} V_n$; we can associate to every $\varepsilon > 0$ an integer $p(\varepsilon)$ such that

(3) $\quad\quad\quad \sum_{n=1}^{\infty} \|x_n^{(p)} - x_n^{(q)}\|_{V_n}^2 \leq \varepsilon \quad \text{when} \quad p, q \geq p(\varepsilon).$

This implies in particular that for every n, $\|x_n^{(p)} - x_n^{(q)}\|_{V_n} \leq \sqrt{\varepsilon}$ and, consequently, that the sequence $x_n^{(p)}$ is a Cauchy sequence in V_n, which converges to an element $y_n \in V_n$ because V_n is complete. For every fixed integer k, it follows from (4) that $\sum_{n=1}^{k} \|x_n^{(p)} - x_n^{(q)}\|_{V_n}^2 \leq \varepsilon$ when $p, q \geq p(\varepsilon)$.

Letting q approach infinity, it follows that

(4) $$\forall k, \quad \sum_{n=1}^{k} \|x_n^{(p)} - y_n\|_{V_n}^2 \leq \varepsilon \quad \text{when} \quad p \geq p(\varepsilon).$$

Since this is true for all k, we can let k approach infinity and deduce that

(5) $$\sum_{n=1}^{\infty} \|x_n^{(p)} - y_n\|_{V_n}^2 \leq \varepsilon \quad \text{when} \quad p \geq p(\varepsilon).$$

Consequently, the sequence $\{x_n^{(p)} - y_n\}_n$ belongs to $\bigoplus_{n=1}^{\infty} V_n$; it follows that $y = \{y_n\}_n$ also belongs to $\bigoplus_{n=1}^{\infty} V_n$. Moreover, (5) also implies that the sequence $x^{(p)}$ converges to y for the norm of $\bigoplus_{n=1}^{\infty} V_n$. Hence the Hilbert sum is complete. ∎

It is clear that the mapping $j_n \in \mathscr{L}(V_n, \bigoplus_{n=1}^{\infty} V_n)$ defined by

$$j_n x_n = \{0, \ldots, 0, x_n, 0, \ldots,\}$$

is an isometry that allows us to *identify* V_n *with a closed subspace of* $\bigoplus_{n=1}^{\infty} V_n$. We remark that

(6) $$\text{if} \quad m \neq n, \quad ((V_m, V_n)) = 0$$

since $((j_m x_m, j_n x_n)) = ((x_m, 0))_{V_m} + ((0, x_n))_{V_n} = 0$.

Consequently, we can write every $x \in \bigoplus_{n=1}^{\infty} V_n$ in the form

(7) $$x = \sum_{n=1}^{\infty} j_n(x_n)$$

since

(8) $$\lim_{k \to \infty} \left\| x - \sum_{n=1}^{k} j_n(x_n) \right\|^2 = \lim_{k \to \infty} \sum_{n=k+1}^{\infty} \|x_n\|_{V_n}^2 = 0.$$

In other words, the (algebraic) sum of the spaces V_n generates a dense space in $\bigoplus_{n=1}^{\infty} V_n$.

Conversely, we prove the following result.

PROPOSITION 2

Let V be a Hilbert space and $\{V_n\}$ a sequence of closed subspaces of V such that

(9) $$\begin{cases} \text{i.} & \forall m, \forall n \text{ such that } m \neq n, ((V_m, V_n))_V = 0. \\ \text{ii.} & \text{the algebraic sum of the } V_n\text{'s is dense in } V. \end{cases}$$

Then there exists an isometry from V onto $\bigoplus_{n=1}^{\infty} V_n$ which, for each V_n, coincides with the isometry j_n. ▲

Proof. Let us set $U_n = j_n(V_n)$, a closed subspace of $U = \bigoplus_{n=1}^{\infty} V_n$. Since j_n is an isometry from V_n onto U_n, it has an inverse g_n from U_n onto V_n. Let F be the algebraic direct sum of the U_n's, defined as the space of *finite sums* of elements $j_n(x_n)$. We then define a linear operator g from F to V by the condition

$$\forall_n, \quad g[j_n(x_n)] = g_n(j_n(x_n)) = x_n.$$

The operator g is an isometry from F *onto* the algebraic sum G of the spaces V_n; we must show that for every finite sequence $n_1, \ldots, n_k$,

$$\left(\left(g\left(\sum_{i=1}^{k} j_{n_i} x_{n_i}\right), g\left(\sum_{j=1}^{k} j_{n_j} y_{n_j}\right)\right)\right)_V = \sum_{i=1}^{k} ((j_{n_i}(x_{n_i}), j_{n_i} y_{n_i}))_{V_{n_i}}.$$

But since $((V_{n_i}, V_{n_j}))_V = 0$ if $i \neq j$, the left-hand side can be written as $\sum_{i=1}^{k} ((x_{n_i}, y_{n_i}))_V$. The right-hand side is also equal to $\sum_{i=1}^{k} ((x_{n_i}, y_{n_i}))_V$, since the j_{n_i}'s are isometries. Therefore g is indeed an isometry.

Furthermore, g has a unique extension by continuity to an isometry $\bar{g}$ from $\bigoplus_{n=1}^{\infty} V_n = U$ to V (by Theorem 1.3.3).

Consequently, $\bar{g}(U)$ is a closed subspace of V; moreover, $\bar{g}(U)$ contains $\bar{g}(F) = g(F) = G$, which is dense in V according to hypothesis (9)ii. Hence $\bar{g}(U) = V$ and $\bar{g}$ is an isometry from $\bigoplus_{n=1}^{\infty} V_n$ onto V, which completes the proof of the proposition. ■

Remark 1

It is this result that justifies the definition of $\bigoplus_{n=1}^{\infty} V_n$ as the *Hilbert sum* of the spaces V_n.

When the space $V_n = \mathbb{R} e_n$ are spaces of dimension 1, condition (9) is equivalent to saying that $\{e_n\}$ is an orthogonal base for V, and we find again Theorem 1.7.1. ■

*9. INTERPOLATION SPACES

Let us consider three Hilbert spaces $U, V,$ and W such that

(1) $U \subset V \subset W$ with the canonical injections being embeddings

We shall identify the transposes of these embeddings with the canonical injections

(2) $$W^* \subset V^* \subset U^*.$$

Let $\theta \in [0,1]$.

DEFINITION 1
*We say that V is an **intermediate space** of order θ between U and W if there exists a constant c such that*

(3) $$\forall x \in U, \quad \|x\|_V \leq c \|x\|_U^{1-\theta} \|x\|_W^{\theta}.$$

*We say that V is an **interpolation space** of order θ between U and V if, in addition, V^* is an intermediate space of order $1-\theta$ between W^* and U^*.* ▲

It is a trivial consequence of the definition that the dual V^* of an interpolation space V of order θ between U and W is an interpolation space of order $1-\theta$ between W^* and U^*.

PROPOSITION 1
Suppose that V is an interpolation space of order θ between U and W. There exist constants c and c^* such that

(4) $$\forall x \in U, \quad c^* \|x\|_U^{1-\theta} \|x\|_W^{\theta} \leq \|x\|_V \leq c \|x\|_U^{1-\theta} \|x\|_W^{\theta}.$$ ▲

Proof. Since V^* is an intermediate space of order $1-\theta$ between W^* and U^*, it follows that there exists $c^* > 0$ such that $\forall f \in W^*$,

$$\|f\|_{W^*} \leq \frac{1}{c^*} \|f\|_{W^*}^{\theta} \|f\|_{U^*}^{1-\theta}.$$

Consequently, $\forall f \in W^*, \forall x \in U$, we obtain

$$c^* \frac{|\langle f, x \rangle|^{1-\theta}}{\|f\|_{U^*}^{1-\theta}} \frac{|\langle f, x \rangle|^{\theta}}{\|f\|_{W^*}^{\theta}} \leq \frac{|\langle f, x \rangle|}{\|f\|_{W^*}}.$$

Taking the supremum as f runs over W^*, we deduce that

$$c^* \|x\|_U^{1-\theta} \|x\|_W^{\theta} \leq \|x\|_V.$$ ∎

The notion of interpolation space is justified by the following proposition.

PROPOSITION 2
Consider two triples of Hilbert spaces satisfying

(5) $$\begin{cases} \text{i.} & U \subset V \subset W; \quad E \subset F \subset G. \\ \text{ii.} & \text{the canonical injections are embeddings.} \end{cases}$$

Suppose that there is a linear operator A from W to G whose restriction from U to E satisfies

(6) $$A \in \mathscr{L}(W, G) \cap \mathscr{L}(U, E).$$

If V and F are interpolation spaces of order θ between U and W and E and G, respectively, then $A \in \mathscr{L}(V, F)$ and there exists a constant d such that

(7) $$\|A\|_{\mathscr{L}(V,F)} \leq d \|A\|_{\mathscr{L}(U,E)}^{1-\theta} \|A\|_{\mathscr{L}(W,G)}^{\theta}. \quad \blacktriangle$$

Proof. For all $x \in U$ we obtain

(8) $$\begin{aligned} \|Ax\|_F &\leq c \|Ax\|_E^{1-\theta} \|Ax\|_G^{\theta} \\ &\leq c \|A\|_{\mathscr{L}(U,E)}^{1-\theta} \|A\|_{\mathscr{L}(W,G)}^{\theta} \|x\|_U^{1-\theta} \|x\|_W^{\theta} \\ &\leq \frac{c}{c^*} \|A\|_{\mathscr{L}(U,E)}^{1-\theta} \|A\|_{\mathscr{L}(W,G)}^{\theta} \|x\|_V. \end{aligned}$$

It follows that A is linear and continuous from U, under the norm of V, with values in E, under the norm of F. Hence A can be extended by density to a continuous linear operator from the Hilbert space V to the Hilbert space F (see Theorem 1.3.1), also denoted by A. Moreover, (8) shows that inequality (7) is satisfied. ∎

Examples

We give some examples of interpolation spaces in Chapter 11, Section 5. The spaces $\hat{H}^s(\mathbb{R}^n)$ and the Sobolev spaces also provide examples of interpolation spaces. (See Chapter 6, Section 3 and Chapter 9, Section 3.)

*10. REPRODUCING KERNELS OF A HILBERT SPACE OF FUNCTIONS

Consider a set X and

(1) a Hilbert space V of functions $x \to \varphi(x)$ defined on X.

Consider, as well, a function $k : X \times X \to \mathbb{R}$ satisfying

(2) i. $\forall x \in X, \quad \forall y \in Y, \quad k(x, y) = k(y, x) \quad$ (symmetry)

ii. for every finite sequence of points $x_i \in X$ and of scalars α_i, we have

$$\sum_{i,j=1}^n \alpha_i \alpha_j k(x_i, x_j) \geq 0 \quad (k \text{ is of positive type}).$$

The function k defines functions $K(x)$ by

(3) $$\forall x \in X, \quad K(x) : y \to K(x)(y) = k(x, y).$$

The question arises whether we associate to every Hilbert space V of functions on X a function k satisfying (2) such that V is generated by the functions $K(x)$ as x runs over X. (It is in this sense that k is called a *reproducing kernel* of V.)

DEFINITION 1
A function $k: X \times X \to \mathbb{R}$ is called a reproducing kernel *of a Hilbert space V of functions on X if every function φ of V satisfies the property*

(4) $$\varphi(x) = ((K(x), \varphi))_V.$$ ▲

(This definition implicitly assumes that the functions $K(x)$ associated with the kernel k belong to V.)

We begin by giving some properties of reproducing kernels.

PROPOSITION 1
Suppose that k is the reproducing kernel of a Hilbert space V of functions. Then the properties (2) along with

(5) $$|\varphi(x)| \leq \sqrt{k(x,x)} \, \|\varphi\|_V$$

and

(6) $\begin{cases} \text{i.} & \forall x \in X, \quad \forall y \in X, \quad ((K(x), K(y)))_V = k(x, y). \\ \text{ii.} & \forall x \in X, \quad \|K(x)\|_V = \sqrt{k(x,x)}. \end{cases}$

are satisfied.

Moreover, the vector space V_0 generated by the functions $K(x)$ is dense in V. ▲

Proof. Taking $\varphi = K(y)$ in (4), we obtain

$$k(y, x) = K(y)(x) = ((K(x), K(y)))_V.$$

Since the scalar product is symmetric, it follows that

$$k(y, x) = ((K(x), K(y)))_V = ((K(y), K(x)))_V = k(x, y).$$

On the other hand, taking $\varphi = \sum_{i=1}^n \alpha_i K(x_i)$, we deduce that

$$0 \leq \|\varphi\|_V^2 = \sum_{i,j=1}^n \alpha_i \alpha_j ((K(x_i), K(x_j)))_V$$
$$= \sum \alpha_i \alpha_j k(x_i, x_j).$$

In particular, we obtain

$$\|K(x)\|_V = \sqrt{((K(x), K(x)))_V} = \sqrt{k(x,x)}.$$

The Cauchy-Schwarz inequality implies that

$$|\varphi(x)| = |((K(x), \varphi))_V| \leq \|K(x)\|_V \|\varphi\|_V$$
$$= \sqrt{k(x,x)} \|\varphi\|_V.$$

To show that the set $V_0 = \{K(x)\}_{x \in X}$ of functions $K(x)$ generates a vector space that in dense in V, it suffices to show that every linear form $f \in V^*$ that vanishes on V_0 is identically zero. (See Theorem 2.2.1.) Let J be the duality operator of V. Then $f = J\varphi$ and, consequently, for every $x \in X$

$$\varphi(x) = ((\varphi, K(x)))_V = \langle J\varphi, K(x) \rangle = \langle f, K(x) \rangle = 0.$$

Hence $\varphi = 0$ and $f = J\varphi = 0$. ∎

We are going to prove successively that every Hilbert space V of functions on X has a reproducing kernel and that, conversely, every function $k : X \times X \to \mathbb{R}$ satisfying (2) is the reproducing kernel of a Hilbert space of functions.

THEOREM 1
Let V be a Hilbert space of functions $\varphi : X \to \mathbb{R}$ satisfying

(7) $\quad \forall \varphi \in V, \quad \forall x \in X, \quad \exists c_x \geq 0 \quad \text{such that} \quad |\varphi(x)| \leq c_x \|\varphi\|_V.$

Then V has a reproducing kernel k. ▲

Proof. We denote by $\rho(x)$ the linear form $\varphi \in V \to \varphi(x) \in \mathbb{R}$. Condition (7) expresses the fact that $\rho(x)$ is a continuous linear form on V, hence an element of V^*.

If J is the duality operator from V onto V^*, we set

(8) $\quad\quad K(x) = J^{-1}\rho(x) \quad \text{and} \quad k(x,y) = K(x)(y).$

Then k is indeed a reproducing kernel of V, since

(9) $\quad\quad \varphi(x) = \langle \rho(x), \varphi \rangle = ((J^{-1}\rho(x), \varphi))_V = ((K(x), \varphi))_V.$ ∎

THEOREM 2
Let $k : X \times X \to \mathbb{R}$ be a symmetric function of positive type [i.e., satisfying (2)]. Then there exists a Hilbert space V of functions on X for which k is the reproducing kernel. ▲

Preliminary Remark

To prove that V is a space of functions on X, it is convenient to introduce the space $\mathbb{R}^X$ of *all* the functions on X, as well as its algebraic dual $\mathbb{R}^{X*}$. We denote by $\mathscr{S}^*(X)$ the vector subspace of $\mathbb{R}^{X\circledast}$ generated by the linear forms

(10) $$\delta(x): \varphi \in \mathbb{R}^X \to \varphi(x) \in \mathbb{R}.$$

First we verify the following.

LEMMA 1
The space $\mathbb{R}^X$ of functions on X is the algebraic dual space of the space $\mathscr{S}^*(X)$. ▲

Proof of the Lemma. Consider a linear form Φ from $\mathscr{S}^*(X)$ to $\mathbb{R}$ and set $\varphi(x) = \Phi(\delta(x))$ for all $x \in X$. Then $\varphi: x \to \varphi(x)$ is indeed a function on X and

$$\Phi\left(\sum_{i=1}^n \alpha_i \delta(x_i)\right) = \sum_{i=1}^n \alpha_i \Phi(\delta(x_i)) = \sum_{i=1}^n \alpha_i \varphi(x_i).$$

Thus the linear form Φ is well defined by the function $\varphi \in \mathbb{R}^X$. ∎

Proof of Theorem 2. Consider the elements

$$m = \sum_{i=1}^p \alpha_i \delta(x_i) \quad \text{and} \quad n = \sum_{j=1}^q \beta_j \delta(y_j)$$

of the vector space $\mathscr{S}^*(X)$.

The function $k: X \times X \to \mathbb{R}$ defines on $\mathscr{S}^*(X)$ the bilinear form

(11) $$((m,n))_* = \sum_{i,j} \alpha_i \beta_j k(x_i, y_j).$$

This bilinear form is symmetric and positive, since k is symmetric and of positive type. It is, therefore, a semiscalar product.

With $((m,n))_*$, $\mathscr{S}^*(X)$ is a nonseparated prehilbert space. We denote by V its dual, which is a Hilbert space (Proposition 7.1). Since every linear form on $\mathscr{S}^*(X)$ is a function $\varphi \in \mathbb{R}^X$ according to Lemma 1, it follows that V is a Hilbert space of functions. It is, in fact, the subspace of functions $\varphi \in \mathbb{R}^X$ satisfying

(12) $$\exists c \geq 0 \quad \text{such that} \quad |\sum \alpha_i \varphi(x_i)| \leq c\sqrt{((m,m))_*}$$
$$= c\sqrt{\sum_{i,j} \alpha_i \alpha_j k(x_i, x_j)} \quad \text{for all } m = \sum \alpha_i \delta(x_i).$$

It remains for us to show that k is the reproducing kernel of V. Let θ be the canonical mapping from the nonseparated prehilbert space $(\mathscr{S}^*(x), ((,))_*)$ to its *Hausdorff completion* V^*. We denote by J^{-1} the canonical isometry from V^* onto V. Then we can write, for all $m = \delta(x) \in \mathscr{S}^*(X)$, for all $\varphi \in V$

(13) $$\varphi(x) = \langle \theta\delta(x), \varphi \rangle = ((J^{-1}\theta\delta(x), \varphi))_V.$$

To show that k is the reproducing kernel of V, it suffices, using (13), to show that $K(x) = J^{-1}\theta\delta(x)$ for all $x \in X$. But according to definition (11) of the scalar product on $\mathscr{S}^*(X)$, we obtain, by setting $\varphi = J^{-1}\theta\delta(y)$,

(14) $$k(x, y) = ((\delta(x), \delta(y)))_* = ((\theta\delta(x), \theta\delta(y)))_{V^*}$$
$$= \langle \theta\delta(x), J^{-1}\theta\delta(y) \rangle = \varphi(x)$$

by formula (13). Hence $\varphi = K(y)$. ∎

CHAPTER 6

L^2 Spaces and Convolution Operators

We present the first (and principal) example of a Hilbert space—namely, the space of square integrable functions—and some examples of continuous linear operators defined on these spaces—namely, convolution operators.

To avoid as much as possible using integration theory, we define the space $L^2(\Omega)$ as the *completion* of the space $\mathscr{C}_0(\Omega)$ of continuous functions with compact support in an open set Ω of $\mathbb{R}^n$ for the scalar product $(x, y) = \int_\Omega x(\omega) y(\omega) d\omega$.

We *admit* that the completion is indeed (isomorphic to) the space of (classes) of square integrable functions, as well as the Lebesgue and Fubini theorems, which we shall use on several occasions.

Since the scalar product (x, y) is the simplest of the scalar products that we can define on $\mathscr{C}_0(\Omega)$, we agree to *choose $L^2(\Omega)$ as a pivot space*. We illustrate the problem of choosing the realization of the dual of a Hilbert space by means of the spaces $L^2(\Omega, a)$ of square integrable functions for the measure $a(\omega) d\omega$ (spaces "with weight" a) where we suppose that a is a continuous strictly positive function. [These spaces are defined as *completions* of $\mathscr{C}_0(\Omega)$ for the scalar products $((x, y))_a = \int_\Omega x(\omega) y(\omega) a(\omega) d\omega$.] We verify that the bilinear form (x, y) is a duality pairing on $L^2(\Omega, 1/a) \times L^2(\Omega, a)$. We then choose to take $L^2(\Omega, 1/a)$ as the realization of the dual of $L^2(\Omega, a)$ [instead of identifying $L^2(\Omega, a)$ with its dual]. This choice is compatible with the preceding, since $L^2(\Omega) = L^2(\Omega, 1)$.

We then give some properties of the spaces $\hat{H}^s(\mathbb{R}^n) = L^2(\mathbb{R}, a_s)$ of square integrable functions for the measure $(1 + \|\omega\|^2)^s d\omega$. Later (Chapter 9, Section 3) we see that these spaces are the images of Sobolev spaces under the Fourier transform.

The first examples of linear operators from $L^2(\mathbb{R}^n)$ to itself are the convolution operators $\lambda *$ by an (integrable) function λ (formally) defined by the convolution product

(1) $$(\lambda * x)(\omega) = \int_{\mathbb{R}^n} x(\omega - \zeta) \lambda(\zeta) d\zeta$$

of λ and x.

We see that this is a continuous operator whose norm is equal to the norm of λ in $L^1(\mathbb{R}^n)$.

To construct this operator, we remark that, first, (1) makes sense when λ and x are continuous with compact support and, second, by extension by continuity, $\lambda * x$ belongs to $L^1(\mathbb{R}^n)$ as λ and x run over $L^1(\mathbb{R}^n)$. Finally we show that $\lambda * x$ belongs to $L^2(\mathbb{R}^n)$ when $\lambda \in L^1(\mathbb{R}^n)$ and $x \in L^2(\mathbb{R}^n)$.

The properties of the convolution product are extremely interesting. The first is that if x is continuous and λ is differentiable, then $\lambda * x$ is also differentiable and

$$(2) \qquad D_i(\lambda * x) = (D_i \lambda) * x \qquad \text{where} \qquad D_i = \frac{\partial}{\partial \omega_i}.$$

The second is the approximation property. If $\lambda \in L^1(\mathbb{R}^n)$ and if we set $\lambda_h(\omega) = (1/h^n)\lambda(x/h)$, we show that $\lambda_h * x$ converges to x in $L^2(\mathbb{R}^n)$.

These two properties allow us to prove that the space $\mathscr{D}(\Omega)$ of infinitely differentiable functions with compact support in Ω is dense in $L^2(\Omega)$.

We conclude this chapter with some examples: that of the mth convolution powers that will be used in the study of approximation by piecewise polynomials (Spline functions) and that of convolution by polynomials.

We briefly define Appell polynomials and, as special cases, Bernouilli and Hermite polynomials.

1. THE SPACE $L^2(\Omega)$ OF SQUARE INTEGRABLE FUNCTIONS

Let us consider an open set Ω of $\mathbb{R}^n$. We denote by

(1) $\quad \mathscr{C}_0(\Omega)$ *the vector space of continuous functions with compact support in* Ω.

This space does not reduce to a single element 0; it is in fact *an* infinite dimensional space.

If $\Omega = \mathbb{R}^n$, it contains in particular the function ρ defined by

$$(2) \qquad \rho(\omega) = \begin{cases} \exp\left(-\dfrac{1}{1 - \|\omega\|^2}\right) & \text{if } \|\omega\| < 1 \\ 0 & \text{if } \|\omega\| \geq 1 \end{cases}$$

whose support is contained in the unit ball of $\mathbb{R}^n$. *This function is in fact infinitely differentiable.*

Since every continuous function with compact support in Ω is Lebesgue integrable with respect to $d\omega$, we can define on $\mathscr{C}_0(\Omega)$ the bilinear form $(.,.)$ given by

$$(3) \qquad (x, y) = \int_\Omega x(\omega) y(\omega) \, d\omega.$$

We recall that $(.,.)$ is a scalar product on $\mathscr{C}_0(\Omega)$ for which $\mathscr{C}_0(\Omega)$ *is not complete.*
[See (AAA), p. 34.] Thus we are led to introduce its completion.

DEFINITION 1
We denote by $L^2(\Omega)$ a completion of $\mathscr{C}_0(\Omega)$ for the scalar product defined by (3), called the space of "square integrable" functions on Ω. ▲

The importance of the role played by this space in analysis is first of all a consequence of the *simplicity* of the scalar product $(.,.)$ defined by (3); it is clearly the simplest of the scalar products that can be defined on a space of functions.

Remark 1

Since $\mathbb{R}^n$ is the space of functions on $\Omega = \{1, 2, \ldots, n\}$, we see that the scalar product $(.,.)$ is the natural generalization of the Euclidean scalar product $(x, y) = \sum_{i=1}^{n} x_i y_i$. ∎

It is for this reason that we agree to *identify the space $L^2(\Omega)$ with its dual*, that is, to take $L^2(\Omega)$ as a *pivot space*.
It is also convenient to use the spaces $L^p(\Omega)$, defined as *completions of the space $\mathscr{C}_0(\Omega)$ for the norm* $\|x\|_p = (\int_\Omega |x(\omega)|^p d\omega)^{1/p} (1 \leq p < +\infty)$.

Remark 2

Having chosen $L^2(\Omega)$ as a pivot space allows us to give another interpretation of the elements of $L^2(\Omega)$: *every element $x \in L^2(\Omega)$ is a linear form on $\mathscr{C}_0(\Omega)$ that is continuous for the norm* $|.|$. (Indeed, $L^2(\Omega)$ is identified with the dual of $L^2(\Omega)$, which is also the dual of the prehilbert space $(\mathscr{C}_0(\Omega), (.,.))$. (See Theorem 1.3.3.) This trivial remark is in fact *fundamental*, because it is at the origin of another way to consider the functions (which will permit us to generalize the notation of function). Consider a function $x : \omega \in \Omega \to x(\omega) \in \mathbb{R}$ that is continuous with compact support belongs to $L^2(\Omega)$.
We shall no longer consider it as an element of $L^2(\Omega)$ in its initial form, but rather as the linear form $y \in \mathscr{C}_0(\Omega) \mapsto (x, y) = \int_\Omega x(\omega) y(\omega) d\omega$, which is continuous on $\mathscr{C}_0(\Omega)$ for the norm $|.|$. It is this latter point of view that we most frequently adopt (almost systematically) in what follows. ∎

Remark 3

The preceding interpretation of the elements of $L^2(\Omega)$ is not sufficiently rich for all the cases we wish to consider. However, making use of integration theory, we can construct realizations of the completions $L^2(\Omega)$ and $L^p(\Omega)$ of $\mathscr{C}_0(\Omega)$ in the following fashion: $L^p(\Omega)$ is the *space of the (classes of) measurable functions on Ω that are pth power integrable (or "p-integrable")*.

Besides the possibility of constructing this realization of $L^p(\Omega)$, integration theory also gives us the Lebesgue-dominated convergence theorem (which we shall accept *without proof*). ■

THEOREM A (LEBESGUE-DOMINATED CONVERGENCE)
Consider a sequence of functions $f_n \in L^p(\Omega)$ that converges pointwise to a function f. Suppose, moreover, that there exists a function $g \in L^p(\Omega)$ such that $\sup_n |f_n(\omega)| \leq g(\omega)$ for every $\omega \in L^p(\Omega)$. Then f_n converges to f in $L^p(\Omega)$. ▲

This result is truly remarkable; *pointwise* convergence (nonuniform, hence very weak), provided that $|f_n(.)|$ is bounded above by $g(.)$, is sufficient to ensure convergence in the mean of order p.

We also admit the Fubini theorem, which we recall.

THEOREM B (FUBINI)
Let $\Omega_1 \subset \mathbb{R}^{n_1}$ and $\Omega_2 \subset \mathbb{R}^{n_2}$ be two open sets and f a function of $L^1(\Omega_1 \times \Omega_2)$. Then, for almost every ω, the function $\zeta \mapsto f(\omega, \zeta)$ is integrable and for almost every ζ, the function $\omega \mapsto f(\omega, \zeta)$ is integrable; moreover, we obtain the following formulas:

$$\int\int_{\Omega_1 \times \Omega_2} f(\omega, \zeta) d\omega\, d\zeta = \int_{\Omega_1} d\omega (\int_{\Omega_2} f(\omega, \zeta) d\zeta) = \int_{\Omega_2} (\int_{\Omega_1} f(\omega, \zeta) d\omega) d\zeta. \quad ▲$$

It is necessary to know *in advance* that the double integral makes sense in order to be able to calculate it by two successive simple integrations. It can happen that

$$\int_{\Omega_1} d\omega (\int_{\Omega_2} f(\omega, \zeta) d\zeta) \quad \text{and} \quad \int_{\Omega_2} d\zeta (\int_{\Omega_1} f(\omega, \zeta) d\omega)$$

are defined and *have different values* (when f is not integrable on $\Omega_1 \times \Omega_2$). For example,

$$\int_0^1 dx \int_0^1 \frac{x^2 - y^2}{(x^2 + y^2)^2} dy = \frac{\pi}{4} \quad \text{and} \quad \int_0^1 dy \int_0^1 \frac{x^2 - y^2}{(x^2 + y^2)^2} dx = -\frac{\pi}{4}.$$

2. THE SPACES $L^2(\Omega, a)$ WITH WEIGHTS

(1) $\begin{cases} \text{Let } \Omega \subset \mathbb{R}^n \text{ be an open set in } \mathbb{R}^n \text{ and } a(.) \text{ be a strictly positive} \\ \text{continuous function on } \Omega, \text{ which we shall call a } weight. \end{cases}$

Then the bilinear form defined on $\mathscr{C}_0(\Omega)$ by

(2) $$((x,y))_a = \int_\Omega x(\omega)y(\omega)a(\omega)\,d\omega$$

is a scalar product.

DEFINITION 1
We denote by $L^2(\Omega, a)$ a completion of $\mathscr{C}_0(\Omega)$ for the scalar product $((.,.))_a$. ▲

Naturally, since $L^2(\Omega, a)$ is a Hilbert space, it can be identified with its dual. However, this would not be a reasonable solution since we have already agreed to use $L^2(\Omega)$ (with weight 1) as a pivot space.

Consider the function $a^{-1}: \omega \mapsto a^{-1}(\omega) = 1/a(\omega)$, which is also a weight, since it is strictly positive and continuous on Ω. We can, therefore, define the space $L^2(\Omega, a^{-1})$.

PROPOSITION 1
The bilinear form $\{x, y\} \mapsto (x, y) = \int_\Omega x(\omega)y(\omega)\,d\omega$ is defined on $L^2(\Omega, a) \times L^2(\Omega, a^{-1})$ and is a duality pairing of $L^2(\Omega, a^{-1})$ and $L^2(\Omega, a)$. The duality operator from $L^2(\Omega, a)$ onto its dual $L^2(\Omega, a^{-1})$ is the multiplication operator by a. ▲

Proof. It suffices to take x and $y \in \mathscr{C}_0(\Omega)$. The Cauchy-Schwarz inequality implies that

$$(x,y) = \int_\Omega x(\omega)\sqrt{a(\omega)}\,y(\omega)\frac{1}{\sqrt{a(\omega)}}\,d\omega$$
$$\leq \left(\int_\Omega |x(\omega)|^2 a(\omega)\,d\omega\right)^{1/2} \left(\int_\Omega |x(\omega)|^2 a^{-1}(\omega)\,d\omega\right)^{1/2}$$
$$= \|x\|_a \|y\|_{a^{-1}}.$$

On the other hand, if $x \in L^2(\Omega, a)$, ax belongs to $L^2(\Omega, a^{-1})$ and $\|ax\|_{a^{-1}} = \|x\|_a$. Hence

$$\|y\|_{a^{-1}} = \sup_{x \in L^2(\Omega, a)} \frac{|(x,y)|}{\|x\|_a}.$$

This implies that $(.,.)$ is a duality pairing. Moreover, the equality

$$((x,y))_a = \int_\Omega x(\omega)y(\omega)a(\omega)\,d\omega = (ax, y)$$

shows that the multiplication operator by a is the duality operator from $L^2(\Omega, a)$ onto its dual $L^2(\Omega, a^{-1})$. ■

Remark 1

This choice is indeed consistent with taking $L^2(\Omega) = L^2(\Omega, 1)$ as a pivot space; the dual of $L^2(\Omega) = L^2(\Omega, 1)$ is identified with $L^2(\Omega, 1^{-1}) = L^2(\Omega)$. ∎

PROPOSITION 2
The multiplication operator by $\sqrt{a}$ is an isometry from $L^2(\Omega, a)$ onto $L^2(\Omega)$. ▲

Proof. (Left as an exercise.)

PROPOSITION 3
Suppose that

(3) $$a(\omega) \geq c > 0 \quad \text{for all} \quad \omega \in \Omega.$$

Then $L^2(\Omega, a)$ is identified with a dense subspace of $L^2(\Omega)$; we have the embeddings

(4) $$L^2(\Omega, a) \subset L^2(\Omega) \subset L^2(\Omega, a^{-1}).$$ ▲

Proof. Indeed, if $x \in \mathscr{C}_0(\Omega)$, the inequality

(5) $$|x|^2 = \int |x(\omega)|^2 \, d\omega \leq \int \frac{|x(\omega)|^2 a(\omega)}{c} \, d\omega = \frac{1}{c} \|x\|_a^2$$

shows that the injection from $(\mathscr{C}_0(\Omega), \|\cdot\|_a)$ to $L^2(\Omega)$ is continuous. Let j be the identity mapping from $(\mathscr{C}_0(\Omega), \|\cdot\|_a)$ to $L^2(\Omega)$, θ be the isometry from $(\mathscr{C}_0(\Omega), \|\cdot\|_a)$ to $L^2(\Omega, a)$ and $\hat{j}$ the mapping from $L^2(\Omega, a)$ to $L^2(\Omega)$ satisfying $j = \hat{j}\theta$. We show that $\hat{j}$ is injective. Let $\hat{x} \in L^2(\Omega, a)$ such that $\hat{j}(\hat{x}) = 0$. Hence $\hat{x} = \lim_{n \to \infty} \theta(x_n)$ and thus, $\{x_n\}_n$ is a Cauchy sequence in $(\mathscr{C}_0(\Omega), \|\cdot\|_a)$. This implies that $\{x_n \sqrt{a}\}_n$ is a Cauchy sequence and, by (5), that $\{x_n\}_n$ is a Cauchy sequence in $L^2(\Omega)$. Then, the sequences x_n and $x_n \sqrt{a}$ converge in $L^2(\Omega)$ to y and $y\sqrt{a}$ respectively. Since $x_n = j(x_n) = \hat{j}\theta(x_n)$ converges to 0, $y = 0$. Therefore, $\hat{x} = 0$ and $\hat{j}(L^2(\Omega, a))$ is a completion of $(\mathscr{C}_0(\Omega), \|\cdot\|_a)$ that is contained in $L^2(\Omega)$. So we can choose $\hat{j}(L^2(\Omega, a))$ as the realization of the completion that is contained in $L^2(\Omega)$. ∎

3. THE SPACE $\hat{H}^s$

We associate to every $s \geq 0$ the weight a_s defined by $a_s(\omega) = (1 + \|\omega\|^2)^s$, bounded below by the constant 1.

DEFINITION 1
If $s > 0$, we denote by $\hat{H}^s = \hat{H}^s(\mathbb{R}^n)$ the space $L^2(\Omega, a_s)$ and by $\hat{H}^{-s} = \hat{H}^{-s}(\mathbb{R}^n)$ the space $L^2(\Omega, a_s^{-1})$. ▲

Proposition 2.3 becomes

PROPOSITION 1
If $s > 0$, the space $\hat{H}^{-s}$ is identified with the dual of $\hat{H}^s$, the duality operator being the multiplication operator by $(1 + \|.\|^2)^s$. Moreover, we have the embeddings

(1) $$\hat{H}^r \subset \hat{H}^s \subset L^2 \subset \hat{H}^{-s} \subset \hat{H}^{-r} \quad \text{if} \quad r > s > 0.$$ ▲

It is also evident that the multiplication operator by $(1 + \|.\|^2)^{(s-r)/2}$ is an isometry from $\hat{H}^s$ onto $\hat{H}^r$ $(r, s \in \mathbb{R})$. If $s = m$ is a positive integer, we have another characterization of the spaces $\hat{H}^m$.

PROPOSITION 2
If m is a positive integer, $\hat{H}^m$ is the subspace of the functions x of $L^2(\mathbb{R}^n)$ for which $\omega^k x(\omega) = \omega_1^{k_1} \omega_2^{k_2} \ldots \omega_n^{k_n} x(\omega_1, \ldots, \omega_n)$ belongs to $L^2(\mathbb{R}^n)$ for every n-tuple $k = (k_1, \ldots, k_n)$ such that $|k| = k_1 + \ldots + k_n \leq m$. ▲

Proof. There exists a constant c such that

(2) $$(1 + \|\omega\|^2)^m \leq \sum_{|k| \leq m} \omega^{2k} \leq c(1 + \|\omega^2\|)^m,$$

which shows that the norms defined by

(3) $$\|x\|_m^2 = \int_\Omega |x(\omega)|^2 (1 + \|\omega\|^2)^m \, d\omega$$

and the norms defined by

(4) $$\|\|x\|\|_m^2 = \sum_{|k| \leq m} \int |x(\omega)|^2 |\omega|^{2k} \, d\omega$$

are equivalent on $\mathcal{C}_0(\mathbb{R}^n)$ and, therefore, on their completion $\hat{H}^m$.

Moreover, the equivalence of these norms implies that if $x \in \hat{H}^m$, then $\omega^k x(\omega) \in L^2(\mathbb{R}^n)$ for all $|k| \leq m$. Conversely, if $x \in L^2 = \hat{H}^0$ is such that $\omega^k x(\omega)$ belongs to L^2 for every $|k| \leq m$, then $(1 + \|\omega\|^2)^{m/2} x(\omega)$ belongs to L^2; that is, x belongs to $\hat{H}^m$. ■

Interpolation Inequalities

PROPOSITION 3
Consider three real numbers r, s, and t such that $r \leq s \leq t$. Let $\theta = (t - s)/(t - r) \in$

[0, 1]. Then if $x \in \mathscr{C}_0(\mathbb{R}^n)$ we obtain the interpolation inequalities
(5) $$\|x\|_s \leq \|x\|_r^\theta \|x\|_t^{1-\theta}.$$ ▲

Proof. We apply the Hölder inequality with $p = 1/\theta$ and $p^* = 1/(1-\theta)$ in the following fashion:

$$\begin{aligned}
\|x\|_s^2 &= \int |x(\omega)|^2 (1 + \|\omega\|^2)^s \\
&= \int |x(\omega)|^{2(1-\theta)} (1 + \|\omega\|^2)^{(1-\theta)t} |x(\omega)|^{2\theta} (1 + \|\omega\|^2)^{\theta r} \\
&\leq [\int |x(\omega)|^2 (1 + \|\omega\|^2)^t]^{1-\theta} [\int |x(\omega)|^2 (1 + \|\omega\|^2)^r]^\theta \\
&= \|x\|_t^{2(1-\theta)} \|x\|_r^{2\theta}
\end{aligned}$$

since $s = (1-\theta)t + \theta r$. ∎

We now give yet another inequality that will have an important consequence (the Sobolev inequality of Proposition 9.3.1).

PROPOSITION 4
If $s > n/2$, $\hat{H}^s(\mathbb{R}^n)$ is contained in $L^1(\mathbb{R}^n)$ with a stronger topology. ▲

Proof. Let $x \in \hat{H}^s(\mathbb{R}^n)$. The Cauchy-Schwarz inequality then implies that

$$\begin{aligned}
\int |x(\omega)| \, d\omega &= \int |x(\omega)| (1 + \|\omega\|^2)^{s/2} (1 + \|\omega\|^2)^{-s/2} \, d\omega \\
&\leq [\int |x(\omega)|^2 (1 + \|\omega\|^2)^s \, d\omega]^{1/2} [\int (1 + \|\omega\|^2)^{-s} \, d\omega]^{1/2} \\
&= \|x\|_s [\int (1 + \|\omega\|^2)^{-s} \, d\omega]^{1/2}
\end{aligned}$$

Therefore it suffices to show that the final integral makes sense if $s > n/2$. This is the object of the following lemma. ∎

LEMMA 1
If $s > n/2$, the integral

(6) $$\int \frac{d\omega}{(1 + \|\omega\|^2)^s} = \prod_{j=1}^n \frac{d\omega_j}{(1 + |\omega_j|^2)^{s-(n-j)/2}}$$

is convergent. ▲

Proof. Writing that $\omega = (\omega_1, \omega_2, \ldots, \omega_n) = (\omega_1, \eta)$ where $\eta = (\omega_2, \ldots, \omega_n)$, we obtain

$$\int \frac{d\omega}{(1 + \|\omega\|^2)^s} = \iint \frac{d\eta \, d\omega_1}{(1 + \|\eta\|^2 + |\omega_1|^2)^s}.$$

Making the change of variable $\omega_1 = t(1 + \|\eta\|^2)^{1/2}$, the integral becomes

$$\int \frac{d\omega}{(1 + \|\omega\|^2)^s} = \int \frac{d\eta}{(1 + \|\eta\|^2)^{s-1/2}} \int \frac{dt}{(1 + t^2)^s}.$$

Repeating this process, we obtain formula (6) by iteration. Since the integral $\int dt/(1+t^2)^\alpha$ converges if $\alpha > \frac{1}{2}$, the integrals appearing in the product on the right-hand side of (6) are convergent if $s - (n-j)/2 > \frac{1}{2}$ for $1 \leq j \leq n$, that is, if $s > n/2$. ∎

4. THE CONVOLUTION PRODUCT FOR FUNCTIONS OF $\mathscr{C}_0(\mathbb{R}^n)$ AND OF $L^1(\mathbb{R}^n)$

First let us consider two continuous functions x and y with compact support in $\mathbb{R}^n$.

DEFINITION 1

The convolution product $x * y$ of two functions x and $y \in \mathscr{C}_0(\mathbb{R}^n)$ is defined by

(1) $$(x * y)(\omega) = \int_{\mathbb{R}^n} x(\omega - \zeta) y(\zeta) \, d\zeta = \int_{\mathbb{R}^n} x(\zeta) y(\omega - \zeta) \, d\zeta.$$ ▲

We see that this is a continuous function *with compact support*; in fact,

(2) $$\operatorname{supp}(x * y) \subset \operatorname{supp} x + \operatorname{supp} y.$$

We also see that the total mass of $x * y$ is the product of the total masses of x and of y:

$$\int_{\mathbb{R}^n} (x * y)(\omega) \, d\omega = \int_{\mathbb{R}^n} d\omega \int_{\mathbb{R}^n} x(\omega - \zeta) y(\zeta) \, d\zeta$$
$$= \int_{\mathbb{R}^n} y(\zeta) \, d\zeta \int_{\mathbb{R}^n} x(\omega - \zeta) \, d\omega$$
$$= \int_{\mathbb{R}^n} y(\zeta) \, d\zeta \int_{\mathbb{R}^n} x(\zeta) \, d\zeta.$$

From this inequality we deduce the *fundamental inequality*

(3) $$\|x * y\|_1 \leq \|x\|_1 \|y\|_1.$$

It is also evident that the convolution product is *commutative and associative*.

If we denote by $\tau_a x$ the function defined by $(\tau_a x)(\omega) = x(\omega - a)$, we see that

(4) $$(\tau_a x) * y = x * (\tau_a y) = \tau_a(x * y).$$

Moreover, the convolution product by a differentiable function is differentiable.

PROPOSITION 1

If $x \in \mathscr{C}_0(\mathbb{R}^n)$ and if $\lambda \in \mathscr{C}_0(\mathbb{R}^n)$ is continuously differentiable, the convolution product $x * \lambda$ is differentiable, and we have

(5) $$D_i(x * \lambda) = x * D_i \lambda$$

where $D_i = \dfrac{\partial}{\partial \omega_i}$ is the *i*th partial derivative. ▲

Proof. Let K be the support of λ. We are going to show that

(6)
$$\frac{(\lambda * x)(\omega + he_i) - (\lambda * x)(\omega)}{h} - (D_i\lambda * x)(\omega)$$
$$= \int_{\mathbb{R}^n} x(\eta)\, d\eta \left(\frac{\lambda(\omega - \eta + he_i) - \lambda(\omega - \eta)}{h} - D_i\lambda(\omega - \eta) \right).$$

converges to zero with h. Since λ is continuously differentiable, we know that we can associate to every $\xi \in K$ an open neighborhood $V(\xi)$ and $r(\xi) > 0$ such that

$$\left| \frac{\lambda(\eta + he_i) - \lambda(\eta)}{h} - D_i\lambda(\xi) \right| \leq \frac{\varepsilon}{2} \quad \text{and} \quad |D_i\lambda(\eta) - D_i\lambda(\xi)| \leq \frac{\varepsilon}{2}$$

when $\eta \in V(\xi)$ and $|h| < r(\xi)$.

Since K is compact, we can cover it with p neighborhoods $V(\xi_j)$. Let $r = \min_j r(\xi_j) > 0$. Then for every $\xi \in K$, we can find ξ_j such that

$$\left| \frac{\lambda(\xi + he_i) - \lambda(\xi)}{h} - D_i\lambda(\xi) \right|$$
$$\leq \left| \frac{\lambda(\xi + he_i) - \lambda(\xi)}{h} - D_i\lambda(\xi_j) \right| + |D_i\lambda(\xi_j) - D_i\lambda(\xi)|$$
$$\leq \frac{2\varepsilon}{2} = \varepsilon \quad \text{when} \quad |h| < r, \text{ independent of } \xi.$$

It follows, therefore, from the mean-value theorem that for every $\varepsilon > 0$, there exists r such that

$$\left| \frac{(\lambda * x)(\omega + he_i) - (\lambda * x)(\omega)}{h} - (D_i\lambda * x)(\omega) \right|$$
$$\leq \varepsilon \int_{\mathbb{R}^n} |x(\omega)|\, d\omega = \varepsilon \|x\|_1 \quad \text{when} \quad |h| \leq r.$$

This expresses the fact that $D_i\lambda * x$ is the partial derivative of $\lambda * x$. ∎

Remark 1

Taking the convolution product of a function $x \in \mathscr{C}_0(\mathbb{R}^n)$ and an infinitely differentiable function with compact support ρ defined by (1.2), we can construct a function $x * \rho$ which is also infinitely differentiable with compact support. ∎

Convolution for Functions of $L^1(\mathbb{R}^n)$

The problem presents itself of extending the convolution product to larger classes of functions. We can reformulate inequality (3) by saying that the bilinear mapping $(x, y) \mapsto x * y$ from $\mathscr{C}_0(\mathbb{R}^n) \times \mathscr{C}_0(\mathbb{R}^n)$ to $L^1(\mathbb{R}^n)$ is continuous when the space $\mathscr{C}_0(\mathbb{R}^n)$ has the norm induced by $L^1(\mathbb{R}^n)$. Then Theorem 1.3.1 on extension by density (which is valid in Banach spaces) implies that this bilinear mapping has a unique extension to a continuous bilinear mapping from $L^1(\mathbb{R}^n) \times L^1(\mathbb{R}^n)$ to $L^1(\mathbb{R}^n)$. Therefore we agree to set $\{x, y\} \mapsto x * y$ as this extended mapping and to say that $x * y$ is the convolution product of the functions x and $y \in L^1(\mathbb{R}^n)$.

THEOREM 1
The mapping $\{x, y\} \mapsto x * y$ has an extension to a continuous bilinear mapping from $L^1(\mathbb{R}^n) \times L^1(\mathbb{R}^n)$ to $L^1(\mathbb{R}^n)$ satisfying

(7) $$\|x * y\|_1 \leq \|x\|_1 \|y\|_1$$

which defines on $L^1(\mathbb{R}^n)$ an internal operation that is commutative and associative. ▲

Example. Convolution Product for Gaussian Probability Densities

Consider the Gaussian probability densities $g_\sigma(\omega)$ defined by

(8) $$g_\sigma(\omega) = \frac{1}{\sigma \sqrt{2\pi}} \exp\left(-\frac{\omega^2}{2\sigma^2}\right)$$

with total mass equal to one (i.e., $\int g_\sigma(\omega)\, d\omega = 1$), with mean value zero (i.e., $\int_\mathbb{R} \omega g_\sigma(\omega)\, d\omega = 0$), and with standard deviation σ (i.e., $(\int \omega^2 g_\sigma(\omega) \cdot d\omega)^{1/2} = \sigma$). ■

PROPOSITION 2
The convolution product of two Gaussian probability densities is again a Gaussian probability density:

(9) $$g_\sigma * g_\tau = g_{\sqrt{\sigma^2 + \tau^2}}.$$
▲

Proof. Indeed, g_σ and g_τ are functions of $L^1(\mathbb{R})$. Moreover,

$$(g_\sigma * g_\tau)(\omega) = \frac{1}{2\pi\sigma\tau} \int_\mathbb{R} \exp\left(-\frac{(\omega - \zeta)^2}{2\sigma^2} - \frac{\zeta^2}{2\tau^2}\right) d\zeta$$

$$= \frac{1}{2\pi\sigma\tau} \int_\mathbb{R} \exp\left(-\frac{\omega^2}{2\sigma^2} + \frac{\omega\zeta}{\sigma^2} - \frac{\zeta^2}{2}\left(\frac{1}{\sigma^2} + \frac{1}{\tau^2}\right)\right) d\zeta$$

$$= \frac{1}{2\pi\sigma\tau} \exp\left(-\frac{\omega^2}{2(\sigma^2 + \tau^2)}\right) \int_\mathbb{R} \exp\left(-\frac{1}{2}\frac{\sigma^2 + \tau^2}{\sigma^2 \tau^2}\left(\zeta - \frac{\tau^2 \omega}{\sigma^2 + \tau^2}\right)^2\right) d\zeta.$$

Making the change of variables

$$\frac{\sqrt{\sigma^2+\tau^2}}{\sigma\tau}\left(\zeta - \frac{\tau^2}{\sigma^2+\tau^2}\omega\right) = u,$$

the final integral becomes

$$\frac{\sigma\tau}{\sqrt{\sigma^2+\tau^2}}\int_{-\infty}^{+\infty}\exp\left(-\frac{u^2}{2}\right)du = \sqrt{2\pi}\,\frac{\sigma\tau}{\sqrt{\sigma^2+\tau^2}},$$

since $\int_{-\infty}^{+\infty}\exp(-u^2/2)\,du = \sqrt{2\pi}$. Consequently, we obtain

$$(g_\sigma * g_\tau)(\omega) = \frac{1}{\sqrt{2\pi}}\frac{1}{\sqrt{\sigma^2+\tau^2}}\exp\left(-\frac{\omega^2}{2(\sigma^2+\tau^2)}\right) = g_{\sqrt{\sigma^2+\tau^2}}.\quad\blacksquare$$

Remark 2

The mth convolution power g_σ^{*m} of a Gaussian probability density g_σ with standard deviation σ is a Gaussian probability density with standard deviation $\sigma\sqrt{m}$. $\blacksquare$

Remark 3

We have shown that the convolution product defined by (1) is defined on $\mathscr{C}_0(\mathbb{R}^n) \times \mathscr{C}_0(\mathbb{R}^n)$ and $L^1(\mathbb{R}^n) \times L^1(\mathbb{R}^n)$. More generally, we can extend the convolution product as a *continuous bilinear mapping from* $L^p(\mathbb{R}^n) \times L^q(\mathbb{R}^n)$ *to* $L^r(\mathbb{R}^n)$ where $1/r = 1/p + 1/q - 1$:

$$\|x * y\|_r \leq \|x\|_p \|y\|_q.$$

(We do not prove this result in this book.) $\blacksquare$

Integral (1) still makes sense if only one of the functions y has compact support. This allows us to define the convolution product of y with the monomials $x_n(\omega) = \omega^n/n!$. We shall see in Section 8 that

(10) $$(x_n * y)(\omega) = \sum_{j=0}^{n}\left(\int y(\zeta)\frac{(-\zeta)^{n-j}}{(n-j)!}d\zeta\right)\frac{\omega^j}{j!},$$

and we shall study the properties of the convolution product for polynomials.

5. CONVOLUTION OPERATORS

We are now going to show that if $\lambda \in L^1(\mathbb{R}^n)$, the operator $x \mapsto \lambda * x$ defines a continuous linear operator from $L^2(\mathbb{R}^n)$ to $L^2(\mathbb{R}^n)$.

THEOREM 1
We can associate to every function $\lambda \in L^1(\mathbb{R}^n)$ the convolution operator $x \in L^2(\mathbb{R}^n) \to \lambda * x \in L^2(\mathbb{R}^n)$, which is a continuous linear operator satisfying

(1) $$|\lambda * x| \leq \|\lambda\|_1 |x|$$ ▲

Proof. According to the theorem on extension by density (Theorem 1.3.3), it suffices to show that inequality (1) holds for all $x \in \mathscr{C}_0(\mathbb{R}^n)$. Indeed, this inequality expresses the fact that the convolution operator $\lambda *$ is continuous from the prehilbert space $(\mathscr{C}_0(\mathbb{R}^n), (.,.))$ to $L^2(\mathbb{R}^n)$ and, consequently, that it has a unique extension to a continuous linear operator from $L^2(\mathbb{R}^n)$ to itself. We also denote this extended operator by $\lambda *$, and we say that $\lambda * x \in L^2(\mathbb{R}^n)$ *is the convolution product of a function of $L^1(\mathbb{R}^n)$ and a function of $L^2(\mathbb{R}^n)$.*

Then consider a function $x \in \mathscr{C}_0(\mathbb{R}^n)$ and let us denote by K its compact support. Since x is a continuous function on the compact set K, it is uniformly continuous: if $\varepsilon > 0$ is fixed and if $\operatorname{mes}(K) = \int_K d\omega$, it follows that $|x(\omega - \zeta) - x(\omega - \eta)| \leq \varepsilon / \sqrt{\operatorname{mes}(K)}$ when $\|\zeta - \eta\| \leq \alpha$. This implies that $\|x(. - \eta) - x(. - \zeta)\|_{L^2(\mathbb{R}^n)} \leq \varepsilon$ when $\|\zeta - \eta\| \leq \alpha$. This inequality expresses the fact that the function $\zeta \mapsto x(\zeta - .)$ is continuous from $\mathbb{R}^n$ to $L^2(\mathbb{R}^n)$. Hence the function z:

(2) $$\zeta \mapsto z(\zeta) = (x(\zeta - .), y) = \int_{\mathbb{R}^n} x(\zeta - \omega) y(\omega) d\omega$$

is also *continuous* and *bounded* as well, since

(3) $$\|z\|_\infty = \sup_\zeta \left| \int_{\mathbb{R}^n} x(\zeta - \omega) y(\omega) d\omega \right|$$
$$\leq \sup_\zeta \left(\int_{\mathbb{R}^n} |x(\zeta - \omega)|^2 d\omega \right)^{1/2} \left(\int_{\mathbb{R}^n} |y(\omega)|^2 d\omega \right)^{1/2}$$
$$= |x| . |y|$$

from the Cauchy-Schwarz inequality.

Consequently, if $\lambda \in L^1(\mathbb{R}^n)$, the mean-value theorem and the Fubini theorem imply that

(4) $$\int_{\mathbb{R}^n} |\lambda(\omega) z(\omega)| d\omega \leq \|\lambda\|_1 \|z\|_\infty \leq \|\lambda\|_1 |x| |y|$$

and that

(5) $$\int_{\mathbb{R}^n} \lambda(\omega) z(\omega) d\omega = \int_{\mathbb{R}^n} \left(\int_{\mathbb{R}^n} x(\zeta - \omega) \lambda(\omega) d\omega \right) y(\zeta) d\zeta.$$

It follows then from (4) and (5) that $y \mapsto \int_{\mathbb{R}^n} (\int_{\mathbb{R}^n} x(\zeta - \omega) \lambda(\omega) d\omega) y(\zeta) d\zeta$ is a continuous linear form on $L^2(\mathbb{R}^n)$, with norm bounded above by $\|\lambda\|_1 |x|$. Since $L^2(\mathbb{R}^n)$ is a pivot space, this continuous linear form is an element of

$L^2(\mathbb{R}^n)$, which coincides (almost everywhere) with the function
$$\zeta \mapsto \int_{\mathbb{R}^n} x(\zeta - \omega)\lambda(\omega)\,d\omega = (\lambda * x)(\zeta). \qquad \blacksquare$$

6. APPROXIMATION BY CONVOLUTION

The convolution operators furnish a means for constructing approximation procedures.

To this end we introduce a function $\lambda \in \mathscr{C}_0(\Omega)$ such that

(1) $\quad \lambda \geq 0, \int_{\mathbb{R}^n} \lambda(\omega)\,d\omega = 1 \quad$ and $\quad \sup \lambda \subset B \quad$ (the unit ball of $\mathbb{R}^n$).

If $h \in\,]0, 1[$, we associate to it the functions λ_h, which are defined by

(2) $$\lambda_h(\omega) = \frac{1}{h^n} \lambda\left(\frac{\omega}{h}\right)$$

and which are always positive continuous functions with compact support contained in a ball of radius h and of total mass equal to one
$$\left(\text{since } \int_{\mathbb{R}^n} \lambda_h(\omega)\,d\omega = \frac{1}{h^n}\int_{\mathbb{R}^n} \lambda\left(\frac{\omega}{h}\right) dx = \int_{\mathbb{R}^n} \lambda(\omega)\,d\omega\right).$$

We shall also show that the convolution operators $\lambda_h *$ converge to the identity, that is, that $\lambda_h * x$ converges to x in $L^2(\mathbb{R}^n)$. To measure the error, we define the *oscillation* $\omega(x, h)$ of a function of $L^2(\mathbb{R}^n)$, which is given by

(3) $$\omega(x, h) = \sup_{\|\zeta\| \leq h} \left(\int |x(\omega - \zeta) - x(\omega)|^2\,d\omega\right)^{1/2}.$$

It is evident that *the oscillation $\omega(x, h)$ approaches zero with h*.

Remark 1

If the function is continuously differentiable with compact support, we obtain the following upper bound for the oscillation:

(4) $$\omega(x, h) \leq ch\left(\sum_{i=1}^{n} |D_i x|^2\right)^{1/2} \qquad \text{where } c \text{ is a constant.} \qquad \blacksquare$$

THEOREM 1
Let $\lambda \in \mathscr{C}_0(\mathbb{R}^n)$ be a function with total mass equal to one. Then the convolution operators λ_{h*} are bounded and converge pointwise to the identity; moreover,

(5) $$|\lambda_h * x - x| \leq \omega(x, h). \qquad \blacktriangle$$

Proof. We know that $\|\lambda_h *\| \leq \|\lambda_h\|_1 = 1$. Let us establish (5). We can write

$$(\lambda_h * x)(\omega) = \frac{1}{h^n} \int_{\mathbb{R}^n} x(\omega - \zeta) \lambda\left(\frac{\zeta}{h}\right) d\zeta = \int x(\omega - \zeta h) \lambda(\zeta) d\zeta$$

and that $x(\omega) = \int x(\omega) \lambda(\zeta) d\zeta$. Therefore, using the Cauchy-Schwarz inequality,

$$|(\lambda_h * x)(\omega) - x(\omega)| = \int_{\mathbb{R}^n} |x(\omega - \zeta h) - x(\omega)| \sqrt{\lambda(\zeta)} \sqrt{\lambda(\zeta)} \, d\zeta$$
$$\leq \int (\int_{\mathbb{R}^n} |x(\omega - \zeta h) - x(\omega)|^2 \lambda(\zeta) \, d\zeta)^{1/2} (\int_{\mathbb{R}^n} \lambda(\zeta) \, d\zeta)^{1/2}$$
$$= (\int_{\mathbb{R}^n} |x(\omega - \zeta h) - x(\omega)|^2 \lambda(\zeta) \, d\zeta)^{1/2}.$$

Consequently, by integrating this inequality and using the Fubini theorem, we obtain

$$\int |\lambda_h * x(\omega) - x(\omega)|^2 \, d\omega \leq \int_{\mathbb{R}^n} d\omega \int_{\mathbb{R}^n} |x(\omega - \zeta h) - x(\omega)|^2 \lambda(\zeta) \, d\zeta$$
$$= \int_{\mathbb{R}^n} d\zeta \lambda(\zeta) \int_{\mathbb{R}^n} |x(\omega - \zeta h) - x(\omega)|^2 \, d\omega.$$

Since ζ runs over the support of λ, which is contained in the unit ball, $\|\zeta h\| \leq h$ and $\int_{\mathbb{R}^n} |x(\omega - \zeta h) - x(\omega)|^2 \, d\omega \leq \omega(x, h)^2$. It follows, therefore, from the mean-value theorem that

$$\int |\lambda_h * x(\omega) - x(\omega)|^2 \, d\omega \leq \omega(x, h)^2 \int_{\mathbb{R}^n} \lambda(\zeta) \, d\zeta = \omega(x, h)^2. \qquad \blacksquare$$

Using Proposition 4.1, this theorem allows us to approximate any function of $L^2(\Omega)$ by a function that is infinitely differentiable with compact support in Ω.

THEOREM 2
The space $\mathscr{D}(\Omega)$ of infinitely differentiable functions with compact support is dense in $L^2(\Omega)$. ▲

Proof. We must approximate a function $x \in L^2(\Omega)$ by a function of $\mathscr{D}(\Omega)$. First of all, for every $\varepsilon > 0$ there exists $x_0 \in \mathscr{C}_0(\Omega)$ such that $|x - x_0| \leq \varepsilon/2$. Let K be the *compact* support of x_0, contained in the open set Ω. There exists h_1 sufficiently small so that $K + B(h) \subset \Omega$ for all $h < h_1$, where $B(h)$ is the ball of radius h. Let ρ_h be a positive definite infinitely differentiable function defined by $\rho_h(\omega) = \frac{1}{h^n} \rho\left(\frac{\omega}{h}\right)$, with support in $B(h)$, where ρ is defined by (1.2), for example. The support of $\sigma_h * x_0$ is contained in $K + B(h) \subset \Omega$ if $h \leq h_1$ and $|x_0 - \rho_h * x_0| \leq \omega(x_0, h) \leq \varepsilon/2$ when $h \leq h_2$, according to Theorem 1.

CH. 6, SEC. 7   CONVOLUTION POWER FOR CHARACTERISTIC FUNCTIONS   135

Thus

$$|x - \rho_h * x_0| \leq \frac{2\varepsilon}{2} = \varepsilon \quad \text{and} \quad \rho_h * x_0 \in \mathscr{D}(\Omega) \qquad \blacksquare$$

according to Proposition 5.1.

*7. EXAMPLE. CONVOLUTION POWER FOR CHARACTERISTIC FUNCTIONS

Let χ be the characteristic function of the interval $[0, 1]$. Then if $x \in L^2(\mathbb{R})$, the function $\chi * x$ is defined by

(1) $$(\chi * x)(\omega) = \int_0^1 x(\omega - \zeta) \, d\zeta = \int_{\omega - 1}^{\omega} x(\zeta) \, d\zeta.$$

Thus we can approximate a function $x \in L^2(\mathbb{R})$ by the functions $\chi_h * x$ defined by

(2) $$(\chi_h * x)(\omega) = \frac{1}{h} \int_0^h x(\omega - \zeta) \, d\zeta = \frac{1}{h} \int_{\omega - h}^{\omega} x(\zeta) \, d\zeta.$$

In particular the following problem occurs: to calculate the powers of convolution by characteristic functions and to study their properties.

These functions play a fundamental role in approximation theory. (See Chapter 8, Sections 5 and 6.) It became clear in effect that the approximation of functions by polynomials led to instable convergence, whereas piecewise approximation provided quasi-optimal approximation procedures.

Consider the functions $\chi, \chi^{*2} = \chi * \chi, \chi^{*3} = \chi * \chi * \chi$. A simple explicit calculation shows that the graphs of these functions are of the following forms:

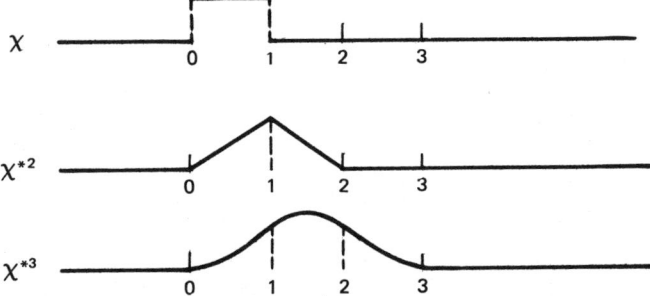

We denote by χ^{*m} the mth convolution power of the characteristic function χ. We then obtain the following result.

PROPOSITION 1
The restriction of $\chi^{*(m+1)}$ to each interval $[k, k+1]$ is a polynomial α_m^k of degree m:

(4) $$\chi^{*(m+1)}(\omega) = \sum_{k=0}^{m} \alpha_m^k(\omega - k)\chi(\omega - k),$$

where

(5) $$\alpha_m^k(\omega) = \sum_{j=0}^{m} a_m(k,j) \frac{\omega^j}{j!}$$

with

(6) $$a_m(k,j) = \sum_{i=0}^{k} (-1)^i \binom{m+1}{i} \frac{(k-i)^{m-j}}{(m-j)!}. \quad \blacktriangle$$

Proof. The proof is accomplished by recursion on m. The formula is true for $m = 0$. Suppose it is true for $m - 1$, and let us establish it for m. Writing that $\chi^{*(m+1)} = \chi * \chi^{*(m)}$ and that $\chi^{*m} = \sum_{k=0}^{m-1} \alpha_{m-1}^k(\omega - k)\chi(\omega - k)$, we see that

(7) $$\chi * (\alpha_{m-1}^k \chi) = \begin{cases} \int_0^\omega \alpha_{m-1}^k(\zeta)\, d\zeta & \text{if} \quad 0 \leq \zeta \leq 1. \\ \int_{\omega-1}^1 \alpha_{m-1}^k(\zeta)\, d\zeta & \text{if} \quad 1 \leq \zeta \leq 2. \end{cases}$$

It then follows that

(8) $$\begin{cases} \text{i.} \quad \alpha_m^0(\omega) = \int_0^\omega \alpha_{m-1}^0(\zeta)\, d\zeta. \\ \text{ii.} \quad \alpha_m^k(\omega) = \int_0^\omega (\alpha_{m-1}^k(\zeta) - \alpha_{m-1}^{k-1}(\zeta))\, d\zeta + \int_0^1 \alpha_{m-1}^{k-1}(\zeta)\, d\zeta \quad \text{if} \quad 1 \leq k \leq m-1. \\ \text{iii.} \quad \alpha_m^m(\omega) = -\int_0^\omega \alpha_{m-1}^{m-1}(\zeta)\, d\zeta + \int_0^1 \alpha_{m-1}^{m-1}(\zeta)\, d\zeta. \end{cases}$$

Hence the coefficients $a_m(k,j)$ of these polynomials satisfy the following recursion relations:

(9) $$\begin{cases} \text{i.} \quad a_m(0,0) = 0, \quad a_m(k,0) = \sum_{j=0}^{m-1} \frac{a_{m-1}(k-1,j)}{(j+1)!} \quad \text{if} \quad 1 \leq k \leq m. \\ \text{ii.} \quad a_m(k,j) = a_{m-1}(k,j-1) - a_{m-1}(k-1,j-1) \\ \qquad \text{if} \quad 1 \leq k \leq m-1, 1 \leq j \leq m. \\ \text{ii.} \quad a_m(0,j) = a_{m-1}(0, j-1) \quad \text{if} \quad 1 \leq j \leq m. \\ \qquad a_m(m,j) = a_{m-1}(m-1, j-1) \quad \text{if} \quad 1 \leq j \leq m. \end{cases}$$

Using the Newton binomial formula $\binom{m+1}{j} = \binom{m}{j} - \binom{m}{j-1}$ and replacing the coefficients $a_{m-1}(k,j)$ by their values defined in the recursion hypothesis, we show that the formulas (6) define the coefficients $a_m(k,j)$. ■

PROPOSITION 2
The coefficients $a_m(k,j)$ satisfy

(10) $\qquad a_m(k,j) = (D^j \alpha_m^k)(0) = (D^j \chi^{*(m+1)})(k).$

The piecewise polynomials (also called Spline functions) $\chi^{*(m+1)}$ are functions that are $(m-1)$ times continuously differentiable. ▲

Proof. (Left as an exercise.)

The following proposition plays an important role in finding upper bounds for the error in the theory of approximation by piecewise polynomials.

PROPOSITION 3
The functions $\chi^{*(m+1)}$ satisfy the following properties: there exist scalars $b^0 = 1, b^1, \ldots, b^m$ such that for every j for which $0 \leq j \leq m$,

(11) $\qquad \sum_{k \in \mathbb{Z}} \frac{k^j}{j!} \chi^{*(m+1)}(\omega - k) = \sum_{p=0}^{j} b^{j-p} \frac{\omega^p}{p!}.$ ▲

Proof. We establish this result by recursion. It is clearly true for $m = 0$. Suppose that there exist $c^0 = 1, \ldots, c^{m-1}$ such that, if $0 \leq j \leq m-1$,

(12) $\qquad \sum_{k \in \mathbb{Z}} \frac{k^j}{j!} \chi^{*(m)}(\omega - k) = \sum_{p=0}^{j} c^{j-p} \frac{\omega^p}{p!}.$

If $0 \leq j \leq m-1$, it follows by taking the convolution product of the two sides of (12) by χ that

$$\sum_{k \in \mathbb{Z}} \frac{k^j}{j!} \chi^{*(m+1)}(\omega - k) = \sum_{p=0}^{j} c^{j-p} \left(\chi * \left(\frac{(\cdot)^p}{p!} \right) \right)$$

$$= \sum_{p=0}^{j} c^{j-p} \sum_{l=0}^{p} d^{p-l} \frac{x^l}{l!}$$

$$= \sum_{l=0}^{j} \frac{x^l}{l!} \left(\sum_{p=l}^{j} c^{j-p} d^{p-l} \right)$$

$$= \sum_{l=0}^{j} \frac{x^l}{l!} \sum_{p=0}^{j-l} c^{j-l-i} d^i$$

$$= \sum_{l=0}^{j} \frac{x^l}{l!} b^{j-l}$$

where we set

(13) $$d^k = \int \chi(\zeta) \frac{(-\zeta)^k}{k!} d\zeta = \frac{(-1)^k}{(k+1)!}$$

and

(14) $$b^k = \sum_{i=0}^{k} c^{k-i} d^i \quad \text{for} \quad 0 \leq k \leq m-1.$$

It remains to show (11) for $j = m$. For this it is sufficient to prove that these two functions take the same value $b^m = \sum_{k \in \mathbb{Z}} k^m/m! \, \chi^{*(m+1)}(-k)$ for $\omega = 0$ and have the same derivatives. These derivatives are, respectively, equal to

$$A = \sum_{k \in \mathbb{Z}} \frac{k^m}{m!} [\chi^{*(m)}(\omega - k) - \chi^{*(m)}(\omega - k - 1)]$$

and

$$B = \sum_{p=0}^{m-1} b^p \frac{\omega^{m-1-p}}{(m-1-p)!} = \sum_{p=0}^{m-1} b^{m-1-p} \frac{\omega^p}{p!}.$$

But we can write A in the following form:

$$A = \sum_{k \in \mathbb{Z}} \left(\frac{k^m}{m!} - \frac{(k-1)^m}{m!} \right) \chi^{*(m)}(\omega - k)$$

$$= \sum_{l=0}^{m-1} \frac{(-1)^{m-l-1}}{(m-l)!} \sum_{k \in \mathbb{Z}} \frac{k^l}{l!} \chi^{*(m)}(\omega - k)$$

$$= \sum_{l=0}^{m-1} d^{m-l-1} \sum_{p=0}^{l} c^{l-p} \frac{\omega^p}{p!}$$

[according to (13) and (12) if $0 \leq l \leq m$]

$$= \sum_{p=0}^{m-1} \frac{\omega^p}{p!} \sum_{l=p}^{m-1} d^{m-l-1} c^{l-p}$$

$$= \sum_{p=0}^{m-1} \frac{\omega^p}{p!} \sum_{i=0}^{m-1-p} d^{m-1-p-i} c^i = \sum_{p=0}^{m-1} \frac{\omega^p}{p!} b^{m-1-p}$$

[according to (14)]. Hence $A = B$, and the proof of the theorem is completed. ∎

*8. EXAMPLE. CONVOLUTION PRODUCT FOR POLYNOMIALS. APPELL POLYNOMIALS

The convolution product of a function $\varphi \in \mathscr{C}_0(\mathbb{R})$ and a polynomial of degree

n is defined. In effect, setting

(1) $$x_n(\omega) = \frac{\omega^n}{n!}$$

the formula for the convolution product gives

$$(\varphi * x_n)(\omega) = \int \varphi(\zeta) x_n(\omega - \zeta)\, d\zeta$$
$$= \sum_{j=0}^{n} \left(\int \varphi(\zeta) \frac{(-\zeta)^{n-j}}{(n-j)!}\, d\zeta \right) \frac{\omega^j}{j!}$$
$$= \sum_{j=0}^{n} \varphi_{n-j} \frac{\omega^j}{j!} \quad \text{where} \quad \varphi_k = (-1)^k \int \varphi(\zeta) \frac{\zeta^k}{k!}\, d\zeta$$

since the Newton binomial formula implies that

$$\frac{(\omega - \zeta)^n}{n!} = \sum_{j=0}^{n} \frac{\omega^j}{j!} \frac{(-\zeta)^{n-j}}{(n-j)!}.$$

In particular, if we take $\varphi = \chi$, the characteristic function of the interval $]0, 1[$, we obtain

(2) $$\chi_k = \frac{(-1)^k}{(k+1)!}; \quad (\chi * x_n)(\omega) = \sum_{j=0}^{n} \frac{(-1)^{n-j}}{(n-j+1)!} \frac{\omega^j}{j!}.$$

*Appell Polynomials

It is useful in certain problems to solve the convolution equations

$$(\varphi * u_n)(\omega) = \frac{\omega^n}{n!}$$

where the u_n's are polynomials (called Appell polynomials of φ).

PROPOSITION 1
Suppose that the function φ satisfies

(3) $$\forall k \in \mathbb{N}, \quad \varphi_k = (-1)^k \int \varphi(\omega) \frac{\omega^k}{k!} < +\infty.$$

Consider the sequence of scalars $\psi_k (k = 0, 1, \ldots)$ defined recursively in the following fashion:

(4) $$\begin{cases} \text{i.} & \varphi_0 \psi_0 = 1. \\ \text{ii.} & \sum_{j=0}^{k} \psi_{k-j} \varphi_j = 0 \quad \text{if} \quad k = 1, \ldots,. \end{cases}$$

The polynomials u_n defined by

$$(5) \qquad u_n(\omega) = \sum_{j=0}^{n} \psi_{n-j} \frac{\omega^j}{j!}$$

are the **Appell polynomials** *of the function* φ. ▲

Proof. Indeed,

$$\begin{aligned}
(\varphi * u_n)(\omega) &= \sum_{j=0}^{n} \psi_{n-j} (\varphi * x_j)(\omega) \\
&= \sum_{j=0}^{n} \psi_{n-j} \sum_{k=j}^{n} \varphi_{j-k} \frac{\omega^k}{k!} \\
&= \sum_{k=0}^{n} \frac{\omega^k}{k!} \sum_{j=k}^{n} \psi_{n-j} \varphi_{j-k} \\
&= \sum_{k=0}^{n} \frac{\omega^k}{k!} \sum_{j=0}^{n-k} \psi_{n-k-j} \varphi_j = \frac{\omega^n}{n!}
\end{aligned}$$

according to (4). ■

PROPOSITION 2
The Appell polynomials satisfy the following properties:

$$(6) \quad \begin{cases} \text{i.} & Du_n = u_{n-1} \quad \left(D = \dfrac{d}{d\omega}\right) \\ \text{ii.} & u_n(\omega + \zeta) = \sum_{k=0}^{n} u_{n-k}(\zeta) \dfrac{\omega^k}{k!} \\ \text{iii.} & u_n(0) = \psi_n. \end{cases}$$

▲

Proof. These properties follow immediately from formula (5). ■

Example. Bernouilli Polynomials

Take as φ the characteristic function χ of the interval $[0,1]$. Consider the sequence of scalars β_k defined recursively by

$$\beta_0 = 1; \qquad \sum_{j=0}^{k} \beta_{k-j} \frac{(-1)^j}{(j+1)!} = 0.$$

The numbers β_k are then rational numbers called *Bernouilli numbers*. The Appell polynomials associated to χ are called *Bernouilli polynomials*; they are defined by

(7) $$b_n(\omega) = \sum_{j=0}^{n} \beta_{n-j} \frac{\omega^j}{j!}.$$

Since $D(\chi * b_n)(\omega) = b_n(\omega) - b_n(\omega - 1)$ (by an easy calculation), it follows that these polynomials satisfy

(8) $$b_n(\omega) - b_n(\omega - 1) = \frac{\omega^{n-1}}{(n-1)!}. \qquad \blacksquare$$

We deduce from this a formula that gives a development analogous to Taylor's, using Appell polynomials instead of the polynomials $\omega^n/n!$.

THEOREM 1
Let f be a function that is m-times continuously differentiable and such that $\varphi * D^k f$ is defined for all $k \leq m$. Then in a neighborhood of zero,

(9) $$f(\omega) = \sum_{k=0}^{m-1} u_k(\omega)(\int \varphi(-\zeta) D^k f(\zeta) \, d\zeta) + R_m(\omega)$$

where the polynomials u_k are Appell polynomials of φ and where

(10) $$R_m(\omega) = \int \varphi(-\zeta) \theta(\zeta) \, d\zeta$$

with θ defined by

(11) $$\theta(\zeta) = \int_0^{\zeta - \omega} u_{m-1}(\omega + \eta) D^m f(\zeta - \eta) \, d\eta. \qquad \blacktriangle$$

Proof. Indeed, we apply the formula for integration by parts $(m-1)$ times to the integral (11) defining $\theta(\zeta)$. Since $D^k u_m = u_{m-k}$, we obtain

(12) $$\theta(\zeta) = \sum_{k=0}^{m-1} u_k(\omega) D^k f(\zeta) - \sum_{k=0}^{m-1} u_k(\zeta) D^k f(\omega).$$

Fixing ω and considering the two sides of this equality as functions of ζ, we obtain by taking the convolution product by φ,

$$(\varphi * \theta)(\xi) = \sum_{k=0}^{m-1} u_k(\omega)(\varphi * D^k f)(\xi) - \sum_{k=0}^{m-1} \frac{\xi^k}{k!} (D^k f)(\omega)$$

since $(\varphi * u_k)(\xi) = \xi^k/k!$.

Taking $\xi = 0$ in this formula, we obtain the desired development since $(\varphi * \psi)(0) = \int \varphi(-\zeta) \psi(\zeta) \, d\zeta$. $\qquad \blacksquare$

We are going to apply this formula to the function $\omega \to e^{\lambda \omega}$. We introduce the function

(13) $$u(\lambda) = \int \varphi(-\zeta) e^{\lambda \zeta} d\zeta$$

which we shall call the *indicator of φ*.

Since $\int \varphi(-\zeta) D^k(e^{\lambda \zeta}) d\zeta = \lambda^k u(\lambda)$, it follows from Theorem 1 that

(14) $$e^{\lambda \omega} = \sum_{k=0}^{m-1} u_k(\omega) \lambda^k u(\lambda) + R_m(\omega)$$

where

(15) $$R_m(\omega) = \lambda^m \int \varphi(-\zeta) d\zeta \int_0^{\zeta-\omega} u_{m-1}(\omega + \eta) e^{\lambda(\zeta-\eta)} d\eta.$$

But since the Taylor series development of $e^{\lambda \omega}$ is absolutely convergent for every λ and every ω, we can in fact write that

(16) $$e^{\lambda \omega} = u(\lambda) \sum_{n=0}^{\infty} \lambda^n u_n(\omega).$$

Example. Hermite Polynomials

We show that the Appell polynomials associated with the function $g = g_1$ (defined by $g(\omega) = (1/\sqrt{2\pi}) \exp(\omega^2/2)$) are the Hermite polynomials H_m (see Chapter 8, Section 2)

(17) $$g * H_m = \frac{\omega^m}{m!}.$$

To verify this we calculate the associated *indicator* u of g defined by

$$u(\lambda) = \frac{1}{\sqrt{2\pi}} \int_{-\infty}^{+\infty} e^{-\zeta^2/2} e^{\lambda \zeta} d\zeta$$

$$= \frac{1}{\sqrt{2\pi}} e^{\lambda^2/2} \int_{-\infty}^{+\infty} e^{-(\zeta-\lambda)^2/2} d\zeta$$

$$= e^{\lambda^2/2}$$

since

$$\int_{-\infty}^{+\infty} e^{-(\zeta-\lambda)^2/2} d\zeta = \int_{-\infty}^{+\infty} e^{-u^2/2} du = \sqrt{2\pi}.$$

Consequently, formula (16) implies that

$$\sum_{n=0}^{\infty} \lambda^n u_n(\omega) = \frac{e^{\lambda \omega}}{u(\lambda)} = e^{-\lambda^2/2} e^{\lambda \omega} = e^{\omega^2/2} \exp\left(-\frac{(\omega-\lambda)^2}{2}\right)$$

Since
$$\exp\left(-\frac{(\omega-\lambda)^2}{2}\right) = \sum_{n=0}^{\infty} \frac{(-\lambda)^n}{n!} D^n(e^{-\omega^2/2})$$
we obtain

(18) $$\sum_{n=0}^{\infty} \lambda^n u_n(\omega) = e^{\omega^2/2} \sum_{n=0}^{\infty} \frac{(-1)^n}{n!} D^n(e^{-\omega^2/2}) \lambda^n.$$

Since these two power series are absolutely convergent, it follows that
$$u_n(\omega) = \frac{1}{n!}(-1)^n e^{\omega^2/2} D^n(e^{-\omega^2/2}) = \frac{1}{n!} H_n(\omega).$$

CHAPTER 7

Sobolev Spaces of Functions of One Variable

In this chapter we present an introduction to the theory of distributions within the framework of Sobolev spaces. To simplify the exposition, we restrict ourselves in this introductory chapter to the case of functions of one variable, postponing to Chapter 9 the study of Sobolev spaces of functions of several variables. This simplification allows us to make a reasonably complete study of Sobolev spaces. The trace theorems (Section 8) are particularly simple in this case. Hence the student can begin at this point the study of boundary-value problems for differential equations (Section 3 of Chapter 13 in the abstract case and Section 4 of Chapter 13 in the concrete case). We also indicate that the extension of the results of the first five sections to functions of several variables poses no difficulty and can be taken up at the end of this chapter. Moreover, as we have already stated, these five sections should be studied at the same time as Section 4 of Chapter 5, which as we pointed out, presents the construction of these spaces abstractly.

The role played by Sobolev spaces is important from two points of view: First, these spaces are the *Hilbert* space analogues of the spaces of the m-times continuously differentiable functions (which are *nonreflexive* Banach spaces, hence "bad"), since we can roughly say that these are the spaces of functions that are m-times "L^2-differentiable." Second, we show that the elements of the dual of a Sobolev space (which we can identify with an overspace of L^2) are *generalized functions*, called distributions, that have properties analogous to those of functions. We shall show how to "differentiate" these distributions (Section 3) and how to form their convolution product with functions of L^1 (Section 9, which can be studied immediately after Section 5). We also see that the duality operator from a Sobolev space onto its dual is a differential operator.

The situation is very good in the case of functions defined on all of $\mathbb{R}$: We give $\mathscr{D}(\mathbb{R})$ the scalar product $((\varphi, \psi))_m = \sum_{k=0}^{m} \int D^k \varphi \, D^k \psi \, d\omega$, and we define the Sobolev space $H^m(\mathbb{R})$ as the completion of $\mathscr{D}(\mathbb{R})$ for this scalar product. This allows us to extend the differential operators $D^k = d^k/d\omega^k$ to operators from $H^m(\mathbb{R})$ to $L^2(\mathbb{R})$. A *distribution* is an element of the dual $H^{-m}(\mathbb{R}) = H^m(\mathbb{R})^*$ of $H^m(\mathbb{R})$. Since $f \in H^{-m}(\mathbb{R})$ is a linear form $\varphi \mapsto (f, \varphi)$ (continuous for the scalar product $((\varphi, \psi))_m$), it is a *generalized function* inasmuch as we have agreed to consider every function f as a linear form $\varphi \mapsto \int_\Omega f(\omega)\varphi(\omega)\,d\omega = (f, \varphi)$. Transposing the operator $D^k \in \mathscr{L}(H^m(\mathbb{R}), L^2(\mathbb{R}))$, we verify that its transpose is the unique extension of the operator $(-1)^k D^k$. We can, therefore, identify it with a continuous operator from $L^2(\mathbb{R})$ to $H^{-m}(\mathbb{R})$ and define the derivative of order k (in the sense of distributions) of a function f of $L^2(\mathbb{R})$ as the distribution $D^k f$ defined by

$$\forall \varphi \in \mathscr{D}(\mathbb{R}), \quad (D^k f, \varphi) = (f, (-1)^k D^k \varphi).$$

Theorem 5.1 allows us to characterize the space $H^m(\mathbb{R})$ as the subspace of functions f of $L^2(\mathbb{R})$ all of whose derivatives (in the sense of distributions) up to order m belong to $L^2(\mathbb{R})$.

This theorem is *no longer true* if we replace $\mathbb{R}$ by a bounded interval $\Omega = \,]a, b[$. Then we denote by $H_0^m(\Omega)$ the completion of $\mathscr{D}(\Omega)$ for the scalar product $((\varphi, \psi))_m$. By analogous methods we consider the dual $H^{-m}(\Omega) = H_0^m(\Omega)^*$ of $H_0^m(\Omega)$ as a space of generalized functions and show that we can define the derivative $D^k f \in H^{-m}(\Omega)$ of a function $f \in L^2(\Omega)$ in the sense of distributions. But then *the Sobolev space $H^m(\Omega)$ of the functions $f \in L^2(\Omega)$ whose derivatives $D^k f$ belong to $L^2(\Omega)$ for $0 \leq k \leq m$ strictly contains $H_0^m(\Omega)$.*

The trace theorem (Section 8) shows that $H_0^m(\Omega)$, which is the closure of $\mathscr{D}(\Omega)$ in $H^m(\Omega)$, is the *subspace of functions f of $H^m(\Omega)$ whose derivatives $D^j f$ vanish on the boundary of $\Omega (0 \leq j \leq m - 1)$.* This theorem is the source of the results concerning boundary-value problems.

1. THE SPACE $H_0^m(\Omega)$ AND ITS DUAL $H^{-m}(\Omega)$

Consider an open interval $\Omega = \,]a, b[$ (either bounded or not) of $\mathbb{R}$. We introduce the space

(1) $\begin{cases} \mathscr{D} = \mathscr{D}(\Omega) \text{ of infinitely differentiable functions with compact} \\ \text{support in } \Omega. \end{cases}$

as well as

(2) $\begin{cases} \text{the pivot space } H = L^2(\Omega), \text{ the completion of } \mathscr{D}(\Omega) \text{ for the} \\ \text{scalar product } (\varphi, \psi) = \int_\Omega \varphi(\omega) \psi(\omega)\,d\omega \end{cases}$

(since $\mathscr{D}(\Omega)$ is dense in $L^2(\Omega)$ according to Theorem 6.6.2). The differential operators

(3) $$D^p : x \to D^p x = \frac{d^p}{d\omega^p} x$$

are linear operators from $\mathscr{D}(\Omega)$ to $\mathscr{D}(\Omega)$ satisfying

(4) $$\int_\Omega D^p\varphi(\omega)\psi(\omega)\,d\omega = (-1)^p \int_\Omega \varphi(\omega) D^p\psi(\omega)\,d\omega$$

according to the formula for integration by parts applied p times. We set

(5) $$D^0\varphi = \varphi.$$

PROPOSITION 1
The family $\mathscr{A}_m$ of operators D^p ($0 \leq p \leq m$) is closed. ▲

Proof. Let us recall (Definition 5.4.1) that this means that if a sequence of elements $\varphi_n \in \mathscr{D}(\Omega)$ converges to zero in quadratic mean and if $D^p\varphi_n$ converges to $f^p \in L^2(\Omega)$ in quadratic mean for all p such that $1 \leq p \leq m$, then the f^p's are identically zero.

Indeed, if $\psi \in \mathscr{D}(\Omega)$ is fixed, the equalities

$$\int_\Omega D^p\varphi_n(\omega)\psi(\omega)\,d\omega = (-1)^p \int \varphi_n(\omega) D^p\psi(\omega)\,d\omega$$

become, by taking limits,

$$\int f^p(\omega)\psi(\omega)\,d\omega = 0 \quad \text{for} \quad p = 1,\ldots,m.$$

As this is true for all $\psi \in \mathscr{D}(\Omega)$ and as $\mathscr{D}(\Omega)$ is dense in $L^2(\Omega)$, it follows that the continuous linear forms f^p on $L^2(\Omega)$ are identically zero. ■

Then we associate to this family $\mathscr{A}_m$ the scalar product defined on $\mathscr{D}(\Omega)$ by

(6) $$((\varphi,\psi))_m = \sum_{p=0}^m (D^p\varphi, D^p\psi) = \sum_{p=0}^m \int_\Omega D^p\varphi(\omega) D^p\psi(\omega)\,d\omega$$

We know that there exists a *realization of the completion of* $\mathscr{D}(\Omega)$ *for* $((.,.))_m$, which is contained in $L^2(\Omega)$. (See Proposition 5.4.1.)

DEFINITION 1
We denote by $H_0^m(\Omega)$ the completion of $\mathscr{D}(\Omega)$ for the scalar product $((.,.))_m$, which is a dense subspace of $L^2(\Omega)$. We denote by $H^{-m}(\Omega)$ its dual. We shall say that $H_0^m(\Omega)$ is the (minimal) Sobolev space of order m and that $H^{-m}(\Omega)$ is the Sobolev space of order $-m$. ▲

2. DEFINITION OF DISTRIBUTIONS

Propositions 5.3.1 and 5.4.1 imply that if $m \geq k$

(1) $$\mathscr{D}(\Omega) \subset H_0^m(\Omega) \subset H_0^k(\Omega) \subset L^2(\Omega) \subset H^{-k}(\Omega) \subset H^{-m}(\Omega) \subset \mathscr{D}^*(\Omega),$$

where $\mathscr{D}^*(\Omega)$ denotes the *algebraic* dual space of $\mathscr{D}(\Omega)$. Moreover, each space is dense in the following spaces (except $\mathscr{D}^*(\Omega)$, which does not have a topology).

The elements of the spaces $H^{-m}(\Omega)$ are not necessarily functions (nor even functions identified with the classes of measurable functions).

DEFINITION 1
The continuous linear forms on the spaces $H_0^m(\Omega)$ are called distributions *(or also, generalized functions) on Ω.* ▲

Remark 1

We can provide the space $\mathscr{D}(\Omega)$ with a topology (unfortunately nonmetrizable) that is stronger than that of the spaces $H_0^m(\Omega)$. Then the injections j from $\mathscr{D}(\Omega)$ to $H_0^m(\Omega)$ are embeddings. We can show that their transposes j^* are also embeddings from $H^{-m}(\Omega)$ into the topological dual space of $\mathscr{D}(\Omega)$. They allow us *to identify the spaces $H^{-m}(\Omega)$ with the (dense) subspaces* of the topological dual space of $\mathscr{D}(\Omega)$, which is called the space of *all distributions*. In fact, in this book we do not need this topological dual space of $\mathscr{D}(\Omega)$. For our purposes it is sufficient to embed all the spaces $H^{-m}(\Omega)$ in the algebraic dual space of $\mathscr{D}(\Omega)$. ■

The inclusions (1) show that every *function* x of $\mathscr{D}(\Omega)$ (or of $H_0^m(\Omega)$) is a *distribution*, which is why distributions are still called generalized functions.

Moreover, since $L^2(\Omega)$ is the pivot space, the duality pairing between $H^{-m}(\Omega)$ and $H_0^m(\Omega)$ is the *unique extension* of the scalar product (x, y) of $L^2(\Omega)$ restricted to $H_0^m(\Omega) \times L^2(\Omega)$. (See Proposition 3.5.4.)

With an abuse of the notation, we sometimes describe the distributions $x \in H^{-m}(\Omega)$ as functions; in particular, if $x \in H^{-m}(\Omega)$ and $\varphi \in H_0^{-m}(\Omega)$, we set

(2) $$(x, \varphi) = \int_\Omega x(\omega)\varphi(\omega)\, d\omega,$$

although we have absolutely no right to write $x(\omega)$ if x is not a function!

Remark 2

This leads us to make the following remark: We have now at our disposal two conceptions of a function $\varphi \in \mathscr{D}(\Omega)$. The first, the classical, is to consider φ as a function $\omega \in \Omega \to \varphi(\omega) \in \mathbb{R}$. The second is to consider (and to use) φ as a distribution of a space $H^{-m}(\Omega)$, that is, as a continuous linear form $\psi \to (\varphi, \psi) = \int_\Omega \varphi(\omega)\psi(\omega)\, d\omega$ on a space $H_0^m(\Omega)$. ■

It is the second point of view (allowing us to extend many usual properties of functions to new mathematical objects, namely, distributions) that has enabled analysis to make considerable progress. This was the work of Laurent Schwartz.

We now see that duality (and, consequently, transposition) is the essential mathematical tool that has made this new conception possible.

We establish this immediately by showing that we can "differentiate" functions that are not differentiable in the usual sense or even "differentiate" distributions by *extending by density* the derivation operators D^p and using the inclusions (1).

3. DIFFERENTIATION OF DISTRIBUTIONS

Since $H_0^m(\Omega)$ is the completion of $\mathscr{D}(\Omega)$ for the initial scalar product, it follows that the operators D^p from $\mathscr{D}(\Omega)$ to $L^2(\Omega)$ *have unique extensions to continuous linear operators* from $H_0^m(\Omega)$ to $L^2(\Omega)$. We can, therefore, write $D^p x$ when x is a function of $H_0^m(\Omega)$, although such a function is not p-times differentiable in the usual sense. We say that $D^p x$ is the pth *weak derivative* (or the pth derivative *in the sense of distributions*) of x.

Hence we have defined $D^p \in \mathscr{L}(H_0^m(\Omega), L^2(\Omega))$. We now extend D^p to $L^2(\Omega)$. Indeed, formulas (1.4) show that the transpose $(-1)^p (D^p)^* \in \mathscr{L}(L^2(\Omega), H^{-m}(\Omega))$ of the operator $(-1)^p (D^p)^* \in \mathscr{L}(H_0^m(\Omega), L^2(\Omega))$ is the *unique extension* of the operator D^p. We can, therefore, identify $(-1)^p (D^p)^*$ to D^p, which allows us to *differentiate in the sense of distributions* the functions of $L^2(\Omega)$,

DEFINITION 1
The pth derivative of a function x of $L^2(\Omega)$ is the distribution $D^p x \in H^{-m}(\Omega)$, defined by

(1) $\qquad \forall \varphi \in \mathscr{D}(\Omega), \quad (D^p x, \varphi) = (-1)^p (x, D^p \varphi).$ ▲

Remark 1

We are going to characterize in another way the derivative Dy in the sense of distributions of a function $y \in H_0^1(\mathbb{R})$. ∎

PROPOSITION 1
Let $y \in H_0^1(\mathbb{R})$. Then $Dy \in L^2(\mathbb{R})$ is the limit in quadratic mean of the difference quotients $\nabla_h y$ defined by

(2) $\qquad \nabla_h y(\omega) = \dfrac{y(\omega) - y(\omega - h)}{h}$

as h approaches zero. ▲

Proof. Let $y \in H_0^1(\mathbb{R})$. Then the convolution product $\chi_h * y$ of y and the functions χ_h defined by

(3) $$\chi_h(\omega) = \frac{1}{h}\chi\left(\frac{\omega}{h}\right) = \begin{cases} \frac{1}{h} & \text{if } \omega \in [0, h]. \\ 0 & \text{if } \omega \notin [0, h]. \end{cases}$$

still belongs to $H_0^1(\mathbb{R})$. We then verify the formula

(4) $$D(\chi_h * y) = \chi_h * Dy = \nabla_h y.$$

Indeed, we calculate $D(\chi_h * y)$ in the sense of distributions using formula (1). For every $\varphi \in \mathcal{D}(\mathbb{R})$, we have

$$(D(\chi_h * y), \varphi) = -(\chi_h * y, D\varphi) = -\int_{-\infty}^{+\infty} D\varphi(\eta)\, d\eta \int_{-\infty}^{+\infty} \chi_h(\eta - \omega) y(\omega)\, d\omega$$

$$= -\int_{-\infty}^{+\infty} y(\omega)\, d\omega \int_{-\infty}^{+\infty} D\varphi(\omega + \eta) \chi_h(\eta)\, d\eta.$$

We have, therefore, obtained the following more symmetric formula:

(5) $$((\chi_h * y), -D\varphi) = \int_{-\infty}^{+\infty}\int_{-\infty}^{+\infty} \chi_h(\eta) y(\omega)(-D\varphi)(\omega + \eta)\, d\omega\, d\eta.$$

Now using the explicit form of χ_h, we obtain

$$((\chi_h * y), -D\varphi) = -\frac{1}{h}\int_{-\infty}^{+\infty} y(\omega)\, d\omega \int_0^h D\varphi(\omega + \eta)\, d\eta$$

$$= \frac{1}{h}\int_{-\infty}^{+\infty} y(\omega)(\varphi(\omega) - \varphi(\omega + h))\, d\omega$$

$$= \frac{1}{h}\int_{-\infty}^{+\infty} (y(\omega) - y(\omega - h))\varphi(\omega)\, d\omega = (\nabla_h y, \varphi),$$

which shows that

(6) $$D(\chi_h * y) = \nabla_h y.$$

Moreover, we can also write that

$$(D(\chi_h * y), \varphi) = -\int_{-\infty}^{+\infty}\int_{-\infty}^{+\infty} \chi_h(\eta) y(\omega) D\varphi(\omega + \eta)\, d\eta\, d\omega$$

$$= -\int_{-\infty}^{+\infty} \chi_h(\eta)\, d\eta \int_{-\infty}^{+\infty} y(\omega) D\varphi(\omega + \eta)\, d\omega$$

$$= \int_{-\infty}^{+\infty} \chi_h(\eta)\, d\eta \int_{-\infty}^{+\infty} (Dy)(\omega)\varphi(\omega + \eta)\, d\omega$$

$$= \int_{-\infty}^{+\infty} \varphi(\eta)\, d\eta \int_{-\infty}^{+\infty} (Dy)(\omega)\chi_h(\eta - \omega)\, d\omega$$

$$= ((\chi_h * Dy), \varphi)$$

Consequently, the approximation Theorem 6.6.1 shows that the function $\nabla_h y = \chi_h * Dy$ converges to Dy in $L^2(\mathbb{R})$. ∎

COROLLARY 1

Let $y \in H_0^m(\mathbb{R})$. Then for every $p = 1, \ldots, m$, $D^p y \in L^2(\mathbb{R})$ is the limit in quadratic mean of the difference quotients $\nabla_h^p y$ of order p defined by

(7) $$\nabla_h^p y(\omega) = h^{-p} \sum_{j=0}^{p} (-1)^j \binom{p}{j} y(\omega - jh).$$ ▲

Proof. We deduce from formula (2) applied p times that

$$D^p(\chi_h^{*p} * y) = \chi_h^{*p} * D^p y = \underbrace{(\chi_h * D) * \ldots (\chi_h * D)}_{p \text{ times}} y = \nabla_h^p y.$$

Since $\chi_h^{*p}(\omega) = \frac{1}{h} \chi^{*p}\left(\frac{\omega}{h}\right)$ and since $\int \chi^{*p}(\omega) d\omega = 1$, Theorem 6.6.1 implies that $\nabla_h^p y = \chi_h^{*p} * D^p y$ converges to $D^p y$ in $L^2(\mathbb{R})$. ∎

Proposition 1 and its corollary justify the notion that the functions of $H_0^m(\mathbb{R})$ are "differentiable": Rather than requiring that $D^p y$ be the limit of $\nabla_h^p y$ *for pointwise convergence* (as in the case of derivatives in the usual sense), we require only that $D^p y$ be the limit of $\nabla_h^p y$ *for quadratic convergence*, which is almost a weaker convergence (i.e., for a weaker metric) by Lebesgue's theorem.

Remark 2

We shall see in studying more especially the properties of the convolution of functions of $H_0^m(\mathbb{R})$ and of distributions of $H^{-m}(\mathbb{R})$ that the derivatives $D^p y \in H^{-m}(\Omega)$ of functions $y \in L^2(\Omega)$ are the limits of difference quotients $\nabla_h^p y$ of order p *in the space* $H^{-m}(\mathbb{R})$ ($1 \leq p \leq m$), that is, for a topology weaker still. (The larger m is, the weaker the topology.) (See Proposition 9.3.) ∎

PROPOSITION 2

The derivation operator D^k has a unique extension to a continuous linear operator from $H^{-m}(\Omega)$ to $H^{-m-k}(\Omega)$ for all $m \geq 0$. ▲

Proof. Indeed, $(-1)^k D^k$ is clearly a continuous operator from $\mathscr{D}(\Omega)$ to $\mathscr{D}(\Omega)$ when these spaces have the scalar products $((\varphi, \psi))_{m+k}$ and $((\varphi, \psi))_m$, respectively. It has an extension to a continuous linear operator $(-1)^k D^k$ from $H_0^{m+k}(\Omega)$ to $H_0^m(\Omega)$. Its transpose $((-1)^k D^k)^* \in \mathscr{L}(H^{-m}(\Omega), H^{-m-k}(\Omega))$ extends the operator D^k according to (1). ∎

This proposition allows us to "differentiate" distributions $f \in H^{-m}(\Omega)$: $D^k f$ is a distribution of $H^{-m-k}(\Omega)$.

We are now going to characterize the dual $H^{-m}(\Omega)$ of $H_0^m(\Omega)$.

PROPOSITION 3
Every distribution $x \in H^{-m}(\Omega)$ can be written (in several ways) as a sum $x = \sum_{k=0}^{m} D^k f_k$ of derivatives of functions f_k of $L^2(\Omega)$. The duality operator from $H_0^m(\Omega)$ onto $H^{-m}(\Omega)$ is the differential operator J_m of order $2m$ defined by

$$(8) \qquad J_m x = \sum_{k=0}^{m} (-1)^p D^{2p} x. \qquad \blacktriangle$$

Proof. We denote by G_0^m the operator from $H_0^m(\Omega)$ to $L^2(\Omega)^{m+1}$ defined by

$$(9) \qquad G_0^m x = \{x, Dx, D^2 x, \ldots, D^p x, \ldots, D^m x\}.$$

It is clear that G_0^m is an isometry from $H_0^m(\Omega)$ to $L^2(\Omega)^{m+1}$. Since $H_0^m(\Omega)$ is complete (by construction), its image is closed. Theorem 4.4.2 implies that its transpose G_0^{m*} is a *surjective* operator from $L^2(\Omega)^{m+1}$ onto $H^{-m}(\Omega)$. It is defined by

$$(G_0^{m*} e, \varphi) = [e, G_0^m \varphi] = \sum_{k=0}^{m} \int_{\Omega} e_k(\omega) D^k \varphi(\omega) \, d\omega$$

for all $\varphi \in H_0^m(\Omega)$ where $e = \{e_0, e_1, \ldots, e_m\} \in L^2(\Omega)$ and where

$$[e^1, e^2] = \sum_{k=0}^{m} \int_{\Omega} e_k^1(\omega) e_k^2(\omega) \, d\omega.$$

Since $\varphi \in H_0^m(\Omega)$ and since we have set $(D^k)^* = (-1)^k D^k$, it follows that

$$(G_0^{m*} e, \varphi) = \sum_{k=0}^{m} (-1)^k \int_{\Omega} D^k e_k(\omega) \varphi(\omega) \, d\omega$$

which leads us to write that

$$(10) \qquad G_0^{m*} e = \sum_{k=0}^{m} (-1)^k D^k e_k.$$

Since G_0^{m*} is surjective, we have shown that every distribution x of $H^{-m}(\Omega)$ can be written as the sum of derivatives of order inferior to m of functions of $L^2(\Omega)$. The representation $x = G_0^{m*} e = \sum_{k=0}^{m} D^k((-1)^k e_k)$ is not unique (since G_0^{m*} is not injective!).

Moreover, if x and $y \in H_0^m(\Omega)$, it follows that

$$(11) \qquad ((x, y))_m = [G_0^m x, G_0^m y] = (G_0^{m*} G_0^m x, y).$$

This shows that the duality operator from $H_0^m(\Omega)$ onto $H^{-m}(\Omega)$ is the operator $J_m = G_0^{m*} G_0^m$ defined by

$$J_m x = \sum_{p=0}^{m} (-1)^p D^{2p} x. \tag{12}$$ ∎

4. RELATIONS BETWEEN $H_0^m(\Omega)$ AND $H_0^m(\mathbb{R})$

Consider the operator π_0 from $\mathscr{D}(\Omega)$ to $\mathscr{D}(\mathbb{R})$ defined by

$$(\pi_0 x)(\omega) = \begin{cases} x(\omega) & \text{if } \omega \in \Omega \\ 0 & \text{if } \omega \notin \Omega \end{cases} \quad \text{if} \quad x \in \mathscr{D}(\Omega). \tag{1}$$

[$\pi_0 x$ does indeed belong to $\mathscr{D}(\mathbb{R})$, since the compact support of x is contained in Ω.]

The operator π_0 is clearly linear and is an isometry:

$$\|\pi_0 x\|_{H_0^m(\mathbb{R})} = \|x\|_{H_0^m(\Omega)}. \tag{2}$$

Hence the isometry π_0 has a unique extension to an *isometry* from $H_0^m(\Omega)$ to $H_0^m(\mathbb{R})$. The transpose π_0^* is, therefore, a *surjective* operator. (See Theorem 4.4.1.)

The restriction of $\pi_0^* \in \mathscr{L}(H^{-m}(\mathbb{R}), H^{-m}(\Omega))$ to $\mathscr{D}(\mathbb{R})$ is defined by

$$\int_\Omega \pi_0^* x(\omega) \varphi(\omega)\, d\omega = \int_\mathbb{R} x(\omega) \pi_0 \varphi(\omega)\, d\omega = 0 + \int_\Omega x(\omega) \varphi(\omega)\, d\omega$$

for every function $\varphi \in H_0^m(\Omega)$.

This shows that $\pi_0^* x$ and x coincide [as continuous linear forms on $H_0^m(\Omega)$, that is, as elements of $H^{-m}(\Omega)$]. Hence $\pi_0^* x$ is the *restriction* of the function x to Ω. Since $\mathscr{D}(\mathbb{R})$ is dense in $H^{-m}(\mathbb{R})$, π_0^* is the *unique extension of the operator ρ of restriction to Ω*.

DEFINITION 1
The isometry π_0 from $H_0^m(\Omega)$ to $H_0^m(\mathbb{R})$ defined by (1) is called the operator of extension by zero. We say that its transpose $\rho = \pi_0^* \in \mathscr{L}(H^{-m}(\mathbb{R}), H^{-m}(\Omega))$ is the operator of restriction to Ω of the distributions of $H^{-m}(\mathbb{R})$. ▲

We sum up the preceding remarks in the statement of the following proposition.

PROPOSITION 1
The operator π_0 of extension by zero is an isometry from $H_0^m(\Omega)$ to $H_0^m(\mathbb{R})$. The space $H^{-m}(\Omega)$ is the space of the restrictions to Ω of the distributions of $H^{-m}(\mathbb{R})$. ▲

PROPOSITION 2 (THE POINCARÉ INEQUALITIES)
If $\Omega = \,]a,b[\,$ is bounded, then for every $x \in H_0^m(\Omega)$, we have

(3) $\qquad |x| \leq c|D^p x| \qquad \forall p \quad \text{such that} \quad 0 \leq p \leq m.$

where c is a constant ▲

Proof. Let $h > b - a$, where $\Omega = \,]a,b[\,$. Consider the function $\varphi = \chi_h^{*p} * \pi_0 x \in H_0^m(\mathbb{R})$. Then

$$D^p \varphi = \chi_h^{*p} * D^p \pi_0 x = \nabla_h^p \pi_0 x.$$

Moreover, if $\omega \in \Omega$,

$$h^p \nabla_h^p \pi_0 x(\omega) = \sum_{j=0}^{p} (-1)^j \binom{p}{j} \pi_0 x(\omega - jh) = \pi_0 x(\omega) = x(\omega)$$

since if $\omega \in \Omega$, $\omega - jh \notin \Omega$ if $1 \leq j \leq m$ and, consequently, $\pi_0 x(\omega - jh) = 0$.
Hence we have shown that

$$h^{-p} x = \chi_h^{*p} * D^p \pi_0 x$$

and, consequently, that

$$h^{-2p} \int_\Omega |x(\omega)|^2 \, d\omega = \int_\Omega |\chi_h^{*p} * D^p \pi_0 x(\omega)|^2 \, d\omega \leq \int_{-\infty}^{+\infty} |\chi_h^{*p} * D^p \pi_0 x(\omega)|^2 \, d\omega$$
$$\leq \|\chi_h^{*p}\|_{L^1}^2 \|D^p \pi_0 x\|_{L^2(\mathbb{R})}^2 = \|D^p \pi_0 x\|_{L^2(\mathbb{R})}^2 = \|D^p x\|_{L^2(\Omega)}^2. \quad \blacksquare$$

COROLLARY 1
If $\Omega = \,]a,b[\,$ is bounded, the norms $\|x\|_m$ and $|D^m x|$ are equivalent on $H_0^m(\Omega)$.
▲

5. THE SOBOLEV SPACE $H^m(\Omega)$

DEFINITION 1
We say that the subspace $H^m(\Omega)$ of $L^2(\Omega)$ defined by

(1) $\quad H^m(\Omega) = \{x \in L^2(\Omega) \quad \text{such that} \quad D^p x \in L^2(\Omega) \quad \text{for} \quad 1 \leq p \leq m\}$

with the scalar product

(2) $\qquad ((x,y))_m = \sum_{k=0}^{m} (D^k x, D^k y) = \sum_{k=0}^{m} \int_\Omega D^k x D^k y \, d\omega$

is the (maximal) Sobolev space of order m. ▲

Proposition 5.4.4 implies that the *Sobolev spaces* $H^m(\Omega)$ are complete.

Consequently, $H_0^m(\Omega)$ is a closed subspace of $H^m(\Omega)$. If Ω is an interval different from $\mathbb{R}$, $H_0^m(\Omega)$ is different from $H^m(\Omega)$.

Consider, for example, the case where $\Omega =]0,1[$. Let χ be the characteristic function of Ω. It is a function of $H^1(\Omega)$; indeed, $\chi \in L^2(\Omega)$ and $D\chi = 0$, since for every $\varphi \in \mathscr{D}(\Omega)$, we have

$$(D\chi, \varphi) = -(\chi, D\varphi) = -\int_0^1 D\varphi(\omega)\,d\omega = \varphi(0) - \varphi(1) = 0 - 0 = 0$$

because φ has compact support in Ω.

Moreover, consider the extension $\pi_0 \chi$ of χ by zero outside of Ω. Then $\pi_0 \chi$ is a function of $L^2(\mathbb{R})$, with compact support in $\mathbb{R}$.

We calculate its derivative in the sense of distributions:

$$(D\pi_0\chi, \varphi) = -(\pi_0\chi, D\varphi) = -\int_{-\infty}^{+\infty} \chi(\omega) D\varphi(\omega)\,d\omega = -\int_0^1 D\varphi(\omega)\,d\omega$$
$$= \varphi(0) - \varphi(1) = (\delta(0), \varphi) - (\delta(1), \varphi),$$

where $\delta(0): \varphi \to \varphi(0)$ and $\delta(1): \varphi \to \varphi(1)$ are the Dirac measures at zero and one. Hence $D\pi_0\chi = \delta(0) - \delta(1)$ belongs to $H^{-1}(\mathbb{R})$ but not to $L^2(\mathbb{R})$. Consequently, $\pi_0\chi$ is not a function of $H^1(\mathbb{R})$. If we had $H_0^1(\Omega) = H^1(\Omega)$, then χ would belong to $H_0^1(\Omega)$ and $\pi_0\chi \in H_0^1(\mathbb{R}) \subset H^1(\mathbb{R})$ according to Proposition 4.1. Since this is impossible, $H_0^1(\Omega) \neq H^1(\Omega)$.

On the other hand, if $\Omega = \mathbb{R}$, the spaces $H_0^m(\mathbb{R})$ and $H^m(\mathbb{R})$ coincide.

THEOREM 1
The spaces $H_0^m(\mathbb{R})$ and $H^m(\mathbb{R})$ coincide. ▲

Proof. We are going to show that we can approximate every function of $H^m(\mathbb{R})$ by a function of $\mathscr{D}(\Omega)$.

Consider a function $\rho \in \mathscr{D}(\mathbb{R})$ that is positive and with total mass equal to one [take ρ defined by formula (6.1.2), for example]. We choose a function $\theta \in \mathscr{D}(\mathbb{R})$ that is equal to one if $|\omega| \leq 1$ and equal to zero if $|\omega| \geq 2$. Let h be a parameter that will approach zero. Let $x \in H^m(\mathbb{R})$. We are going to approximate it by the functions

(3) $$\varphi_h = \rho_h * (\theta(\cdot h) x(\cdot)),$$

where $\rho_h(\omega) = (1/h)\rho(\omega/h)$. Since $\theta(\omega h) = 0$ if $|\omega| \geq 2/h$, φ_h has compact support.

Since ρ is infinitely differentiable, φ_h is also infinitely differentiable. We must show that for every $p \leq m$, $D^p\varphi_h$ converges to $D^p x$ in $L^2(\mathbb{R})$. But

$$|D^p\varphi_h - D^p x| = |\rho_h * D^p(\theta(\cdot h)x(\cdot)) - \rho_h * D^p x| + |\rho_h * D^p x - D^p x|$$
$$\leq \|\rho_h\|_1 |D^p(\theta(\cdot h)x(\cdot)) - D^p x| + |\rho_h * D^p x - D^p x|.$$

Since we known that $\rho_h * D^p x$ converges to $D^p x$ in $L^2(\mathbb{R})$ as h approaches zero (Theorem 6.6.1) and since $\|\rho_h\|_1 = \|\rho\| = 1$, it remains to show that $D^p \theta(\cdot h)x(\cdot)$ converges to $x(\cdot)$ in $L^2(\mathbb{R})$. The Leibniz formula implies that

(4)
$$\begin{cases} \dfrac{D^p}{p!}(\theta(\omega h)x(\omega) - x(\omega)) = (\theta(\omega h) - 1)\dfrac{D^p x(\omega)}{p!} \\ \qquad + \displaystyle\sum_{j=0}^{p-1} \dfrac{h^{p-j}}{(p-j)!}(D^{p-j}\theta)(\omega h)\dfrac{D^j x(\omega)}{j!}. \end{cases}$$

Since

$$\int_{-\infty}^{+\infty} |\theta(\omega h) - 1|^2 |D^p x(\omega)|^2 \, d\omega \leq \int_{|x| \leq 2/h} |D^p x(\omega)|^2 \, d\omega$$

converges to zero and since

$$\int_{-\infty}^{+\infty} |D^{p-j}\theta(\omega h)|^2 |D^j x(\omega)|^2 \, d\omega \leq \|D^{p-j}\theta\|_\infty^2 |D^j x|^2,$$

formula (4) shows that $D^p \theta(\cdot h)x(\cdot)$ converges to $D^p x$ in $L^2(\mathbb{R})$. ∎

Properties of the Functions of $H^m(\Omega)$

We are going to give some results that are *only true for functions of one variable*. (See Proposition 9.3.1 for the case of functions of several variables.)

PROPOSITION 1
Every function of $H^1(\Omega)$ is uniformly continuous and bounded. The injection from $H^1(\Omega)$ to $\mathscr{C}_\infty(\bar{\Omega})$ is continuous; there exists a constant $c > 0$ such that

(5) $$\sup_\omega |x(\omega)| \leq c \|x\|_1. \qquad \blacktriangle$$

Proof. **a.** Indeed, if $x \in H^1(\Omega)$, we can write

$$|x(\eta) - x(\xi)| \leq \int_\xi^\eta |Dx(\omega)| \, d\omega \leq (\xi - \eta)^{1/2} |D\xi|$$

according to the Cauchy-Schwarz inequality. Hence x is Hölder continuous and, consequently, uniformly continuous. It has a unique extension to a uniformly continuous function on $\bar{\Omega}$.

b. Consider the case where $\Omega = \mathbb{R}$. We take $\theta \in \mathscr{D}(\mathbb{R})$ such that $\theta(0) = 1$. We can then write

$$x(\omega) = \int_{-\infty}^{\omega} \frac{d}{d\eta}(\theta(\omega - \eta)x(\eta))\, d\eta$$

$$= -\int_{-\infty}^{\omega} D\theta(\omega - \eta)x(\eta)\, d\eta + \int_{-\infty}^{\omega} \theta(\omega - \eta)Dx(\eta)\, d\eta.$$

The Cauchy-Schwarz inequality implies that

(6) $\qquad |x(\omega)| \leq |D\theta|\,|x| + |\theta|\,|Dx| \leq \|\theta\|_1 \|x\|_1.$

c. Consider that case where $\Omega = \,]a, b[$. We choose a function $\theta_1 \in \mathscr{D}(\mathbb{R})$ equal to one on a neighborhood of a and equal to zero on a neighborhood of b, and we set $\theta_2 = 1 - \theta_1$. We can then write $x = x_1 + x_2$ where $x_i = \theta_i x$ ($i = 1, 2$). Moreover,

$$x_1(\omega) = -\int_{\omega}^{b} \frac{d}{d\eta}(x(\eta)\theta_1(\eta))\, d\eta = -\int_{\omega}^{b} Dx(\eta)\theta_1(\eta)\, d\eta - \int_{\omega}^{b} x(\eta)D\theta_1(\eta)\, d\eta$$

and

$$x_2(\omega) = \int_{a}^{\omega} \frac{d}{d\eta}(x(\eta)\theta_2(\eta))\, d\eta = \int_{a}^{b} Dx(\eta)\theta_2(\eta)\, d\eta + \int_{a}^{b} x(\eta)D\theta_2(\eta)\, d\eta.$$

It follows that for $i = 1, 2$,

$$|x_i(\omega)| \leq Dx\|\theta_i| + |x|\,|D\theta_i| \leq (|Dx|^2 + |x|^2)^{1/2}(|\theta_i|^2 + |D\theta_i|^2)^{1/2}$$
$$= \|x\|_1 \|\theta_i\|_1.$$

Consequently, for all $\omega \in [a, b]$,

$$|x(\omega)| \leq |x_1(\omega)| + |x_2(\omega)| \leq \|x\|_1(\|\theta_1\|_1 + \|\theta_2\|_1),$$

that is,

(7) $\qquad \|x\|_\infty \leq c\|x\|_1.$ ∎

We denote by $\mathscr{C}_\infty^{(k)}(\bar{\Omega})$ the space of functions defined on a neighborhood of $\bar{\Omega}$ whose derivatives up to order k exist and are continuous and bounded on $\bar{\Omega}$. We deduce from Proposition 1:

COROLLARY 1
Let Ω be an open interval of $\mathbb{R}$, either bounded or not. The injection from $H^{k+1}(\Omega)$ to $\mathscr{C}_\infty^{(k)}(\bar{\Omega})$ is continuous. ▲

*Compactness Conditions

We use in Chapter 11 compactness properties of the following type: If the interval Ω is *bounded*, the unit ball of $H^1(\Omega)$ not only is bounded in $\mathscr{C}_\infty(\bar\Omega)$ but is in fact relatively compact.

PROPOSITION 2
If Ω is bounded, the unit ball of $H^{k+1}(\Omega)$ is relatively compact in $\mathscr{C}_\infty^{(k)}(\bar\Omega)$. ▲

Proof. It suffices to establish the result for $k = 0$. Let B be the unit ball of $H^1(\Omega)$. The inequalities (6) show that the functions of B form an equicontinuous set of $\mathscr{C}_\infty(\bar\Omega)$, and inequality (7) implies that for every $\omega \in \bar\Omega$, $B(\omega) = \{x(\omega)\}_{x \in B}$ is relatively compact. Hence the Ascoli theorem [Theorem 4.6.2 of (AAA) p. 153.] allows us to conclude that B is relatively compact in $\mathscr{C}_\infty(\bar\Omega)$. ■

We deduce from this the following interesting result.

PROPOSITION 3
Let Ω be a bounded interval of $\mathbb{R}$. If $m \geq k + 1$, the unit ball of $H^m(\Omega)$ is relatively compact in $H^k(\Omega)$. ▲

Proof. The unit ball B of $H^m(\Omega)$ is bounded in $H^{k+1}(\Omega)$, hence relatively compact in $\mathscr{C}_\infty^{(k)}(\bar\Omega)$ according to Proposition 2. Since the injection from $\mathscr{C}_\infty^{(k)}(\bar\Omega)$ to $H^k(\Omega)$ is clearly continuous, B remains relatively compact in $H^k(\Omega)$. ■

(This result generalizes to functions of several variables.)

6. RELATIONS BETWEEN $H^m(\Omega)$ AND $H^m(\mathbb{R})$

We have seen that π_0, the operator of extension by zero, is not an operator from $H^m(\Omega)$ to $H^m(\mathbb{R})$ [since $\pi_0 \chi \notin H^1(\mathbb{R})$ when $\chi \in H^1(\Omega)$, where χ is the characteristic function of $\Omega = \,]0, 1[\,.$] We are going to construct another extension operator π that is an isomorphism from $H^m(\Omega)$ onto its closed image in $H^m(\mathbb{R})$.

Let $\Omega = \,]a, b[\,.$ We can find m functions $\alpha_p \in H^m(\mathbb{R}) \, (0 \leq p \leq m)$ with compact support in $\,]a - \varepsilon, a + \varepsilon[\,$ and m functions $\beta_p \in H^m(\mathbb{R}) \, (0 \leq p \leq m)$ with compact support in $\,]b - \varepsilon, b + \varepsilon[\,$ such that

(1) $\quad D^k \alpha_p(a) = \begin{cases} 0 & \text{if } p \neq k, \\ 1 & \text{if } p = k, \end{cases} \quad D^k \beta_p(b) = \begin{cases} 0 & \text{if } p \neq k, \\ 1 & \text{if } p = k. \end{cases}$

(For example, we can take the *Hermite interpolation polynomials* at the nodes $a - \varepsilon, a, a + \varepsilon$, and $b - \varepsilon, b, b + \varepsilon$.) Naturally we choose ε such that $a + \varepsilon < b - \varepsilon$.

We associate to every function $x \in H^m(\Omega)$ the function πx defined by

(2) $$\pi x(\omega) = \begin{cases} x(\omega) & \text{if } a \leq \omega \leq b, \\ \sum_{j=0}^{m-1} D^j x(a) \alpha_j(\omega) & \text{if } \omega \leq a, \\ \sum_{j=0}^{m-1} D^j x(b) \beta_j(\omega) & \text{if } b \leq \omega. \end{cases}$$

THEOREM 1
The operator π defined by (2) is a continuous linear operator from $H^m(\Omega)$ to $H^m(\mathbb{R})$ that has the operator ρ of restriction to Ω as a left inverse. Moreover, for every $k \leq m$,

(3) $$\rho D^k \pi x = D^k x \quad \text{if} \quad x \in H^m(\Omega).$$

If $x \in H_0^m(\Omega)$, then $\pi x = \pi_0 x$ coincides with the extension of x by zero outside of Ω. ▲

Proof. We shall show that π is a linear operator from $H^m(\Omega)$ to $H^m(\mathbb{R})$. Indeed, for all $p \leq m - 1$,

$$D^k \pi x(a) = D^k x(a) \quad \text{and} \quad D^k \pi x(b) = x(b).$$

Hence for every function $\varphi \in \mathscr{D}(\Omega)$,

(4)
$$(D^k \pi x, \varphi) = (-1)^k \int_{-\infty}^{+\infty} \pi x(\omega) D^k \varphi(\omega) \, d\omega$$
$$= (-1)^k \sum_{j=0}^{m-1} D^j x(a) \int_{-\infty}^{a} \alpha_j(\omega) D^k \varphi(\omega) \, d\omega$$
$$+ (-1)^k \int_a^b x(\omega) D^k \varphi(\omega) \, d\omega$$
$$+ (-1)^k \sum_{j=0}^{m-1} D^j x(b) \int_b^{+\infty} \beta_j(\omega) D^k \varphi(\omega) \, d\omega.$$

But, according to (1) and by using integration by parts,

$$\int_{-\infty}^{a} \alpha_j(\omega) D^k \varphi(\omega) \, d\omega = (-1)^k \int_{-\infty}^{a} (D^k \alpha_j(\omega)) \varphi(\omega) \, d\omega$$
$$+ \sum_{q=0}^{k-1} (-1)^q D^{k-1-q} \alpha_j(\omega) D^q \varphi(\omega) \Big]_{-\infty}^{a}$$
$$= (-1)^k \int_{-\infty}^{a} D^k \alpha_j(\omega) \varphi(\omega) \, d\omega + (-1)^{k-1-j} D^{k-1-j} \varphi(a).$$

Similarly,

$$\int_a^b x(\omega)D^k\varphi(\omega)\,d\omega = (-1)^k \int_a^b D^k x(\omega)\varphi(\omega)\,d\omega$$
$$+ \sum_{q=0}^{k-1}(-1)^q D^{k-1-q}x(b)D^q\varphi(b)$$
$$- \sum_{q=0}^{k-1}(-1)^q D^{k-1-q}x(a)D^q\varphi(a)$$

and

$$\int_b^\infty \beta_j(\omega)D^k\varphi(\omega) = (-1)^k \int_b^\infty D^k\beta_j(\omega)\varphi(\omega)\,d\omega - (-1)^{k-1-j}D^{k-1-j}\varphi(b).$$

Hence, evaluating all the integrals in formula (4), we obtain

(5) $\quad (D^k\pi x, \varphi) = \sum_{j=0}^{m-1} D^j x(a) \int_{-\infty}^a D^k\alpha_j(\omega)\varphi(\omega)\,d\omega$
$$+ \int_a^b D^k x(\omega)\varphi(\omega)\,d\omega + \sum_{j=0}^{m-1} D^j x(b) \int_b^\infty D^k\beta_j(\omega)\varphi(\omega)\,d\omega.$$

This last formula shows that if $k \leq m$, $D^k\pi x$ is the function of $L^2(\mathbb{R})$ defined by

(6) $\quad D^k\pi x(\omega) = \begin{cases} D^k x(\omega) & \text{if} \quad a \leq \omega \leq b. \\ \sum_{j=0}^{m-1} D^j x(a)D^k\alpha_j(\omega) & \text{if} \quad \omega \leq a. \\ \sum_{j=0}^{m-1} D^j x(b)D^k\beta_j(\omega) & \text{if} \quad \omega \geq b. \end{cases}$

Moreover, Corollary 5.1 shows that

(7) $\quad \begin{cases} |D^j x(a)| \leq c|D^{j+1}x| \\ |D^j x(b)| \leq c|D^{j+1}x|. \end{cases}$

Consequently,

(8) $\quad \begin{cases} |D^k\pi x|^2 \leq \left(\sum_{j=0}^{m-1} |D^j x(a)|^2 + |D^j x(b)|^2 + |D^k x|^2\right). \\ \left(\sum_{j=0}^{m-1} |D^k\alpha_j|^2 + |D^k\beta_j|^2 + 1\right) \leq c\|x\|_m^2. \end{cases}$

Hence π is a *continuous* linear operator from $H^m(\Omega)$ to $H^m(\mathbb{R})$.

The operator ρ of restriction to Ω is clearly a continuous linear operator from $H^m(\mathbb{R})$ to $H^m(\Omega)$, which is a left inverse of π. Indeed, formula (6) shows that

(9) $$\rho D^k \pi x = D^k x.$$

Moreover, if $x \in H_0^m(\Omega)$, we have

$$D^j x(a) = D^j x(b) = 0 \quad \text{for} \quad 0 \leq j \leq m-1,$$

since x is the limit of functions of $\mathscr{D}(\Omega)$. Hence

$$\pi x(\omega) = 0 \quad \text{if} \quad \omega \notin [a,b]. \qquad \blacksquare$$

The operator of extension π is very useful. It allows us, for example, to obtain an approximation theorem.

THEOREM 2
Let $\mathscr{E}(\bar{\Omega})$ be the space of restrictions to Ω of functions infinitely differentiable on $\mathbb{R}$. Then $\mathscr{E}(\bar{\Omega})$ is dense in $H^m(\Omega)$. ▲

Proof. Let $x \in H^m(\Omega)$. Then $\pi x \in H^m(\mathbb{R}) = H_0^m(\mathbb{R})$ and is the limit [in $H^1(\mathbb{R})$] of functions $y_h \in \mathscr{D}(\mathbb{R})$. Hence the restrictions $x_h = \rho y_h$ to Ω belong to $\mathscr{E}(\bar{\Omega})$, and since ρ is continuous, x_h converges to $\rho \pi x = x$ in $H^m(\Omega)$. ∎

*7. CHARACTERIZATION OF THE DUAL OF $H^m(\Omega)$

We are going to characterize the dual of the space $H^m(\Omega)$. To this end we denote by $D^k \delta(a)$ and $D^k \delta(b)$ the linear forms defined by

(1) $\quad \varphi \mapsto (D^k \delta(a), \varphi) = (-1)^k (D^k \varphi)(a) \quad$ and $\quad (D^k \delta(b), \varphi) = (-1)^k (D^k \varphi)(b)$

Corollary 5.1 implies that these linear forms are continuous on $H^m(\Omega)$ if $0 \leq k \leq m-1$.

THEOREM 1
Every continuous linear form x on $H^m(\Omega)$ can be written as the sum of the restriction to Ω of a distribution $f \in H^{-m}(\mathbb{R})$ and of a linear combination of linear forms $D^k \delta(a)$ and $D^k \delta(b)$ when $0 \leq k \leq m-1$. ▲

Proof. Since π is left invertible, its transpose π^* is right invertible and, consequently, is a surjective operator from $H^{-m}(\mathbb{R})$ onto the dual $H^m(\Omega)^*$ of $H^m(\Omega)$. Hence every $x \in H^m(\Omega)^*$ can be written as $x = \pi^* f$ where $f \in H^{-m}(\mathbb{R})$. Moreover, formula (6) allows us to calculate the transpose π^* of π; for every $\varphi \in H^m(\Omega)$,

(2)
$$\begin{cases} (x, \varphi) = (\pi^* f, \varphi) = \int_a^b f(\omega)\varphi(\omega)\,d\omega \\ + \sum_{j=0}^{m-1} \left\langle \left(\int_{-\infty}^a \alpha_j(\omega) f(\omega)\,d\omega \right) D^j \delta(a) + \left(\int_b^\infty b_j(\omega) f(\omega)\,d\omega \right) D^j \delta(b), \varphi \right\rangle \end{cases}$$

Consequently,

(3) $\quad x = \pi^* f = \rho f + \sum_{j=0}^{m-1} \left[\left(\int_{-\infty}^a \alpha_j f\,d\omega \right) D^j \delta(a) + \left(\int_b^\infty \beta_j f\,d\omega \right) D^j \delta(b) \right].$ ∎

We are able to give explicitly the duality operator from $H^1(\Omega)$ onto its dual and to compare it with the duality operator $-D^2 + 1$ from $H_0^1(\Omega)$ onto $H^{-1}(\Omega)$. (See Proposition 3.2.)

PROPOSITION 1

The duality operator from $H^1(\Omega)$ onto its dual is defined by

(4) $\qquad Jx = -D^2 x + x + (Dx(b))\delta(b) - (Dx(a))\delta(a).$ ▲

Proof. Since $\mathscr{E}(\bar{\Omega})$ is dense in $H^1(\Omega)$, we are going to restrict the scalar product $((x, y))$ to $\mathscr{E}(\bar{\Omega})$. Therefore integration by parts gives

$$((x, y)) = \int_a^b Dx(\omega) Dy(\omega)\,d\omega + \int_a^b x(\omega)y(\omega)\,d\omega$$
$$= \int_a^b (-D^2 x(\omega) + x(\omega))y(\omega)\,d\omega + Dx(b)y(b) - Dx(a)y(a)$$
$$= \langle -D^2 x(\omega) + x(\omega) + Dx(b)\delta(b) - Dx(a)\delta(a), y \rangle$$
$$= \langle Jx, y \rangle. \qquad ∎$$

Remark 1

Since $\mathscr{D}(\Omega)$ is not dense in $H^m(\Omega)$ when Ω is bounded, $H^m(\Omega)$ is not a "normal" space; consequently, the dual of $H^m(\Omega)$ cannot be identified with a subspace of the *algebraic* dual space $\mathscr{D}^*(\Omega)$ of the space $\mathscr{D}(\Omega)$. ∎

8. TRACE THEOREMS

Let $\Omega =]a, b[$ be a bounded interval of $\mathbb{R}$. Corollary 5.1 implies that we can define the "traces" $\{D^k x(a), D^k x(b)\}$ of the functions $x \in H^m(\Omega)$ on the boundary $\{a, b\}$ of Ω (for $0 \leq k \leq m - 1$). They vanish if $x \in H_0^m(\Omega)$.

The trace theorem states that $H_0^m(\Omega)$ is the space of functions all of whose derivatives of order inferior to $m - 1$ vanish on the boundary of Ω.

We set

(1) $$\gamma_k x = \{D^k x(a), D^k x(b)\} \in \mathbb{R}^2$$

and we denote by γ the operator $\gamma_0 \times \ldots \times \gamma_{m-1}$ defined by

(2) $$\gamma x = \{\gamma_0 x, \ldots, \gamma_{m-1} x\} \in \mathbb{R}^{2m}.$$

THEOREM 1

The operator γ is continuous, linear, and surjective from $H^m(\Omega)$ onto $\mathbb{R}^{2m}$. Its kernel is the subspace $H_0^m(\Omega)$. ▲

Proof. Corollary 5.1 implies that γ is continuous and linear. It is *surjective*: to this end consider the functions α_j and $\beta_j (0 \leq j \leq m - 1)$ satisfying the following conditions:

$$\begin{cases} \text{i.} & D^k \alpha_j(a) = \begin{cases} 0 & \text{if} \quad k \neq j \\ 1 & \text{if} \quad k = j \end{cases} \quad \text{where } 0 \leq j, \ k \leq m - 1. \\ \text{ii.} & D^k \alpha_j(b) = 0 \quad \text{if} \quad 0 \leq j, \ k \leq m - 1 \end{cases}$$

and

$$\begin{cases} \text{i.} & D^k \beta_j(a) = 0 \quad \text{if} \quad 0 \leq j, \ k \leq m - 1. \\ \text{ii.} & D^k \beta_j(b) = \begin{cases} 0 & \text{if} \quad k \neq j. \\ 1 & \text{if} \quad k = j. \end{cases} \end{cases}$$

(We can take, for example, the Hermite interpolation polynomials.)

We associate to the vector $c = \{c^j\}_{0 \leq j \leq m} \in \mathbb{R}^{2m}$ (where $c^j = (c_a^j, c_b^j)$) the restriction to Ω of

(3) $$\sigma c = \sum_{j=0}^{m-1} (c_a^j \alpha_j + c_b^j \beta_j),$$

It is clear that σc belongs to $H^m(\Omega)$ and that $\gamma \sigma c = c$.

Moreover, it is evident that $\gamma x = 0$ for all $x \in \mathscr{D}(\Omega)$. Since $\mathscr{D}(\Omega)$ is dense in $H_0^m(\Omega)$ and since γ is continuous, it follows that $\gamma x = 0$ for all $x \in H_0^m(\Omega)$.

It remains for us to show the converse. We take $x \in H^m(\Omega)$ such that $\gamma x = 0$. It suffices to approximate x by functions $x_k \in H^m(\Omega)$ with compact support.

Let $\theta \in \mathscr{D}(\mathbb{R})$ be a positive function such that $\theta(\omega) = 1$ on a neighborhood of a and $\theta(\omega) = 0$ on a neighborhood of b. We can then write that $x(\omega) = x_1(\omega) + x_2(\omega)$ where $x_1(\omega) = \theta(\omega) x(\omega)$ (respectively, $x_2(\omega) = (1 - \theta(\omega)) x(\omega)$) vanishes on a neighborhood of b (respectively, a) and equals

$x(\omega)$ on a neighborhood of a (respectively, b). Thus it suffices to approximate x_1 by functions that vanish on a neighborhood of a, since the same argument will show that x_2 can be approximated by functions that vanish on a neighborhood of b.

This is the object of the following lemma.

LEMMA 1
Let $x \in H^m(0, 1)$ vanish on a neighborhood of 1 and suppose that $D^k x(0) = 0$ for all $k \leq m - 1$. Then x is the limit of functions $x_\varepsilon \in H^m(0, 1)$ with compact support in $]0, 1[$. ▲

Proof. Since $D^k x(0) = 0$ if $0 \leq k \leq m - 1$, the extension πx of x coincides with the extension $\pi_0 x$ of x by zero outside of Ω. [See formula (6) in Section 6.] Theorem 6.1 implies, therefore, that $\pi_0 x \in H^m(\mathbb{R})$ and has its support contained in an interval $[0, b]$ with $b < 1$. Hence for all $\varepsilon < 1 - b$, the functions x_ε defined by $x_\varepsilon(\omega) = \pi_0 x(\omega - \varepsilon)$ have compact support $[\varepsilon, b + \varepsilon] \subset]0, 1[$ and converge to $\pi_0 x$ in $H^m(\mathbb{R})$ and, consequently, to x in $H^m(0, 1)$. ■

9. CONVOLUTION OF DISTRIBUTIONS

Let $\varphi \in L^1(\mathbb{R})$. We have seen that the convolution operator $\varphi *$ by φ is a continuous operator from $L^2(\mathbb{R})$ to itself.

PROPOSITION 1
Let $\varphi \in L^1(\mathbb{R})$. The convolution operator by φ is continuous from $H^m(\mathbb{R})$ to $H^m(\mathbb{R})$ for every $m \in \mathbb{Z}$ and

(1) $$\|\varphi *\|_{\mathscr{L}(H^m(\mathbb{R}), H^m(\mathbb{R}))} \leq \|\varphi\|_{L^1} \quad \text{for every} \quad m \in \mathbb{Z}.$$ ▲

Proof. **a.** First we consider the case where $m \geq 0$. We are going to show that if $x \in \mathscr{D}(\mathbb{R})$,

(2) $$D^k(\varphi * x) = \varphi * D^k x \in L^2(\mathbb{R}) \quad \text{for all} \quad k \leq m.$$

This implies that $\varphi *$ sends $\mathscr{D}(\mathbb{R})$ to $H^m(\mathbb{R})$ and that $\|\varphi *\| \leq \|\varphi\|_{L^1}$, since

$$\|\varphi * x\|_m^2 = \sum_{k=0}^m |D^k(\varphi * x)|^2 = \sum_{k=0}^m |\varphi * D^k x|^2$$
$$\leq \|\varphi\|_{L^1}^2 \sum_{k=0}^m |D^k x|^2 = \|\varphi\|_{L^1}^2 \|x\|_m^2.$$

Then $\varphi *$ has an extension to a continuous operator from $H^m(\mathbb{R})$ to $H^m(\mathbb{R})$.

Therefore let us establish (2): for every $\psi \in \mathcal{D}(\mathbb{R})$, we have

$$(D^k(\varphi * x), \psi) = (-1)^k(\varphi * x, D^k\psi) = (-1)^k \int d\omega \int d\eta D^k\psi(\omega)x(\omega - \eta)\varphi(\eta)$$
$$= \int d\eta \varphi(\eta) \int d\omega \psi(\omega)(D^k x)(\omega - \eta) = \int d\omega \psi(\omega) \int d\eta D^k x(\omega - \eta)\varphi(\eta)$$
$$= (D^k x * \varphi, \psi).$$

b. Now let us establish the proposition for the case of the spaces H^{-m}. We denote by $\check{\varphi}$ the symmetric function of φ defined by

(3) $$\check{\varphi}(\omega) = \varphi(-\omega).$$

Then since $\check{\varphi} *$ is a continuous linear operator from $H^m(\mathbb{R})$ to $H^m(\mathbb{R})$, its transpose $(\check{\varphi}*)^*$ is a continuous linear operator from $H^{-m}(\mathbb{R})$ to $H^{-m}(\mathbb{R})$. Its restriction to $\mathcal{D}(\mathbb{R})$ coincides with the operator $\varphi *$ of convolution by φ; indeed, if

$$x \in \mathcal{D}(\mathbb{R}), \quad ((\check{\varphi}*)^*x, y) = (x, \check{\varphi}*y) = \int x(\omega)\, d\omega \int \check{\varphi}(\omega - \eta) y(\eta)\, d\eta$$
$$= \int x(\omega)\, d\omega \int \varphi(\eta - \omega) y(\eta)\, d\eta$$
$$= \int d\eta\, y(\eta) \int \varphi(\eta - \omega) x(\omega)\, d\omega$$
$$= (\varphi * x, y) \quad \text{for all} \quad y \in \mathcal{D}(\Omega).$$

Thus

(4) $$(\check{\varphi}*)^*x = \varphi * x \quad \text{for all} \quad x \in \mathcal{D}(\Omega).$$

Since $\mathcal{D}(\Omega)$ is dense in $H^{-m}(\mathbb{R})$, $(\check{\varphi}*)^*$ is the unique extension of $\varphi *$ to $H^{-m}(\mathbb{R})$. ∎

Since we can define $\varphi * x$ for every $x \in H^{-m}(\mathbb{R})$, we say that $\varphi * x$ is the *convolution product of the function* φ $L^1(\mathbb{R})$ *and the distribution* $x \in H^{-m}(\mathbb{R})$.

We are going to use the convolution product to approximate distributions by smooth functions.

PROPOSITION 2
Let $\lambda \in \mathscr{C}_0(\mathbb{R})$ be a positive function with total mass equal to one. Then for every $m \in \mathbb{Z}$ the convolution operators $\lambda_h * \in \mathscr{L}(H^m(\mathbb{R}), H^m(\mathbb{R}))$ are bounded and converge pointwise to the identity mapping from $H^m(\mathbb{R})$ to $H^m(\mathbb{R})$ (and, in fact, uniformly on every compact set of $H^m(\mathbb{R})$.) ▲

Proof. We know that $\|\lambda_h *\|_{\mathscr{L}(H^m(\mathbb{R}), H^m(\mathbb{R}))} \leq 1$. The operators $\lambda_h *$ therefore form a bounded set of linear operators. Theorem 4.1.2 implies that if $\lambda_h * \varphi$ converges to φ for every $\varphi \in \mathcal{D}(\mathbb{R})$, which is dense in $H^m(\mathbb{R})$, then $\lambda_h *$ converges to the identity mapping uniformly on every compact set.

If $m \geq 0$, $D^k(\lambda_h * \varphi) = \lambda_h * D^k\varphi$ converges to $D^k\varphi$ in $L^2(\mathbb{R})$ for all $k \leq m$

according to Theorem 6.6.1. Thus $\lambda_h * \varphi$ converges to φ in $H^m(\mathbb{R})$. If $m < 0$, Theorem 6.6.1 implies that $\lambda_h * \varphi$ converges to φ in $L^2(\mathbb{R})$, which is contained in $H^m(\mathbb{R})$ with a finer topology and dense. Hence $\lambda_h * \varphi$ converges to φ in $H^m(\mathbb{R})$. ∎

This result allows us to give an interpretation to the derivative in the sense of distributions.

PROPOSITION 3
If $x \in L^2(\mathbb{R})$, then $D^k x \in H^{-k}(\mathbb{R})$ is the limit in $H^{-k}(\mathbb{R})$ of the difference quotients $\nabla_h^k x$ (for $k \geq 0$).

Proof. Indeed we can write that
$$\nabla_h^k x = D^k(\chi_h^{*k} * x) = \chi_h^{*k} * D^k x.$$
Since χ^{*k} is a function with total mass equal to 1, $\nabla_h^k x = \chi_k^{*k} * D^k x$ converges to x in $H^{-k}(\mathbb{R})$ according to the preceding Proposition 2. ∎

CHAPTER 8

Some Approximation Procedures in Spaces of Functions

The problem of approximating an arbitrary function by *simpler* or *more convenient* functions has long been a fundamental problem of analysis and, more recently, of numerical analysis. We devote this chapter to a study of three types of constructive procedures for approximating functions: by polynomials (Sections 1 and 2), by trigonometric polynomials (Section 3), and by piecewise polynomials (Sections 4, 5, and 6).

We already know that $\mathscr{D}(\Omega)$ is dense in $L^2(\Omega)$, which result we have obtained by a method of approximation by convolution. This type of procedure is mainly theoretic, since the functions $\rho_h * f$, though smoother than f, are not "simpler" than f from the point of view of calculations.

For a long time the "simple" functions to manipulate have been polynomials and trigonometric polynomials. The Stone-Weierstrass theorem shows that every continuous function can be approximated uniformly by polynomials, and the Bernstein theorem permits us to construct these polynomials. (See [AAA], p. 185.)

With a Hilbert space $L^2(\Omega, a)$, the Gramm-Schmidt orthogonalization process gives a method of constructing an orthogonal sequence of polynomials called *orthogonal polynomials* and, therefore, a polynomial approximation procedure. With the space $L^2(-1, +1)$ these orthogonal polynomials are the Legendre polynomials. For the space $L^2(-\frac{1}{2}, +\frac{1}{2})$ (the complex space of complex valued functions) we use the density of the trigonometric polynomials to approximate (in quadratic mean) by trigonometric polynomials every function $x \in L^2(-\frac{1}{2}, +\frac{1}{2})$ and to represent it by its Fourier series $\sum_{k=-\infty}^{+\infty} c_k \exp(2i\pi k\omega)$.

The use of computers instead of direct calculation has led to replacing polynomial approximation by approximation by piecewise polynomial functions of fixed degree m (also called *Spline functions*). These approximation procedures play an important role in the approximation of the solutions of boundary value problems by the method of "finite elements," as we see in Chapter 13, Section 5.

1. APPROXIMATION BY ORTHOGONAL POLYNOMIALS

Suppose that Ω is an interval of the real line $\mathbb{R}$. We are going to construct orthogonal bases for the *weighted spaces $L^2(\Omega, a)$ consisting of polynomials*.

In order for the polynomials to belong to the space $L^2(\Omega, a)$ we shall suppose that the weight a satisfies

(1) $$\forall n \in \mathbb{N}, \quad \int_\Omega \omega^n a(\omega)\, d\omega < +\infty.$$

Since the monomials $\omega \mapsto \omega^n$ are linearly independent, we can make this system orthogonal by using the Gramm-Schmidt orthogonalization process: If we denote by x_n the monomial defined by $x_n(\omega) = \omega^n$, we define by recursion the nth element of the associated orthogonal system by

(2) $$e_0 = x_0; \quad e_n = x_n - t_{V_{n-1}} x_n$$

where $t_{V_{n-1}}$ is the orthogonal projector onto the space V_{n-1} generated by $\{x_0, \ldots, x_{n-1}\}$, which is the space of polynomials of degree $n-1$. (See Proposition 1.7.2.) Then the element e_n is a polynomial of degree n whose term of highest degree is ω^n. The polynomials e_n are called the *orthogonal polynomials associated with the weight a on Ω*.

The study of these associated orthogonal polynomials for various examples of the weight has been systematically undertaken in the framework of the theory of special functions. We give one complete example only. Before this we establish some general properties of orthogonal polynomials.

Remark 1

When Ω is bounded, $\mathscr{C}_\infty(\bar{\Omega})$ is dense in $L^2(\Omega, a)$, and the Stone-Weierstrass theorem (see Theorem 5.2.2 of [AAA] p. 185) implies that the space of polynomials is dense in $\mathscr{C}_\infty(\bar{\Omega})$. Therefore the space of polynomials is dense in $L^2(\Omega, a)$; consequently, the orthogonal polynomials form an orthogonal base for $L^2(\Omega, a)$. If Ω is not bounded, there exist weights for which the sequence of orthogonal polynomials does not form a base; in this case it is necessary to verify in each example whether or not the orthogonal polynomials form a base. ∎

Remark 2

If the orthogonal polynomials e_n form an orthogonal base for $L^2(\Omega, a)$, every function $x \in L^2(\Omega, a)$ is the limit in quadratic mean weighted by a of the sequence of polynomials

(3) $$\sum_{k=0}^{n} \left(\int_\Omega x(\omega) e_n(\omega) a(\omega)\, d\omega \right) \frac{e_n(\omega)}{\|e_n\|_a^2}.$$

This sequence does not always converge uniformly (nor even pointwise) to the function x. ∎

PROPOSITION 1 (RECURSION RELATION)
There exist two sequences of real numbers $\lambda_n, \mu_n > 0$ such that, for every $n \geq 2$,

(3) $$e_n = (\omega + \lambda_n)e_{n-1} - \mu_n e_{n-2}.$$ ▲

Proof. Since $e_{n-1} = x_{n-1} - t_{V_{n-2}} x_{n-1}$, we deduce that
$$\omega e_{n-1}(\omega) = \omega^n - \omega t_{V_{n-2}} x_{n-1}.$$

Thus
$$e_n(\omega) - \omega e_{n-1}(\omega) = \omega t_{V_{n-2}} x_{n-1} - t_{V_{n-1}} x_n$$

is a polynomial of degree $n - 1$. We can, therefore, write

(4) $$e_n - \omega e_{n-1} = \sum_{i=0}^{n-1} c_i e_i.$$

Since the polynomials are orthogonal, we obtain, by taking the scalar product of (4) with e_i,

(5) $$-((\omega e_{n-1}, e_i))_a = c_i \|e_i\|_a^2 \quad \text{if} \quad 0 \leq i \leq n-1$$

and by taking the scalar product with e_n,

(6) $$\|e_n\|_a^2 - ((\omega e_{n-1}, e_n))_a = 0.$$

Moreover, since $((\omega e_{n-1}, e_i))_a = ((e_{n-1}, \omega e_i))_a$, this scalar product is zero if $i + 1 \leq n - 2$. Hence (5) implies that the coefficients c_i are zero for $i \leq n - 3$. For $i = n - 2$, we deduce from (6), replacing n by $n - 1$, that $c_{n-2} \|e_{n-2}\|_a^2 = ((\omega e_{n-1}, e_{n-2}))_a = -\|e_{n-1}\|_a^2$. Therefore $c_{n-2} < 0$. Hence $e_n - \omega e_{n-1} = c_{n-2} \omega_{n-2} + c_{n-1} \omega_{n-1}$, which implies (3) with $\lambda_n = c_{n-1}$ and $\mu_n = -c_{n-2}$. ∎

PROPOSITION 2
Every orthogonal polynomial of degree n has its n distinct real roots in the interval Ω. ▲

Proof. Since e_n and $e_0 = 1$ are orthogonal, we deduce that $\int_\Omega e_n(\omega) a(\omega) \, d\omega = 0$. Since $a(\omega) > 0$, the polynomial $e_n(\omega)$ can be neither strictly positive nor strictly negative and must, therefore, change sign at at least one point of Ω.

More generally, let $\{\omega_1, \ldots, \omega_k\}$ be the sequence of roots of e_n belonging to Ω and at which e_n changes sign. We have seen that $k \geq 1$, and we want

CH. 8, SEC. 2   LEGENDRE, LAGUERRE, AND HERMITE POLYNOMIALS   169

to show that $k = n$. Since $k \leq n$, it suffices to exclude the case where $k < n$. To this end we set $y(\omega) = (\omega - \omega_1) \ldots (\omega - \omega_k)$. By construction the polynomial $e_n y$ has constant sign on Ω. Moreover, if $k < n$, $((e_n, y))_a = \int_\Omega e_n(\omega) y(\omega) a(\omega) \, d\omega = 0$. This is a contradiction. Hence $k = n$. ∎

2. LEGENDRE, LAGUERRE, AND HERMITE POLYNOMIALS

Legendre Polynomials

Legendre polynomials are the orthogonal polynomials in the case where $\Omega = \,]-1, +1[$ and $a(\omega) = 1$; they therefore form an *orthogonal base* for the space $L^2(-1, +1)$. We shall verify that the Legendre polynomials are defined by

(1) $$l_n(\omega) = D^n([\omega^2 - 1]^n).$$

Suppose that $m > n$. Integrating by parts, we obtain

$$\int_{-1}^{+1} D^m[(\omega^2-1)^m] D^n[(\omega^2-1)^n] \, d\omega = [D^{m-1}[(\omega^2-1)^m] D^n[(\omega^2-1)^n]]_{-1}^{+1}$$
$$- \int_{-1}^{+1} D^{m-1}[(\omega^2-1)^m] D^{n+1}[(\omega^2-1)^n] \, d\omega.$$

But at the points ± 1, the function $(\omega^2 - 1)^m = (\omega - 1)^m (\omega + 1)^m$ vanishes along with its derivatives of order less than $m - 1$. Hence the first term of the second member of this equality is zero. Continuing, we find

$$(-1)^n \int_{-1}^{+1} D^{m-n}[(\omega^2-1)^m] D^{2n}[(\omega^2-1)^n] \, d\omega$$
$$= (-1)^n (2n)! \int_{-1}^{+1} D^{m-n}[(\omega^2-1)^m] \, d\omega$$
$$= (-1)^n (2n)! [D^{m-n-1}((\omega^2-1)^m)]_{-1}^{+1} = 0 \quad \text{since} \quad m > n.$$

Calculating now the norm of l_n:

$$\|l_n\|^2 = \int_{-1}^{+1} (D^n[(\omega^2-1)^n])^2 \, d\omega$$

(2) $$= (-1)^n (2n)! \int_{-1}^{+1} (\omega^2 - 1)^n \, d\omega$$

$$= (2n)! \int_{-1}^{+1} (1 - \omega^2)^n \, d\omega.$$

This integral is in turn calculated by integration by parts:

$$\int_{-1}^{+1} (1-\omega^2)^n \, d\omega = \int_{-1}^{+1} (1-\omega)^n (1+\omega)^n \, d\omega$$

$$= \left[(1-\omega)^n \frac{(1+\omega)^{n+1}}{n+1} \right]_{-1}^{+1} - \int_{-1}^{+1} n(1-\omega)^{n-1} \frac{(1+\omega)^{n+1}}{n+1} \, d\omega$$

$$= \frac{1}{n+1} \int_{-1}^{+1} (1-\omega)^{n-1}(1+\omega)^{n+1} \, d\omega = \ldots$$

$$= \frac{n!}{(n+1)\ldots(2n)} \int_{-1}^{+1} (1-\omega)^{2n} \, d\omega = \frac{(n!)^2}{(2n)!} \frac{2^{2n+1}}{(2n+1)},$$

Hence

(3) $$\|l_n\| = 2^n n! \sqrt{\frac{2}{(2n+1)!}}.$$ ∎

Remark 1

It is usual to normalize the Legendre polynomials by imposing the condition $\bar{l}_n(1) = 1$. This gives then the polynomials $\bar{l}_n$ defined by $\bar{l}_n = \frac{1}{2^n n!} l_n$. We can also show that they satisfy the recursion relation

(4) $$n\bar{l}_n = (2n-1)\bar{l}_{n-1} - (n-1)\bar{l}_{n-2}$$

and the differential equations

(5) $$D[(1-\omega^2)D\bar{l}_n] + n(n+1)\bar{l}_n = 0.$$ ∎

Laguerre Polynomials

Laguerre polynomials L_n are the orthogonal polynomials of the space $L^2(]0, \infty[; e^{-\omega})$ where $\Omega =]0, \infty[$ and $a(\omega) = e^{-\omega}$. We shall verify that they are defined by

(6) $$L_n(\omega) = e^{\omega} D^n [\omega^n e^{-\omega}].$$

Suppose that $m > n$. To show that $\int_0^\infty L_m(\omega) L_n(\omega) e^{-\omega} \, d\omega = 0$, we are going to verify that $\int_0^\infty L_m(\omega) \omega^k e^{-\omega} \, d\omega = 0$ for all $k < m$. Integrating by parts k times, we obtain

$$\int_0^\infty L_m(\omega) \omega^k e^{-\omega} \, d\omega = \int_0^\infty \omega^k D^m [\omega^m e^{-\omega}] \, d\omega = (-1)^k k! \int_0^\infty D^{m-k} [\omega^m e^{-\omega}] \, d\omega = 0$$

since $k < m$. Moreover,

$$\int_0^\infty e^{-\omega}L_n^2(\omega)\,d\omega = \int_0^\infty D^n[\omega^n e^{-\omega}] \sum_{k=0}^n (-1)^k \binom{n}{k} n(n-1)\ldots(k+1)\omega^k\,d\omega$$
$$= \int_0^\infty (-1)^n \omega^n D^n(\omega^n e^{-\omega})\,d\omega = n! \int_0^\infty \omega^n e^{-\omega}\,d\omega = (n!)^2.$$

Thus

$$\|L_n\|_a = n!$$

Remark 2

We can verify that the Laguerre polynomials form a topological base for the space $L^2(]0, \infty[\,;e^{-\omega})$.

It is customary to normalize the Laguerre polynomials by imposing the condition $\bar{L}_n(0) = 1$; we then obtain

(7) $$\bar{L}_n(\omega) = \frac{1}{n!}L_n(\omega) = \frac{e}{n!}D^n[\omega^n e^{-\omega}].$$

They satisfy the differential equations

(8) $$\omega D^2 \bar{L}_n + (1 - \omega)D\bar{L}_n + n\bar{L}_n = 0 \qquad \blacksquare$$

Hermite Polynomials

The Hermite polynomials H_n are the orthogonal polynomials of the space $L^2(]-\infty, +\infty[\,;e^{-\omega^2/2})$ where $\Omega =]-\infty, +\infty[$ and $a(\omega) = e^{-\omega^2/2}$. They are defined by the formulas

(9) $$H_n(\omega) = (-1)^n e^{\omega^2/2} D^n[e^{-\omega^2/2}].$$

We verify, as in the case of the Legendre polynomials and the Laguerre polynomials, that

(10) $$\int_{-\infty}^{+\infty} H_m(\omega)H_n(\omega)e^{-\omega^2/2}\,d\omega = 0 \qquad \text{if} \qquad m > n.$$

Indeed, we show that for every $k < m$,

$$\int_{-\infty}^{+\infty} H_n(\omega)\omega^k e^{-\omega^2/2}\,d\omega = (-1)^n \int_{-\infty}^{+\infty} D^n[e^{-\omega^2/2}]\omega^k\,d\omega = 0$$

by integrating by parts k times. We also obtain

(11) $$\|H_n(\omega)\|_a = (n!\sqrt{2\pi})^{1/2}.$$

We verify as well that the Hermite polynomials satisfy the recursion relations

(12) $H_n = \omega H_{n-1} - (n-1)H_{n-2}$, $H_0 = 1$, $H_1(\omega) = 2\omega$

and are solutions of the differential equations

(13) $D^2 H_n - \omega D H_n + n H_n = 0$ and $DH_n = nH_{n-1}$.

They form an orthogonal base for $L^2(]-\infty, +\infty[; e^{-\omega^2/2})$. We recall that they are also the Appell polynomials of the Gaussian probability densities. (See Chapter 6, Section 8.)

3. FOURIER SERIES

Consider the open interval $\Omega =]-\frac{1}{2}, +\frac{1}{2}[$ and the *complex* Hilbert space $L^2(-\frac{1}{2}, +\frac{1}{2})$. We are going to show that the sequence $\{e_k\}_{k\in\mathbb{Z}}$ of functions

(1) $$e_k(\omega) = \exp(2i\pi k\omega)$$

forms an orthogonal base for the complex Hilbert space $L^2(-\frac{1}{2}, +\frac{1}{2})$.

First of all it is clear that the sequence $\{e_k\}_{k\in\mathbb{Z}}$ is orthogonal, since

$$(e_k, e_l) = \int_{-1/2}^{+1/2} \exp(2i\pi k\omega) \exp(-2i\pi l\omega)\, d\omega$$

$$= \int_{-1/2}^{+1/2} \exp(2i\pi(k-l)\omega)\, d\omega = \begin{cases} 0 & \text{if } k \neq l. \\ 1 & \text{if } k = l. \end{cases}$$

Consequently, if P_n is the subspace generated by the functions e_k for $|k| \leq n$, the orthogonal projector t_n onto P_n can be written

$$t_n x = \sum_{|k| \leq n} (x, e_k) e_k$$

(2) $$= \sum_{|k| \leq n} \left(\int_{-1/2}^{+1/2} x(\omega) \exp(-2i\pi k\omega)\, d\omega \right) e_k.$$

We are going to show that the orthogonal sequence $\{e_n\}_{n\in\mathbb{Z}}$ is a *base*.

THEOREM 1
The functions $\omega \to \exp(2i\pi k\omega)$ (where $k \in \mathbb{Z}$) form an orthonormal base for $L^2(-\frac{1}{2}, +\frac{1}{2})$. ▲

Proof. We must show that the space $\mathscr{A}$ generated by the functions e_k when $k \in \mathbb{Z}$ is dense in $L^2(-\frac{1}{2}, +\frac{1}{2})$.

Since $\mathscr{D}(-\frac{1}{2}, +\frac{1}{2})$ is dense in $L^2(-\frac{1}{2}, +\frac{1}{2})$ and since $\mathscr{D}(-\frac{1}{2}, +\frac{1}{2})$ is contained in the space $\mathscr{C}_\pi(-\frac{1}{2}, +\frac{1}{2})$ of continuous functions f on $[-\frac{1}{2}, +\frac{1}{2}]$ for which $f(-\frac{1}{2}) = f(+\frac{1}{2})$, we have as a result that $\mathscr{C}_\pi(-\frac{1}{2}, +\frac{1}{2})$ is dense in $L^2(-\frac{1}{2},+\frac{1}{2})$. It suffices, therefore, to show that $\mathscr{A}$ is dense in $\mathscr{C}_\pi(-\frac{1}{2},+\frac{1}{2})$. But if ρ is the function associating to $\omega \in [-\frac{1}{2},+\frac{1}{2}]$ the complex number $\exp(2i\pi\omega)$ of the unit circle $\chi = \{z \in \mathbb{C}$ such that $|z| = 1\}$, the application T that associates to $f \in \mathscr{C}(\chi)$ the function $f\rho \in \mathscr{C}_\pi(-\frac{1}{2},+\frac{1}{2})$ is an isomorphism that maps the algebra $\mathscr{B}$ generated by the monomials $1, z_k$, and $\bar{z}^k$ onto the algebra $\mathscr{A}$. Hence it is sufficient to show that $\mathscr{B}$ is dense in $\mathscr{C}(\chi)$. Since χ is compact, since $\mathscr{B}$ contains the constants, and since the function $z \to z$ separates the points of χ, the theorem of Stone-Weierstrass implies precisely that $\mathscr{B}$ is dense in $\mathscr{C}(\chi)$. (See Theorem 5.2.1 of [AAA], p. 182.) ∎

The functions $t_n x$ are the *partial sums* of the *Fourier series* of x. Theorem 1 implies that the sequence $t_n x$ converges in *quadratic mean* to x. This allows us to say that every function $x \in L^2(-\frac{1}{2},+\frac{1}{2})$ can be written as

(3) $$x = \sum_{k \in \mathbb{Z}} \left(\int_{-1/2}^{+1/2} x(\omega) \exp(-2i\pi k\omega)\, d\omega \right) e_k.$$

Estimate of the Error

Consider now the prehilbert space $\mathscr{C}_\pi(-\frac{1}{2},+\frac{1}{2})$ with the scalar product

$$((x,y))_1 = (x,y) + (Dx, Dy),$$

We remark that

(4) $$((x, e_0))_1 = (x, e_0)$$

and that

(5) $$\begin{aligned}((x, e_k))_1 &= \int_{-1/2}^{+1/2} x(\omega) \bar{e}_k(\omega)\, d\omega + \int_{-1/2}^{+1/2} Dx(\omega) D\bar{e}_k(\omega)\, d\omega \\ &= \int_{-1/2}^{+1/2} x(\omega) [\bar{e}_k(\omega) - D^2 \bar{e}_k(\omega)\, d\omega]\, d\omega \\ &= \alpha_k \int_{-1/2}^{+1/2} x(\omega) \bar{e}_k(\omega)\, d\omega = \alpha_k(x, e_k)\end{aligned}$$

where

(6) $$\alpha_k = 1 + (2\pi k)^2.$$

This implies that *the functions* $(1/\sqrt{\alpha_k}) e_k$ (*for* $|k| \le n$) *form an orthonormal sequence in the prehilbert space* $\mathscr{C}_\pi(-\frac{1}{2},+\frac{1}{2})$ [*under the scalar product* $((x,y))_1$]. Consequently, Theorem 1.7.1 implies that

(7) $$\sum_{k \in \mathbb{Z}} \left|\left(\left(x, \frac{e_k}{\sqrt{\alpha_k}}\right)\right)_1\right|^2 = \sum_{k \in \mathbb{Z}} \alpha_k |(x, e_k)|^2 \le \|x\|_1^2.$$

PROPOSITION 1
For every $x \in \mathscr{C}^{(1)}(-\frac{1}{2}, +\frac{1}{2})$ the Fourier series of x is convergent in quadratic mean. Moreover,

$$(8) \qquad |x - t_{n-1} x| \leq \frac{1}{\sqrt{1 + (2\pi n)^2}} \|x\|_1. \qquad \blacktriangle$$

Proof. Indeed, since $\alpha_k \geq \alpha_n$ if $|k| \geq n$, we obtain

$$|x - t_{n-1} x|^2 \leq \sum_{|k| \geq n} \frac{\alpha_k}{\alpha_n} |(x, e_k)|^2 = \frac{1}{\alpha_n} \sum_{|k| \geq n} \alpha_k |(x, e_k)|^2$$

$$\leq \frac{1}{\alpha_n} \sum_{k \in \mathbb{Z}} \alpha_k |(x, e_k)|^2 \leq \frac{1}{\alpha_n} \|x\|_1^2. \qquad \blacksquare$$

Remark 1

The study of the convergence of Fourier series for other topologies forms an important part of *harmonic analysis*. ∎

4. APPROXIMATION BY STEP FUNCTIONS

Consider the space $L^2(\mathbb{R}^n)$. Its *discrete analogue* is the space $l^2(\mathbb{Z}^n)$ of square summable sequences $u = \{u^j\}_{j \in \mathbb{Z}^n}$, where $\mathbb{Z}^n$ is the set of n-tuples $j = (j_1, j_2, \ldots, j_n)$ whose components are positive or negative integers.

We are going to associate in *as simple a fashion as possible* the sequences with the functions and the functions with the sequences.

Let $h \in \,]0, 1[$ be a parameter (denoting the "step" or "mesh"). We associate with h the grid formed of the points jh as j runs over $\mathbb{Z}^n$.

The Extension Operator p_h^0

Let θ_{jh} be the characteristic function of the "cube" $[jh, (j+1)h] = \prod_{k=1}^{n} [j_k h, (j_k + 1)h]$. The operator $p_h^0 \in \mathscr{L}(l^2(\mathbb{Z}^n), L^2(\mathbb{R}^n))$ associates to every sequence $u \in l^2(\mathbb{Z}^n)$ the step function $p_h^0 u$ defined by

$$(1) \qquad p_h^0 u = \sum_{j \in \mathbb{Z}^n} u^j \theta_{jh}.$$

It is clear that

$$(2) \qquad |p_h^0 u| = h^{n/2} \left(\sum_{j \in \mathbb{Z}^n} |u^j|^2 \right)^{1/2}.$$

We set $|u|_h = |p_h^0 u|$.

Remark 1

In approximation theory we use sequences of scalars (those that appear on the "listing" of a computer, for example). These sequences of scalars do not have an intrinsic meaning; they are useful only when they are "interpreted" (as functions, for example) by means of operators (called extensions) such as p_h^0, which associates to every sequence a step function. ■

The Restriction Operator r_h

We can associate to every function λ satisfying

(3) $\quad \lambda \in L^\infty(\mathbb{R}^n) \quad$ with compact support, positive, $\quad \int \lambda(\omega)\,d\omega = 1$

an operator $r_h \in \mathcal{L}(L^2(\mathbb{R}^n), l^2(\mathbb{Z}^n))$ defined in the following fashion:

(4) $$(r_h x)^j = \frac{1}{h^n} \int x(\omega) \lambda\left(\frac{\omega}{h} - j\right) d\omega.$$

The Cauchy-Schwarz inequality implies that

$$|(r_h x)^j|^2 \leq \frac{1}{h^n}\left[\int |x(\omega)|^2 \lambda\left(\frac{\omega}{h} - j\right) d\omega\right] \int \lambda\left(\frac{\omega}{h} - j\right) d\omega$$

$$= \frac{1}{h^n} \int |x(\omega)|^2 \lambda\left(\frac{\omega}{h} - j\right) d\omega.$$

Consequently,

$$|p_h^0 r_h x| = h^n \sum_{j \in \mathbb{Z}^n} |(r_h x)^j|^2 = \int |x(\omega)|^2 \left(\sum_{j \in \mathbb{Z}^n} \lambda\left(\frac{\omega}{h} - j\right)\right) d\omega.$$

If we set $c^2 = \sup_\omega (\sum_{j \in \mathbb{Z}^n} \lambda(\omega - j))$, we derive the inequality

(5) $$|p_h^0 r_h x| \leq c |x|.$$

The Operators $p_h^0 r_h$

The operator $p_h^0 r_h$, therefore, associates to every function $x \in L^2(\mathbb{R}^n)$ a step function $p_h^0 r_h x$ on a grid of mesh h. As one would expect, $p_h^0 r_h x$ converges to x in $L^2(\mathbb{R}^n)$.

PROPOSITION 1

For every $x \in L^2(\mathbb{R}^n)$ we obtain the inequality

(6) $$|x - p_h r_h x| \leq \omega(x, ch)$$

where c is a constant independent of h and of x and $\omega(x, ch)$ is the oscillation of x. (See Section 6.6.) ▲

Proof. Since $\sum_j \theta_h^j(\omega) = 1$ and since $x(\omega) = 1/h^n \int x(\omega) \lambda(\zeta/h - j)\, d\zeta$, we can write

$$|x(\omega) - p_h r_h x(\omega)|^2 = \sum_{j \in \mathbb{Z}^n} \theta_h^j(\omega) \frac{1}{h^{2n}} \left| \int (x(\omega) - x(\zeta)) \lambda\left(\frac{\zeta}{h} - j\right) d\zeta \right|^2$$

$$\leq \sum_{j \in \mathbb{Z}^n} \theta_h^j(\omega) \left[\frac{1}{h^n} \int |x(\omega) - x(\zeta)|^2 \lambda\left(\frac{\zeta}{h} - j\right) d\zeta \right]$$

$$\times \left[\frac{1}{h^n} \int \lambda\left(\frac{\zeta}{h} - j\right) d\zeta \right]$$

$$\leq \sum_{j \in \mathbb{Z}^n} \theta_h^j(\omega) \frac{1}{h^n} \int |x(\omega) - x(\zeta)|^2 \lambda\left(\frac{\zeta}{h} - j\right) d\zeta$$

according to the Cauchy-Schwarz inequality and the fact that $\int \lambda(\omega) \, d\omega = 1$. Integrating this inequality on $\mathbb{R}^n$, we obtain

$$\int |x(\omega) - p_h r_h x(\omega)|^2 \, d\omega = \frac{1}{h^n} \sum_j \int_{\mathbb{R}^n} \int_{\mathbb{R}^n} \lambda\left(\frac{\zeta}{h} - j\right) \theta_h^j(\omega) |x(\omega) - x(\zeta)|^2 \, d\zeta \, d\omega$$

$$= \sum_j \int_{\mathbb{R}^n} \int_{\mathbb{R}^n} \lambda(\eta) \theta_h^j(\omega) |x(\omega) - x(\eta h - jh)|^2 \, d\eta \, d\omega.$$

But η runs over the support K of λ, which is assumed to be contained in a ball of radius d. Hence

$$\sum_j \int_{\mathbb{R}^n} \theta_h^j(\omega) |x(\omega) - x(\eta h + jh)|^2 \, d\omega \leq \sup_{\|\zeta\| \leq (d+1)h} \int_{\mathbb{R}^n} |x(\omega) - x(\omega + \zeta)|^2 \, d\omega$$

$$= \omega(x, (d+1)h)^2.$$

Consequently,

$$\int |x(\omega) - p_h r_h x(\omega)|^2 \, d\omega \leq \omega(x, (d+1)h)^2 \int \lambda(\eta) \, d\eta = \omega(x, (d+1)h)^2. \quad \blacksquare$$

5. APPROXIMATION BY PIECEWISE POLYNOMIAL FUNCTIONS

Naturally, approximations by step functions are the simplest, but they are also the least smooth. Nevertheless, it can be interesting to approximate functions by smoother functions, at the price, of course, of greater complexity.

Since the convolution product by a function ρ is at least as many times differentiable as ρ, we can consider approximating a function x by $\rho_h * p_h^0 r_h x$ where $\rho_h(\omega) = (1/h^n)\rho(\omega/h)$ when

(1) $\qquad \rho \in \mathscr{C}_0(\mathbb{R}^n), \qquad \rho \geq 0 \quad$ and $\quad \int \rho(\omega)\, d\omega = 1.$

PROPOSITION 1
Every $x \in L^2(\mathbb{R}^n)$ can be approximated by the functions

$$\rho_h * p_h^0 r_h x = \sum_j (r_h x)^j (\rho_h * \theta_h^j)(\omega). \qquad \blacktriangle$$

CH. 8, SEC. 5   APPROXIMATION BY PIECEWISE POLYNOMIAL FUNCTIONS   177

Proof. We write that $|x - \rho_h * p_h^0 r_h x| \leq |x - \rho_h * x| + |\rho_h * (x - p_h^0 r_h x)| \leq |x - \rho_h * x| + \|\rho_h\|_1 |x - p_h^0 r_h x| = |x - \rho_h * x| + |x - p_h^0 r_h x|$. These two terms converge to zero according to Theorem 6.6.1 and Proposition 4.1. ∎

The most natural choice for a smoothing function ρ is $\rho = \chi^{*m}$ (where χ denotes the characteristic function of $[0, 1]$.)

To simplify the discussion we restrict ourselves to the case of functions of one variable ($n = 1$). This leads us to define the *extension operator* $p_h^m \in \mathscr{L}(l^2(\mathbb{Z}), L^2(\mathbb{R}))$ by

$$(2) \qquad p_h^m u = \sum_{j \in \mathbb{Z}^n} u^j \chi_h^{*m} * \theta_{jh}.$$

We see (by a simple calculation) that

$$(3) \qquad \chi_h^{*m} * \theta_{jh}(\omega) = \chi^{*(m+1)}\left(\frac{\omega}{h} - j\right)$$

where the functions $\chi^{*(m+1)}$ (studied in Chapter 6, Section 7) are written in the form

$$(4) \qquad \chi^{*(m+1)}(\omega) = \sum_{k=0}^{m} \alpha_m^k(\omega - k)\chi(\omega - k)$$

(See Proposition 6.7.1). It follows that

$$(5) \qquad p_h^m u = \sum_{j \in \mathbb{Z}} \theta_{jh}(\omega) \left(\sum_{k=0}^{m} u^{j-k} \alpha_m^k \left(\frac{x}{h} - j \right) \right)$$

since (3) and (4) imply that

$$(6) \qquad \chi_h^{*m} * \theta_{jh}(\omega) = \sum_{k=0}^{m} \alpha_m^k \left(\frac{\omega}{h} - k - j \right) \theta_{(j+k)h}(\omega).$$

Hence the restriction of $p_h^m u$ to each interval $]jh, (j+1)h[$ is a polynomial of degree m and $p_h^m u$ is $(m-1)$ times continuously differentiable.

This formula indeed shows that smoothness is obtained at the price of complexity (the degree of the piecewise polynomials).

These approximations provide the simplest examples of "approximation by finite elements." Certain authors denote the functions $p_h^m u$ by the term *Spline functions*.

*Estimate of the Error

We shall show that the smoothness of the approximation is tied to the speed of convergence. To this end let us recall that the functions $\chi^{*(m+1)}$

satisfy the conditions

$$\sum_{k\in\mathbb{Z}} \frac{k^j}{j!} \chi^{*(m+1)}(\omega - k) = \sum_{p=0}^{j} b^{j-p} \frac{\omega^p}{p!} \tag{7}$$

if $0 \le j \le m$ (Proposition 6.7.3). We associate to these scalars $b^0 = 1, \ldots, b^m$ the scalars $d^0 = 1, d^1, \ldots, d^m$ defined by

$$d^p + \sum_{j=1}^{p} d^{p-j} b^j = 0. \tag{8}$$

We choose a function $\lambda \in \mathscr{C}_0(\mathbb{R}^n)$ that is positive and of total mass equal to one such that

$$\int \lambda(\omega) \frac{\omega^p}{p!} d\omega = d^p \quad \text{for} \quad 0 \le p \le m. \tag{9}$$

Let us consider the restriction operators r_h associated with λ by (1.4). We can then establish the following estimate of the error.

THEOREM 1
Suppose that the function λ satisfies (8) and (9). If the derivative of order $(m+1)$ of the function $x \in L^2(\mathbb{R})$ exists and belongs to $L^2(\mathbb{R})$, then

$$|x - p_h^m r_h x| \le ch^{m+1} |D^{m+1} x|. \tag{10}$$ ▲

Proof. We are going to use the Taylor expansion of the function x on each interval $[jh, (j+1)h]$; we write

$$x(\omega) = A_{jh}(\omega) + B_{jh}(\omega) \quad \text{if} \quad jh \le \omega \le (j+1)h \tag{11}$$

where

$$A_{jh}(\omega) = \sum_{q=0}^{m} h^q (D^q x)(jh) \frac{\left(\frac{\omega}{h} - j\right)^q}{q!} \tag{12}$$

and

$$B_{jh}(\omega) = \int_{jh}^{\omega} \frac{(\omega - \eta)^m}{m!} (D^{m+1} x)(\eta) \, d\eta. \tag{13}$$

As a first step we are going to show that

$$A_{jh}(\omega) = (p_h r_h A_{jh})(\omega) \quad \text{if} \quad jh \le \omega \le (j+1)h. \tag{14}$$

For this it suffices [due to (5)] to show that

$$\frac{\left(\frac{\omega}{h} - j\right)^q}{q!} = p_h r_h \frac{\left(\frac{\omega}{h} - j\right)^q}{q!} \quad \text{if} \quad jh \le \omega \le (j+1)h, \quad 0 \le q \le m. \tag{15}$$

CH. 8, SEC. 5 APPROXIMATION BY PIECEWISE POLYNOMIAL FUNCTIONS

Calculating the right-hand side of this last equation gives

(16) $$p_h r_h \frac{\left(\frac{\omega}{h} - j\right)^q}{q!} = \sum_{k \in \mathbb{Z}} \left(\int \lambda(\omega) \frac{(\omega + k - j)^q}{q!} d\omega \right) \chi^{*(m+1)}\left(\frac{\omega}{h} - k\right).$$

Using the Newton binomial formula, we obtain

(17) $$\int \lambda(\omega) \frac{(\omega + k - j)^q}{q!} d\omega = \sum_{p=0}^{q} d^{q-p} \frac{(k-j)^p}{p!}.$$

Hence, by the choice of the d^p's [in (8), and by property (7), which was established in Proposition 6.7.3], we obtain

$$\sum_{k \in \mathbb{Z}} \frac{(k-j)^p}{p!} \chi^{*(m+1)}\left(\frac{\omega}{h} - k\right) = \sum_{k \in \mathbb{Z}} \frac{k^p}{p!} \chi^{*(m+1)}\left(\frac{\omega}{h} - k - j\right)$$

$$= \sum_{i=0}^{p} b^{p-i} \frac{\left(\frac{\omega}{h} - j\right)^i}{i!}.$$

Consequently, these relations imply that

$$p_h r_h \frac{\left(\frac{\omega}{h} - j\right)^q}{q!} = \sum_{p=0}^{q} \sum_{i=0}^{p} d^{q-p} b^{p-i} \frac{\left(\frac{\omega}{h} - j\right)^i}{i!}$$

$$= \sum_{i=0}^{q} \frac{\left(\frac{\omega}{h} - j\right)^i}{i!} \sum_{p=i}^{q} d^{q-p} b^{p-i}$$

$$= \sum_{i=0}^{q} \frac{\left(\frac{\omega}{h} - j\right)^i}{i!} \sum_{r=0}^{q-i} d^{q-i-r} b^r = \frac{\left(\frac{\omega}{h} - j\right)^q}{q!}.$$

We therefore derive from (14) that

$$(1 - p_h^m r_h) x(\omega) = (1 - p_h^m r_h) B_{jh}(\omega) \quad \text{if} \quad jh \leq \omega \leq (j+h)h.$$

Thus we can write that

(18) $$|(1 - p_h^m r_h) x|^2 = \sum_{j \in \mathbb{Z}} U_{jh}$$

where

(19) $$U_{jh} = \int_{jh}^{(j+1)h} |(1 - p_h^m r_h) B_{jh}(\omega)|^2 \, d\omega.$$

We are going to show that $2(V_{jh} + W_{jh})$ is an upper bound for U_{jh} where

(20) $$V_{jh} = \int_{jh}^{(j+1)h} |B_{jh}(\omega)|^2 \, d\omega$$

and

(21) $$W_{jh} = \int_{jh}^{(j+1)h} |p_h^m r_h B_{jh}(\omega)|^2 \, d\omega,$$

To find an upper bound for V_{jh} we use the Cauchy-Schwarz inequality:

(22) $$|B_{jh}(\omega)|^2 \leq \frac{(\omega - jh)^{2(m+1)-1}}{m!(2(m+1)-1)} \int_{jh}^{\omega} |D^{m+1}x(\omega)|^2 \, d\omega.$$

Integrating this inequality from jh to $(j+1)h$, we obtain

(23) $$V_{jh} \leq ch^{2(m+1)} \int_{jh}^{(j+1)h} |D^{m+1}x(\omega)|^2 \, d\omega.$$

Now to find an upper bound for W_{jh}, we use the fact that the Cauchy-Schwarz inequality implies

$$|p_h^m r_h B_{jh}(\omega)|^2$$

(24) $$\leq \left(\sum_{p \in \mathbb{Z}} |(r_h B_{jh})^p|^2 \chi^{*(m+1)}\left(\frac{\omega}{h} - p\right) \right) \left(\sum_{p \in \mathbb{Z}} \chi^{*(m+1)}\left(\frac{\omega}{h} - p\right) \right)$$

$$= \sum_{p \in \mathbb{Z}} |(r_h B_{jh})^p|^2 \chi^{*(m+1)}\left(\frac{\omega}{h} - p\right)$$

since

$$\sum_{p \in \mathbb{Z}} \chi^{*(m+1)}(\omega - p) = 1.$$

Moreover, if $[a, b]$ denotes the support of λ, the Cauchy-Schwarz inequality also implies that

$$|(r_h B_{jh})^p|^2 \leq c \frac{1}{h} \int_{(a+p)h}^{(b+p)h} |B_{jh}(\omega)|^2 \, dx.$$

Integrating $p_h^m r_h B_{jh}$ on $[jh, (j+1)h]$, we derive from inequality (24) the following estimate

$$W_{jh} \leq \sum_{p=j-m}^{j} h|(r_h B_{jh})^p|^2 \leq c \sum_{p=j-m}^{j} \int_{(a+p)h}^{(b+p)h} |B_{jh}(\omega)|^2 \, dx$$

since the support of $\chi^{*(m+1)}(\omega/h - p)$ contains the interval $[jh, (j+1)h]$ only if $j - m \leq p \leq j$.

Now using the upper bound (22) for $|B_{jh}(\omega)|^2$, it follows that

$$W_{jh} \leq dh^{2(m+1)} \int_{(j-r)h}^{(j+r)h} |D^{m+1}x(\omega)|^2 \, d\omega$$

where d and r are constants independent of h and j. Consequently,

$$U_{jh} \leq c'h^{2(m+1)} \int_{(j-r)h}^{(j+r)h} |D^{m+1}x(\omega)|^2 \, d\omega,$$

which completes the proof. ∎

6. APPROXIMATION IN SOBOLEV SPACES

Consider the Sobolev space $H^k(\mathbb{R})$. If $m \geq k$, the functions $p_h^m u = \chi_h^{*m} * p_h^0 u$ defined by (2) in Section 5 belong to $H^k(\mathbb{R})$, since if $j \leq k \leq m$,

(1) $$D^j p_h^m u = D^j \chi_h^{*j} * \chi_h^{*(m-j)} * p_h^0 u = \nabla_h^j p_h^{m-j} u \in L^2(\mathbb{R}).$$

We associate to every function

(2) $\lambda \in L^1(\mathbb{R})$ with compact support, positive, and $\int \lambda(\omega) \, d\omega = 1$

the operator $r_h \in \mathscr{L}(L^2(\mathbb{R}), l^2(\mathbb{Z}))$ defined by

(3) $$(r_h x)^l = \frac{1}{h} \int x(\omega) \lambda\left(\frac{\omega}{h} - l\right) d\omega.$$

PROPOSITION 1
If $m \geq k$, every function $x \in H^k(\mathbb{R})$ can be approximated by the functions $p_h^m r_h x$ as h approaches zero. ▲

Proof. We already know that $p_h^m r_h x$ converges to x in $L^2(\mathbb{R})$. We must show that if $j \leq k$, $D^j p_h^m r_h x = \nabla_h^j p_h^{m-j} r_h x$ converges to $D^j x$ in $L^2(\mathbb{R})$. We denote by r_h^j the operator defined by

(4) $$(r_h^j x)^l = \frac{1}{h} \int x(\omega) (\check{\chi}^{*j} * \lambda)\left(\frac{\omega}{h} - l\right) d\omega$$

where the function $\check{\chi}^{*j} * \lambda \in L^1(\mathbb{R})$ is positive and has total mass equal to one.
  We are going to show that

(5) $$D^j p_h^m r_h x = p_h^{m-j} r_h^j D^j x.$$

[Proposition 5.1 will then imply that $p_h^{m-j} r_h^j D^j x$ converges to $D^j x$ in $L^2(\mathbb{R})$.] Hence let us establish (5).

First we are going to show that

(6) $$\nabla_h^j p_h^{m-j} r_h x = p_h^{m-j} r_h \nabla_h^j x.$$

Indeed, by definition we have

(7) $$\nabla_h^j(p_h^{m-j} r_h x)(\omega) = h^{-j} \sum_{q=0}^{j} (-1)^q \binom{j}{q} (p_h^{m-j} r_h x)(\omega - qh).$$

Moreover,

$$\begin{aligned}(p_h^{m-j} r_h x)(\omega - qh) &= \sum_{l \in \mathbb{Z}} (r_h x)^l \chi_h^{*(m-j)} * \theta_{lh}(\omega - qh) \\ &= \sum_{l \in \mathbb{Z}} (r_h x)^l \chi_h^{*(m-j)} * \theta_{(l+q)h}(\omega) \\ &= \sum_{l \in \mathbb{Z}} (r_h x)^{l-q} \chi_h^{*(m-j)} * \theta_{lh}(\omega) \\ &= \sum_{l \in \mathbb{Z}} \frac{1}{h} \left(\int_\Omega x(\omega) \lambda\left(\frac{\omega}{h} - l + q\right) d\omega \right) \chi_h^{*(m-j)} * \theta_{lh}(\omega) \\ &= \sum_{l \in \mathbb{Z}} \frac{1}{h} \left(\int x(\omega - qh) \lambda\left(\frac{\omega}{h} - l\right) d\omega \right) \chi_h^{*(m-j)} * \theta_{lh}(\omega).\end{aligned}$$

Thus (6) follows from (7) and from this equality. Furthermore,

$$\begin{aligned}(r_h \nabla_h^j x)^l &= r_h(\chi_h^{*j} * D^j x)^l \\ &= \frac{1}{h} \int \lambda\left(\frac{\omega}{h} - l\right) d\omega \int \check{\chi}_h^{*j}(\omega - \eta) D^j x(\eta) \, d\eta \\ &= \frac{1}{h} \int (\lambda * \check{\chi}^{*j})\left(\frac{\omega}{h} - l\right) D^j x(\omega) \, d\omega.\end{aligned}$$

This implies

(8) $$r_h \nabla_h^j x = r_h^j D^j x.$$

Formula (5) follows, therefore, from (1), (6), and (8). ∎

Approximation in the Spaces $H^k(\Omega)$

Let us consider now an interval $\Omega =]a, b[$, along with the restriction operator ρ from $H^k(\mathbb{R})$ onto $H^k(\Omega)$ and the extension operator π, which is continuous from $H^k(\Omega)$ to $H^k(\mathbb{R})$, defined in Chapter 7, Section 6 and having the properties stated in Theorem 7.6.1.

PROPOSITION 2

If $m \geq k$, every function $x \in H^k(\Omega)$ can be approximated by the functions $\rho p_h^m r_h \pi x$ as h approaches zero ▲

Proof. Indeed, according to Proposition 1, $\rho p_h^m r_h \pi x$ converges in $H^k(\Omega)$ to $\rho \pi x = x$. ∎

We can, therefore, approximate $x \in H^k(\Omega)$ by piecewise polynomial functions of degree $m \geq k$.

Estimate of the Error

We choose the function λ_m defining r_h such that

(9) $$\forall j = 0, \ldots, m-1 \quad \int \lambda_m(\omega) \frac{\omega^j}{j!} d\omega = d_m^j$$

when the sequence d_m^j satisfies

(10) $$\sum_{k=0}^{j} d_m^{j-k} b_m^k = 0 \quad \text{if} \quad j > 0, \quad d_m^0 = 1,$$

where the scalars b_m^k are those that have the property

(11) $$\sum_{l \in \mathbb{Z}} \frac{l^j}{j!} \chi^{*(m+1)}(\omega - l) = \sum_{p=0}^{j} b_m^{p-j} \frac{\omega^p}{p!}.$$

Theorem 5.1 implies that $|x - p_h^m r_h x| \leq ch^{m+1} \|x\|_{m+1}$. We are going to establish an analogous upper bound for the error.

THEOREM 1
Suppose that the function λ_m satisfies (9). If $x \in H^{m+1}(\Omega)$, then

(12) $$\|x - \rho p_h^m r_h \pi x\|_k \leq ch^{m+1-k} \|x\|_{m+1}. \qquad \blacktriangle$$

Proof. We are going to verify that

(13) $$|D^j(x - \rho p_h^m r_h \pi x)| \leq ch^{m+1-j} |D^{m+1} \pi x| \quad \text{if} \quad j \leq k.$$

To this end we know that $D^j \rho p_h^m r_h \pi x = \rho p_h^{m-j} r_h^j D^j \pi x$ where r_h^j is the restriction operator associated with the function $\lambda * \check{\chi}^{*j}$. It is then necessary to verify that its moments $d_{m+j}^q = \int (\lambda * \check{\chi}^{*j})(\omega)(\omega^q/q!) d\omega$ satisfy the relations

(14) $$\sum_{q=0}^{l} b_{m-j}^{l-q} d_{m+j}^q = 0 \quad \text{if} \quad q \geq 1,$$

since, in this case, Theorem 5.1 implies the upper estimate

$$|D^j(x - \rho p_h^m r_h \pi x)| = |D^j x - \rho p_h^{m-j} r_h^j D^j \pi x| \leq ch^{m+1-j} |D^{m+1} \pi x|.$$

The definition of the d_{m+j}^q's shows that

(15) $$d_{m+j}^q = \sum_{k=0}^{q} c_j^k \int \lambda(\omega) \frac{\omega^{q-k}}{(q-k)!} d\omega = \sum_{k=0}^{q} c_j^k d_m^{q-k}$$

where we have set

(16) $$c_j^k = \int \chi^{*j}(\omega) \frac{\omega^k}{k!} d\omega.$$

Taking the convolution product by χ^{*j} of the equality

(17) $$\sum_{l \in \mathbb{Z}} \frac{l^p}{p!} \chi^{*(m-j+1)}(\omega - l) = \sum_{q=0}^{p} b_{m-j}^{p-q} \frac{\omega^q}{q!}$$

and making use of (11), we derive the relations

(18) $$b_m^p = \sum_{q=0}^{p} b_{m-j}^{p-q} c_j^q.$$

Consequently,

$$\sum_{q=0}^{l} b_{m-j}^{l-q} d_{m+j}^q = \sum_{q=0}^{l} b_{m-j}^{l-q} \sum_{k=0}^{q} c_j^k d_m^{q-k} = \sum_{q=0}^{l} b_m^{l-q} d_m^q = 0 \quad \text{if} \quad l > 0,$$

which completes the proof of the theorem. ∎

CHAPTER 9

Sobolev Spaces of Functions of Several Variables and the Fourier Transform

This chapter is devoted to the study of Sobolev spaces of functions defined on an open set Ω of $\mathbb{R}^n$ and to an introduction to the Fourier Transform.

We construct successively the spaces $H_0^m(\Omega)$, $H^{-m}(\Omega)$ and $H^m(\Omega)$ as in the case of functions of one variable. (See Chapter 7, Sections 1 to 5.) We also show that $H_0^m(\mathbb{R}^n) = H^m(\mathbb{R}^n)$.

For a bounded open set we characterize the space $H_0^m(\Omega)$ as the kernel of a trace operator, that is, as the subspace of the functions $x \in H^m(\Omega)$ whose "traces" $\gamma_j x$ on the boundary Γ of the normal derivatives of order $j \leq m-1$ are zero. This time the boundary of Ω does not reduce to two points, as in the case of an interval, but to a manifold Γ of dimension $n-1$. Hence the traces $\gamma_j x$ are functions on Γ whose properties need to be studied.

The essential tool that we use is the Fourier transform. We devote Sections 2 and 3, therefore, to an introduction to the Fourier transform F, which associates to every function $\varphi \in \mathscr{D}(\mathbb{R})$ the function $F\varphi$ defined by

$$F\varphi(\xi) = \int_{-\infty}^{+\infty} e^{-2i\pi\xi\eta} \varphi(\eta) \, d\eta.$$

The importance of the role played by this operator is due to the fact that it exchanges the differential operators D^k and the multiplication operators by the monomials ω^k, the translation operators and the multiplication operators by the exponentials $\exp(ia\omega)$, the products of functions and the convolution products. Furthermore, the Fourier transform is an isometry from $L^2(\mathbb{R})$ onto $L^2(\mathbb{R})$, and its inverse is the inverse Fourier transform $\bar{F}$ defined by

$$\bar{F}\psi(\eta) = \int_{-\infty}^{+\infty} e^{2i\pi\xi\eta} \psi(\xi) \, d\xi.$$

We also establish that F is an isomorphism from the Sobolev spaces

$H^m(\mathbb{R}^n)$ onto the spaces $\hat{H}^m(\mathbb{R}^n)$ of square summable functions for the weight $(1 + \|\omega\|^2)^m$. This allows us to "transport" the properties of these spaces to the Sobolev spaces. (For example, we obtain in this fashion the Sobolev and the interpolation inequalities.) Above all, this allows us to state and prove in a precise way the trace theorem and to construct an extension operator π from $H^m(\Omega)$ to $H^m(\mathbb{R}^n)$.

1. THE SOBOLEV SPACES $H_0^m(\Omega)$, $H^m(\Omega)$, AND $H^{-m}(\Omega)$

Consider an open set Ω of $\mathbb{R}^n$. We introduce the space

(1) $\quad \mathscr{D} = \mathscr{D}(\Omega)$ of functions infinitely differentiable with compact support in Ω

as well as the pivot space

(2) $\quad \begin{cases} H = L^2(\Omega), \text{ the completion of } \mathscr{D}(\Omega) \text{ for the} \\ \text{scalar product } (\varphi, \psi) = \int_\Omega \varphi(\omega)\psi(\omega)\,d\omega. \end{cases}$

We introduce the family $\mathscr{A}_m$ of differential operators

(3) $\qquad\qquad D^p = D_1^{p_1} \ldots D_n^{p_n} = \dfrac{\partial^{|p|}}{\partial \omega_1^{p_1} \ldots \partial \omega_n^{p_n}}$

of order $|p| = p_1 + \ldots + p_n$ less than m, which are linear operators from $\mathscr{D}(\Omega)$ to $\mathscr{D}(\Omega)$ satisfying

(4) $\qquad \int_\Omega D^p \varphi(\omega)\psi(\omega)\,d\omega = (-1)^{|p|} \int_\Omega \varphi(\omega) D^p \psi(\omega)\,d\omega$

according to the formula for integration by parts. We prove, *as in the case of functions of one variable, that the family $\mathscr{A}_m$ of operators D^p for $|p| \leq m$ is closed.* Since $\mathscr{A}_m$ contains the identity (identified with D^0), the family $\mathscr{A}_m$ is *collectively injective*.

Hence there exists, according to Proposition 5.4.1, a realization of the completion of $\mathscr{D}(\Omega)$ for the scalar product

(5) $\qquad ((\varphi, \psi))_m = \sum_{|p| \leq m} (D^p \varphi, D^p \psi) = \sum_{|p| \leq m} \int_\Omega D^p \varphi(\omega) D^p \psi(\omega)\,d\omega,$

which is contained in $L^2(\Omega)$.

DEFINITION 1

We denote by $H_0^m(\Omega)$ the completion of $\mathscr{D}(\Omega)$ for the scalar product $((\cdot,\cdot))_m$ which is a dense subspace of $L^2(\Omega)$ and by $H^{-m}(\Omega)$ its dual. We say that $H_0^m(\Omega)$ is the (minimal) Sobolev space of order m and that $H^{-m}(\Omega)$ is the Sobolev space of order $-m$. ▲

Propositions 5.3.1 and 5.4.1 imply that

(6) $\quad \mathscr{D}(\Omega) \subset H_0^m(\Omega) \subset H_0^k(\Omega) \subset L^2(\Omega) \subset H^{-k}(\Omega) \subset H^{-m}(\Omega) \subset \mathscr{D}^*(\Omega)$

(when $m \geq k$), each space being *dense* in the following spaces [except $\mathscr{D}^*(\Omega)$].

The elements of $H^{-m}(\Omega)$ are called *distributions* or *generalized functions*. The spaces $H^{-m}(\Omega)$ are bigger as m is larger.

We make the following abuse of the notation: If $x \in H^{-m}(\Omega)$ and $\varphi \in H_0^m(\Omega)$, we denote by

(7) $\quad\quad\quad\quad (x, \varphi) = \int_\Omega x(\omega)\varphi(\omega)\, d\omega$

the duality pairing (which extends the $L^2(\Omega)$ scalar product), even though we have no right at all to write $x(\omega)$ if x is not a function from Ω to $\mathbb{R}$, but only a continuous linear form on $H_0^m(\Omega)$.

If $|p| \leq m$, the *differential operator* D^p has a unique extension to a continuous linear operator from $H_0^m(\Omega)$ to $L^2(\Omega)$.

If $x \in H_0^m(\Omega)$, we say that $D^p x \in L^2(\Omega)$ is the *weak derivative* (or the *derivative in the sense of distributions*) of order p of x.

We remark that the transpose $((-1)^{|p|} D^p)^* \in \mathscr{L}(L^2(\Omega), H^{-m}(\Omega))$ of the operator $(-1)^{|p|} D^p \in \mathscr{L}(H_0^m(\Omega), L^2(\Omega))$ is, according to (4), an extension of the linear operator D^p from $\mathscr{D}(\Omega)$ to $\mathscr{D}(\Omega) \subset H^{-m}(\Omega)$. Since $\mathscr{D}(\Omega)$ is dense in $L^2(\Omega)$ and since $((-1)^{|p|} D^p)^*$ is continuous from $L^2(\Omega)$ to $H^{-m}(\Omega)$, this extension is *unique*. Hence we set

(8) $\quad\quad D^p = ((-1)^{|p|} D^p)^* \in \mathscr{L}(L^2(\Omega), H^{-m}(\Omega))$ $\quad$ if $\quad$ $|p| \leq m$.

For every $x \in L^2(\Omega)$, we say that the distribution $D^p x \in H^{-m}(\Omega)$ defined by

(9) $\quad\quad\quad\quad (D^p x, \varphi) = (-1)^{|p|} (x, D^p \varphi) \quad\quad \forall \varphi \in H_0^m(\Omega)$

is the *weak derivative* (or the *derivative in the sense of distributions*) of order p of x.

Moreover, since $(-1)^{|k|} D^k$ is, of course, continuous from $H_0^{m+|k|}(\Omega)$ to $H_0^m(\Omega)$ if $m \geq 0$, it follows by transposition that it has a unique extension to a continuous linear operator D^k from $H^{-m}(\Omega)$ to $H^{-m-|k|}(\Omega)$ for every $m \geq 0$. Formula (9) allows us, therefore, to differentiate the distributions of $H^{-m}(\Omega)$ for all $m \geq 0$.

DEFINITION 2

We say that the subspace $H^m(\Omega)$ of $L^2(\Omega)$ defined by

(10) $\quad\quad H^m(\Omega) = \{x \in L^2(\Omega) \text{ such that } D^p x \in L^2(\Omega) \text{ for all } |p| \leq m\}$

with the scalar product

(11) $\quad\quad ((x, y))_m = \sum_{|p| \leq m} (D^p x, D^p y) = \sum_{|p| \leq m} \int_\Omega D^p x(\omega) D^p y(\omega)\, d\omega$

is the (*maximal*) *Sobolev space of order m.* ▲

Proposition 5.4.4 implies that the *Sobolev spaces* $H^m(\Omega)$ *are complete.*

We have seen that in the case of functions of one variable, $H_0^m(\Omega)$ can be a proper subspace of $H^m(\Omega)$. However, where $\Omega = \mathbb{R}^n$, they coincide.

THEOREM 1
The spaces $H_0^m(\mathbb{R}^n)$ *and* $H^m(\mathbb{R}^n)$ *coincide.* ▲

Proof. The proof is analogous to that of Theorem 6.5.1 (with functions of one variable). We approximate a function x through "cutting and smoothing," by associating to a parameter h, destined to approach zero, the function $\varphi_h \in \mathcal{D}(\mathbb{R}^n)$ defined by

(12) $$\varphi_h = \rho_h * (\theta(\cdot h) x(\cdot))$$

where $\theta \in \mathcal{D}(\mathbb{R}^n)$ is equal to one if $\|\omega\| \leq 1$ and to zero if $\|\omega\| \geq 2$ and where $\rho \in \mathcal{D}(\mathbb{R}^n)$ is positive and has a total mass equal to one. ∎

2. THE FOURIER TRANSFORM OF INFINITELY DIFFERENTIABLE AND RAPIDLY DECREASING FUNCTIONS

When we use the Fourier transform it is convenient to assume that the functions considered are complex valued, since, as we shall see, the Fourier transform of a function (real valued or complex valued) is complex valued. In this chapter we assume that the spaces of functions we use are spaces of complex-valued functions.

As usual we first define the Fourier transform on the functions $\varphi \in \mathcal{D}(\mathbb{R}^n)$ and then extend it by density to different spaces of functions and of distributions.

DEFINITION 1
Let $\varphi \in \mathcal{D}(\mathbb{R}^n)$. *We denote by* $F\varphi$ *and* $\bar{F}\varphi$ *the functions defined by*

(1)
i. $(F\varphi)(\xi) = \int_{\mathbb{R}^n} e^{-2i\pi \langle \xi, \eta \rangle} \varphi(\eta) \, d\eta$

ii. $(\bar{F}\varphi)(\xi) = \int_{\mathbb{R}^n} e^{2\pi i \langle \xi, \eta \rangle} \varphi(\eta) \, d\eta$

where $\xi = \{\xi_1, \ldots, \xi_n\}, \eta = \{\eta_1, \ldots, \eta_n\}$, *and where* $\langle \xi, \eta \rangle = \sum_{i=1}^n \xi_i \eta_i$. *We say that* $F\varphi$ *is the Fourier transform of* φ *and that the mappings* $\varphi \mapsto F\varphi$ *and* $\varphi \mapsto \bar{F}\varphi$ *are the direct and inverse* Fourier *transforms.* ▲

Abuses of notation are numerous in this chapter. For example, we denote by $\omega^k\varphi$ the function $\omega \mapsto \omega^k\varphi(\omega)$.

PROPOSITION 1
Let $\varphi \in \mathcal{D}(\mathbb{R}^n)$. Its Fourier transform is infinitely differentiable and

(2) $$D^k F\varphi = F((-2i\pi\omega)^k \varphi)$$

and is "rapidly decreasing" in the sense that for all k,

(3) $$(2i\pi\xi)^k F(\varphi)(\xi) = F[D^k \varphi].$$

Moreover, for every n-tuple of integers k and for every $\xi \in \mathbb{R}^n$,

(4) $$|2\pi\xi|^{|k|}|F\varphi(\xi)| \leq \|D^k\varphi\|_{L^1}$$

and

(5) $$|D^k(F\varphi)(\xi)| \leq \| |2\pi\omega|^{|k|}\varphi \|_{L^1}. \qquad \blacktriangle$$

Proof. Indeed, because the functions $\xi \mapsto e^{-2i\pi\langle \xi, \eta \rangle}$ are infinitely differentiable and the functions $\eta^k \varphi(\eta)$ are continuous and bounded (on the *compact* support of φ), we can differentiate $F\varphi(\cdot)$ under the $\int$ sign and obtain

$$D^k F\varphi(\xi) = \int_{\mathbb{R}^n} (D^k_\xi e^{-2i\pi\langle \xi, \eta \rangle}) \varphi(\eta) \, d\eta$$
$$= (-2i\pi\xi_1)^{k_1} \ldots (-2i\pi\xi_n)^{k_n} \int_{\mathbb{R}^n} e^{-2i\pi\langle \xi, \eta \rangle} \varphi(\eta) \, d\eta = (-2i\pi\xi)^k (F\varphi)(\xi).$$

Moreover, since $|e^{-2i\pi\langle \xi, \eta \rangle}| = 1$, we obtain inequality (5) because

$$|D^k F\varphi(\xi)| \leq |(2i\pi\xi)^k| \int_{\mathbb{R}^n} |\varphi(\eta)| \, d\eta \leq |(2i\pi\xi)^k| \, \|\varphi\|_{L^1}.$$

We now calculate the Fourier transform of $D^k\varphi$. Integrating by parts, we obtain

$$F(D^k\varphi)(\xi) = \int_{\mathbb{R}^n} e^{-2i\pi\langle \xi, \eta \rangle} D^k\varphi(\eta) \, d\eta = \int_{\mathbb{R}^n} (-1)^{|k|} (D^k_\eta e^{-2i\pi\langle \xi, \eta \rangle}) \varphi(\eta) \, d\eta$$
$$= \int_{\mathbb{R}^n} (-1)^{|k|} (-2i\pi\eta)^k e^{-2i\pi\langle \xi, \eta \rangle} \varphi(\eta) \, d\eta$$
$$= \int_{\mathbb{R}^n} (2i\pi\eta)^k e^{-2i\pi\langle \xi, \eta \rangle} \varphi(\eta) \, d\eta$$
$$= F((2i\pi\eta)^k \varphi(\cdot)).$$

This formula implies inequality (4). $\blacksquare$

This proposition shows that the Fourier transforms of the functions of $\mathcal{D}(\mathbb{R}^n)$ belong to the vector space $\mathcal{S}(\mathbb{R}^n)$ defined as follows:

DEFINITION 2
We denote by $\mathscr{S}(\mathbb{R}^n)$ the space of functions φ infinitely differentiable on $\mathbb{R}^n$ such that

(6) $$\forall k \in \mathbb{N}^n, \quad \sup_{\omega \in \mathbb{R}^n} |\omega^k \varphi(\omega)| < +\infty.$$

We say that $\mathscr{S}(\mathbb{R}^n)$ is the space of "rapidly decreasing" functions. ▲

If we look again at the proof of Proposition 1, we observe that we need suppose only that $\varphi \in \mathscr{S}(\mathbb{R}^n)$. Hence we can state the following proposition.

PROPOSITION 2
The Fourier transforms F and $\bar{F}$ are linear operators from $\mathscr{S}(\mathbb{R}^n)$ to $\mathscr{S}(\mathbb{R}^n)$. ▲

The important example of a function of $\mathscr{S}(\mathbb{R}^n)$ that does not have compact support is that of the Gaussian probability density g defined by

(7) $$g(\omega) = \exp(-\pi \|\omega\|^2)$$

(whose standard deviation is $\sigma = 1/\sqrt{2\pi}$; see Chapter 6, Section 4).
It is evident that g belongs to $\mathscr{S}(\mathbb{R}^n)$. Let us recall that

(8) $$\int_{-\infty}^{+\infty} g(\omega)\, d\omega = 1.$$

Elementary Properties of the Fourier Transform

We now prove some elementary properties of the Fourier transform.

PROPOSITION 3
Let $\varphi \in \mathscr{S}(\mathbb{R}^n)$. Then

(9) $$F(\tau_a \varphi) = e^{-2i\pi\langle \cdot, a\rangle} F\varphi \quad \text{where} \quad \tau_a \varphi(\omega) = \varphi(\omega - a)$$

and

(10) $$F(\varphi_h) = (F\varphi)(\cdot h) \quad \text{and} \quad F(\varphi(\cdot h)) = (F\varphi)_h \quad \text{where} \quad \varphi_h(\omega) = \frac{1}{h^n} \varphi\left(\frac{\omega}{h}\right).$$

If the functions φ and ψ belong to $\mathscr{D}(\mathbb{R}^n)$, we obtain

(11) $$F(\varphi * \psi) = F(\varphi) F(\psi)$$

and if φ and ψ belong to $\mathscr{S}(\mathbb{R}^n)$

(12) $$(F\varphi, \psi) = (\varphi, \bar{F}\psi). \quad ▲$$

Proof.

$$F(\tau_a\varphi)(\xi) = \int e^{-2i\pi\langle\xi,\eta\rangle}\varphi(\eta - a)\,d\eta = e^{-2i\pi\langle\xi,a\rangle}\int e^{-2i\pi\langle\xi,\eta\rangle}\varphi(\eta)\,d\eta$$
$$= e^{-2i\pi\langle\xi,a\rangle}F\varphi(\xi)$$

and

$$F(\varphi_h)(\xi) = \int e^{-2i\pi\langle\xi,\eta\rangle}\varphi\left(\frac{\eta}{h}\right)\frac{d\eta}{h^n} = \int e^{-2i\pi\langle\xi,\eta h\rangle}\varphi(\eta)\,d\eta = F(\varphi)(\xi h).$$

With $e^{-2i\pi\langle\xi,\eta\rangle} = e^{-2i\pi\langle\xi,\eta-\omega\rangle}e^{-2i\pi\langle\xi,\omega\rangle}$, we obtain

$$F(\varphi * \psi)(\xi) = \int e^{-2i\pi\langle\xi,\eta\rangle}\int\varphi(\eta - \omega)\psi(\omega)\,d\omega\,d\xi$$
$$= \int e^{-2i\pi\langle\xi,\eta-\omega\rangle}\varphi(\eta - \omega)\int e^{-2i\pi\langle\xi,\omega\rangle}\psi(\omega)\,d\omega\,d\eta = (F\varphi)(\xi)(F\psi)(\xi).$$

The last formula is evident:

$$(F\varphi, \psi) = \int\left(\int e^{-2i\pi\langle\xi,\eta\rangle}\varphi(\eta)\,d\eta\right)\overline{\psi(\xi)}\,d\xi$$
$$= \int\varphi(\eta)\,d\eta\left(\int e^{-2i\pi\langle\xi,\eta\rangle}\overline{\psi(\xi)}\,d\xi\right)$$
$$= \int\varphi(\eta)\,d\eta\overline{\left(\int e^{+2i\pi\langle\xi,\eta\rangle}\psi(\xi)\,d\xi\right)}$$
$$= \int\varphi(\eta)\,\overline{\bar{F}\psi(\eta)}\,d\eta = (\varphi, \bar{F}\psi). \quad\blacksquare$$

The following proposition is going to play an important role.

PROPOSITION 4
The Gaussian probability density g defined by $g(\omega) = \exp(-\pi\|\omega\|^2)$ is a fixed point for the Fourier transform. ▲

Proof. If $n = 1$ the function g is a solution of the linear differential equation

$$y' + 2\pi\omega y = 0$$

since $g'(\omega) = -2\pi\omega g(\omega)$. Applying the Fourier transform, we see that Fg is a solution of the same differential equation, since $DFg = F(-2i\pi\omega g) = iF(Dg) = -2\pi\omega F(g)$. Hence $g(\omega) = \alpha Fg(\omega)$. To determine α, we write that $1 = g(0) = \alpha Fg(0) = \alpha\int g(\eta)\,d\eta = \alpha$, since $\int_{-\infty}^{+\infty} e^{-\pi x^2}\,dx = 1$.
If $n > 1$, we write $g(\omega) = g(\omega_1)\ldots g(\omega_n)$. Then

$$F(g)(\xi) = \int_{-\infty}^{+\infty} e^{-2\pi i \sum_{k=1}^{n}\xi_k\omega_k}g(\omega_1)\ldots g(\omega_n)\,d\omega_1\ldots d\omega_n$$
$$= Fg(\xi_1)Fg(\xi_2)\ldots Fg(\xi_n) = g(\xi_1)\ldots g(\xi_n) = g(\xi). \quad\blacksquare$$

The Inversion Theorem

We are now going to prove the fundamental theorem concerning the Fourier transform, that is, that F is bijective from $\mathscr{S}(\mathbb{R}^n)$ onto $\mathscr{S}(\mathbb{R}^n)$ and that $\bar{F}$ is its inverse.

THEOREM 1 (INVERSION THEOREM)
The transform $\bar{F}$ is the inverse of the Fourier transform F, which is a linear operator from $\mathscr{S}(\mathbb{R}^n)$ onto $\mathscr{S}(\mathbb{R}^n)$. ▲

Proof. Let $\varphi \in \mathscr{S}(\mathbb{R}^n)$. Successively using formulas (9) and (10), Proposition 4, and formula (10), we obtain

$$\int_{-\infty}^{+\infty} e^{-2i\pi\langle\xi,a\rangle} F\varphi(\xi) g(\xi h)\, d\xi = \int_{-\infty}^{+\infty} F(\tau_{-a}\varphi)(\xi) g(\xi h)\, d\xi$$

$$= \int_{-\infty}^{+\infty} (\tau_{-a}\varphi)(\xi) F(g(\cdot h))(\xi)\, d\xi$$

$$= \int_{-\infty}^{+\infty} (\tau_{-a}\varphi)(\xi)(Fg)_h(\xi)\, d\xi$$

$$= \int_{-\infty}^{+\infty} (\tau_{-a}\varphi)(\xi) g_h(\xi)\, d\xi = \int_{-\infty}^{+\infty} (\tau_{-a}\varphi)(\xi h) g(\xi)\, d\xi$$

$$= \int_{-\infty}^{+\infty} \varphi(\xi h + a) g(\xi)\, d\xi.$$

Hence

(13) $$\int_{-\infty}^{+\infty} e^{2i\pi\langle\xi,a\rangle} F\varphi(\xi) g(\xi h)\, d\xi = \int_{-\infty}^{+\infty} \varphi(\xi h + a) g(\xi)\, d\xi.$$

Letting h approach zero, it follows that

(14) $$g(0) \int_{-\infty}^{+\infty} e^{2i\pi\langle a,\xi\rangle} F\varphi(\xi)\, d\xi = \varphi(a) \int_{-\infty}^{+\infty} g(\xi)\, d\xi,$$

that is $\varphi(a) = (\bar{F}F\varphi)(a)$, since $g(0) = \int_{-\infty}^{+\infty} g(\xi)\, d\xi = 1$.
We prove, similarly, that $F\bar{F}\varphi = \varphi$. Hence $\bar{F}$ is the inverse of F. ■

The inversion theorem implies, in particular, the Parseval-Plancherel formula.

PROPOSITION 5 (PARSEVAL-PLANCHEREL)
Let φ and $\psi \in \mathscr{S}(\mathbb{R}^n)$. Then

(15) $$(\varphi, \psi) = (F\varphi, F\psi)$$

and, consequently,

(16) $$|\varphi|^2 = |F\varphi|^2.$$ ▲

CH. 9, SEC. 2   THE FOURIER TRANSFORM OF FUNCTIONS

Proof. Using the inversion theorem, we obtain

(17) $$(\varphi, \psi) = (\varphi, \bar{F}F\psi) = (F\varphi, F\psi).$$ ∎

*Application: The Poisson Formula

We are going to prove the Poisson summation formula.

PROPOSITION 6 (POISSON SUMMATION FORMULA)
For every $\varphi \in \mathscr{S}(\mathbb{R})$, we have

(18) $$\sum_{n=-\infty}^{+\infty} \varphi(n) = \sum_{n=-\infty}^{+\infty} (F\varphi)(n).$$ ▲

Proof. Consider the function f defined by

(19) $$f(\omega) = \sum_{n=-\infty}^{+\infty} \varphi(\omega + n).$$

Since φ is rapidly decreasing, the series $\sum_{n=1}^{+\infty} D^k \varphi(\omega + n)$ are absolutely convergent. This implies that f is infinitely differentiable and periodic of period 1. Hence we can write

(20) $$f(\omega) = \sum_{n=-\infty}^{+\infty} c_k e^{-2i\pi k \omega} \qquad [\text{in the sense of } L^2(-\tfrac{1}{2}, +\tfrac{1}{2})],$$

since the functions $e^{-2i\pi k \omega}$ form an orthonormal base for $L^2(-\tfrac{1}{2}, +\tfrac{1}{2})$. Calculating the coefficients c_k:

$$c_k = \int_{-1/2}^{+1/2} f(\omega) e^{-2i\pi k\omega} d\omega = \sum_{n=-\infty}^{+\infty} \int_{-1/2}^{+1/2} \varphi(\omega + n) e^{-2i\pi k\omega} d\omega$$

$$= \sum_{n=-\infty}^{+\infty} \int_{n-1/2}^{n+1/2} \varphi(\omega) e^{-2i\pi k\omega} d\omega = \int_{-\infty}^{+\infty} \varphi(\omega) e^{-2i\pi\omega} d\omega = (F\varphi)(k).$$

But since $F\varphi \in \mathscr{S}(\mathbb{R})$, the series $\sum_{k=-\infty}^{+\infty} (F\varphi)(k) e^{-2i\pi k\omega}$ converges absolutely. Thus we can write

(21) $$f(\omega) = \sum_{n=-\infty}^{+\infty} \varphi(\omega + n) = \sum_{k=-\infty}^{+\infty} (F\varphi)(k) e^{-2i\pi k\omega}.$$

Taking $\omega = 0$, we obtain the desired formula. ∎

Remark 1

We apply this formula to the function φ defined by $\varphi(\omega) = (1/h)g(\omega/h) = g_h(\omega)$.

Then $F\varphi(\zeta) = (Fg_h)(\zeta) = (Fg)(\zeta h) = g(\zeta h)$. Hence we obtain

(22) $$\frac{1}{h}\sum_{n=-\infty}^{+\infty} e^{-\pi n^2/h^2} = \sum_{n=-\infty}^{+\infty} e^{-\pi n^2 h^2}.$$

This important formula (the functional equation of the θ functions) plays an important role in the theory of elliptic functions. ∎

3. THE FOURIER TRANSFORM OF SOBOLEV SPACES

Since $\mathscr{D}(\mathbb{R}^n)$ (and, consequently, $\mathscr{S}(\mathbb{R}^n)$) is dense in $L^2(\mathbb{R}^n)$, there are unique extensions of the isometries F and $\bar{F}$ of the prehilbert spaces $\mathscr{S}(\mathbb{R}^n)$ considered with the norm of $L^2(\mathbb{R}^n)$.

THEOREM 1
The Fourier transforms F and $\bar{F}$ have unique extensions to mutually inverse isometries from $L^2(\mathbb{R}^n)$ onto $L^2(\mathbb{R}^n)$. ▲

Remark 1

We continue the abuse of notation by writing

$$F\varphi(\xi) = \int_{-\infty}^{+\infty} e^{-2i\pi\langle\xi,\eta\rangle}\varphi(\eta)\,d\eta \qquad \text{when} \qquad \varphi \in L^2(\mathbb{R}^n).$$

This integral has no meaning because a square integrable function is not necessarily integrable. ∎

Similarly, inequality (1.5) of Proposition 2.1 shows that F is continuous from $\mathscr{D}(\mathbb{R}^n)$ with the norm $\|\varphi\|_{L^1(\mathbb{R}^n)}$ to the space $\mathscr{C}_\infty(\mathbb{R}^n)$ of continuous bounded functions on $\mathbb{R}^n$. The Fourier transform has a unique extension to a continuous operator from $L^1(\mathbb{R}^n)$ to $\mathscr{C}(\mathbb{R}^n)$.

THEOREM 2
The Fourier transformations F and $\bar{F}$ have unique extensions to continuous linear operators from $L^1(\mathbb{R}^n)$ to $\mathscr{C}_\infty(\mathbb{R}^n)$. ▲

Remark 2

In fact the Fourier transform maps $L^1(\mathbb{R}^n)$ to the closure of $\mathscr{S}(\mathbb{R}^n)$ in $\mathscr{C}_\infty(\mathbb{R}^n)$, which consists of functions $\xi \mapsto \varphi(\xi)$ *that approach zero as* $\|\xi\| \mapsto \infty$.

We are now going to prove that the Fourier transform of the functions of $H^m(\mathbb{R}^n)$ belong to the space $\hat{H}^m(\mathbb{R}^n)$ of square summable functions for the weight $(1 + \|\cdot\|^2)^m$. ∎

THEOREM 3
The Fourier transforms F and $\bar{F}$ have unique extensions to mutually inverse isomorphisms from $H^m(\mathbb{R}^n)$ onto $\hat{H}^m(\mathbb{R}^n)$. ▲

Proof. Let us recall that Proposition 6.3.2 affirms that $\hat{H}^m(\mathbb{R}^n)$ is the space of functions $\psi(\cdot) \in L^2(\mathbb{R}^n)$ such that $\xi^k \psi(\xi) \in L^2(\mathbb{R}^n)$ for every n-tuple of integers k such that $|k| = k_1 + \ldots + k_n \leq m$, considered with the equivalent norm defined by

$$\|\psi\|_{\hat{H}^m(\mathbb{R}^n)} = \left(\sum_{|k| \leq m} |2\pi \xi^k \psi|^2 \right)^{1/2}. \tag{1}$$

Now Propositions 2.1 and 2.5 show that if φ and ψ belong to $\mathscr{S}(\mathbb{R}^n)$, we have

$$\|F\varphi\|_{\hat{H}^m(\mathbb{R}^n)} = \|\varphi\|_{H^m(\mathbb{R}^n)} \tag{2}$$

and

$$\|\bar{F}\psi\|_{H^m(\mathbb{R}^n)} = \|\psi\|_{\hat{H}^m(\mathbb{R}^n)}. \tag{3}$$

Hence F and $\bar{F}$ have unique extensions to isometries from $H^m(\mathbb{R}^n)$ to $\hat{H}^m(\mathbb{R}^n)$ and from $\hat{H}^m(\mathbb{R}^n)$ to $H^m(\mathbb{R}^n)$, respectively. Since $\bar{F}F\varphi = \varphi$ and $F\bar{F}\psi = \psi$ when φ and ψ run over $\mathscr{S}(\mathbb{R}^n)$, it follows that $\bar{F}$ is the inverse of F. ∎

Since we can define the spaces $\hat{H}^s(\mathbb{R}^n)$ for arbitrary real numbers s, we can define the Sobolev spaces $H^s(\mathbb{R}^n)$ in the following fashion.

DEFINITION 1
Let $s \in [0, \infty[$. We call the Sobolev space of order $s \geq 0$ the image $H^s(\mathbb{R}^n) = \bar{F}(\hat{H}^s(\mathbb{R}^n))$ under the Fourier transform $\bar{F}$ of the space $\hat{H}^s(\mathbb{R}^n)$ of square summable functions for the weight $(1 + \|\cdot\|^2)^s$, considered with the norm

$$\|\varphi\|_s = \|F\varphi\|_{H^s(\mathbb{R}^n)}. \tag{4}$$

We denote by $H^{-s}(\mathbb{R}^n)$ the dual of $H^s(\mathbb{R}^n)$. ▲

Propositions 6.3.3 and 6.3.4, therefore, imply the following results.

PROPOSITION 1 (SOBOLEV AND INTERPOLATION INEQUALITIES)
If $s > n/2$, the Sobolev space $H^s(\mathbb{R}^n)$ is contained in the space of continuous bounded functions with a stronger topology.

If $r \leq s \leq t$ and if $\theta = \dfrac{t-s}{t-r} \in [0,1]$, then

$$\|\varphi\|_s \leq \|\varphi\|_r^\theta \|\varphi\|_t^{1-\theta} \qquad \forall \varphi \in H^t(\mathbb{R}^n). \tag{5}$$
▲

Proof. Let $\varphi \in H^s(\mathbb{R}^n)$. Then $\bar{F}\varphi \in \hat{H}^s(\mathbb{R}^n)$, which is contained in $L^1(\mathbb{R}^n)$ according to Proposition 6.3.4. Hence $\varphi = F\bar{F}\varphi$ is continuous and bounded according to Theorem 2; moreover,

$$(6) \qquad \|\varphi\|_{\mathscr{C}_\infty(\mathbb{R}^n)} \leq c \|\bar{F}\varphi\|_{L^1(\mathbb{R}^n)} \leq c' \|\bar{F}\varphi\|_{\hat{H}^s(\mathbb{R}^n)} = c' \|\varphi\|_{H^s(\mathbb{R}^n)}.$$

The second assertion follows trivially from Proposition 6.3.3. ∎

Fourier Transforms of Distributions

Since the Fourier transform $\bar{F}$ is an isomorphism from $\hat{H}^s(\mathbb{R}^n)$ onto $H^s(\mathbb{R}^n)$, its transpose $\bar{F}^*$ is an isomorphism from the dual $H^{-s}(\mathbb{R}^n)$ of $H^s(\mathbb{R}^n)$ onto the dual $\hat{H}^{-s}(\mathbb{R}^n)$ of $\hat{H}^s(\mathbb{R}^n)$. Since $\mathscr{D}(\mathbb{R}^n)$ [and, therefore, $\mathscr{S}(\mathbb{R}^n)$] is dense in $H^{-s}(\mathbb{R}^n)$, $\bar{F}^*$ is the *unique extension* of its restriction to $\mathscr{S}(\mathbb{R}^n)$, which coincides with F, since, if $\varphi \in \mathscr{S}(\mathbb{R}^n)$,

$$(7) \qquad (\bar{F}^*\varphi, \psi) = (\varphi, \bar{F}\psi) = (F\varphi, \psi) \qquad \forall \psi \in \mathscr{S}(\mathbb{R}^n)$$

according to formula (12) of Proposition 2.3.

Similarly, the transpose $F^* \in \mathscr{L}(\hat{H}^{-s}(\mathbb{R}^n), H^{-s}(\mathbb{R}^n))$ of F is the unique extension of the Fourier transform $\bar{F}$. Hence we have proved the following fundamental result:

THEOREM 4
The Fourier transforms F and $\bar{F}$ have unique extensions to mutually inverse isomorphisms from $H^{-s}(\mathbb{R}^n)$ onto $\hat{H}^{-s}(\mathbb{R}^n)$ (for every $s \geq 0$). ▲

Hence if $x \in H^{-s}(\mathbb{R}^n)$ is a distribution, we say that $Fx \in \hat{H}^{-s}(\mathbb{R}^n)$ is its Fourier transform.

We are going to extend to Sobolev spaces property (11) of Proposition 2.3.

PROPOSITION 2
The bilinear mapping $\{\lambda, \varphi\} \mapsto \lambda * \varphi$ has a unique extension to a continuous bilinear mapping from $L^1(\mathbb{R}^n) \times H^s(\mathbb{R}^n)$ to $H^s(\mathbb{R}^n)$ for every real number s. In this case we obtain the formula

$$(8) \qquad F(\lambda * \varphi) = F(\lambda) F(\varphi). \qquad ▲$$

Proof. It is clear that if $\mu \in \mathscr{C}_\infty(\mathbb{R}^n)$ and $\psi \in \hat{H}^s(\mathbb{R}^n)$, then $\mu\psi \in \hat{H}^s(\mathbb{R}^n)$ and $\|\mu\psi\|_{\hat{H}^s(\mathbb{R}^n)} \leq \|\mu\|_{\mathscr{C}_\infty(\mathbb{R}^n)} \|\psi\|_{\hat{H}^s(\mathbb{R}^n)}$, according to the mean value theorem (for arbitrary $s \in \mathbb{R}$). Consequently, if $\lambda \in L^1(\mathbb{R}^n)$ and $\varphi \in H^s(\mathbb{R}^n)$, then $F(\lambda) \in \mathscr{C}_\infty(\mathbb{R}^n)$ and $F(\varphi) \in \hat{H}^s(\mathbb{R}^n)$, and hence $F(\lambda)F(\varphi) \in \hat{H}^s(\mathbb{R}^n)$. This implies that $\bar{F}(F(\lambda)F(\varphi)) \in H^s(\mathbb{R}^n)$. Therefore the bilinear mapping $\{\lambda, \varphi\} \in L^1(\mathbb{R}^n) \times H^s(\mathbb{R}^n) \to \bar{F}(F(\lambda)F(\varphi))$ is bilinear and continuous (of norm 1), since

$$\|\bar{F}(F(\lambda)F(\varphi)\|_{H^s(\mathbb{R}^n)} = \|F(\lambda)F(\varphi)\|_{\hat{H}^s(\mathbb{R}^n)}$$
$$\leq \|F(\lambda)\|_{\mathscr{C}(\mathbb{R}^n)} \|F(\varphi)\|_{\hat{H}^s(\mathbb{R}^n)} \leq \|\lambda\|_{L^1(\mathbb{R}^n)} \|\varphi\|_{H^s(\mathbb{R}^n)}.$$

Moreover, it is clear that if λ and φ belong to $\mathscr{D}(\mathbb{R}^n)$, $\lambda * \varphi = \bar{F}(F(\lambda)F(\varphi))$, since, according to (11), Section 2, $F(\lambda * \varphi) = F(\lambda)F(\varphi)$. Thus $\{\lambda, \varphi\} \to \bar{F}(F(\lambda)F(\varphi))$ is the unique extension by density of $\{\lambda, \varphi\} \to \lambda * \varphi$. We therefore set $\lambda * \varphi = \bar{F}(F(\lambda)F(\varphi))$ when $\lambda \in L^1(\mathbb{R}^n)$ and $\varphi \in H^s(\mathbb{R}^n)$, which naturally implies (8).

4. THE TRACE THEOREM FOR THE SPACES $H^m(\mathring{\mathbb{R}}^n_+)$

We are going to prove the trace theorem in the case of the open set $\Omega = \mathring{\mathbb{R}}^n_+$ whose boundary is $\Gamma = \mathbb{R}^{n-1}$. To this end we first define and study the properties of trace operators for functions defined on $\mathbb{R}^n$, after which we consider the case of $\Omega = \mathring{\mathbb{R}}^n_+$ using an extension theorem that we prove. Finally we characterize $H^m(\mathbb{R}^{n-1})$ as the kernel of the trace operator.

Trace Theorem in $\mathbb{R}^n$

We are going to consider $\mathbb{R}^n$ as the product $\mathbb{R}^{n-1} \times \mathbb{R}$ and write $\omega = \{\omega_1, \ldots, \omega_{n-1}, \omega_n\}$ as $\{\alpha, \tau\}$ where $\alpha = \{\alpha_1, \ldots, \alpha_{n-1}\} = \{\omega_1, \ldots, \omega_{n-1}\} \in \mathbb{R}^{n-1}$ and $\tau = \omega_n \in \mathbb{R}$.

We thus consider every function $x \in \mathscr{S}(\mathbb{R}^n)$ as the function $\{\alpha, \tau\} \to x(\alpha, \tau)$ defined on $\mathbb{R}^{n-1} \times \mathbb{R}$.

DEFINITION 1
We call a trace operator of order j the operator γ_j from $\mathscr{S}(\mathbb{R}^n)$ to $\mathscr{S}(\mathbb{R}^{n-1})$ defined by

(1) $$\gamma_j x(\alpha) = (D_\tau^j x)(\alpha, 0).$$

We denote by $\gamma = \gamma_0 \times \ldots \times \gamma_{m-1}$ the operator from $\mathscr{S}(\mathbb{R}^n)$ to $\mathscr{S}(\mathbb{R}^{n-1})^m$ defined by

(2) $$\gamma x = \{\gamma_0 x, \ldots, \gamma_{m-1} x\}.$$ ▲

THEOREM 1
If $s > m - \frac{1}{2}$, the operator γ defined by (1) and (2) has a unique extension to a surjective continuous linear operator from $H^s(\mathbb{R}^n)$ onto $\prod_{j=0}^{m-1} H^{s-j-1/2}(\mathbb{R}^{n-1})$. ▲

Proof. The theorem is the consequence of the following five lemmas. We denote by F_n, F_{n-1}, and F the Fourier transforms of the functions of $n, n-1$, and one variable(s), respectively: We can write $F_n = F_{n-1}F = FF_{n-1}$, since

$$(F_n\varphi)(\beta,\sigma) = \int\int_{\mathbb{R}^{n-1}\times\mathbb{R}} d\alpha\,d\tau\, e^{-2i\pi\langle\beta,\alpha\rangle} e^{-2i\pi\tau\sigma}\varphi(\alpha,\tau)$$
$$= F_{n-1}[F\varphi(\cdot,\sigma)](\beta) = F[F_{n-1}\varphi(\alpha,\cdot)](\sigma)$$

according to the Fubini theorem. ∎

LEMMA 1
We have the relation

(3) $$(F_{n-1}\gamma_j x)(\beta) = \int_{-\infty}^{+\infty}(-2i\pi\sigma)^j(F_n x)(\beta,\sigma)\,d\sigma.$$ ▲

Proof. Indeed, using the inversion theorem, Theorem 2.1, we can write that

$$\gamma_j x(\alpha) = [\bar{F}F[D_t^j x(\alpha,.)]](0) = [\bar{F}(-2i\pi\sigma)^j(Fx)(\alpha,\sigma)](0)$$
$$= \int_{-\infty}^{+\infty}(-2i\pi\sigma)^j(Fx)(\alpha,\sigma)\,d\sigma$$

Applying the Fourier transformation F_{n-1} to the members of this equality, we obtain (3). ∎

LEMMA 2
If $s - j - \frac{1}{2} > 0$, we obtain the inequality

(4) $$\|F_{n-1}\gamma_j x\|_{\hat{H}^{s-j-1/2}(\mathbb{R}^n)} \le c_s \|F_n x\|_{\hat{H}^s(\mathbb{R}^n)}.$$ ▲

Proof. Let us remark that we can write

(5) $$\int_{-\infty}^{+\infty}\frac{\sigma^{2j}d\sigma}{(1+\|\beta\|^2+\sigma^2)^s} = \frac{1}{(1+\|\beta\|^2)^{s-j-1/2}}C_{s,j}^2,$$

where

$$C_{s,j} = \left(\int\frac{t^{2j}dt}{(1+t^2)^s}\right)^{1/2} < +\infty \quad\text{if}\quad s > j + \tfrac{1}{2}.$$

(it suffices to make the change of variable $\sigma = (1+\|\beta\|^2)^{1/2}t$.) We can then rewrite formula (3) of Lemma 1 in the form

$$(F_{n-1}\gamma_j x)(\beta) = (-2i\pi)^j\int_{-\infty}^{+\infty}(F_n x)(\beta,\sigma)(1+\|\beta\|^2+\sigma^2)^{s/2}\frac{\sigma^j d\sigma}{(1+\|\beta\|^2+\sigma^2)^{s/2}}$$

Using the Cauchy-Schwarz formula, we obtain the following upper bound:

$$|(F_{n-1}\gamma_j x)(\beta)|^2 \leq c \left(\int_{-\infty}^{+\infty} |F_n x(\beta,\sigma)|^2 (1+\|\beta\|^2+\sigma^2)^s d\sigma \right) \int_{-\infty}^{+\infty} \frac{\sigma^{2j} d\sigma}{(1+\|\beta\|^2+\sigma^2)^s}$$

$$\leq cC_{s,j}^2 \int_{-\infty}^{+\infty} |F_n x(\beta,\sigma)|^2 (1+\|\beta\|^2+\sigma^2)^s d\sigma \frac{1}{(1+\|\beta\|^2)^{s-j-1/2}}$$

Hence

$$\|F_{n-1}\gamma_j\|_{\hat{H}^{s-j-1/2}(\mathbb{R}^{n-1})}^2 = \int |(F_{n-1}\gamma_j x)(\beta)|^2 (1+\|\beta\|^2)^{s-j-1/2} d\beta$$
$$\leq cC_{s,j}^2 \iint |(F_n x)(\beta,\sigma)|^2 (1+\|\beta\|^2+\sigma^2)^s d\sigma\, d\beta$$
$$= cC_{s,j}^2 \|F_n x\|_{\hat{H}^s(\mathbb{R}^n)}^2 \quad \blacksquare$$

LEMMA 3
If $s - j - \frac{1}{2} > 0$, the operator γ_j has a unique extension to a continuous operator from $H^s(\mathbb{R}^n)$ to $H^{s-j-1/2}(\mathbb{R}^{n-1})$. ▲

Proof. Lemma 2 and Definition 3.1 of the Sobolev spaces H^s imply that

$$\|\gamma_j x\|_{H^{s-j-1/2}(\mathbb{R}^{n-1})} \leq \|F_{n-1}\gamma_j x\|_{\hat{H}^{s-j-1/2}(\mathbb{R}^{n-1})} \leq c'\|F_n x\|_{\hat{H}^s(\mathbb{R}^n)}$$
$$\leq c''\|x\|_{H^s(\mathbb{R}^n)}$$

for all $x \in \mathscr{S}(\mathbb{R}^n)$. Therefore γ_j is continuous from $\mathscr{S}(\mathbb{R}^n)$ with the norm induced by $H^s(\mathbb{R}^n)$ to $H^{s-j-1/2}(\mathbb{R}^{n-1})$ and has a unique extension to a continuous operator from $H^s(\mathbb{R}^n)$ to $H^{s-j-1/2}(\mathbb{R}^{n-1})$. ∎

We are now going to show that γ_j possesses a continuous right inverse σ_j. To construct it, it suffices for us to show that $\gamma_j \sigma_j \varphi = \varphi$ for every $\varphi \in \mathscr{S}(\mathbb{R}^{n-1})$ and that σ_j is continuous from $\mathscr{S}(\mathbb{R}^{n-1})$ with the norm induced by $H^{s-j-1/2}(\mathbb{R}^{n-1})$ to $H^s(\mathbb{R}^n)$. For this we take a function $\theta_j \in \mathscr{D}(\mathbb{R})$ such that

(6) $$(D^k \theta_j)(0) = \begin{cases} 1 & \text{if } j = k. \\ 0 & \text{if } j \neq k. \end{cases}$$

We then define

(7) $$(\sigma_j \varphi)(\alpha, \tau) = \bar{F}_{n-1}\left[F_{n-1}\varphi(\beta) \frac{\theta_j(\tau\sqrt{1+\|\beta\|^2})}{\sqrt{(1+\|\beta\|^2)^j}} \right](\alpha).$$

LEMMA 4
The operator σ_j defined by (7) satisfies

(8) $$\gamma_k \sigma_j \varphi = \begin{cases} 0 & \text{if } j \neq k. \\ \varphi & \text{if } j = k. \end{cases}$$

If $s > j + \frac{1}{2}$, the operator σ_j has an extension to a continuous operator from $H^{s-j-1/2}(\mathbb{R}^{n-1})$ to $H^s(\mathbb{R}^{n-1})$. ▲

Proof. Differentiating k times with respect to τ, we obtain

$$(D_\tau^k \sigma_j \varphi)(\alpha, \tau) = \left[(\bar{F}_{n-1} \left[F_{n-1} \varphi(\beta) \frac{D^k \theta_j(\tau \sqrt{1 + \|\beta\|^2})}{\sqrt{(1 + \|\beta\|^2)^{j-k}}} \right] \right](\alpha).$$

Now taking $\tau = 0$, it follows that

(9) $$(\gamma_k \sigma_j \varphi)(\alpha) = (\bar{F}_{n-1} \left[F_{n-1} \varphi(\beta) \frac{D^k \theta_j(0)}{\sqrt{(1 + \|\beta\|^2)^{j-k}}} \right](\alpha).$$

If $j \neq k$, we deduce that $\gamma_k \sigma_j \varphi = 0$, since $D^k \theta_j(0) = 0$. If $j = k$, it follows from (9) that

$$(\gamma_j \sigma_j \varphi)(\alpha) = (\bar{F}_{n-1} F_{n-1} \varphi)(\alpha) = \varphi(\alpha).$$

It remains for us to show that σ_j is continuous from $\mathscr{S}(\mathbb{R}^{n-1})$ with the norm induced by $H^{s-j-1/2}(\mathbb{R}^{n-1})$ to $H^s(\mathbb{R}^n)$. For this we calculate $F_n \sigma_j \varphi$:

$$F_n \sigma_j \varphi = FF_{n-1} \sigma_j \varphi = F\left[F_{n-1} \varphi(\beta) \frac{\theta_j(.\sqrt{1 + \|\beta\|^2})}{\sqrt{(1 + \|\beta\|^2)^j}} \right]$$

$$= F_{n-1} \varphi(\beta) \frac{1}{\sqrt{(1 + \|\beta\|^2)^{j+1}}} (F\theta_j)\left(\frac{\tau}{\sqrt{(1 + \|\beta\|^2)}} \right).$$

(We use formula (10) of Proposition 2.3.)

Hence

$$\|F_n \sigma_j \varphi\|^2_{\hat{H}^s(\mathbb{R}^n)}$$

$$= \int_{-\infty}^{+\infty} d\sigma \int_{\mathbb{R}^{n-1}} d\beta (1 + \|\beta\|^2 + \sigma^2)^s (1 + \|\beta\|^2)^{-j-1} |F_{n-1}\varphi(\beta)|^2$$

$$\cdot \left| F\theta_j\left(\frac{\sigma}{\sqrt{1 + \|\beta\|^2}} \right) \right|^2$$

$$= \int_{-\infty}^{+\infty} d\tau \int_{\mathbb{R}^{n-1}} d\beta (1 + \|\beta\|^2)^{s-j-1/2} (1 + \tau^2)^s |F_{n-1}\varphi(\beta)|^2 |F\theta_j(\tau)|^2$$

$$= c^2 \|F_{n-1}\varphi\|^2_{\hat{H}^{s-j-1/2}(\mathbb{R}^{n-1})}$$

making the change of variable $\sigma = \tau\sqrt{1 + \|\beta\|^2}$ and setting $c^2 = \int (1 + \tau^2)^s |F\theta_j(\tau)|^2 d\tau = \|\theta_j\|^2_{\hat{H}^s(\mathbb{R})}$. It follows that $\|\sigma_j \varphi\|_{H^s(\mathbb{R}^n)} \leq c\|\varphi\|_{H^{s-j-1/2}(\mathbb{R}^{n-1})}$ for every $\varphi \in \mathscr{S}(\mathbb{R}^{n-1})$, which is what we needed to prove. ∎

LEMMA 5

If $s > m - \frac{1}{2}$, the operator $\gamma = \gamma_0 \times \gamma_1 \times \ldots \times \gamma_{m-1}$ is a surjective continuous linear operator from $H^s(\mathbb{R}^n)$ onto $\prod_{j=0}^{m-1} H^{s-j-1/2}(\mathbb{R}^{n-1})$. ▲

Proof. According to Lemma 3, we know that γ is a continuous linear operator from $H^s(\mathbb{R}^n)$ onto $\prod_{j=0}^{m-1} H^{s-j-1/2}(\mathbb{R}^{n-1})$.

Let $\sigma_j \in \mathscr{L}(H^{s-j-1/2}(\mathbb{R}^{n-1}), H^s(\mathbb{R}^n))$ be the right inverse of γ_j defined by (7). We set $\sigma\{\varphi_0, \ldots, \varphi_{m-1}\}(\alpha, \tau) = \sum_{j=1}^{m} \sigma_j \varphi_j(\alpha, \tau)$. Lemma 4 implies that σ is continuous from $\prod_{j=0}^{m-1} H^{s-j-1/2}(\mathbb{R}^{n-1})$ to $H^s(\mathbb{R}^n)$ and that σ is a right inverse of γ, since, according to (8),

$$\gamma_j \sigma\{\varphi_0, \ldots, \varphi_{m-1}\} = \sum_{k=1}^{m} \gamma_j \sigma_k \varphi_k = \varphi_j \quad \text{for all} \quad j. \qquad \blacksquare$$

Trace Theorem and Extension Theorem in $\mathring{\mathbb{R}}^n_+$

In fact we can define the trace operators γ_j for the functions $x \in \mathscr{D}(\mathring{\mathbb{R}}^n_+)$, which are infinitely differentiable with compact support contained in $\mathring{\mathbb{R}}^n_+$

(10) $$(\gamma_j x)(\alpha) = D^j_\tau x(\alpha, 0)$$

and the operator $\gamma = \gamma_0 \times \ldots \times \gamma_{m-1}$ from $\mathscr{D}(\mathring{\mathbb{R}}^n_+)$ to $\mathscr{D}(\mathbb{R}^{n-1})^m$.

THEOREM 2
The operator γ has a unique extension to a surjective continuous linear operator from $H^m(\mathring{\mathbb{R}}^n_+)$ onto $\prod_{j=0}^{m-1} H^{m-j-1/2}(\mathbb{R}^{n-1})$. ▲

Proof. We are going to construct an extension operator $\bar{\pi} \in \mathscr{L}(H^m(\mathring{\mathbb{R}}^n_+), H^m(\mathbb{R}^n))$ such that $\gamma_j x = \gamma_j \bar{\pi} x$ for every $j \leq m-1$. Then Theorem 2 will follow from Theorem 1 in an obvious way.

We define $\bar{\pi}$ in the following fashion:

(11) $$\bar{\pi} x(\alpha, \tau) = \begin{cases} x(\alpha, \tau) & \text{if} \quad \tau > 0. \\ \sum_{j=1}^{m} \alpha_j x(\alpha, -j\tau) & \text{if} \quad \tau < 0. \end{cases}$$

where the α_j's are the solutions of the Van der Monde system

(12) $$\sum_{j=1}^{m} \alpha_j(-j)^k = 1 \quad \text{if} \quad 0 \leq k \leq m-1.$$

We calculate the derivative $D^k = D^p_\alpha D^q_\tau$ (where $|p| + q \leq m$) of $\bar{\pi} x$ in the sense of distributions: for every $\varphi \in \mathscr{D}(\mathbb{R}^n)$ we have

(13) $$(\bar{\pi} x, D^k \varphi) = \sum_{j=1}^{m} \alpha_j \int_{-\infty}^{0} d\tau \int_{\mathbb{R}^{n-1}} x(\alpha, -j\tau) D^k \varphi(\alpha, \tau) \, d\alpha$$
$$+ \int_{0}^{\infty} d\tau \int_{\mathbb{R}^{n-1}} x(\alpha, \tau) D^k \varphi(\alpha, \tau) \, d\alpha.$$

Integrating by parts each of the integrals, we obtain, since φ has compact support in $\mathbb{R}^n$,

$$\int_0^\infty \int_{\mathbb{R}^{n-1}} x(\alpha,\tau) D^k \varphi(\alpha,\tau) d\alpha\, d\tau = (-1)^{|k|} \int_0^\infty \int_{\mathbb{R}^{n-1}} D^k x(\alpha,\tau) \varphi(\alpha,\tau) d\alpha\, d\tau$$
$$+ \sum_{r=0}^{q-1} (-1)^{q-r} \int_{\mathbb{R}^{n-1}} (\gamma_{q-1-r} D_\alpha^p x)(\alpha)(\gamma_r \varphi)(\alpha)\, d\alpha$$

and

$$\int_{-\infty}^0 \int_{\mathbb{R}^{n-1}} x(\alpha, -j\tau) D^k \varphi(\alpha,\tau)\, d\alpha\, d\tau$$
$$= (-1)^{|k|}(-j)^q \int_{-\infty}^0 \int_{\mathbb{R}^{n-1}} (D^k x)(\alpha, -j\tau) \varphi(\alpha,\tau)\, d\alpha\, d\tau$$
$$- \sum_{r=0}^{q-1} (-1)^{q-r} \int_{\mathbb{R}^{n-1}} (-j)^{q-1-r}(\gamma_{q-1-r} D_\alpha^p x)(\alpha)(\gamma_r \varphi)(\alpha)\, d\alpha.$$

When we take the sum of the integrals, the relations (12) imply that the integrals on $\mathbb{R}^{n-1}$ do not appear and that

(14) $$D^k \bar{\pi} x(\alpha,\tau) = \begin{cases} D^k x(\alpha,\tau) & \text{if } \tau > 0. \\ \sum_{j=1}^m (-j)^q \alpha_j D^k(\alpha, -j\tau) & \text{if } \tau < 0. \end{cases}$$

Hence $\bar{\pi} x \in H^m(\mathbb{R}^n)$. It is also clear that (14) implies that $\bar{\pi}$ is a continuous linear operator from $H^m(\mathring{\mathbb{R}}^n_+)$ to $H^m(\mathbb{R}^n)$. ∎

We are now going to use the trace operators γ_j defined on $H^m(\mathring{\mathbb{R}}^n_+)$ to define the operator π from $H^m(\mathring{\mathbb{R}}^n_+)$ to $L^2(\mathbb{R}^n)$ by

(15) $$(\pi x)(\alpha,\tau) = \begin{cases} x(\alpha,\tau) & \text{if } \tau \geq 0 \\ \sum_{j=0}^{m-1} (\sigma_j \gamma_j x)(\alpha,\tau) & \text{if } \tau \leq 0 \end{cases}$$

where the γ_j's are the trace operators of order j and the σ_j's are their right inverses defined by (7).

THEOREM 3
The operator π defined by (15) is a continuous linear operator from $H^m(\mathring{\mathbb{R}}^n_+)$ to $H^m(\mathbb{R}^n)$ whose left inverse is the operator ρ of restriction to $\mathring{\mathbb{R}}^n_+$. Moreover, for every $|k| \leq m$.

(16) $$\rho D^k \pi x = D^k x \quad \text{if} \quad x \in H^m(\mathring{\mathbb{R}}^n_+).$$

If $\gamma_j x = 0$ for $0 \leq j \leq m-1$, then $\pi x = \pi_0 x$ coincides with the extension of x by zero outside of $\mathring{\mathbb{R}}^n_+$. ▲

Proof. We are going to show that for every n-tuple of integers k such that $|k| \leq m$,

(17) $$(D^k \pi x)(\alpha, \tau) = \begin{cases} (D^k x)(\alpha, \tau) & \text{if } \tau > 0. \\ \sum_{j=1}^{m-1} (D^k \sigma_j \gamma_j x)(\alpha, \tau) & \text{if } \tau \leq 0. \end{cases}$$

For this we write that $k = (p, q)$ where $p = (k_1, \ldots, k_{n-1})$ and $q = k_n$. For every $\varphi \in \mathscr{D}(\mathbb{R}^n)$ we have by definition

$$(D^k \pi x, \varphi) = (-1)^{|k|} (\pi x, D^k \varphi) = \sum_{j=0}^{m-1} \int_0^\infty \int_{\mathbb{R}^{n-1}} x(\alpha, \tau)(-1)^{|p|+q} D_\alpha^p D_\tau^q \varphi(\alpha, \tau) \, d\alpha \, d\tau$$

$$+ \int_{-\infty}^0 \int_{\mathbb{R}^{n-1}} \left(\sum_{j=1}^m \sigma_j \gamma_j x \right)(\alpha, \tau)(-1)^{|p|+q} D_\alpha^p D_\tau^q \varphi(\alpha, \tau) \, d\alpha \, d\tau.$$

We are going to integrate each of these integrals by parts. First of all, we see that

$$\int_0^\infty \int_{\mathbb{R}^{n-1}} x(\alpha, \tau)(-1)^{|p|+q} D_\alpha^p D_\tau^q \varphi(\alpha, \tau) \, d\alpha \, d\tau = \int_0^\infty \int_{\mathbb{R}^{n-1}} D_\alpha^p D_\tau^q x(\alpha, \tau) \varphi(\alpha, \tau) \, d\alpha \, d\tau$$

$$+ \int_{\mathbb{R}^{n-1}} d\alpha \left(\sum_{r=0}^{q-1} (-1)^{q-r-1} \gamma_{q-1-r} D_\alpha^p x(\alpha) \gamma_r \varphi(\alpha) \right).$$

Similarly, for every $j = 0, \ldots, m-1$ we obtain

$$\int_{-\infty}^0 \int_{\mathbb{R}^{n-1}} (\sigma_j \gamma_j x)(\alpha, \tau)(-1)^{|p|+q} D_\alpha^p D_\tau^q \varphi(\alpha, \tau) \, d\alpha \, d\tau$$

$$= \int_{-\infty}^0 \int_{\mathbb{R}^{n-1}} D_\alpha^p D_\tau^q (\sigma_j \gamma_j x)(\alpha, \tau) \varphi(\alpha, \tau) \, d\alpha \, d\tau$$

$$+ \int_{\mathbb{R}^{n-1}} d\alpha \left(\sum_{r=0}^{q-1} (-1)^{q-r} \gamma_{q-1-r} D_\alpha^p \sigma_j \gamma_j x(\alpha) \gamma_r \varphi(\alpha) \right)$$

since $\gamma_{q-1-r} D^p \sigma_j \gamma_j x(\alpha) = D^p \gamma_{q-1-r} \sigma_j \gamma_j x(\alpha) = 0$ if $r \neq q - j - 1$ and is equal to $D^p \gamma_j x(\alpha)$ if $r = q - j - 1$.

Consequently, taking the sum of these terms, we see that the integrals on $\mathbb{R}^{n-1}$ do not appear and that $(-1)^{|k|} (\pi x, D^k \varphi) = (D^k \pi x, \varphi)$ where $D^k \pi x$ is defined by (17).

This shows that $D^k \pi x \in L^2(\mathbb{R}^n)$ and that $\rho D^k \pi x = D^k x$ when $|k| \leq m$. Moreover, it is clear that π is a continuous linear operator from $H^m(\mathring{\mathbb{R}}^n_+)$ to $H^m(\mathbb{R}^n)$. Formula (17) implies as well that $\pi x(\alpha, \tau) = 0$ when $\tau \leq 0$ and when the traces $\gamma_j x$ are zero for $0 \leq j \leq m - 1$. In this case $\pi x = \pi_0 x$ coincides with the extension of x by zero outside of $\mathring{\mathbb{R}}^n_+$. ∎

Characterization of $H_0^m(\mathring{\mathbb{R}}_+^n)$

It is clear that $H_0^m(\mathring{\mathbb{R}}_+^n)$ is contained in the kernel of γ. We are going to show that $H_0^m(\mathring{\mathbb{R}}_+^n)$ is the subspace of $H^m(\mathring{\mathbb{R}}_+^n)$ whose first m traces are zero.

THEOREM 4
The minimal Sobolev space $H_0^m(\mathring{\mathbb{R}}_+^n)$ is the kernel of the operator $\gamma = \gamma_0 \times \ldots \times \gamma_{m-1}$ from $H^m(\mathring{\mathbb{R}}_+^n)$ onto $\prod_{j=0}^{m-1} H^{m-j-1/2}(\mathbb{R}^{n-1})$. ▲

Proof. Theorem 3 implies that $\pi_0 x = \pi x \in H^m(\mathbb{R}^n)$ when $x \in \operatorname{Ker}\gamma$. Hence the functions x_ε defined by $x_\varepsilon(\alpha, \tau) = \pi_0 x(\alpha, \tau - \varepsilon)$ belong to $H^m(\mathbb{R}^n)$ and have their support contained in $\mathbb{R}^{n-1} \times [\varepsilon, \infty[$. They converge to x in $H^m(\mathbb{R}^n)$. Moreover, every function $x_\varepsilon \in H^m(\mathbb{R}^n)$ whose support is contained in $\mathbb{R}^{n-1} \times [\varepsilon, \infty[$ can be approximated by a function of $\mathscr{D}(\mathbb{R}^n)$ whose compact support is contained in $\mathbb{R}^{n-1} \times [\varepsilon/2, \infty[\subset \mathring{\mathbb{R}}_+^n$. (See the proof of Lemma 7.8.1.) Thus this shows that $x \in H_0^m(\mathring{\mathbb{R}}_+^n)$. ∎

5. THE TRACE THEOREM FOR THE SPACES $H^m(\Omega)$

Before proving the trace theorem in the case of open sets Ω with boundary Γ, we must construct the spaces $H^s(\Gamma)$ of functions defined on the boundary.

DEFINITION 1
We say that Ω is regular if it is a bounded open set of $\mathbb{R}^n$, if its boundary Γ is an infinitely differentiable manifold of dimension $n-1$, and if Ω is on one side of Γ. ▲

Consequently, if Ω is regular, there exist a covering of Γ by a finite number of bounded open sets $\theta_j (1 \leq j \leq J)$ of $\mathbb{R}^n$ and mappings Φ_j and Ψ_j such that

(1)
$\begin{cases}
\text{i.} & \Phi_j \text{ sends } \theta_j \text{ to the subset } Q \text{ of } \mathbb{R}^n \text{ defined by} \\
& Q = \{\{\alpha, \tau\} \in \mathbb{R}^{n-1} \times \mathbb{R} \text{ such that } \|\alpha\| < 1 \text{ and } |\tau| < 1\}. \\
\text{ii.} & \Phi_j \text{ sends } \theta_j \cap \Omega \text{ to } Q_+ = \{\{\alpha, \tau\} \in Q \text{ such that } \tau > 0\}. \\
\text{iii.} & \Phi_j \text{ sends } \theta_j \cap \Gamma \text{ to } Q_0 = \{\{\alpha, \tau\} \in Q \text{ such that } \tau = 0\}. \\
\text{iv.} & \Psi_j \text{ is the inverse of } \Phi_j. \\
\text{v.} & \Phi_j \text{ and } \Psi_j \text{ are infinitely differentiable and their Jacobians are} \\
& \text{strictly positive.}
\end{cases}$

There also exists an open set $\theta_0 \subset \Omega$ such that the open sets $\theta_0, \theta_1, \ldots, \theta_j$ form a covering of Ω.

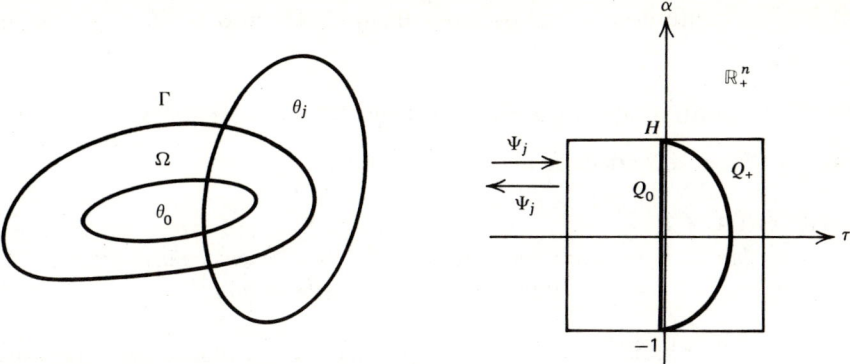

PROPOSITION 1
There exists an infinitely differentiable partition of unity subordinate to the covering $\{\theta_0, \ldots, \theta_J\}$ of Ω. ▲

Proof. Consider an open covering $\{\Xi_0, \ldots, \Xi_j, \ldots, \Xi_J\}$ of Ω such that $\bar{\Xi}_j \subset \theta_j$. (See Proposition 5.3.3 of [AAA], p. 197.) Hence there exists a continuous partition of unity $\{\beta_0, \beta_1, \ldots, \beta_j, \ldots, \beta_J\}$ subordinate to the covering $\{\Xi_0, \ldots, \Xi_J\}$. Since $\bar{\Xi}_j \subset \theta_j$ there exists a compact neighborhood K such that $\Xi_j + K \subset \theta_j$ for all j. We take $\rho \in \mathscr{D}(\mathbb{R}^n)$ for which the support is contained in K and such that $\int \rho(\omega)\,d\omega = 1$. Since $\rho * 1 = 1$ (because $(\rho * 1)(\omega) = \int \rho(\omega - \xi)\,d\xi = \int \rho(\omega)\,d\omega = 1$), it follows that the functions $\alpha_j = \rho * \beta_j$ form an infinitely differentiable partition of unity. It is subordinate to the covering $\{\theta_0, \ldots, \theta_J\}$, since $\text{supp}(\beta_j * \rho) \subset \text{supp}\,\beta_j + K \subset \theta_j$ for all j. ∎

If x is a function defined on Ω and y is a function defined on Q, we set

(2) $\quad \begin{cases} \text{i.} & (\Phi_j^* \cdot y)(\omega) = y[\Phi_j(\omega)]. \\ \text{ii.} & (\Psi_j^* \cdot x)(\xi) = x[\Psi_j(\xi)]. \end{cases}$

Every function x defined on Ω can, therefore, be written

(3) $\quad x = \sqrt{\alpha_0}\sqrt{\alpha_0}\,x + \sum_{j=1}^{J} \sqrt{\alpha_j}\,\Phi_j^* \Psi_j^* \cdot (\sqrt{\alpha_j}\,x) = \Phi^* \Psi^* \cdot x.$

where

(4) $\quad \begin{cases} \text{i.} & \Psi^* x = \{\sqrt{\alpha_0}\,x, \ldots, \Psi_j^* \cdot (\sqrt{\alpha_j}\,x), \ldots, \}. \\ \text{ii.} & \Phi^*(y_0, \ldots, y_J) = \sqrt{\alpha_0}\,y_0 + \sum_{j=1}^{J} \sqrt{\alpha_j}\,\Phi_j^* \cdot y_j. \end{cases}$

Since the functions Φ_j, Ψ_j, and α_j are infinitely differentiable, it is clear that

(5) Ψ^* is a continuous linear operator from $H^m(\Omega)$ onto $H^m(\mathbb{R}) \times (H^m(\mathring{\mathbb{R}}^n_+))^J$

and that

(6) Φ^* is a continuous linear operator from $H^m(\mathbb{R}^n) \times (H^m(\mathring{\mathbb{R}}^n_+))^J$ to $H^m(\Omega)$.

Since $\Phi^*\Psi^* = 1$, we deduce the following result. ∎

PROPOSITION 2
Suppose that Ω is regular. The operator Ψ^* is an isomorphism from $H^m(\Omega)$ onto its closed image in $H^m(\mathbb{R}^n) \times (H^m(\mathring{\mathbb{R}}^n_+))^J$ for all $m \geq 0$. ▲

On the other hand, let us consider the operator Ψ_0^* that maps every function x defined on the boundary Γ of Ω to a sequence

(7) $$(\Psi_0^* x)(\alpha) = \{\psi_j^* \cdot (\sqrt{\alpha_j} x)(\alpha, 0)\}_{j=1,\ldots,J}$$

of functions defined on $\mathbb{R}^{n-1}$ and its left inverse defined by

(8) $$\Phi_0^*(y_1, \ldots, y_J) = \sum_{j=1}^{J} \sqrt{\alpha_j} \phi_j^* \cdot y_j.$$

DEFINITION 2
We denote by $H^s(\Gamma)$ the space of functions x such that $\Psi_0^* \cdot x$ belongs to $H^s(\mathbb{R}^{n-1})^J$ with the norm

(9) $$\|x\|_{H^s(\Gamma)} = \left(\sum_{j=1}^{J} \|\psi_j^* \cdot (\sqrt{\alpha_j} x)(\alpha, 0)\|_{H^s(\mathbb{R}^{n-1})}^2 \right)^{1/2}.$$ ▲

In other words, we identify the space $H^s(\Gamma)$ with a closed subspace of $H^s(\mathbb{R}^{n-1})^J$ by making use of the isomorphism Ψ_0^*. Of course, this definition appears to depend on the choice of the system of local coordinates Φ_j defining the manifold Γ. Quite the contrary, for it is easy, though tedious, to verify that the norms on $\mathcal{S}(\Gamma)$ associated with different equivalent systems of local coordinates are equivalent.

Trace Theorem and Extension Theorem in Ω

Let Ω be a "regular" open set of $\mathbb{R}^n$ with boundary Γ. We define the trace operators of order j of the functions of $H^m(\Omega)$ by means of the trace operators γ_j of order j defined on the spaces $H^m(\mathring{\mathbb{R}}^n_+)$ in the following fashion:

(10) $$\gamma_j^\Omega = \Phi_0^* \gamma_j \Psi^* x.$$

Similarly, we define the extension operator $\pi^\Omega \in \mathcal{L}(H^m(\Omega), H^m(\mathbb{R}^n))$ by means of the extension operator $\pi \in \mathscr{L}(H^m(\mathring{\mathbb{R}}^n_+), H^m(\mathbb{R}^n))$ by the following formula:

(11) $$\pi_\Omega x = \Phi^* \pi \Psi^* x.$$

Theorems 4.2, 4.3, and 4.4 then imply the following results.

THEOREM 1
Suppose that Ω is regular. The operator $\gamma^\Omega = \gamma_0^\Omega \times \ldots \times \gamma_{m-1}^\Omega$ is a continuous linear operator from $H^m(\Omega)$ onto $\prod_{j=0}^{m-1} H^{m-j-1/2}(\Gamma)$ whose kernel is $H_0^m(\Omega)$. ▲

THEOREM 2
Suppose that Ω is regular. The operator π^Ω is a continuous linear operator from $H^m(\Omega)$ to $H^m(\mathbb{R}^n)$ whose left inverse is the operator of restriction to Ω. Moreover, for every $|k| \leq m$,

(12) $$\rho D^k \pi x = D^k x \quad \text{when} \quad x \in H^m(\Omega).$$

If $x \in H_0^m(\Omega)$, $\pi^\Omega x$ coincides with the extension of x by zero outside of Ω. ▲

We also obtain the following result.

THEOREM 3
Suppose that Ω is regular. Then $\mathscr{E}(\bar{\Omega})$ is dense in $H^m(\Omega)$. ▲

Proof. (The proof is analogous to that of Theorem 7.6.2.) ■

6. THE COMPACTNESS THEOREM

We are going to prove that the canonical injection from a Sobolev space to another is compact when Ω is bounded.

THEOREM 1
Suppose that Ω is regular (hence, bounded). For every $m > k \geq 0$, the unit ball of $H^m(\Omega)$ is relatively compact in $H^k(\Omega)$. ▲

Proof. Let B_m be the unit ball of $H^m(\Omega)$ and $\{x_n\}_n$ a sequence of elements of B_m. We are going to extract a convergent subsequence in $H^k(\Omega)$. Before beginning the proof proper, we use the extension operator π^Ω from $H^m(\Omega)$ to $H^m(\mathbb{R}^n)$ constructed in (ii), Section 5. Let $\theta \in \mathscr{D}(\mathbb{R}^n)$ be a function that is equal to one on a neighborhood of $\bar{\Omega}$. Then the functions $y_n = \theta \pi^\Omega x_n$ are in a ball $B_m(a) \subset H^m(\mathbb{R}^n)$ of radius $a > 1$ and their supports are contained in the support K of θ. Since the restriction of y_n to Ω is x_n, it suffices to extract a convergent subsequence in $H^k(\mathbb{R}^n)$ from the sequence $\{y_n\}_n$. Hence we fix

$\varepsilon > 0$ and associate to ε the constant $M = M(\varepsilon)$ such that $a(1 + M^2)^{k-m} \leq \varepsilon/4$. We then obtain the inequality

(1)
$$\begin{aligned}&\int_{\|\xi\| \geq M} (1 + \|\xi\|^2)^k |Fy_n(\xi) - Fy_p(\xi)|^2 \, d\xi \\ &\leq (1 + M^2)^{k-m} \int_{\|\xi\| \geq M} (1 + \|\xi\|^2)^m |Fy_n(\xi) - Fy_p(\xi)|^2 \, d\xi \\ &\leq (1 + M^2)^{k-m} \|y_n - y_p\|_{H^m(\mathbb{R}^n)} \leq 2a(1 + M^2)^{k-m} \leq \frac{\varepsilon}{2}.\end{aligned}$$

Now consider the sequence of Fourier transforms Fy_n. We are going to show that they form an equicontinuous and bounded set of continuous functions.

Indeed, let $\varphi \in \mathscr{D}(\mathbb{R}^n)$ be a function equal to one on the compact support K of θ. We can write

$$Fy_n(\xi_1) - Fy_n(\xi_2) = \int y_n(\eta)\varphi(\eta)(e^{-2i\pi\langle \xi_1, \eta\rangle} - e^{-2i\pi\langle \xi_2, \eta\rangle}) \, d\eta.$$

Since $|e^{-2i\pi\langle \xi_1\, \eta\rangle} - e^{-2i\pi\langle \xi_2, \eta\rangle}| \leq \alpha$ when $\|\xi_1 - \xi_2\| \leq \beta(\alpha)$ uniformly as η runs over the compact set K, the Cauchy-Schwarz inequality implies that

(2) $\quad |Fy_n(\xi_1) - Fy_n(\xi_2)| \leq \alpha \|y_n\|_{L^2(\mathbb{R}^n)} \|\varphi\|_{L^2(\mathbb{R}^n)} \leq a\alpha \|\varphi\|_{L^2(\mathbb{R}^n)}$
$\quad$ when $\quad \|\xi_1 - \xi_2\| \leq \beta(\alpha)$,

that is, that the Fy_n's are equicontinuous. Similarly, since $Fy_n(\xi) = \int y_n(\eta)\varphi(\eta)e^{-2i\pi\langle \xi, \eta\rangle} \, d\eta$, the Cauchy-Schwarz inequality implies that $|Fy_n(\xi)| \leq \|y_n\|_{L^2(\mathbb{R}^n)} \|\varphi\|_{L^2(\mathbb{R}^n)} \leq a \|\varphi\|_{L^2(\mathbb{R}^n)}$.

Hence for every integer M, the restriction of the sequence Fy_n to the ball of radius M forms a relatively compact set of continuous functions according to the Ascoli theorem (Theorem 4.6.2 of [AAA], p. 153.) Thus we can extract a subsequence such that Fy_{n_p} converges uniformly to z for $\|\xi\| \leq M$. Consequently, for every $\varepsilon > 0$, there exists an integer p such that for all $p, q \geq p$,

(3) $$\sup_{\|\xi\| \leq M} |Fy_{n_p}(\xi) - Fy_{n_q}(\xi)|^2 \leq \frac{\varepsilon}{2(1 + M^2)^k N},$$

where N is the measure of the ball of radius M. Therefore

(4)
$$\begin{cases}\int_{\|\xi\| \leq M} (1 + \|\xi\|^2)^k |Fy_{n_p} - Fy_{n_q}(\xi)|^2 \, d\xi \\ \leq (1 + M^2)^k \int_{\|\xi\| \leq M} |Fy_{n_p}(\xi) - Fy_{n_q}(\xi)|^2 \, d\xi \leq \frac{\varepsilon}{2}.\end{cases}$$

Inequalities (1) and (4) show that the subsequence Fy_{n_p} is a Cauchy sequence in $\hat{H}^k(\mathbb{R}^n)$. Hence the subsequence y_{n_p} is a convergent subsequence of $H^k(\mathbb{R}^n)$. ∎

CHAPTER 10

Elementary Convex Analysis

In this chapter we present the principal results of convex analysis that are essential to the study of optimization theory. In Section 1 we associate to every function $f: U \mapsto]-\infty, +\infty]$ its conjugate function $f^*: U^* \mapsto]-\infty, +\infty]$ defined by

$$f^*(p) = \sup_{x \in U}[\langle p, x \rangle - f(x)]$$

and its biconjugate function $f^{**}: U \mapsto]-\infty, +\infty]$ defined by

$$f^{**}(x) = \sup_{p \in U^*}[\langle p, x \rangle - f^*(p)].$$

The functions f^* and f^{**} are clearly convex and lower semicontinuous. The principal result of Section 1 establishes that *f is convex and lower semicontinuous if and only if $f = f^{**}$*.

In Section 2 we review various concepts of differentiability, and we define the subdifferential

$$\partial f(x) = \{p \in U^* \text{ such that } f(x) - f(y) \leq \langle p, x - y \rangle \text{ for all } y \in U\}$$

of a function $f: U \mapsto]-\infty, +\infty]$ (which may be empty). If f is convex and lower semicontinuous, we show that $\partial f^*(p)$ is the set of solutions $x \in U$ that minimize $y \mapsto f(y) - \langle p, y \rangle$. This subdifferential generalizes the concept of differentiability in convex functions. We see that if f possesses a gradient $Df(x) \in U^*$, then $\partial f(x)$ reduces to the single element $Df(x)$. These concepts are used to study minimization problems of the form

$$v = \inf_{x \in U} F(x, Lx) \quad \text{and} \quad v^* = \inf_{p \in V^*} F^*(-L^*p, p)$$

where F is a function defined on the product $X \times Y$ of closed convex subsets of U and of V, $L \in \mathscr{L}(U, V)$, F^* is the conjugate function of F, and L^* is the transpose of L.

We say that $\bar{p}$ is a Lagrange multiplier if $\bar{p}$ is a minimal solution of the problem v^* *and if* $-v = v^*$. It is related to the optimal solution $\bar{x}$ of the problem v by the extremality relations $\{-L^*\bar{p}, \bar{p}\} \in \partial F(\bar{x}, L\bar{x})$. We give in Section 4 an existence theorem for a Lagrange multiplier; Section 5 is devoted to some supplementary results on this problem.

1. CONJUGATE FUNCTIONS

We can justify the introduction of the conjugate function f^* of a function f defined on a subset X of a Hilbert space U in the following fashion. Rather than studying simply the minimization problem $\alpha = \inf_{x \in X} f(x)$, we study the perturbed problems $-f^*(p) = \inf_{x \in X}[f(x) - \langle p, x \rangle]$ where the function f is perturbed by the continuous linear forms $x \mapsto \langle p, x \rangle$ (which form the class of the "simplest" perturbations that can be considered). This defines the function $p \mapsto f^*(p)$.

Furthermore, we can take into account the domain X on which f is defined by associating to it the function $f_X : U \mapsto]-\infty, +\infty]$ defined by

(1) $$f_X(x) = \begin{cases} f(x) & \text{if } x \in X \\ +\infty & \text{if } x \notin X. \end{cases}$$

Conversely, we can associate to every function $f : U \mapsto]-\infty, +\infty]$ its domain $X = \text{Dom}\, f = \{x \in U \text{ such that } f(x) < \infty\}$ on which it is finite. We suppose, therefore, that the functions f that we use are all functions from U to $]-\infty, +\infty]$.

Observe that if $f : U \mapsto]-\infty, +\infty]$ has a nonempty domain and if $x_0 \in \text{Dom}\, f$, then for all $p \in U^*$,

(2) $$-\infty < \langle p, x_0 \rangle - f(x_0) \leq \sup_{x \in X}[\langle p, x \rangle - f(x)].$$

DEFINITION 1

Let f be a function from a Hilbert space U to $]-\infty, +\infty]$ whose domain is nonempty. The function $f^* : U^* \mapsto]-\infty, +\infty]$ associating to every $p \in U^*$

(3) $$f^*(p) = \sup_{x \in U}[\langle p, x \rangle - f(x)] \in]-\infty, +\infty]$$

is called the **conjugate function** *of f.*

Similarly, if g is a function from U^* to $]-\infty, +\infty]$ whose domain is nonempty, we define $g^* : U \mapsto]-\infty, +\infty]$ by

(4) $$g^*(x) = \sup_{p \in U^*}[\langle p, x \rangle - g(p)] \in]-\infty, +\infty]$$

and, consequently, the biconjugate f^{**} of f defined by $f^{**} = (f^*)^*$. ▲

Remark 1

If we interpret the vector space U as a *commodity space* and accordingly its dual U^* as a *price space* and if we interpret $f : U \mapsto \,]-\infty, +\infty]$ as a *cost function* that associates to every commodity $x \in U$ its cost $f(x) \in \,]-\infty, +\infty]$, then the conjugate function f^* can be interpreted as a *profit function* that associates to every price $p \in U^*$ the maximum profit $f^*(p) = \sup_{x \in U}[\langle p, x \rangle - f(x)]$ (since $\langle p, x \rangle$ is the value of x when the price system p prevails). ∎

Let us remark first of all that

(5) $$-f^*(0) = \inf_{x \in U} f(x)$$

and that the biconjugate function satisfies

(6) $$f^{**}(x) \leq \sup_{p \in U^*}[\langle p, x \rangle - [\langle p, x \rangle - f(x)]] \leq f(x).$$

We also obtain the *Fenchel inequality*:

(7) $$\langle p, x \rangle \leq f(x) + f^*(p) \quad \forall x \in U, \quad \forall p \in U^*.$$

Since f^* is the pointwise supremum of the family of continuous affine functions $p \mapsto \langle p, x \rangle - f(x)$,

(8) the functions f^* and f^{**} are convex and lower semicontinuous.

Hence a necessary condition for f to equal its biconjugate is that f be convex and lower semicontinuous. This condition turns out to be sufficient as well.

THEOREM 1
Let U be a Hilbert space. A function $f : U \mapsto \,]-\infty, +\infty]$ whose domain is nonempty is convex and lower semicontinuous if and only if $f = f^{**}$. ▲

Proof. Since f is convex and lower semicontinuous, its epigraph $\mathscr{E}p(f) = \{\{x, \alpha\} \in U \times \mathbb{R} \text{ such that } f(x) - \alpha \leq 0\}$ is a closed convex subset of $U \times \mathbb{R}$ (See Proposition 3.8.1 of [AAA], p. 109.)

a. We assume that $R < f(x)$. Since the pair $\{x, R\}$ does not belong to the epigraph $\mathscr{E}p(f)$ of f, there exists a continuous linear form $\{p, \alpha\} \in U^* \times \mathbb{R}$ that strictly separates $\{x, R\}$ from the closed convex set $\mathscr{E}p(f)$ (see Theorem 2.3.1.): there exists $\varepsilon > 0$ such that

(9) $\forall y \in \text{Dom } f, \quad \forall \lambda \geq f(y), \quad \langle p, y \rangle - \alpha\lambda \leq \langle p, x \rangle - \alpha R - \varepsilon$

b. We note that $\alpha \geq 0$; if not, by taking $y = x_0 \in \text{Dom } f$ (which is nonempty by assumption) and $\lambda = f(x_0) + n$, we would have $-\lambda\alpha n \leq \langle p, x \rangle - \alpha R -$

$\langle p, x_0 \rangle + \alpha f(x_0) - \varepsilon$, which is impossible if $\alpha < 0$ and n is large enough.

c. We consider the case where $x \in \text{Dom } f$. By taking $y = x$ and $\lambda = f(x)$, we deduce from (9) that $\alpha(f(x) - R) \geq \varepsilon$. Hence $\alpha > 0$; by dividing by $\alpha > 0$ and setting $\bar{p} = p/\alpha$, we obtain

$$(10) \qquad \forall y \in \text{Dom } f, \quad \langle \bar{p}, y \rangle - f(y) \leq \langle \bar{p}, x \rangle - R - \frac{\varepsilon}{\alpha}$$

By taking the supremum with respect to λ, we deduce that

$$(11) \qquad f^*(\bar{p}) \leq \langle \bar{p}, x \rangle - R - \frac{\varepsilon}{\alpha}$$

This implies that $\bar{p} \in \text{Dom } f^*$ and also, that $R \leq f^{**}(x)$ for any $R < f(x)$. Hence $f(x) = f^{**}(x)$.

d. We consider the case where $x \notin \text{Dom } f$. If $\alpha > 0$, assertions (10) and (11) still hold true and we deduce that $R \leq f^{**}(x)$ for all R. Hence $f^{**}(x) = +\infty$.

It remains to consider the case where $\alpha = 0$. Inequalities (9) imply that

$$(12) \qquad \forall y \in \text{Dom } f, \quad \langle p, y - x \rangle + \varepsilon \leq 0$$

Let us take $\bar{p} \in \text{Dom}(f^*)$ (which we proved to be nonempty). By multiplying inequality (12) by $n \geq 0$ and adding it to inequality $\langle \bar{p}, y \rangle - f^*(\bar{p}) - f(y) \leq 0$, we obtain

$$(13) \qquad \forall y \in \text{Dom } f, \quad \langle \bar{p} + np, y \rangle - n \langle p, x \rangle + n\varepsilon - f^*(\bar{p}) - f(y) \leq 0$$

It implies, by taking the supremum with respect to $y \in \text{Dom } f$,

$$(14) \qquad f^*(\bar{p} + np) \leq n \langle p, x \rangle - n\varepsilon + f^*(\bar{p}).$$

Therefore

$$\langle \bar{p}, x \rangle + n\varepsilon - f^*(\bar{p}) \leq \langle \bar{p} + np, x \rangle - f^*(\bar{p} + np) \leq f^{**}(x).$$

By letting $n \to \infty$, we deduce that $f^{**}(x) = +\infty$. ∎

Example 1

Let X be a nonempty subset of U. We denote by $\Psi_X : U \mapsto \,]-\infty, +\infty]$ its *indicator function* defined by

$$(15) \qquad \Psi_X(x) = \begin{cases} 0 & \text{if} \quad x \in X \\ \infty & \text{if} \quad x \notin X. \end{cases}$$

This function is convex if and only if X is convex, lower semicontinuous if and only if X is closed, continuous at x if x is in the interior of X.

Moreover, the conjugate function of Ψ_X is the support function σ_X defined by

(16) $$\sigma_X(p) = \sup_{x \in X} \langle p, x \rangle = \sup_{x \in U}[\langle p, x \rangle - \Psi_X(x)] = \Psi_X^*(p).$$ ∎

We have, therefore, established the following result.

PROPOSITION 1
The support function of a set X is the conjugate function of its indicator function. Conversely, if X is closed and convex, $\Psi_X = \sigma_X^*$. ▲

PROPOSITION 2
The function $p \mapsto \frac{1}{2}\|p\|_*^2$ is the conjugate of the function $x \mapsto \frac{1}{2}\|x\|^2$. ▲

Proof. Indeed, it follows from the Cauchy-Schwarz inequality that $\langle p, x \rangle \leq \|p\|_* \|x\| \leq \frac{1}{2}\|p\|_*^2 + \frac{1}{2}\|x\|^2$ and that for $p = Jx, \langle Jx, x \rangle = \|x\|^2 = \frac{1}{2}\|Jx\|_*^2 + \frac{1}{2}\|x\|^2$. Hence we have $\frac{1}{2}\|p\|_*^2 = \sup_{x \in U}[\langle p, x \rangle - \frac{1}{2}\|x\|^2]$. ∎

Elementary Properties of Conjugate Functions

PROPOSITION 3
If $f \leq g$, then $g^* \leq f^*$.
If $g(x) = f(x - x_0) + \langle p_0, x \rangle + \alpha$, then $g^*(p) = f^*(p - p_0) + \langle p, x_0 \rangle - (\langle p_0, x_0 \rangle + \alpha)$. ▲

Proof. The first statement is obvious. We verify the second:

$$\sup_{x \in U}[\langle p, x \rangle - g(x)] = \sup_{x \in U}[\langle p - p_0, x \rangle - f(x - x_0)] - \alpha$$
$$= \sup_{x \in U}[\langle p - p_0, x - x_0 \rangle - f(x - x_0)] - \alpha + \langle p - p_0, x_0 \rangle$$
$$= f^*(p - p_0) + \langle p, x_0 \rangle - \alpha - \langle p_0, x_0 \rangle.$$ ∎

2. THE GRADIENT

Let us recall some of the common notions of differentiability.

DEFINITION 1
Let $f: U \mapsto]-\infty, +\infty]$ be a function with nonempty domain. Let x_0 be an element of the interior of $\text{Dom} f$. If the limit

(1) $$Df(x_0)(x) = \lim_{\theta \to 0^+} \frac{f(x_0 + \theta x) - f(x_0)}{\theta}$$

exists, we say that $Df(x_0)(x)$ is the derivative from the right of f at x_0 in the direction of x. If $Df(x_0)(x)$ exists for every direction $x \in U$, we say that f is differentiable from the right at x_0. If, moreover,

$$(2) \qquad x \to Df(x_0)(x) = \langle Df(x_0), x \rangle$$

is continuous and linear, we say that $Df(x_0)$ is the gradient of f at x_0 and that f is differentiable or Gâteaux-differentiable at x_0. ▲

Observe that $x \to Df(x_0)(x)$ is positively homogeneous.

The principal justification of differentiability in the context of optimization theory is the *variational principle*.

PROPOSITION 1

Let $f = g + h$ be the sum of a convex function g and a function h which is Gâteaux-differentiable on a convex subset X. If $\bar{x} \in X$ minimizes f on X, then

$$(3) \qquad \forall x \in X, \quad \langle Dh(\bar{x}), \bar{x} - x \rangle + g(\bar{x}) - g(x) \leq 0.$$

In particular, if $g = 0$ and if $\bar{x}$ belongs to the interior of X, this condition implies that $Df(\bar{x}) = 0$. ▲

Proof. Since X is convex, $y = \bar{x} + \theta(x - \bar{x})$ belongs to X when $\theta \in \,]0, 1]$. If $\bar{x}$ minimizes f on X, then $[f(\bar{x} + \theta(x - \bar{x})) - f(\bar{x})]/\theta \geq 0$. Since g is convex, $[h(\bar{x} + \theta(x - \bar{x})) - h(\bar{x})]/\theta \geq g(\bar{x}) - g(x)$. Then (3) follows by taking the limit as $\theta \to 0$. ■

First of all we are going to study the existence of a derivative from the right for a convex function.

LEMMA 1

Let $f: U \mapsto \,]-\infty, +\infty]$ be a convex function of nonempty domain. Let x_0 and x be two elements such that the segment $[x_0 - x, x_0 + x] = \{x_0 + \theta x$ where $\theta \in [-1, +1]\}$ is contained in the domain of f. Then $Df(x_0)(x)$ exists and satisfies

$$(4) \qquad f(x_0) - f(x_0 - x) \leq Df(x_0)(x) \leq f(x_0 + x) - f(x_0).$$

Moreover,

$$(5) \qquad x \to Df(x_0)(x) \quad \text{is convex.} \qquad ▲$$

Proof. Let $\theta \in \,]0, 1]$; we first remark that

$$(6) \qquad f(x_0) - f(x_0 - x) \leq \frac{f(x_0 + \theta x) - f(x_0)}{\theta}.$$

Indeed, $x_0 = \frac{1}{1+\theta}(x_0 + \theta x) + \frac{\theta}{1+\theta}(x_0 - x)$. Hence $f(x_0) \leq \frac{1}{1+\theta} f(x_0 + \theta x) + \frac{\theta}{1+\theta} f(x_0 - x)$ because of the convexity of f. This inequality implies (6).

Moreover, the function
$$\theta \to \frac{f(x_0 + \theta x) - f(x_0)}{\theta}$$
is increasing; indeed, if $\theta_1 \leq \theta_2 \leq 1$, then
$$f(x_0 + \theta_1 x) - f(x_0) = f\left(\frac{\theta_1}{\theta_2}(x_0 + \theta_2 x) + \left(1 - \frac{\theta_1}{\theta_2}\right)x_0\right) - f(x_0)$$
$$\leq \frac{\theta_1}{\theta_2} f(x_0 + \theta_2 x) + \left(1 - \frac{\theta_1}{\theta_2}\right) f(x_0) - f(x_0),$$

Hence

(7) $$\frac{f(x_0 + \theta_1 x) - f(x_0)}{\theta_1} \leq \frac{f(x_0 + \theta_2 x) - f(x_0)}{\theta_2} \leq f(x_0 + x) - f(x_0).$$

Consequently, we obtain

$$f(x_0) - f(x_0 - x) \leq \inf_{\theta > 0} \frac{f(x_0 + \theta x) - f(x_0)}{\theta}$$

(8) $$= \lim_{\theta \to 0^+} \frac{f(x_0 + \theta x) - f(x_0)}{\theta}$$
$$= Df(x_0)(x) \leq f(x_0 + x) - f(x_0).$$

Finally, since

$$\frac{f(x_0 + \theta[\lambda x_1 + (1-\lambda)x_2]) - f(x_0)}{\theta} \leq \lambda \frac{f(x_0 + \theta x_1) - f(x_0)}{\theta}$$
$$+ (1-\lambda) \frac{f(x_0 + \theta x_2) - f(x_0)}{\theta}$$

it follows by letting θ approach zero that

(9) $\quad Df(x_0)(\lambda x_1 + (1-\lambda)x_2) \leq \lambda Df(x_0)(x_1) + (1-\lambda) Df(x_0)(x_2).$ ∎

THEOREM 1

Let $f: U \mapsto]-\infty, +\infty]$ *be a convex function whose domain has a nonempty interior. Then f is differentiable from the right at every point x_0 interior to its domain. If, in addition, f is continuous at x_0, then*

(10) $\qquad\qquad x \to Df(x_0)(x)$ *is continuous.* ▲

Proof. If x_0 is the center of a ball of radius η contained in Dom f, the hypothesis of Lemma 1 is satisfied with x replaced by $\eta(x/\|x\|)$. Hence $Df(x_0)(x)$ exists for all x. Let $x \in \eta B$.

Since $Df(x_0)(x) \leq f(x_0 + x) - f(x_0)$ and since f is continuous at x_0, it follows that $Df(x_0)(x)$ is bounded above on a ball of radius α by a number $a > 0$. Hence

$$(11) \qquad Df(x_0)(x) \leq \frac{a}{\alpha} \|x\| \qquad \text{for all} \qquad x \in U,$$

which implies that $Df(x_0)(\cdot)$ is continuous. (See [AAA], Theorem 3.8.2, p. 111.) ∎

3. THE SUBDIFFERENTIAL

We are going to characterize the set of elements that minimize a function f by means of the notion of subdifferential, which we define.

DEFINITION 1
Let $f : U \mapsto \,]-\infty, +\infty]$ be a function with nonempty domain. If $x_0 \in \text{Dom } f$, the "subdifferential $\partial f(x_0)$ of f at x_0" is the subset (which may be empty) of U^* defined by

$$(1) \qquad \partial f(x_0) = \{p \in U^* \text{ such that } f(x_0) - f(x) \leq \langle p, x_0 - x \rangle \text{ for all } x \in U\}.$$

The elements $p \in \partial f(x_0)$ are also called **subgradients**. ▲

Clearly an equivalent definition is

$$(2) \qquad p_0 \in \partial f(x_0) \qquad \text{if and only if} \qquad x_0 \text{ minimizes } x \to f(x) - \langle p_0, x \rangle \text{ on } U$$

or indeed,

$$(3) \qquad p_0 \in \partial f(x_0) \qquad \text{if and only if} \qquad \langle p_0, x_0 \rangle = f^*(p_0) + f(x_0).$$

Note that $\partial f(x_0)$ *is a closed convex subset*.
An initial justification of this notion is given by the following result:

PROPOSITION 1
Let $f : U \mapsto \,]-\infty, +\infty]$ be a function with nonempty domain. Then

$$(4) \qquad x_0 \text{ minimizes } x \mapsto f(x) - \langle p_0, x \rangle \text{ on } U$$

if and only if

$$(5) \qquad \begin{cases} \text{i.} & x_0 \in \partial f^*(p_0). \\ \text{ii.} & f(x_0) = f^{**}(x_0). \end{cases}$$

Proof. Suppose that (4) is the case. The inequalities
$$f(x_0) - \langle p_0, x_0 \rangle \leq f(x) - \langle p_0, x \rangle$$
imply

(6) $$f^*(p_0) = \langle p_0, x_0 \rangle - f(x_0).$$

Since for every $p \in U^*$
$$f^*(p) \geq \langle p, x_0 \rangle - f(x_0),$$
it follows that

(7) $$f^*(p_0) - f^*(p) \leq \langle p_0 - p, x_0 \rangle$$

(i.e., that $x_0 \in \partial f^*(p_0)$). Moreover, (7) implies that $f^{**}(x_0) = \langle p_0, x_0 \rangle - f^*(p_0)$. Hence, comparing this with (6), we deduce that $f(x_0) = f^{**}(x_0)$.

Conversely, let us suppose that (5) holds. Since $x_0 \in \partial f^*(p_0)$, we obtain from the inequalities
$$f^*(p_0) - \langle p_0, x \rangle \leq f^*(p) - \langle p, x_0 \rangle$$
that
$$f^{**}(x_0) - \langle p_0, x_0 \rangle = -f^*(p_0) = \inf_{x \in U} [f(x) - \langle p_0, x \rangle].$$

Since $f(x_0) = f^{**}(x_0)$, we have proved that x_0 minimizes $x \mapsto f(x) - \langle p_0, x \rangle$ on U. ∎

For convex lower semicontinuous functions this result becomes

PROPOSITION 2
Let $f : U \mapsto \,]-\infty, +\infty]$ be a convex lower semicontinuous function with nonempty domain. The set of points that minimize $x \mapsto f(x) - \langle p, x \rangle$ is the subdifferential $\partial f^*(p)$. ▲

This is equivalent to saying that

(8) $\quad\quad p_0 \in \partial f(x_0) \quad$ if and only if $\quad x_0 \in \partial f^*(p_0).$

Example 1

Let X be a nonempty subset. If Ψ_X is its indicator function, we verify that $p \in \partial \Psi_X(x)$ if and only if

(9) $$\begin{cases} \text{i.} & x \in X. \\ \text{ii.} & \langle p, x - y \rangle \geq 0 \quad \text{for all} \quad y \in X. \end{cases}$$

If $x \in \text{Int}(X)$, $\partial \Psi_X(x) = 0$.

We say that $\partial \Psi_X(x)$ (when $x \in X$) is the "*normal cone to X at x*". The study of normal cones is continued in Chapter 15, Section 4.

Moreover, $\partial \sigma_X(p)$ is the set of $x \in X$ such that $\langle p, x \rangle = \sigma_X(p)$. If X is closed and convex, Proposition 2 implies that *p is in the normal cone to X at $x \in X$ if and only if* $\langle p, x \rangle = \sigma_X(p)$. We say that $\partial \sigma_X(p)$ is the *support zone of X at p*. ∎

We are now going to examine the relations that exist between the notions of subdifferential and derivative from the right.

PROPOSITION 3
Let $f : U \mapsto \,]-\infty, +\infty]$ be a convex function. Then
(10) $\qquad \partial f(x_0) = \{p \in U^* \text{ such that } \langle p, x \rangle \leq Df(x_0)(x) \quad \forall x \in U\}.$ ▲

Proof. We denote by M the set that is the right-hand side of equality (10). If $p \in \partial f(x_0)$, it follows, if $\theta \in \,]0, 1[$, that

$$\langle p, x \rangle = \frac{1}{\theta} \langle p, x_0 + \theta x - x_0 \rangle \leq \frac{1}{\theta}(f(x_0 + \theta x) - f(x_0)).$$

Hence, letting θ approach zero, we obtain
(11) $\qquad\qquad\qquad \langle p, x \rangle \leq Df(x_0)(x).$

Conversely, inequality (11) implies that

$$\langle p, x \rangle \leq Df(x_0)(x) \leq f(x_0 + x) - f(x_0) \qquad \text{for all} \qquad x \in U,$$

which shows that $p \in \partial f(x_0)$. ∎

PROPOSITION 4
Let $f : U \mapsto \,]-\infty, +\infty]$ be a convex function. Suppose that f is differentiable from the right at x_0 and that $x \to Df(x_0)(x)$ is lower semicontinuous. Then f is subdifferentiable at x_0 and
(12) $\qquad\qquad\qquad Df(x_0)(x) = \sigma(\partial f(x_0); x).$ ▲

Proof. Since $Df(x_0)(\cdot)$ is convex, lower semicontinuous, and positively homogeneous, it is the conjugate function of the indicator function of the subset

$$M = \{p \in U^* \quad \text{such that} \quad \langle p, x \rangle - Df(x_0)(x) \leq 0 \quad \forall x \in U\}.$$

According to Proposition 3, $M = \partial f(x_0)$. ∎

Thus we obtain the following important theorem on subdifferentiability.

THEOREM 1
Let $f: U \mapsto \,]-\infty, +\infty]$ be a convex function. Suppose that f is continuous at a point x_0. Then f is subdifferentiable and differentiable from the right at x_0; moreover, $x \to Df(x_0)(x) = \sigma(\partial f(x_0); x)$ is continuous and $\partial f(x_0)$ is a convex, closed, and bounded subset. ▲

Proof. According to Theorem 2.1, f is differentiable from the right at x_0 and $Df(x_0)(\cdot)$ is continuous. Proposition 3 implies, therefore, that $\partial f(x_0) \neq \emptyset$ and that

$$\sigma(\partial f(x_0); x) = Df(x_0)(x) \leq c\|x\| = c\sigma(B_*; x), \tag{13}$$

since $\|x\| = \sup_{p \in B_*} \langle p, x \rangle$ where B_* is the unit ball of the dual space. Hence $\partial f(x_0) \subset cB_*$ is bounded. ■

PROPOSITION 5
Let $f: U \mapsto \,]-\infty, +\infty]$ be a convex function. Suppose that the function f is both subdifferentiable and differentiable in the sense of Gâteaux at a point $\hat{x}$ of the interior of its domain. Then

$$\partial f(\hat{x}) = \{Df(\hat{x})\}. \tag{14}$$
▲

Proof. If $p \in \partial f(\bar{x})$, Proposition 3 implies that $\langle p, x \rangle \leq \langle Df(\bar{x}), x \rangle$ for all $x \in U$ and, consequently, that $p = Df(\bar{x})$. ■

PROPOSITION 6
If f is convex and continuous at x_0 and $\partial f(x_0)$ contains exactly one element, then f is differentiable at x_0. ▲

Proof. If f is continuous at x_0 and if $\partial f(x_0) = \{p_0\}$ contains the unique element p_0, Proposition 4 implies that $Df(x_0)(x) = \sigma\{p_0\}(x) = \langle p_0, x \rangle$ for all $x \in U$, that is, that $p_0 = Df(x_0)$ is the gradient of f at x_0. ■

Suppose we are given a convex lower semicontinuous function whose domain $\text{Dom } f$ is nonempty. If its interior is nonempty, we know that as a consequence of Baire's theorem (see Theorem 5.3.4 of [AAA], p. 189) f is continuous on $\text{Int}(\text{Dom } f)$. Hence Theorem 1 implies that f is subdifferentiable and that

$$\text{Int}(\text{Dom } f) \subset D(\partial f) \subset \text{Dom } f \tag{15}$$

where $D(\partial f) = \{x \in U \text{ such that } \partial f(x) \neq \emptyset\}$.

In fact we are going to show that $D(\partial f) \neq \emptyset$ and is, also, dense in $\text{Dom } f$ even when $\text{Int}(\text{Dom } f) = \emptyset$.

Let J be the duality operator from U onto U^*.

THEOREM 2
Let $f : U \mapsto]-\infty, +\infty]$ be a convex lower semicontinuous function. Then $D(\partial f) \neq \emptyset$ and

(16) $\qquad\qquad\qquad D(\partial f)$ is dense in Dom f.

Moreover, for every $k > 0$, the mapping $J + k\partial f$ is surjective in the sense that

(17) $\qquad \forall p \in U^*, \quad \exists x_k \in D(\partial f) \quad \text{such that} \quad p \in Jx_k + k\partial f(x_k).$ ▲

Proof. We fix $k > 0$ and set

(18) $$f_k(x) = \inf_{y \in U}\left[f(y) + \frac{1}{2k}\|y - x\|^2 \right].$$

a. First we show that there exists $x_k \in U$ such that

(19) $$f_k(x) = f(x_k) + \frac{1}{2k}\|x_k - x\|^2.$$

To this end we consider a minimizing sequence of elements $y^n \in U$ that satisfies $f(y^n) + (1/2k)\|y^n - x\|^2 \leq f_k(x) + 1/n$. We verify that this is a Cauchy sequence (as in the proof of Theorem 1.4.1 of projectors). Indeed,

$$\|y^n - y^m\|^2 = 2\|y^n - x\|^2 + 2\|y^m - x\|^2 - 4\left\|\frac{y_n + y_m}{2} - x\right\|^2$$

$$\leq 4k\left[\frac{1}{n} + \frac{1}{m} + 2f_k(x) - f(y^n) - f(y^m)\right] + 8k\left[f\left(\frac{y^n + y^m}{2}\right) - f_k(x)\right]$$

$$\leq 4k\left[\frac{1}{n} + \frac{1}{m} + 2f\left(\frac{y^n + y^m}{2}\right) - f(y^n) - f(y^m)\right]$$

$$\leq 4k\left(\frac{1}{n} + \frac{1}{m}\right),$$

since f is convex. Hence y^n converges to an element x_k. Since f is lower semicontinuous, it follows that

$$f(x_k) + \frac{1}{2k}\|x_k - x\|^2 \leq \lim_{n \to \infty} \inf f(y^n) + \lim_{n \to \infty} \frac{1}{2k}\|y^n - x\|^2 \leq f_k(x).$$

b. We deduce that $J + k\partial f$ is surjective. Proposition 2.1 where g is replaced by f and where $h(y) = 1/(2k)\|y - x\|^2$ implies that any minimum x_k is a solution of

(20) $\qquad f(x_k) - f(y) + \left\langle \frac{1}{k}J(x_k - x), x_k - y \right\rangle \leq 0 \qquad \forall y \in U,$

since $Dh(y) = (1/k)J(y - x)$. This shows that

$$\text{(21)} \quad \frac{1}{k}J(x-x_k) \in \partial f(x_k).$$

If $p \in U^*$ and if $x = J^{-1}p$, it follows that $p \in Jx_k + k\,\partial f(x_k)$, that is, that the mapping $J + k\,\partial F$ is surjective (and, therefore, that $x_k \in D(\partial f)$, which shows that $D(\partial f) \neq \varnothing$).

c. *The solution x_k is unique.* If $\bar{x}_k$ were another minimum, it would follow from (20) with $y = \bar{x}_k$ that

$$f(x_k) - f(\bar{x}_k) + \frac{1}{k}\langle J(x_k - x), x_k - \bar{x}_k \rangle \leq 0.$$

Interchanging the roles of x_k and $\bar{x}_k$, we obtain

$$f(\bar{x}_k) - f(x_k) + \frac{1}{k}\langle J(\bar{x}_k - x), \bar{x}_k - x_k \rangle \leq 0.$$

Adding these two inequalities, we deduce that $\|\bar{x}_k - x_k\| \leq 0$, that is, that $\bar{x}_k = x_k$.

d. *$D(\partial f)$ is dense in Dom f.* We take $x \in \text{Dom } f$ and choose $p \in \text{Dom } f^*$, which is nonempty. Since

$$\text{(22)} \quad \frac{1}{2k}\|x_k - x\|^2 + f(x_k) = f_k(x) \leq f(x)$$

and since $-f(x_k) \leq f^*(p) - \langle p, x_k \rangle$, it follows from inequality $\langle p, x - x_k \rangle \leq \|p\|_* \|x - x_k\| \leq k\|p\|_*^2 + \frac{1}{4k}\|x_k - x\|^2$ that

$$\frac{1}{2k}\|x_k - x\|^2 \leq f(x) + f^*(p) - \langle p, x \rangle + \langle p, x - x_k \rangle$$

$$\leq \frac{1}{4k}\|x - x_k\|^2 + f(x) + f^*(p) - \langle p, x \rangle + k\|p\|_*^2.$$

This implies that $\|x - x_k\|^2 \leq 4k(f(x) + f^*(p) - \langle p, x \rangle + k\|p\|_*^2)$. Hence $x_k \in D(\partial f)$ converges to $x \in \text{Dom } f$, which shows that $\text{Dom } f \subset \overline{D(\partial f)}$. ∎

In addition, (22) and $f(x) \leq \liminf_{n \to \infty} f(x_k)$ imply that

$$\text{(23)} \quad \forall x \in \text{Dom } f, \quad f(x) = \lim f_k(x).$$

We set

$$\text{(24)} \quad \begin{cases} \text{i.} & R_k(x) = x_k = (J + k\partial f)^{-1}(Jx), \text{ the unique solution of the minimization problem (18).} \\ \text{ii.} & P_k(x) = \frac{1}{k}(Jx - JR_k(x)). \end{cases}$$

Relations (18) and (21) become

(25) $\begin{cases} \text{i.} & f_k(x) = f(R_k(x)) + \dfrac{k}{2} \| P_k(x) \|_*^2. \\ \text{ii.} & P_k(x) \in \partial f(R_k(x)). \end{cases}$

The mappings $R_k : U \to U$ and $P_k : U \to U^*$ are Lipschitz.

PROPOSITION 7
The mappings R_k and P_k are Lipschitz with constant 1 and $1/k$, respectively. ▲

Proof. Since $kP_k = J - JR_k$, we obtain

(26) $\begin{aligned} \| x - y \|^2 &= \| kP_k(x) - kP_k(y) + JR_k(x) - JR_k(y) \|_*^2 \\ &= k^2 \| P_k(x) - P_k(y) \|_*^2 + \| R_k(x) - R_k(y) \|_*^2 \\ &\quad + 2k \langle P_k(x) - P_k(y), R_k(x) - R_k(y) \rangle \end{aligned}$

Since $P_k(x) \in \partial f(R_k(x))$ and $P_k(y) \in \partial f(R_k(y))$, it follows that

(27) $\langle P_k(x) - P_k(y), R_k(x) - R_k(y) \rangle \geq 0,$

by adding the inequalites

$$f(R_k(x)) - f(R_k(y)) \leq \langle P_k(x), R_k(x) - R_k(y) \rangle$$

and

$$f(R_k(y)) - f(R_k(x)) \leq \langle P_k(y), R_k(y) - R_k(x) \rangle.$$

Hence (26) and (27) imply

(28) $\| x - y \|^2 \geq k^2 \| P_k(x) - P_k(y) \|^2 + \| R_k(x) - R_k(y) \|^2.$ ■

We are going to show that $P_k(x)$ is the gradient of f_k at x. This, together with (23), provides an *approximation procedure for a convex lower semi-continuous function by convex continuously differentiable functions*.

PROPOSITION 8
Let $f : U \mapsto \,]-\infty, +\infty]$ be a convex lower semicontinuous function. Then $\forall k > 0$, the functions f_k are convex and differentiable, $Df_k(x) = P_k(x)$ is Lipschitz with respect to x and $f_k(x)$ converges to $f(x)$ for every $x \in \text{Dom } f$. ▲

Proof. It is easy to verify that f_k is convex. Since $P_k(x) \in \partial f(R_k(x))$, it follows from (25)i that $P_k(x) \in \partial f_k(x)$. Indeed,

$$f_k(x) - f_k(y) = f(R_k(x)) - f(R_k(y)) + \frac{k}{2}\|P_k(x)\|_*^2 - \frac{k}{2}\|P_k(y)\|_*^2$$

$$\leq \langle P_k(x), R_k(x) - R_k(y)\rangle + \frac{k}{2}\|P_k(x)\|_*^2 - \frac{k}{2}\|P_k(y)\|_*^2$$

$$\leq \langle P_k(x), x - y \rangle$$

$$- k\langle P_k(x), J^{-1}P_k(x) - J^{-1}P_k(y)\rangle + \frac{k}{2}\|P_k(x)\|_*^2 - \frac{k}{2}\|P_k(y)\|_*^2$$

since $(1 - R_k = kJ^{-1}P_k)$

$$= \langle P_k(x), x - y\rangle - k(\tfrac{1}{2}\|P_k(x)\|_*^2 + \tfrac{1}{2}\|P_k(y)\|_*^2$$
$$- \langle P_k(x), J^{-1}P_k(y)\rangle)$$

$$= \langle P_k(x), x - y\rangle - \frac{k}{2}\|P_k(x) - P_k(y)\|_*^2$$

$$\leq \langle P_k(x), x - y\rangle).$$

Moreover, since $P_k(y) \in \partial f_k(y)$ for all $y \in U$, we obtain the inequalities

$$f_k(x) - f_k(y) \geq \langle P_k(y), x - y\rangle = \langle P_k(x), x - y\rangle + \langle P_k(y) - P_k(x), x - y\rangle$$

$$\geq \langle P_k(x), x - y\rangle - \|P_k(y) - P_k(x)\|_* \|x - y\|$$

$$\geq \langle P_k(x), x - y\rangle - \frac{1}{k}\|x - y\|^2$$

according to Proposition 7. Hence

(29) $$\left|\frac{f_p(x) - f_k(y) - \langle P_k(x), x - y\rangle}{\|x - y\|}\right| \leq \frac{1}{k}\|x - y\|,$$

which implies that $P_k(x) = Df_k(x)$. We have already proved that $f_k(x)$ converges to $f(x)$ for all $x \in \text{Dom } f$. ∎

4. EXTREMALITY CONDITIONS FOR A MINIMIZATION PROBLEM

We are now going to use the concepts of conjugate function and subdifferential to study the properties of minimization problems.

Consider the following situation:

$$X \subset U \xrightarrow{L} V \supset Y$$
$$U^* \xleftarrow{L^*} V^*$$

where

(1) $\begin{cases} \text{i.} & U \text{ and } V \text{ are Hilbert spaces.} \\ \text{ii.} & L \in \mathscr{L}(U, V) \text{ is a continuous linear operator.} \\ \text{iii.} & X \subset U \text{ and } Y \subset V \text{ are closed convex subsets.} \end{cases}$

Suppose we are given

(2) a convex lower semicontinuous function F whose domain is $X \times Y$,

and consider its conjugate function F^* defined on $U^* \times V^*$ by

(3) $\qquad F^*(q, p) = \sup_{x \in X} \sup_{y \in Y} [\langle q, x \rangle + \langle p, y \rangle - F(x, y)].$

We propose to compare the minimization problems

(4) $$v = \inf_{x \in U} F(x, Lx)$$

and

(5) $$v^* = \inf_{p \in V^*} F^*(-L^*p, p).$$

First of all it is clear that

(6) $$v + v^* \geq 0,$$

since for all $x \in U$ and $p \in V^*$, we have

(7) $\qquad 0 = \langle -L^*p, x \rangle + \langle p, Lx \rangle \leq F(x, Lx) + F^*(-L^*p, p).$

DEFINITION 1
We say that $\bar{p} \in V^*$ is a Lagrange multiplier *for the problem* (4) *or a solution of its dual problem if*

(8) $\begin{cases} \text{i.} & v = -v^*. \\ \text{ii.} & v^* = F^*(-L^*\bar{p}, \bar{p}). \end{cases}$ ▲

The Lagrange multipliers are related to the minimal solutions of problem (4) by the following *extremality relations*.

PROPOSITION 1
Suppose that F is convex and lower semicontinuous. Then $\bar{p}$ is a Lagrange multiplier and $\bar{x}$ minimizes $x \to F(x, Lx)$ if and only if

(9) $$\{-L^*\bar{p}, \bar{p}\} \in \partial F(\bar{x}, L\bar{x}).$$ ▲

Proof. Indeed, the pair $\{-L^*\bar{p}, \bar{p}\}$ belongs to $\partial F(\bar{x}, L\bar{x})$ if and only if

(10) $\quad 0 = \langle -L^*\bar{p}, \bar{x}\rangle + \langle \bar{p}, L\bar{x}\rangle = F(\bar{x}, L\bar{x}) + F^*(-L^*\bar{p}, \bar{p}).$

Since inequality (7) always holds, it follows that

$$F(\bar{x}, L\bar{x}) = \min_x F(x, Lx), \quad F^*(-L^*\bar{p}, \bar{p}) = \min_p F^*(-L^*p, p)$$

and that $v + v^* = 0$. Conversely, if $v = F(\bar{x}, L\bar{x}), v^* = F^*(-L^*\bar{p}, \bar{p})$ and $v + v^* = 0$, equality (10) is clearly true. ∎

The relations (9) generalize the *Euler-Lagrange equations* of calculus of variations in the convex case.

Example 1

Consider the case where

(11) $\quad\quad\quad\quad F(x, y) = f(x) + g(y),$

where f and g are functions with domains X and Y, respectively. Then $F^*(q, p) = f^*(q) + g^*(p)$, and we have

(12) $\begin{cases} \text{i.} & v = \inf_x [f(x) + g(Lx)]. \\ \text{ii.} & v^* = \inf_p [f^*(-L^*p) + g^*(p)]. \end{cases}$

The extremality relations become

(13) $\quad\quad\quad -L^*\bar{p} \in \partial f(\bar{x}) \quad\quad \text{and} \quad\quad \bar{p} \in \partial g(L\bar{x}),$

In a more symmetric form we obtain

(14) $\quad\quad\quad L\bar{x} \in \partial g^*(\bar{p}) \quad\quad \text{and} \quad\quad L^*\bar{p} \in -\partial f(\bar{x}).$

(This is the Hamiltonian form; see Proposition 5.1 below in the general case.) ∎

Example 2

If $F(x, y) = f(x) + \Psi_Y(y)$, the problem becomes

(15) $\quad\quad\quad\quad v = \inf_{\substack{x \in X \\ Lx \in Y}} f(x)$

and the conjugate problem,

(16) $\quad\quad\quad\quad v^* = \inf_p [f^*(-L^*p) + \sigma_Y(p)].$

The extremality relations are

(17) $\quad -L^*\bar{p}\in\partial f(\bar{x})$ and $\bar{p}$ belongs to the normal cone to Y at $L\bar{x}$. ▲

In the case of Example 1 we can rewrite the extremality conditions in such a way as to obtain $\bar{p}$ and from this $\bar{x}$ as a function of $\bar{p}$.

PROPOSITION 2
Let f and g be two convex lower semicontinuous functions defined on X and Y, respectively. Then $\bar{x}$ minimizes $x \to f(x) + g(Lx)$ and $\bar{p}$ is a Lagrange multiplier if and only if

(18) $\quad \begin{cases} \text{i.} & \bar{x}\in\partial f^*(-L^*\bar{p})\cap L^{-1}\partial g^*(\bar{p}) \quad \text{where } \bar{p} \text{ is a solution of} \\ \text{ii.} & 0\in\partial g^*(\bar{p}) - L\partial f^*(-L^*\bar{p}). \end{cases}$ ▲

Proof. Suppose that $\bar{x}$ and $\bar{p}$ satisfy the extremality conditions (14). Then $\bar{x}\in L^{-1}\partial g^*(\bar{p})$ and $\bar{x}\in\partial f^*(-L^*\bar{p})$. Hence $L\bar{x}\in\partial g^*(\bar{p})$ and $L\bar{x}\in L\partial f^*(-L^*\bar{p})$, which implies (18)ii.

Conversely, if $\bar{p}$ is a solution of (18)ii, there exists $y\in\partial g^*(\bar{p})\cap L\partial f^*L^*(-\bar{p})$. Hence there exists $\bar{x}\in\partial f^*(-L^*\bar{p})$ such that $y = L\bar{x}\in\partial g^*(\bar{p})$. The extremality relations are, therefore, satisfied. ∎

Relations (18) become simpler under supplementary hypotheses.

PROPOSITION 3
Suppose that f and g are two convex lower semicontinuous functions and that

(19) $\quad f^*$ is Gâteaux-differentiable on $-L^*$ Dom g^*.

The extremality relations become

(20) $\bar{x} = Df^*(-L^*\bar{p})\quad$ where $\quad \bar{p}\in$ Dom $g^*\quad$ satisfies

$\forall p\in$ Dom $g^*, \quad \langle -LDf^*(-L^*\bar{p}), \bar{p} - p\rangle + g^*(\bar{p}) - g^*(p) \leq 0,$ ▲

Proof. Indeed, hypothesis (19) implies that $\partial f^*(-L^*\bar{p})$ consists of a single element $Df^*(-L^*\bar{p})$, according to Theorem 3.1. Hence (18)i amounts to saying that $\bar{x} = Df^*(-L^*p)$ and (18)ii that $LDf^*(-L^*\bar{p})\in\partial g^*(\bar{p})$, that is, conditions (20). ∎

Example 3

If $g = \Psi_Y$, then $g^* = \sigma_Y$ and conditions (20) become $\bar{x} = Df^*(-L^*\bar{p})$ where $\bar{p}$ satisfies

(21) $\quad \langle -LDf^*(-L^*\bar{p}), \bar{p} - p\rangle + \sigma_Y(\bar{p}) - \sigma_Y(p) \leq 0 \quad \forall p.$

In particular, if $Y = \{y\}$ reduces to a single point, then $\sigma_Y(p) = \langle p, y \rangle$, and (21) is equivalent to

(22) $$-LDf^*(-L^*\bar{p}) = y.$$

When $f(x) = \frac{1}{2}\|x - u\|^2$, then $f^*(p) = \frac{1}{2}\|p\|_*^2 + \langle p, u \rangle$ (according to Propositions 1.2 and 1.3) and $Df^*(p) = J^{-1}p + u$. Equation (22) becomes $LJ^{-1}L^*\bar{p} + Lu = y$, and equations (20) become

(23) $$\bar{x} = u - J^{-1}L^*\bar{p}, \quad \text{where}$$
$$\bar{p} = (LJ^{-1}L^*)^{-1}(y - Lu).$$

[These equations were obtained in Chapter 4, Section 7 (Proposition 4.7.1).] ∎

We are now going to establish the existence of a Lagrange multiplier $\bar{p}$.

THEOREM 1 (FENCHEL)
Let U and V be two Hilbert spaces, $L \in \mathscr{L}(U, V)$ and F a convex lower semicontinuous function with nonempty domain $X \times Y$. Suppose that

(24) $$0 \in \text{Int}(L(X) - Y).$$

Then there exists a Lagrange multiplier $\bar{p}$ for the problem $v = \inf_x F(x, Lx)$. ▲

Proof. We are going to restrict ourselves to the case where the Hilbert space V is finite dimensional. (We assume, without proof, the result for infinite dimensional Hilbert spaces.) Thus the proof is analogous to that of the duality theorem, Theorem 2.6.1.

We introduce the mapping Φ from $X \times Y$ to $\mathbb{R} \times V$ defined by

(25) $$\Phi(x, y) = \{F(x, y), L(x) - y\}$$

as well as

(26) $$\begin{cases} \text{i.} & \text{the vector } \{v, 0\} \text{ of } \mathbb{R} \times V. \\ \text{ii.} & \text{the cone } Q =]0, \infty[\times \{0\} \subset \mathbb{R} \times V. \end{cases}$$

It is clear that the convexity of F and the linearity of L imply that

(27) $$\Phi(X \times Y) + Q \text{ is a convex subset of } \mathbb{R} \times V.$$

(See lemma 2.6.1 of Chapter 2, Section 6.) Moreover, we see that

(28) $$\{v, 0\} \notin \Phi(X \times Y) + Q.$$

(Indeed, if $\{v, 0\} \in \Phi(X \times Y) + Q$, there would exist $x \in X$ and $y \in Y$ such that $Lx - y = 0$ and $v > F(x, y) = F(x, Lx)$, which contradicts the definition of v.)

Since V is a *finite dimensional* space, we can use the theorem of separation for $\{v, 0\}$ and $\Phi(X \times Y) + Q$. (See Theorem 2.4.1.) There exists a linear form, not identically zero, $\{\alpha, p\} \in \mathbb{R} \times V^*$ such that

$$\begin{aligned}(29) \quad \alpha v &= \{\langle \alpha, p \rangle, \{v, 0\}\} \\ &= \inf_{x \in X, y \in Y} [\alpha F(x, y) + \langle p, Lx - y \rangle] + \inf_{\theta > 0} \alpha \theta. \end{aligned}$$

This implies, first of all, that $\alpha \geq 0$ and that $\inf_{\theta > 0} \alpha \theta = 0$. It is not possible that $\alpha = 0$, for in that case the inequality would become

$$(30) \quad 0 \leq \inf_{x \in X, y \in Y} \langle p, Lx - y \rangle = \inf_{u \in L(X) - Y} \langle p, u \rangle.$$

Since $0 \in \text{Int}(L(X) - Y)$, this would imply that $p = 0$ and, consequently, that $\{\alpha, p\} = 0$, which is impossible. Hence, dividing the two members of (29) by $\alpha > 0$ and setting $\bar{p} = p/\alpha$, it follows that

$$v = \inf_{x \in X} \inf_{y \in Y} [F(x, y) - \langle -L^* \bar{p}, x \rangle - \langle \bar{p}, y \rangle] = -F^*(-L^* \bar{p}, \bar{p})$$

Since $F^*(-L^* \bar{p}, \bar{p}) \geq v^* \geq -v$, we obtain that $\bar{p}$ is a Lagrange multiplier. ■

*5. HAMILTONIAN AND LAGRANGIAN OF A MINIMIZATION PROBLEM

It is convenient to rewrite the extremality relations by introducing the Hamiltonian of problem (4), Section 4.

DEFINITION 1
The function H from $U \times V^*$ to $]-\infty, +\infty]$ defined by

$$(1) \quad H(x, p) = \sup_{y \in Y} [\langle p, y \rangle - F(x, y)]$$

is called the **Hamiltonian** *associated with F.* ▲

Formulas (1) and (3), Section 4, show that

$$(2) \quad F^*(q, p) = \sup_{x \in U} [\langle q, x \rangle + H(x, p)]$$

and if convexity hypothesis (4.2) is satisfied, that

$$(3) \quad F(x, y) = \sup_{p \in V^*} [\langle p, y \rangle - H(x, p)],$$

since $p \mapsto H(x, p)$ is precisely the conjugate function of $y \mapsto F(x, y)$.

It is clear that the Hamiltonian H satisfies

(4) $\begin{cases} \text{i.} & p \mapsto H(x, p) \quad \text{is convex and lower semicontinuous.} \\ \text{ii.} & x \mapsto H(x, p) \quad \text{is concave (when } F \text{ is convex).} \end{cases}$

[We verify this latter point: Consider a convex combination $\sum_{i=1}^{n} \alpha^i x_i$ of points of U. We can associate to every $\varepsilon > 0$ points $y_i \in V$ such that $H(x_i, p) \leq \langle p, y_i \rangle - F(x_i, y_i) + \varepsilon$. Multiplying these inequalities by α^i, adding them, and using the convexity of F, we obtain

$$\sum_{i=1}^{n} \alpha^i H(x_i, p) \leq \langle p, y \rangle - F(x, y) + \varepsilon \leq H(x, p) + \varepsilon$$

where $x = \sum_{i=1}^{n} \alpha^i x_i$ and $y = \sum_{i=1}^{n} \alpha^i y_i$. The result is obtained by letting ε approach zero.]

We denote by $\partial_p H(\bar{x}, \bar{p})$ the subdifferential of $p \mapsto H(\bar{x}, p)$ at $\bar{p}$ and by $\bar{\partial}_x H(\bar{x}, \bar{p})$ the *overdifferential* of $x \mapsto H(x, \bar{p})$ at $\bar{x}$, defined by

(5) $\qquad\qquad\qquad \bar{\partial}_x H(\bar{x}, \bar{p}) = - \partial_x(- H(\bar{x}, \bar{p})).$

PROPOSITION 1
Suppose that F is convex and lower semicontinuous. The extremality relations $\{-L^*\bar{p}, \bar{p}\} \in \partial F(\bar{x}, L\bar{x})$ are equivalent to

(6) $\qquad\qquad L\bar{x} \in \partial_p H(\bar{x}, \bar{p}) \quad \text{and} \quad L^*\bar{p} \in \bar{\partial}_x H(\bar{x}, \bar{p}).$ ▲

Proof. The relation $\{-L^*\bar{p}, \bar{p}\} \in \partial F(\bar{x}, L\bar{x})$ can be written

(7) $\qquad F(\bar{x}, L\bar{x}) - F(x, y) \leq \langle -L^*\bar{p}, \bar{x} - x \rangle + \langle \bar{p}, L\bar{x} - y \rangle.$

Taking $x = \bar{x}$, this implies that

(8) $\qquad H(\bar{x}, \bar{p}) = \sup_{y} [\langle \bar{p}, y \rangle - F(\bar{x}, y)] = \langle \bar{p}, L\bar{x} \rangle - F(\bar{x}, L\bar{x}).$

The definition of H implies that

(9) $\qquad\qquad H(\bar{x}, \bar{p}) - H(\bar{x}, p) \leq \langle \bar{p} - p, L\bar{x} \rangle$

(i.e., that $L\bar{x} \in \partial_p H(\bar{x}, \bar{p})$). Moreover, writing (7) in the form

(10) $\quad \langle \bar{p}, y \rangle - F(x, y) \leq \langle L^*\bar{p}, x \rangle - F(\bar{x}, L\bar{x}) = \langle L^*\bar{p}, x - \bar{x} \rangle + H(\bar{x}, \bar{p})$

[making use of (8)] and taking the supremum on V, we obtain

(11) $\qquad\qquad H(x, \bar{p}) - H(\bar{x}, \bar{p}) \leq \langle -L^*\bar{p}, \bar{x} - x \rangle$

(i.e., $L^*\bar{p} \in \bar{\partial}_x H(\bar{x}, \bar{p})$).

Conversely, suppose that $L\bar{x} \in \partial_p H(\bar{x}, \bar{p})$ and that $L^*\bar{p} \in \bar{\partial}_x H(\bar{x}, \bar{p})$. Since $F(x, y) = \sup_p [\langle p, y \rangle - H(x, p)]$, according to (3), and since $\langle \bar{p}, L\bar{x} \rangle =$

$F(\bar{x}, L\bar{x}) + H(\bar{x}, \bar{p})$, it follows that

(12) $\quad F(\bar{x}, L\bar{x}) - F(x, y) + \langle \bar{p}, y - L\bar{x} \rangle \leq H(x, \bar{p}) - H(\bar{x}, \bar{p})$.

Moreover, since $L^*\bar{p} \in \bar{\partial}_x H(\bar{x}, \bar{p})$, we deduce that

(13) $\quad F(\bar{x}, L\bar{x}) - F(x, y) + \langle \bar{p}, y - L\bar{x} \rangle \leq \langle -L^*\bar{p}, \bar{x} - x \rangle$.

Hence $\{-L^*\bar{p}, \bar{p}\} \in \partial F(\bar{x}, L\bar{x})$. ■

We indicate, in conclusion, another characterization of minimal solutions and Lagrange multipliers.

DEFINITION 2
We say that the function $l: U \times V^* \mapsto]-\infty, +\infty]$ defined by $l(x, p) = \langle p, Lx \rangle - H(x, p)$ is the Lagrangian of the problem $v = \inf_x F(x, Lx)$. ▲

With this definition we obtain

PROPOSITION 2
We can always write

(14) $$-v^* = \sup_{p \in V^*} \inf_{x \in U} l(x, p).$$

If F is convex and lower semicontinuous, then

(15) $$v = \inf_{x \in U} \sup_{p \in V^*} l(x, p).$$

In this case the Lagrange multipliers are the max-inf's of the Lagrangian. ▲

Proof. Indeed,

$$\inf_x l(x, p) = \inf_x [\langle p, Lx \rangle - H(x, p)] = -\sup_x [\langle -L^*p, x \rangle + H(x, p)]$$
$$= -F^*(-L^*p, p),$$

according to (2). Hence

$$-v^* = -\inf_p F^*(-L^*p, p) = \sup_p \inf_x l(x, p).$$

Similarly, if F is convex and lower semicontinuous, property (3) implies that

$$\sup_p l(x, p) = \sup_p [\langle p, Lx \rangle - H(x, p)] = F(x, Lx).$$

Hence $v = \inf_x \sup_p l(x, p)$. Finally, saying that $\bar{p}$ is a Lagrange multiplier is the same as saying that $\bar{p}$ is a max-inf of the Lagrangian, since

(16) $$\inf_{x \in U} l(x, \bar{p}) = \inf_{x \in U} \sup_{p \in V^*} l(x, p) = v = -v^*.$$ ■

Remark 1

It is worth remarking that

(17) $y \mapsto F(x, Lx + y)$ is the conjugate function of $p \mapsto -l(x, p)$,

since

$$F(x, Lx + y) = \sup_{p \in V^*}[\langle p, y + Lx \rangle - H(x, p)] = \sup_{p \in V^*}[\langle p, y \rangle - (-l(x, p))].$$

This implies that

(18) $$v(y) = \inf_{x \in U} F(x, Lx + y) = \inf_{x \in U} \sup_{p \in V^*}[\langle p, y \rangle + l(x, p)]$$

and that

(19) $$v = v(0) = \inf_{x \in U} \sup_{p \in V^*} l(x, p). \qquad \blacksquare$$

PROPOSITION 3
Every Lagrange multiplier $\bar{p}$ belongs to $\partial v(0)$.   ▲

Proof. Indeed, for every $y \in V$, we have

$$v(0) + \langle \bar{p}, y \rangle = \inf_{x \in U} l(x, \bar{p}) + \langle \bar{p}, y \rangle \leq \inf_{x \in U} \sup_{p \in V^*}[l(x, p) + \langle p, y \rangle] = v(y),$$

which shows that $v(0) - v(y) \leq \langle \bar{p}, 0 - y \rangle$.   ∎

In addition, one can prove that the converse is true when the function $y \to v(y)$ satisfies $v(0) = v^{**}(0)$.

CHAPTER 11

Elementary Spectral Theory

We devote this chapter to the spectral theory of *compact* symmetric operators and to some applications.

We begin by studying in Section 1 the *elementary* properties of *compact* operators from a Hilbert space to another (which transform bounded sets to relatively compact sets).

In Section 2 we take up the study of the perturbation of an isomorphism $K \in \mathscr{L}(V, V^*)$ by a compact operator $J \in \mathscr{L}(V, V^*)$, where K and J are self-transposed. We then prove the well-known Riesz-Fredholm theorem, which guarantees the existence of an orthonormal base $\{e_n\}_n$ of U and of a sequence λ_n of scalars approaching zero such that

$$\forall n \quad Je_n = \lambda_n K e_n.$$

Making use of this theorem, we characterize the compact operators $A \in \mathscr{L}(U, F)$ in Section 3; there exist orthonormal bases $\{e_n^*\}$ and $\{f_n\}$ of U^* and F such that A can be written

$$\forall x \in U, Ax = \sum_{n=1}^{\infty} \lambda_n \langle e_n^*, x \rangle f_n.$$

We deal with the first application in Section 4 using the Riesz-Fredholm theorem to prove that the operators $A + \lambda$ are isomorphisms for every value of λ different from a sequence of numbers μ_n that approaches $-\infty$ when

(1) $\begin{cases} \text{i.} & U \subset H = H^* \subset U^*. \\ \text{ii.} & \text{the injection from } U \text{ to } H \text{ is compact.} \\ \text{iii.} & A \in \mathscr{L}(U, U^*) \text{ is } U\text{-elliptic.} \end{cases}$

In Section 5 we associate with two spaces U and H, which satisfy the first two of the preceding conditions, a base $\{e_n\}$ that is orthogonal both in U and in H, such that $\{e_n\}$ is orthonormal in H and $\{e_n/\sqrt{\lambda_n}\}$ is orthonormal in U (where $\lambda_n \geq 0$ approaches zero). The spaces U^s generated by the base $\{e_n/\sqrt{\lambda_n^s}\}$ are interpolation spaces between U and H when $s \in [0, 1]$.

We then show that such a base defines the *best approximation process* in a sense that is made precise in Section 6.

Finally, Section 7 is devoted to the study of the perturbations of an isomorphism $K \in \mathscr{L}(V, F)$ by a compact operator $J \in \mathscr{L}(V, F)$.

The exposition of general spectral theory can be made in an elegant fashion within the framework of the theory of normed Hilbert algebras, but this theory is beyond the limits of the present book.

1. COMPACT OPERATORS

Let U and F be two prehilbert spaces.

DEFINITION 1
An operator $A \in \mathscr{L}(U, F)$ is called compact *if the image of the unit ball of U under the operator is a relatively compact set of F.*

We say that A is a finite rank operator if the image of U under A is finite dimensional. ▲

It is clear that *every operator of finite rank is compact*, since the image $AB_U(I)$ of the unit ball of U is a bounded set of a finite dimensional space, which is relatively compact.

Remark 1

If U is infinite dimensional, the identity operator from U to itself *is not compact* in view of Theorem 1.1.1, which affirms that the unit ball of U is not compact. (It is, moreover, in connection with the theory of compact operators that F. Riesz proved this theorem.) ■

Remark 2

We have already encountered examples of compact operators: the canonical injections from $H^m(\Omega)$ to $H^k(\Omega)$ are compact if $m > k > 0$, according to Proposition 7.5.3 if Ω is a bounded interval and to Theorem 9.6.1 if Ω is a regular bounded open set of $\mathbb{R}^n$. We also recall that Theorem 4.1.4 is true if the injection from U to V is compact. ■

We take up in Chapter 12, Section 1, the fundamental example of Fredholm integral operators, whose study is, moreover, at the origin of spectral theory. These operators benefit from a richer structure than that of the compact operators; they are the Hilbert-Schmidt operators, which will be studied in Chapter 12.

It is clear that the set $\mathscr{L}_0(U, F)$ of compact operators is a vector subspace of $\mathscr{L}(U, F)$.

*PROPOSITION 1
Let U and F be two Hilbert spaces. The subspace $\mathscr{L}_0(U, F)$ of compact operators is closed in the Banach space $\mathscr{L}(U, F)$. ▲

Proof. We consider the limit $A \in \mathscr{L}(U, F)$ of a sequence of compact operators $A_n \in \mathscr{L}(U, F)$. We want to show that the image $A(B)$ of the unit ball B of U is relatively compact. Since F is *complete*, it suffices to show that for every $\varepsilon > 0$, $A(B)$ can be covered by a finite number of balls of radius ε. However, there exists N such that if $n \geq N$, $\|Ax - A_n x\| \leq \|A - A_n\| \leq \varepsilon/2$ for all $x \in B$. Since $A_n(B)$ is relatively compact, it can be covered by a finite number of balls $B_F(y_j, \varepsilon/2)$ of F. Hence for every $x \in B$, there exists y_j such that $\|A_n x - y_j\|_F \leq \varepsilon/2$ and, consequently, such that

$$\|Ax - y_j\|_F \leq \|Ax - A_n x\|_F + \|A_n x - y_j\|_F \leq 2\frac{\varepsilon}{2} = \varepsilon.$$ ■

The following result is obvious.

PROPOSITION 2
Consider $M \in \mathscr{L}(U_1, U)$ and $N \in \mathscr{L}(F, F_1)$ where U, U_1, F, and F_1 are prehilbert spaces. If $A \in \mathscr{L}(U, F)$ is compact, the operator $NAM \in \mathscr{L}(U_1, F_1)$ is also compact. ▲

Proof. (Left as an exercise.) ■

Hence if we compose a compact operator on the right or on the left by a continuous operator, we obtain a compact operator.

We are now going to show that the transpose of a compact operator is compact.

PROPOSITION 3
Suppose that U and F are Hilbert spaces. Then the transpose $A^* \in \mathscr{L}(F^*, U^*)$ of a compact operator $A \in \mathscr{L}(U, F)$ is compact. ▲

Proof. Let B and B^* be the unit balls of U and F^*, respectively, $K = \overline{A(B)}$ the closure of $A(B)$, which is compact, since A is compact. We associate to every $f \in B^*$ the function $\hat{f} \in \mathscr{C}_\infty(K)$ defined by $\hat{f}(z) = \langle f, z \rangle$. The set $\hat{B}^*$ of these functions is an equicontinuous set, since

$$\forall y, z \in K, \quad |\hat{f}(y) - \hat{f}(z)| = |\langle f, y - z \rangle| \leq \|f\|_{F^*} \|y - z\| \leq \|y - z\|.$$

Hence $\hat{B}^*$ is relatively compact in $\mathscr{C}_\infty(K)$ according to the Ascoli theorem. (See Theorem 4.6.2 of [AAA], p. 153.) It can, therefore, be covered by a *finite* number of balls with centers $\hat{f}_j$ and of a given radius $\varepsilon > 0$. Hence for all $f \in B^*$ there exists $f_i \in B^*$ such that

$$\|A^*f - A^*f_i\|_{U^*} = \sup_{x \in B} |\langle f - f_i, Ax \rangle| \leq \sup_{y \in K} |\langle f - f_i, y \rangle|$$
$$\leq \sup_{y \in K} |\hat{f}(y) - \hat{f}_i(y)| \leq \varepsilon.$$

Hence $A^*(B^*)$ can be covered by a finite number of balls with center A^*f_i and of given radius $\varepsilon > 0$, which implies that $A^*(B^*)$ is relatively compact. ∎

The compact operators are going to play an important role in the theory of perturbations of an isomorphism. We pose the problem now and resolve it in a particular case in the following section, leaving until the end of the chapter some remarks concerning the general case.

Perturbation of an Isomorphism by a Compact Operator

Let U and F be two Hilbert spaces. We consider

(1) $\begin{cases} \text{i.} & \text{an } \textit{isomorphism } K \text{ from } U \text{ onto } F. \\ \text{ii.} & \textit{a compact operator } J \text{ from } U \text{ to } F. \end{cases}$

The theory of F. Riesz shows that the perturbations

(2) $$A_\lambda = \lambda K - J$$

of the isomorphism λK by the compact operator J are isomorphisms when λ is not an "eigenvalue". Moreover, there is at most a countable sequence of such eigenvalues.

DEFINITION 2
We say that $\lambda \in \mathbb{R}$ is an eigenvalue *of J (with respect to K) if*

(3) $\qquad N_\lambda = \operatorname{Ker} A_\lambda = \{x \in U \text{ such that } \lambda Kx = Jx\} \neq \{0\}.$

The kernel N_λ of A_λ is called the eigensubspace *associated with the eigenvalue λ, and its nonzero elements are called* eigenvectors. *The dimension of N_λ is called the* order of multiplicity *of λ.* ▲

We now prove the following theorem.

THEOREM 1
Suppose that the hypotheses (1) are satisfied. Then for every $\lambda \neq 0$, the kernel N_λ of A_λ is finite dimensional. ▲

Proof. Let us show that N_λ, is finite dimensional. If $x \in N_\lambda$, then $x = (1/\lambda)K^{-1}Jx$. Since J is compact, the unit ball B of N_λ is equal to $(1/\lambda)K^{-1}J(B)$, which is compact. This implies that the vector space N_λ is finite dimensional. (See Theorem 1.1.1.) ∎

2. THE THEORY OF RIESZ-FREDHOLM

We suppose now that

(1) $\begin{cases} K \text{ is the duality operator from a Hilbert space } U \text{ onto} \\ \text{its dual space } F = U^*. \end{cases}$

and that the operator J is defined by

(2) J is the duality operator from a Hilbert space V onto its dual space V^*

where

(3) $\begin{cases} \text{i.} & U \text{ is a dense subspace of } V. \\ \text{ii.} & \text{the injection from } U \text{ to } V \text{ is a } \textit{compact} \text{ operator.} \end{cases}$

We identify V^* with a dense subspace of U^*. Proposition 1.4 implies that the injection from V^* to U^* is *compact*. Hence we deduce from Proposition 1.3 that J is a *compact operator from V to U^* that is self-transposed*.

THEOREM 1 (F. RIESZ AND FREDHOLM)
Suppose that the embedding from U into V is compact. Let $K \in \mathcal{L}(U, U^*)$ and $J \in \mathcal{L}(V, V^*)$ be the duality operators from U onto U^* and from V onto V^*, respectively. Then there exist an orthonormal base $\{e_n\}$ of vectors of U and a sequence λ_n of positive scalars satisfying

(4) $\begin{cases} \text{i.} & \forall n \geq 1, \quad Je_n = \lambda_n Ke_n. \\ \text{ii.} & \lambda_1 \geq \lambda_2 \geq \ldots \geq \lambda_n \ldots; \quad \lim_{n \to \infty} \lambda_n = 0 \\ & \text{when } U \text{ is } \textit{infinite dimensional}. \\ \text{iii.} & ((e_n, e_k))_U = ((e_n, e_k))_V = 0 \quad \text{if} \quad n \neq k. \\ \text{iv.} & \|e_n\|_U = 1; \quad \|e_n\|_V = \sqrt{\lambda_n}. \end{cases}$ ▲

Before proving this theorem we establish the following lemma, which we shall need.

LEMMA 1
Suppose that the injection from U to V is compact. Let P be a closed subspace

of U and
$$\lambda = \sup_{\substack{x \in P \\ \|x\|_U \leq 1}} \|x\|_V^2.$$

Then there exists $e \in P$ such that

(5) $$Je - \lambda Ke \in P^\perp. \qquad \blacktriangle$$

Proof. Since the injection from U to V is compact, the set $C = \{x \in P$ such that $\|x\|_U \leq 1\}$ is compact in V. Hence there exists an element $e \in C$ maximizing on C the continuous function $x \to \|x\|_V^2$. We set $\lambda = \|e\|_V^2$. On the other hand, for all $y \in P$ and $\theta \in \mathbb{R}$, we have

$$\frac{\|e+\theta y\|_V^2}{\|e+\theta y\|_U^2} - \frac{\|e\|_V^2}{\|e\|_U^2} \geq 0.$$

Consequently, it follows that the derivative at zero of the function l defined by

$$l(\theta) = \frac{u(\theta)}{v(\theta)} \quad \text{(where } u(\theta) = \|e+\theta y\|_V^2 \quad \text{and} \quad v(\theta) = \|e+\theta y\|_U^2\text{)}$$

is identically zero. But

$$\frac{du}{d\theta}(0) = \langle Je, y \rangle \quad \text{and} \quad \frac{dv}{d\theta}(0) = \langle Ke, y \rangle.$$

Hence

$$\frac{dl}{d\theta}(0) = \frac{1}{\|e\|_U^2}\left[\langle Je, y \rangle - \frac{\|e\|_V^2}{\|e\|_U^2}\langle Ke, y \rangle\right] = \frac{\langle Je - \lambda Ke, y \rangle}{\|e\|_U^2} = 0.$$

Since this is true for all $y \in P$, we obtain (5). $\blacksquare$

Proof of Theorem 1. We are going to construct the sequence $\{\lambda_n\}$ of eigenvalues and the sequence $\{e_n\}$ of eigenvectors by recursion. For $n = 1$ we take $P_1 = U$ and $P_1^\perp = \{0\}$. Lemma 1 implies the existence of $e_1 \in U$, satisfying

(6) $$\|e_1\|_U = 1, \qquad \lambda_1 = \|e_1\|_V^2, \qquad Je_1 = \lambda Ke_1.$$

Suppose that the eigenvalues $\lambda_1, \lambda_2, \ldots, \lambda_{n-1}$ and the eigenvectors $e_1, \ldots, e_{n-1}$ satisfying the conditions (4) have been constructed. We denote by $P_n^\perp = \{Ke_1, \ldots, Ke_{n-1}\}$ the closed vector subspace of U^* generated by the vectors $Ke_1, \ldots, Ke_{n-1}$. We take $P_n = (P_n^\perp)^\perp$, the orthogonal complement of $P_n^\perp$ in U. Lemma 1 implies the existence of $e_n \in P_n$ and of $(n-1)$ scalars

$a_k (1 \leq k \leq n-1)$ satisfying

(7) $\quad \|e_n\|_U = 1; \quad \lambda_n = \|e_n\|_V^2, \quad Je_n - \lambda_n Ke_n = \sum_{k=1}^{n-1} a_k Ke_k \in P_n^\perp$

where

$$\lambda_n = \sup_{\substack{x \in P_n \\ \|x\|_U = 1}} \|x\|_V^2 > 0 \quad \text{if} \quad P_n^\perp \neq U^*.$$

Since $e_n \in P_n$, it follows that for all k between 1 and $n-1$

$$((e_n, e_k))_U = \langle e_n, Ke_k \rangle = 0.$$

Consequently, taking the scalar product of $Je_n - \lambda_n Ke_n$ with e_k, we obtain

$$\langle Je_n, e_k \rangle - \lambda_n \langle Ke_n, e_k \rangle = \langle e_n, Je_k - \lambda_k Ke_k \rangle = \langle e_n, (\lambda_k - \lambda_n) Ke_k \rangle$$
$$= (\lambda_k - \lambda_n)((e_n, e_k))_U = 0.$$

Moreover, since $Je_n - \lambda_n Ke_n = \sum_{j=1}^{n-1} a_j Ke_j$ and since $((e_j, e_k))_U = \langle Ke_j, e_k \rangle = 0$ for all $j \neq k$, j and k ranging from 1 to $n-1$, we obtain the equality

$$0 = \langle Je_n - \lambda_n Ke_n, e_k \rangle = \sum_{j=1}^{n-1} a_j \langle Ke_j, e_k \rangle = a_k \|e_k\|_U^2 = a_k.$$

Hence the scalars a_k are zero; consequently, e_n satisfies

(8) $\quad \|e_n\|_U = 1, \quad \lambda_n = \|e_n\|_V^2, \quad Je_n = \lambda_n Ke_n, \quad \text{and} \quad ((e_n, e_k))_U = 0$

for all $k \leq n$, and the sequence λ_n is decreasing. It is stationary if U is finite dimensional. Let us show that the sequence λ_n converges to zero if U is infinite dimensional. If this were not the case, there would exist $c > 0$ such that $\lambda_n \geq c$ for all n. Since $\|e_n\|_U = 1$ and since the injection from U to V is compact, a subsequence e_{n_k} is convergent and, consequently, a Cauchy sequence in V. This contradicts the fact that for n, m,

$$\|e_n - e_m\|_V^2 = \|e_n\|_V^2 + \|e_m\|_V^2 - 2((e_n, e_m))_V$$
$$= \|e_n\|_V^2 + \|e_m\|_V^2 = \lambda_n + \lambda_m \geq 2c.$$

Finally, it is clear that the orthogonal family $\{e_n\}$ generates a dense subspace in U (and in V). Indeed, let $f = Kx$ be a linear form that is identically zero on the space generated by the e_k's. This implies that $\langle f, e_k \rangle = \langle x, Ke_k \rangle = 0$ for all k, that is, that $x = K^{-1} f \in \bigcap_{n=0}^\infty P_n$. We can suppose that $\|x\|_U = 1$. Since $x \in P_n$, it follows that

$$\|x\|_V^2 \leq \lambda_n = \sup_{\substack{x \in P_n \\ \|x\|_U = 1}} \|x\|_V^2.$$

Furthermore, since λ_n converges to zero, we obtain that $\|x\|_V^2 = 0$ and, consequently, that $f = Kx = 0$. Hence Theorem 2.2.1 implies that $\{e_k\}_k$ generates a dense subspace. ∎

3. CHARACTERIZATION OF COMPACT OPERATORS FROM A HILBERT SPACE TO ANOTHER

We are going to show that every compact operator from a Hilbert space to another is the limit of a sequence of operators of finite rank.

THEOREM 1
Let A be a compact operator from a Hilbert space U to a Hilbert space F. The operator A can be expressed in the form

(1) $$Ax = \sum_{n=1}^{\infty} \lambda_n \langle e_n^*, x \rangle f_n$$

where
$\{e_n^*\}$ *and* $\{f_n\}$ *are the orthonormal bases of U^* and of F and where λ_n is either a finite sequence or a decreasing sequence of positive numbers that converges to zero. Conversely, every operator A of the form* (1) *where the sequences* $\{e_n^*\}, \{f_n\}$, *and* $\{\lambda_n\}$ *satisfy the proceeding properties is a compact operator.* ▲

Remark 1

We state in Chapter 12 an analogous result characterizing Hilbert-Schmidt operators. ∎

Proof that the Condition is Sufficient

To show that A is compact, it suffices to verify that A is the uniform limit of the operators of finite rank A_k defined by

$$A_k(x) = \sum_{n=1}^{k} \lambda_n \langle e_n^*, x \rangle f_n.$$

Indeed,
$$\|Ax - A_k x\|_F^2 = \left\| \sum_{n=k+1}^{\infty} \lambda_n \langle e_n^*, x \rangle f_n \right\|_F^2$$
$$= \sum_{n=k+1}^{\infty} \lambda_n^2 |\langle e_n^*, x \rangle|^2 \leq \sup_{n \geq k+1} \lambda_n^2 \sum_{n=k+1}^{\infty} |\langle e_n^*, x \rangle|^2$$
$$\leq \sup_{n \geq k+1} \lambda_n^2 \|x\|_U^2.$$

Since the sequence λ_n approaches zero, it follows that $\|A - A_k\|_{\mathscr{L}(U,F)} \leq \sup_{n \geq k+1} \lambda_n$ approaches zero. ∎

Proof that the Condition is Necessary

We can suppose that we are in the case where $A \in \mathscr{L}(U, F)$ is injective. (If necessary, we replace U by $\tilde{U} = (\operatorname{Ker} A)^{\oplus}$ and A be its restriction $\tilde{A}$ to $\tilde{U}$.)

Then we consider the norm $\|u\|_V = \|Au\|_F$ on U, and we introduce

(2) \qquad the completion V of U for the norm $\|u\|_V$.

We identify A to its extension $\hat{A} \in \mathscr{L}(V, F)$.

Since A is compact, *the embedding from U into V is compact*. Moreover, if L denotes the duality operator from F onto F^*, the duality operator J from V onto V^* is equal to $J = A^* L A$. (See Proposition 5.1.1.)

Theorem 2.1 of Riesz-Fredholm therefore implies the existence of an orthonormal base $\{e_n\}$ in U and of a decreasing sequence of positive scalars λ_n^2 converging to zero such that

(3) $$A^* L A e_n = J e_n = \lambda_n^2 K e_n = \lambda_n^2 e_n^*,$$

where the $e_n^* = K e_n$ form the dual orthonormal base of U^*. Hence we define the operator $B \in \mathscr{L}(U, U)$ by

(4) $$Bx = \sum_{n=1}^{\infty} \lambda_n \langle e_n^*, x \rangle e_n$$

whose transpose $B^* \in \mathscr{L}(U^*, U^*)$ is defined by

(5) $$B^* f = \sum_{n=1}^{\infty} \lambda_n \langle f, e_n \rangle e_n^*.$$

Using (3), we then obtain

(6) $$A^* L A = B^* K B$$

as well as

(7) $$\begin{cases} ((Ax, Ay))_F = \langle LAx, Ay \rangle = \langle A^* LAx, y \rangle \\ = \langle B^* KBx, y \rangle = \langle KBx, By \rangle = ((Bx, By))_U. \end{cases}$$

Moreover, B is a *compact* operator (according to the sufficient condition that we have already proved). Consider the image $U_1 = B(U) \subset U$ of U. We define the operator C from U_1 to F by

(8) \qquad $C(Bx) = Ax$ \qquad for all \qquad $x \in U$,

which makes sense, since A is injective. This operator is an isometry, since,

according to (7),
$$((C(Bx), C(By)))_F = ((Ax, Ay))_F = ((Bx, By))_U.$$

The operator C, therefore, maps the orthonormal base $\{e_n\}$ to an orthonormal family $\{f_n\}$ of F by setting $Ce_n = f_n$. Hence we deduce from formulas (7) and (8) that

$$Ax = C(Bx) = \sum_{n=1}^{\infty} \lambda_n \langle e_n^*, x \rangle f_n.$$ ∎

Remark 2

Theorem 1 suggests, more generally, *classifying* the compact operators from a Hilbert space U to a Hilbert space V by means of the properties of the sequence λ of the λ_n's of decomposition (1).

If $\lambda \in l^2(\mathbb{N})$ we obtain the *Hilbert-Schmidt* operators. (See Chapter 12.) The compact operators for which the sequence $\lambda \in l^1(\mathbb{N})$ are the *nuclear operators*, which we shall not study in this book. ∎

Remark 3

If $U = F$ is a pivot space, then $K = L = 1$. Then it follows that $B = B^*$ and that $A^*A = B^2$. If we suppose that A is self-transposed, we find that $A^2 = B^2$ and, consequently, that the isometry C is defined by $Cy = y$ or $-y$. ∎

THEOREM 2
Let H be a pivot space and $A \in \mathscr{L}(H, H)$ a compact self-transposed operator. The operator A can be written in the form

(9) $$Ax = \sum_{n=1}^{\infty} \mu_n (e_n, x) e_n$$

where $\{e_n\}$ is an orthonormal base of H and where μ_n is a finite sequence or an infinite sequence such that $|\mu_n|$ approaches zero. ▲

4. THE FREDHOLM ALTERNATIVE

We consider an infinite dimensional pivot space H and a Hilbert space U embedded in H and dense in H. It will be convenient to embed H in U^*. (See Proposition 3.5.4.)

(1) $$U \subset H \subset U^*.$$

Let A be a continuous linear operator from U onto U^*.

DEFINITION 1

We say that an operator $A \in \mathscr{L}(U, U^*)$ is (U, H)-coercive if there exists $c > 0$ and $\mu_0 \in \mathbb{R}$ such that

(2) $$\forall x \in U, \quad \langle Ax, x \rangle + \mu_0 |x|^2 \geq c \|x\|_U^2,$$

where $|x| = \sqrt{(x, x)}$ denotes the norm of H. ▲

In other words, A is (U, H)-coercive if and only if there exists a constant μ_0 such that the operator $A + \mu_0 : x \in U \to Ax + \mu_0 x \in U^*$ is U-elliptic (see Chapter 3, Section 6) because $\langle (A + \mu_0)x, x \rangle = \langle Ax, x \rangle + \mu_0 |x|^2$. Since $A + \mu$ is U-elliptic for μ sufficiently large, it is an isomorphism (according to Theorem 3.6.1). We are going to prove from the Riesz-Fredholm theorem, Theorem 2.1, that when A is self-transposed, the operators $A + \mu$ are all isomorphisms except for a countable sequence of scalars that approach $-\infty$.

THEOREM 1

Suppose that

(3) the injection from U to H is compact

and that the operator $A \in \mathscr{L}(U, U^*)$ satisfies

(4) $\begin{cases} \text{i.} & A \text{ is self-transposed.} \\ \text{ii.} & \exists c > 0 \quad \text{and} \quad \mu_0 \in \mathbb{R} \quad \text{such that} \\ & \langle Ax, x \rangle + \mu_0 |x|^2 \geq c \|x\|_U^2 \quad \forall x \in U. \end{cases}$

Then there exists a countable sequence of numbers μ_n that approach $-\infty$ and an orthonormal base $\{e_n\}$ such that $(A + \mu_n)e_n = 0$.

The following alternative holds

(5) If $\forall n, \mu \neq \mu_n$, $A + \mu$ is an isomorphism from U onto U^*.

If $\mu = \mu_{n_0}$, we denote by

(6) $\begin{cases} \text{i.} & N(n_0) = \{n \in \mathbb{N} \text{ such that } \mu_n = \mu_{n_0}\}. \\ \text{ii.} & U_{n_0} = \text{the space generated by the } e_n\text{'s when } n \in N(n_0). \\ \text{iii.} & F_{n_0} = U_{n_0}^\perp = \text{the closure in } U^* \text{ of the space generated by the } e_n\text{'s when } n \notin N(n_0). \end{cases}$

Then

(7) If $\mu = \mu_{n_0}$, $A + \mu$ is an isomorphism from U/U_{n_0} onto F_{n_0}.

In fact,

(8) $\quad$ if $\forall n, \mu \neq \mu_n$, $\quad (A+\mu)^{-1}f = \sum_{n=1}^{\infty} \frac{\mu_0 - \mu_n}{\mu - \mu_n} \langle f, e_n \rangle$

and

(9) if $\mu = \mu_{n_0}$ and $f \in F_{n_0}$, $\quad (A+\mu)^{-1}f = \sum_{n \notin N(n_0)} \frac{\mu_0 - \mu_n}{\mu - \mu_n} \langle f, e_n \rangle + \sum_{n \in N(n_0)} a_n e_n$

where the a_n's ($n \in N(n_0)$) are arbitrary. ▲

Proof. We apply Theorem 2.1 with $K = A + \mu_0$ and $J = 1$. Indeed, since $A + \mu_0$ is self-transposed and U-elliptic, it can be considered as the duality operator of the space U with the scalar product $\langle Ax + \mu_0 x, y \rangle$, which is equivalent to the initial scalar product of U. We take $V = H$. Then there exists an orthonormal base $\{e_n\}_n$ of U and a sequence $\{\lambda_n\}$ of positive eigenvalues approaching zero such that

(10) $\quad e_n = \lambda_n(A + \mu_0)e_n = \lambda_n K e_n.$

Setting

$$\mu_n = \mu_0 - \frac{1}{\lambda_n},$$

we can write

(11) $\quad (A + \mu_n)e_n = 0 \quad$ for all n.

Consequently, if

$$x = \sum_{n=1}^{\infty} ((x, e_n))_U e_n,$$

we can write

(12) $\quad (A + \mu)x = \sum_{n=1}^{\infty} ((x, e_n))_U (\mu - \mu_n) e_n.$

Moreover, since

$$\|e_n\|_{U^*} = \|K^{-1}e_n\|_U = \|\lambda_n e_n\|_U = \lambda_n$$

the sequence $\left\{\frac{1}{\lambda_n} e_n\right\}_n$ forms an orthonormal base of U^*. Hence every $f \in U^*$

can be written in the form

(13)
$$\begin{cases} f = \sum_{n=1}^{\infty} \left(\left(f, \frac{1}{\lambda_n} e_n\right)\right)_{U^*} \frac{1}{\lambda_n} e_n \\ = \sum_{n=1}^{\infty} \left\langle f, \frac{1}{\lambda_n} K^{-1} e_n \right\rangle \frac{1}{\lambda_n} e_n = \sum_{n=1}^{\infty} \frac{1}{\lambda_n} \langle f, e_n \rangle e_n \\ = \sum_{n=1}^{\infty} (\mu_0 - \mu_n) \langle f, e_n \rangle e_n. \end{cases}$$

Then consider the equation

(14) $$(A + \mu)x = f,$$

which is equivalent to

(15) $$\forall n, \quad (\mu - \mu_n)((x, e_n))_U = (\mu_0 - \mu_n)\langle f, e_n \rangle.$$

Hence if $\mu \neq \mu_n$, it follows that

$$((x, e_n))_U = \frac{\mu_0 - \mu_n}{\mu - \mu_n} \langle f, e_n \rangle,$$

and if $\mu = \mu_n$, it follows that $((x, e_n))_U$ is arbitrary.

Consequently, if $\forall n, \mu \neq \mu_n$, we deduce that

(16) $$(A + \mu)^{-1} = \sum_{n=1}^{\infty} \frac{\mu_0 - \mu_n}{\mu - \mu_n} \langle f, e_n \rangle.$$

If, on the contrary, $\mu = \mu_{n_0}$ and if $N(n_0) = \{n$ such that $\mu_n = \mu_{n_0}\}$, and if $f \in F_{n_0}$, we obtain

(17) $$(A + \mu)^{-1} f = \sum_{n \notin N(n_0)} \frac{\mu_0 - \mu_n}{\mu - \mu_n} \langle f, e_n \rangle + \sum_{n \in N(x_0)} a_n e_n. \quad \blacksquare$$

*5. APPLICATIONS: CONSTRUCTION OF INTERMEDIATE SPACES

We consider two Hilbert spaces U and V such that

(1) the embedding from U into V is compact.

We are going to construct a family of spaces U^s indexed by $s \in \mathbb{R}$ such that $U^0 = V, U^1 = U, U \subset U^s \subset V$ if $s \in [0, 1]$.

DEFINITION 1
We denote by U^s the vector subspace of elements $x \in V$ that satisfy

$$\|x\|_s = \left(\sum_{j=1}^{\infty} \frac{1}{\lambda_j^s} |((x, f_j))_V|^2 \right)^{1/2} < +\infty$$

CH. 11, SEC. 5 APPLICATIONS: CONSTRUCTION INTERMEDIATE SPACES

where $\{f_j\}$ is the orthonormal base of V composed of the eigenvectors of $K^{-1}J$:

(2) $$Jf_n = \lambda_n Kf_n, \quad \|f_n\|_V = 1, \quad \|f_n\|_U = \frac{1}{\sqrt{\lambda_n}}.$$

It is clear that the operator

(3) $$x \in U^s \mapsto \left\{ \frac{1}{\sqrt{\lambda_j^s}} ((x, f_j))_V \right\}_{1 \le j \le \infty} \in l^2$$

is a bijection from U^s onto the Hilbert space l^2. Consequently, U^s is a Hilbert space for the scalar product

(4) $$((x, y))_s = \sum_{j=1}^{\infty} \frac{1}{\lambda_j^s} ((x, f_j))_V ((y, f_j))_V.$$

It is also clear that

$$U^1 = U \quad \text{and} \quad U^0 = V.$$

We define the operator K^s formally by

(5) $$K^s x = \sum_{j=1}^{\infty} \frac{1}{\lambda_j^s} ((x, f_j))_V f_j$$

and then verify that

(6) $\quad K^{s/2}$ is an isometry from U^s onto V,

since

$$((x, y))_s = ((K^{s/2} x, K^{s/2} y))_V.$$

It is obvious that if $m \ge s$, U^m is embedded in U^s. In fact, we show that U^m is an intermediate space between U^k and U^s if $k \le m \le s$. (See Definition 5.9.1.)

PROPOSITION 1
Suppose that $k \le m \le s$. Then for all $x \in U^s$

(7) $$\|x\|_m \le \|x\|_s^{1-\theta} \|x\|_k^{\theta} \quad \text{where} \quad \theta = \frac{s-m}{s-k}.$$

Proof. Indeed, we can write

$$\|x\|_m^2 = \sum_{j=1}^{\infty} \frac{1}{\lambda_j^m} |((x, f_j))_V|^2$$

$$= \sum_{j=1}^{\infty} \left[\left(\frac{1}{\lambda_j} \right)^{s(1-\theta)} |((x, f_j))_V|^{2(1-\theta)} \right] \left[\left(\frac{1}{\lambda_j} \right)^{k\theta} |((x, f_j))_V|^2 \right],$$

since $s(1-\theta) + k\theta = m$. Applying the Hölder inequality (See [AAA] p. 41.) with

$$p = \frac{1}{1-\theta} \quad \text{and} \quad p^* = \frac{1}{\theta}$$

(which is possible, since $\theta \in [0,1]$), it follows that

$$\|x\|_m^2 \leq \left[\sum_{j=1}^{\infty} \frac{1}{\lambda_j^s} |((x,f_j))_V|^2\right]^{(1-\theta)} \left[\sum_{j=1}^{\infty} \frac{1}{\lambda_j^k} |((x,f_j))_V|^2\right]^{\theta}$$
$$= \|x\|_s^{2(1-\theta)} \|x\|_k^{2\theta}.$$

In particular, if $s \in [0,1]$, we deduce that

(8) $$\|x\|_s \leq \|x\|_U^s \|x\|_V^{1-s}.$$

When V is a pivot space, we obtain the embeddings

(9) $$U^s \subset U^0 = U^{0*} \subset (U^s)^*.$$

Thus we verify that the *duality operator from U^s onto its dual space is equal to K^s*, since

$$((x,y))_s = \sum_{j=1}^{\infty} \left(x, \frac{1}{\lambda_j^s} f_j\right)(y,f_j)$$
$$= \sum_{j=1}^{\infty} (x, K^s f_j)(y, f_j)$$
$$= \sum_{j=1}^{\infty} (K^s x, f_j)(y, f_j)$$
$$= (K^s x, y).$$

Consequently, the scalar product on $(U^s)^*$ is defined by

(10) $$((x,y))_{s^*} = ((K^s)^{-1} x, y) = \sum_{j=1}^{\infty} \lambda_j^s (x,f_j)(y,f_j).$$

Therefore if $s > 0$, we are led to set

(11) $$\begin{cases} \text{i.} & ((x,y))_{-s} = \sum_{j=1}^{\infty} \lambda_j^s (x,f_j)(y,f_j) = ((x,y))_{s^*}. \\ \text{ii.} & \|x\|_{-s} = \|x\|_{s^*} = \left(\sum_{j=1}^{\infty} \lambda_j^s |(x,f_j)|^2\right)^{1/2}. \\ \text{iii.} & U^{-s} = (U^s)^*. \end{cases}$$

Hence we have defined a family of spaces U^s, $s \in \mathbb{R}$, such that

(12) $$\begin{cases} \text{i.} & \text{If } s \geq t, U^s \subset U^t \text{ and } K^{(s-t)/2} \text{ is an isometry from } U^s \text{ onto } U^t. \\ \text{ii.} & \text{if } s \geq t, \text{ the injection from } U^s \text{ to } U^t \text{ is an embedding.} \\ \text{iii.} & \text{for all } s, \{\sqrt{\lambda_j^s} f_j\}_j \text{ is an orthonormal base of } U^s. \end{cases}$$

In particular, it follows that the spaces U^m are *interpolation spaces* between U^s and U^k if $k \leq m \leq s$. ∎

*6. APPLICATION: BEST APPROXIMATION PROCESSES

A "process" of approximation of a Hilbert space V by subspaces of finite dimension is defined by a family of projectors $t_n \in \mathscr{L}(V, V)$ on the subspaces of dimension n that satisfy

(1) $$\forall x \in V, \quad \lim_{n \to \infty} \|x - t_n x\| = 0.$$

We may suppose that the projectors t_n are orthogonal projectors; indeed, for a given subspace P_n, the orthogonal projector t_n onto P_n is better than any other projector s_n onto P_n in the sense that

$$\forall v \in V, \quad \|v - t_n v\| \leq \|v - s_n v\|.$$

We cannot compare the projectors t_n by means of their norms since the norm $\|1 - t_n\|_{\mathscr{L}(V,V)}$ of every orthogonal projector is equal to one.

To be able to measure the speed of convergence of a process of approximation and thereby to compare two methods, we introduce

(2)   a Hilbert space U embedded in V

and the *error function* $e_U^V(t_n)$ of a projector t_n,

(3) $$e_U^V(t_n) = \|1 - t_n\|_{\mathscr{L}(U,V)} = \sup_{x \in U} \frac{\|x - t_n x\|_V}{\|x\|_U}.$$

Theorem 4.1.4 clearly implies the following result:

PROPOSITION 1
Suppose that the orthogonal projectors t_n from V to V satisfy

(4) $\begin{cases} \text{i.} & \text{the embedding from } U \text{ into } V \text{ is compact.} \\ \text{ii.} & \forall x \in U, \lim_{n \to \infty} \|x - t_n x\|_V = 0. \end{cases}$

Then

(5) $$\lim_{n \to \infty} e_U^V(t_n) = 0. \quad\blacktriangle$$

Our aim now is to look for the *best orthogonal projectors t_n* for the subspaces of dimension n in the sense that

(6) $$e_U^V(t_n) = \min_{s_n} e_U^V(s_n)$$

and to calculate $e_U^V(t_n)$.

To do this it is useful to introduce another notion of approximation theory, namely, that of the *stability function* $s_U^V(t_n)$ defined by

$$(7) \qquad s_U^V(t_n) = \sup_{x \in U} \frac{\|t_n x\|_U}{\|t_n x\|_V}.$$

Observe that this makes sense, since the restrictions $\|t_n x\|_U$ and $\|t_n x\|_V$ of the norms U and V to the subspace of finite dimension $P_n = t_n(V)$ are equivalent.

The following lemma allows us to estimate the product of the error function and the stability function.

LEMMA 1
Consider two projectors s_n and t_{n+1} onto subspaces Q_n and P_{n+1} of dimensions n and $n+1$, respectively. Then

$$(8) \qquad 1 \leq e_U^V(s_n) s_U^V(t_{n+1}).$$

Proof. Let $Q_n = s_n(V)$ and $P_{n+1} = t_{n+1}(V)$ be the images of the projectors s_n and t_{n+1}. Since the dimension of Q_n is less than that of P_{n+1}, we have

$$(9) \qquad P_{n+1} \cap Q_n^{\oplus} \neq \{0\}.$$

Indeed, the n elements e_j of a base of Q_n define n linear forms $x \mapsto ((e_j, x))$, which are linearly independent on the space P_{n+1} of dimension $n+1$. Hence there exists a nonzero element $x \in P_{n+1}$ satisfying

$$((e_j, x)) = 0 \quad \text{for} \quad 1 \leq j \leq n.$$

In other words, there exists x satisfying

$$(10) \qquad x \in P_{n+1}, \quad x \neq 0, \quad \|x\|_V = \inf_{y \in Q_n} \|x - y\| = e_U^V(s_n) \|x\|_U.$$

For such an x we have

$$1 \leq s_U^V(t_{n+1}) \frac{\|x\|_V}{\|x\|_U} = s_U^V(t_{n+1}) e_U^V(s_n). \qquad \blacksquare$$

From this lemma we are going to obtain the construction of orthogonal projectors t_n that minimize the error function $e_U^V(s_n)$.

THEOREM 1
Suppose that

$$(11) \qquad \text{the embedding from } U \text{ into } V \text{ is compact.}$$

Consider the eigenvectors e_n and the decreasing sequence consisting of the eigenvalues λ_n of the operator $K^{-1}J$ that converge to zero:

(12) $$Je_n = \lambda_n Ke_n, \quad \|e_n\|_U = 1, \quad \|e_n\|_V = \sqrt{\lambda_n}$$

and also the projectors t_n defined by

(13) $$t_n x = \sum_{j=1}^{n} ((x, e_j))_U e_j = \sum_{j=1}^{n} \frac{1}{\lambda_j}((x, e_j))_V e_j.$$

Then

(14) $\begin{cases} \text{i.} & e_U^V(t_n) = \sqrt{\lambda_{n+1}} \\ \text{ii.} & s_U^V(t_n) = \dfrac{1}{\sqrt{\lambda_n}}, \end{cases}$

and these projectors t_n are optimal in the sense that

(15) $$e_U^V(t_n) = \min_{s_n} e_U^V(s_n). \quad \blacktriangle$$

Proof. **a.** First of all let us show that $e_U^V(t_n) = \sqrt{\lambda_{n+1}}$. Indeed, since

$$\lambda_j \leq \lambda_{n+1} \quad \text{if} \quad j \geq n+1,$$

we obtain

$$\|x - t_n x\|_V^2 = \sum_{j=n+1}^{\infty} \frac{1}{\lambda_j}|((x, e_j))_V|^2 = \sum_{j=n+1}^{\infty} \lambda_j |((x, e_j))_U|^2$$

$$= \lambda_{n+1} \sum_{j=n+1}^{\infty} \frac{\lambda_j}{\lambda_{n+1}}|((x, e_j))_U|^2 \leq \lambda_{n+1} \sum_{j=n+1}^{\infty} |((x, e_j))_U|^2$$

$$\leq \lambda_{n+1} \left(\sum_{j=1}^{\infty} |((x, e_j))_U|^2 \right) = \lambda_{n+1} \|x\|_U^2.$$

Moreover,

$$\|e_{n+1} - t_n e_{n+1}\|_V^2 = \|e_{n+1}\|_V^2 = \lambda_{n+1} \|e_{n+1}\|_U^2.$$

Hence

(16) $$e_U^V(t_n) = \max_x \frac{\|x - t_n x\|_V}{\|x\|_U} = \sqrt{\lambda_{n+1}}.$$

b. Next let us show that

$$s_U^V(t_n) = \frac{1}{\sqrt{\lambda_n}}.$$

Indeed,

$$\|t_n x\|_U^2 = \sum_{j=1}^n |((x, e_j))_U|^2 = \sum_{j=1}^n \frac{1}{\lambda_j} \left|\left(\left(x, \frac{e_j}{\sqrt{\lambda_j}}\right)\right)_V\right|^2$$

$$= \frac{1}{\lambda_n} \sum_{j=1}^n \frac{\lambda_n}{\lambda_j} \left|\left(\left(x, \frac{e_j}{\sqrt{\lambda_j}}\right)\right)_V\right|^2 \leq \frac{1}{\lambda_n} \sum_{j=1}^n \left|\left(\left(x, \frac{e_j}{\sqrt{\lambda_j}}\right)\right)_V\right|^2$$

$$= \frac{1}{\lambda_n} \|tx\|_V^2.$$

Moreover, taking $x = e_n$, we have $t_n e_n = e_n$ and, consequently,

$$\frac{\|t_n e_n\|_U}{\|t_n e_n\|_V} = \frac{1}{\sqrt{\lambda_n}}.$$

Therefore

(17) $$s_U^V(t_n) = \max_x \frac{\|t_n x\|_U}{\|t_n x\|_V} = \frac{1}{\sqrt{\lambda_n}}.$$

c. Finally we establish that

$$e_U^V(t_n) = \inf_{s_n} e_U^V(s_n),$$

that is, that the approximation by the orthogonal projectors t_n is optimal. It follows from the estimates (16) and (17) that

(18) $$1 = \frac{1}{\sqrt{\lambda_{n+1}}} \sqrt{\lambda_{n+1}} = s_U^V(t_{n+1}) e_U^V(t_n).$$

In addition, Lemma 1 implies that

(19) $$1 \leq s_U^V(t_{n+1}) e_U^V(s_n)$$

for every projector s_n onto a subspace of dimension n. Hence (18) and (19) imply that

$$e_U^V(t_n) = \frac{1}{s_U^V(t_{n+1})} \leq e_U^V(s_n). \quad \blacksquare$$

PROPOSITION 2
Consider a sequence of projectors s_n from V onto n-dimensional subspaces that satisfies

(20) $$\sup_n e_U^V(s_n) s_U^V(s_n) = M < +\infty.$$

Then the error functions satisfy

(21) $$e_U^V(s_n) \leq M\sqrt{\lambda_n}.$$

Proof. Indeed,

$$e_U^V(s_n) \leq \frac{M}{s_U^V(s_n)} \leq M e_U^V(t_{n-1}) = M\sqrt{\lambda_n}$$

Remark 1

The function $e_U^V(t_n) = \sqrt{\lambda_{n+1}}$ is called the *n*th *width* of the unit ball of U in V.

Remark 2

When we know that the product $e_U^V(s_n) \cdot s_U^V(s_n)$ of the error function and the stability function of a sequence of projectors is bounded, it follows that the speed of convergence of s_n is of the same order as the optimal speed of convergence, without need of a finer upper estimate of $e_U^V(s_n)$ than (20).

*7. PERTURBATION OF AN ISOMORPHISM BY A COMPACT OPERATOR

We are going to pursue the study of the perturbations $A_\lambda = \lambda K - J$ when

(1) $\begin{cases} \textbf{i.} & K \text{ is an isomorphism from } U \text{ onto } F. \\ \textbf{ii.} & J \text{ is a compact operator from } U \text{ to } F. \end{cases}$

We already know that for every $\lambda \neq 0$, the kernel N_λ of A_λ is of finite dimensional (Theorem 1 of Section 1.)

THEOREM 1
Suppose that (1) holds. If $\lambda \neq 0$, the image F_λ of V under A_λ is closed.

Proof. Consider a sequence $f_n = A_\lambda x_n = \lambda K x_n - J x_n$ that converges to f, and let us show that $f = A_\lambda x = \lambda K x - J x$ for a suitable x.
a. If the sequence x_n is bounded in U, the sequence $J x_n$ is relatively compact in F and a subsequence $J x_m$ converges to an element g. Hence $x_m = \lambda^{-1} K^{-1}(f_m + J x_m)$ converges to $x = \lambda^{-1} K^{-1}(f + g)$ and, consequently, $f = A_\lambda x$.
b. Now suppose that the sequence $\|x_n\|_U$ is not bounded. We introduce the

distance $\alpha_n = \inf_{y \in N_\lambda} \|x_n - y\|$ from x_n to N_λ and elements $y_n \in N_\lambda$ such that

(2) $$\alpha_n \leq \|x_n - y_n\| \leq \alpha_n\left(1 + \frac{1}{n}\right).$$

The elements $\tilde{x}_n = x_n - y_n$ satisfy $A_\lambda(\tilde{x}_n) = A_\lambda(x_n) = f_n$. If the sequence α_n is bounded, the sequence $\|\tilde{x}_n\|$ is bounded as well; the preceding argument shows that a subsequence $\tilde{x}_m$ converges to $\tilde{x}$ and that $A_\lambda(\tilde{x}_m) = f_m$ converges to $A_\lambda(\tilde{x}) = f$ and, consequently, that the image is closed.

c. This will always be the case because the sequence α_n is always bounded. Otherwise we would introduce the elements $z_n = \dfrac{x_n - y_n}{\|x_n - y_n\|}$, which form a bounded sequence in U such that $A_\lambda z_n = \dfrac{f_n}{\|x_n - y_n\|}$ converges to zero (since $\|x_n - y_n\| \geq \alpha_n$ approaches infinity). Hence it follows that a subsequence $z_m = \lambda^{-1} K^{-1}\left(\dfrac{f_m}{\|x_m - y_m\|} - Jz_m\right)$ converges to an element z_0 for which $A_\lambda(z_0) = \lim_m A_\lambda(z_m) = 0$ (i.e., which belongs to N_λ). Then

$$\|x_n - y_n - z_0(\|x_n - y_n\|)\| = \|x_n - y_n\| \|z_n - z_0\| \geq \alpha_n.$$

since $y_n + z_0 \|x_n - y_n\| \in N_\lambda$. Therefore, according to (2), we obtain

$$\|z_m - z_0\| \geq \frac{\alpha_m}{\|x_m - y_m\|} \geq \frac{m}{m+1},$$

which contradicts the fact that z_m converges to z_0. ∎

Before going further, we need the following lemma.

LEMMA 1
Let F be a Hilbert space and G a closed subspace. There exists then $f \in F$ satisfying

(3) $$\|f\| = 1, \quad \inf_{g \in G} \|f - g\| \geq 1.$$

Proof. Let t be the orthogonal projector from F onto G. We take $h \notin G$. Then $f = (h - th)/(\|h - th\|)$ satisfies $\|f\| = 1$ and

$$\|f - g\| = \frac{1}{\|h - th\|} \|h - th - g\|h - th\|\| \geq \frac{\|h - th\|}{\|h - th\|} = 1,$$

since $th + g\|h - th\|$ belongs to G as g runs over G. ∎

THEOREM 2

Suppose that (1) *holds. If λ is not an eigenvalue, then*

(4) $\qquad A_\lambda$ *is an isomorphism from V onto F.* ▲

Proof. A_λ is an injective operator from U to F. According to the Banach theorem (Theorem 5.2.3), it will follows that A is an isomorphism from the fact that it is surjective. Hence let us show that A_λ is surjective. For this we suppose that $F_1 = A_\lambda(V) \neq F$. We are going to show that this leads to a contradiction of the fact that J is compact. Set $V_1 = K^{-1}F_1 \neq V$. Then $F_2 = A_\lambda(V_1)$ is different from F_1, since A_λ is injective. Hence we construct in this way a sequence of closed subspaces $F_n = A_\lambda(V_{n-1})$ and $V_n = K^{-1}F_n$. According to the lemma, there exists a sequence $\{f_n\}$ of elements of F satisfying

(5) $\qquad f_n \in F_n, \quad \|f_n\|_F = 1, \quad \inf_{g \in F_{n+1}} \|f_n - g\| \geq 1.$

Then if $y_n = K^{-1}f_n$, we obtain

$$\frac{1}{\lambda}(Jy_m - Jy_n) = f_m - (f_n + (A_\lambda(y_m) - A\lambda(y_n))) = f_m - g_m$$

where $g_m = f_n + A_\lambda(y_m) - A_\lambda(y_n)$ belongs to F_{m+1} when $n \geq m+1$. Then it follows from (5) that

(6) $\qquad \frac{1}{\lambda}\|Jy_m - Jy_n\|_F = \|f_m - g_n\|_F \geq 1 \qquad \forall n \geq m+1.$

But the sequence Jy_m, being relatively compact, has a Cauchy subsequence, which is impossible. ∎

THEOREM 3

Suppose that the hypotheses (1) *are satisfied. Then the sequence of distinct eigenvalues either is finite or is a countable sequence that converges to zero.* ▲

Proof. Suppose there exists an infinite sequence of distinct eigenvalues λ_n such that $\lim \lambda_n \neq 0$. We take eigenvectors $e_n \in U$ for which

(7) $\qquad \|e_n\| = 1, \quad Je_n = \lambda_n Ke_n.$

To arrive at a contradiction, we consider the subspaces $x_n = \{e_1, \ldots, e_n\}$ generated by the first n eigenvectors (which are linearly independent). According to Lemma 1, we can construct

(8) $\qquad x_n \in X_n, \quad \|x_n\| = 1, \quad \inf_{x \in X_{n-1}} \|x_n - x\| \geq 1.$

If $n \geq m+1$, we can write

(9) $$\lambda_n^{-1} K^{-1} J x_n - \lambda_m^{-1} K^{-1} J x_m = x_n - z.$$

But $z = x_m + \lambda_n^{-1} K^{-1} A_{\lambda_n} x_n - \lambda_m^{-1} K^{-1} A_{\lambda_m} x_m$ belongs to X_{n-1}, since, if $x_n = \sum_{j=1}^n \beta_j e_j$,

$$\lambda_n^{-1} K^{-1} A_{\lambda_n} x_n = x_n - \lambda_n^{-1} K^{-1} J x_n$$
$$= \sum_{j=1}^n (\beta_j - \lambda_n^{-1} \lambda_j \beta_j) e_j = \sum_{j=1}^{n-1} (\beta_j - \lambda_n^{-1} \lambda_j \beta_j) e_j$$

belongs to X_{n-1} and since x_m and $\lambda_m^{-1} K^{-1} A_{\lambda_m} x_m$ belongs to $X_m \subset X_{n-1}$. Hence

(10) $$\| \lambda_n^{-1} K^{-1} J x_n - \lambda_m^{-1} K^{-1} J x_m \| = \| x_n - z \| \geq 1$$

Since the sequence $\lambda_n^{-1} K^{-1} J x_n$ is relatively compact, it has a Cauchy subsequence, which contradicts equality (10). ∎

We conclude this section with the proof of

THEOREM 4
Suppose that (1) *is the case. Then λ is an eigenvalue of J with respect to K if and only if λ is an eigenvalue of J^* with respect to K^*. Moreover,*

(11) $$N_\lambda = (F_\lambda^*)^\perp \quad \text{and} \quad N_\lambda^* = (F_\lambda)^\perp$$

where $F_\lambda^* = \operatorname{Im} A_\lambda^*$ and $N_\lambda^* = \operatorname{Ker} A_\lambda^*$. ▲

Proof. Suppose that λ is not an eigenvalue of J^*. Then $F_\lambda = F$ according to Theorem 2 and, consequently, $N_\lambda^* = \operatorname{Ker} A_\lambda^* = (\operatorname{Im} F_\lambda)^\perp = 0$. Hence λ is not an eigenvalue of J^*.

Interchanging the roles of J and J^*, we show, similarly, that if λ is not an eigenvalue of J^*, it is not an eigenvalue of J, since J^* is compact (according to Proposition 1.3) and since K^* is an isomorphism. ∎

CHAPTER 12

Hilbert-Schmidt Operators and Tensor Products

The space $\mathscr{L}(U, F)$ of continuous linear operators from a separable Hilbert space U to a separable Hilbert space F is a Banach space. We show in Section 1 that it contains a Hilbert space $\mathscr{L}_2(U, F)$ consisting of operators called *Hilbert-Schmidt operators*, which are defined in the following fashion: We verify that if $\{e_n\}$ and $\{f_m^*\}$ are orthonormal bases of U and of F^* and if the series

$$\|A\|_2 = \left(\sum_{n,m=1}^{\infty} |\langle f_n^*, Ae_m \rangle|^2 \right)^{1/2}$$

is convergent, then the sum of this series does not depend on the choice of the bases. Thus we say that A is a Hilbert-Schmidt operator if $\|A\|_2$ is finite. These operators are compact and can be represented by "infinite matrices" $\{a_{m,n}\} \in l^2(\mathbb{N} \times \mathbb{N})$ in the following fashion:

$$Ax = \sum_{m,n=1}^{\infty} a_{m,n} \langle e_n^*, x \rangle f_m$$

where $\{e_n^*\}$ and $\{f_m\}$ are orthonormal bases of V^* and of F.

The most important example is that of the integral operators $\tilde{K}$ associated with the functions $K \in L^2(\Omega_1 \times \Omega_2)$ by the formula

$$(\tilde{K}x)(\omega_2) = \int_{\Omega_1} K(\omega_1, \omega_2) x(\omega_1) \, d\omega_1.$$

These integral operators, therefore, form a class of compact operators. Moreover, we establish in Section 6 that every Hilbert-Schmidt operator $\tilde{K}$ from $L^2(\Omega_1)$ to $L^2(\Omega_2)$ is defined by a *kernel* $K \in L^2(\Omega_1 \times \Omega_2)$.

Then we show in Section 2 that if V, F, and G are three Hilbert spaces, $\mathscr{L}_2(\mathscr{L}_2(V, F), G)$ is isometric to $\mathscr{L}_2(V^*, \mathscr{L}_2(F, G))$.

This theorem is then going to play a fundamental role. In fact, it will allow us to interpret in Section 3 the space $\mathscr{L}_2(E^*, F)$ as the "tensor product"

$E \hat{\otimes} F$ of the Hilbert spaces E and F in the sense that every bilinear Hilbert-Schmidt operator on $E \times F$ is associated in a one-to-one fashion with a Hilbert-Schmidt operator on $E \hat{\otimes} F$. We then verify that this tensor product is associative and commutative and that $\mathbb{R}$ is the neutral element for it.

In Section 4 we study the concept of the tensor product of continuous linear operators. This allows us to characterize the tensor product $D(A) \hat{\otimes} E$ of the domain of an operator A by E as the domain of an operator.

We show successively that $l^2 \hat{\otimes} E$ is isometric to $l^2(E)$ and that $L^2 \hat{\otimes} E$ is isometric to $L^2(E)$ (where $l^2(E)$ denotes the space of square summable sequences with values in E and $L^2(E)$ the space of square integrable functions with values in E).

Finally, we characterize the Hilbert tensor product $H^m \hat{\otimes} F$ as the Sobolev space $H^m(F)$ of functions with values in a suitably defined F. Then we obtain a *kernel theorem* implying that every Hilbert-Schmidt operator $\tilde{K}$ from $H^m(\Omega_1)^*$ to $H^p(\Omega_2)$ is associated with a kernel $K \in L^2(\Omega_1 \times \Omega_2)$ whose derivatives $D_1^k D_2^l K$ belong to $L^2(\Omega_1 \times \Omega_2)$ when $|k| \leq m$ and $|l| \leq p$: for every $f \in H^m(\Omega_1)^*$, $\tilde{K}f(\omega_2) = (f, K(., \omega_2))$.

These theorems will play a very important role in systems theory when it is a question of constructing "external and internal representations" of systems. The first part of Section 5 in Chapter 14 can be studied when the present chapter is completed.

1. THE HILBERT SPACE OF HILBERT-SCHMIDT OPERATORS

Let V and F be two *separable* Hilbert spaces. We are going to construct a Hilbert space $\mathscr{L}_2(V, F)$ of continuous linear operators from V to F, operators that are called *Hilbert-Schmidt*.

Before defining them, we need to prove the following lemma.

LEMMA 1

Let V and F be two separable Hilbert spaces. If the series

$$\sum_{n=1}^{\infty} \|Ae_n\|_F^2$$

converges for an orthonormal base $\{e_n\}$ of V, then

(1) $\quad \displaystyle\sum_{n=1}^{\infty} \|Ae_n\|_F^2 = \sum_{n=1}^{\infty} \|Ae_n'\|_F^2 = \sum_{n,m=1}^{\infty} |\langle Ae_n, f_m^* \rangle|^2 = \sum_{m=1}^{\infty} \|A^* f_m^*\|^2$

no matter what orthonormal bases $\{e_n'\}$ of V and $\{f_m^*\}$ of F^* are chosen. ▲

CH. 12, SEC. 1   HILBERT SPACE OF HILBERT-SCHMIDT OPERATORS   257

Proof. Indeed, if $\{f_m^*\}$ is an orthonormal base of F^*, we can write

$$\|Ae_n\|^2 = \sum_{m=1}^{\infty} |\langle f_m^*, Ae_n\rangle|^2 = \sum_{m=1}^{\infty} |\langle A^*f_m^*, e_n\rangle|^2.$$

Consequently,

(2)
$$\sum_{n=1}^{\infty} \|Ae_n\|_F^2 = \sum_{m=1}^{\infty} \sum_{n=1}^{\infty} |\langle A^*f_m^*, e_n\rangle|^2$$
$$= \sum_{m=1}^{\infty} \|A^*f_m^*\|_{V^*}^2.$$

does not depend on the choice of the orthonormal base $\{e_n\}$.   ∎

DEFINITION 1
We say that a continuous linear operator $A \in \mathscr{L}(V, F)$ is a Hilbert-Schmidt operator if the series

$$\|A\|_2 = \left(\sum_{n=1}^{\infty} \|Ae_n\|_F^2\right)^{1/2}$$

converges for at least one orthonormal base $\{e_n\}$ of V.   ▲

Example 1

The operators of finite rank

(3)
$$A_p : x \mapsto \sum_{m,n=1}^{p} a_{m,n}\langle e_n^*, x\rangle f_m$$

are Hilbert-Schmidt operators, since

$$\|A\|_2^2 = \sum_{m,n=1}^{p} |a_{m,n}|^2$$

More generally, consider the square summable double sequence

$$\{a_{m,n}\}_{m,n=1,\ldots,\infty} \in l^2(\mathbb{N} \times \mathbb{N})$$

The operators A_p defined by (3) form a Cauchy sequence in $\mathscr{L}(V, F)$, since if $p \leq q$,

$$\|A_p x - A_q x\|_F^2 \leq \sum_{\substack{m \text{ and } n \leq q \\ m \text{ or } n \geq p}} |a_{m,n}|^2 |\langle e_n^*, x\rangle|^2$$
$$\leq \|x\|^2 \sum_{\substack{m \text{ and } n \leq q \\ m \text{ or } n \geq p}} |a_{m,n}|^2$$

converges to zero when p and q approach infinity. Hence the operators A_p

converge to the operator A defined by

$$(4) \qquad Ax = \sum_{m,n=1}^{\infty} a_{m,n} \langle e_n^*, x \rangle f_m,$$

which satisfies

$$(5) \qquad \|A\|_2^2 = \sum_{m,n=1}^{\infty} |a_{m,n}|^2 < +\infty. \qquad \blacktriangle$$

The following theorem shows that all the Hilbert-Schmidt operators are of the form (4).

THEOREM 1
The space $\mathscr{L}_2(V, F)$ of Hilbert-Schmidt operators is a Hilbert space for the norm

$$(6) \qquad \left\{ \|A\|_2 = \left(\sum_{m=1}^{\infty} \|A e_n\|_F^2 \right)^{1/2} \text{ where } \{e_n\} \text{ is an orthonormal base of } V \right.$$

and the scalar product

$$(7) \qquad ((A, B)) = \sum_{n,m=1}^{\infty} \langle f_m^*, A e_n \rangle \langle f_m^*, B e_n \rangle$$

where $\{e_n\}$ and $\{f_m^*\}$ are orthonormal bases of V and of F^*, respectively. The double sequence of Hilbert-Schmidt operators $e_n^* \hat{\otimes} f_m$ defined by

$$(8) \qquad e_n^* \otimes f_m : x \in V \mapsto \langle e_n^*, x \rangle f_m \in F$$

forms an orthonormal base of $\mathscr{L}_2(V, F)$. $\blacktriangle$

Proof. Let $A \in \mathscr{L}_2(V, F)$, let $\{e_n\}$ be an orthonormal base of V and $\{f_m^*\}$ an orthonormal base of F^*. Then

$$\|A\|_2 = \left(\sum_{n=1}^{\infty} \|A e_n\|_F^2 \right)^{1/2} = \sum_{n,m=1}^{\infty} |\langle f_m^*, A e_n \rangle|^2 < +\infty$$

and does not depend on the choice of these bases, according to the preceding lemma.

Consequently, the mapping $\theta : \mathscr{L}_2(V, F) \mapsto l^2(\mathbb{N} \times \mathbb{N})$ defined by

$$(9) \qquad A \mapsto \{ \langle f_m^*, A e_n \rangle \}_{n,m=1,\ldots,\infty} = \theta(A)$$

is a bijection from $\mathscr{L}_2(V, F)$ onto $l^2(\mathbb{N} \times \mathbb{N})$.

Hence $\mathscr{L}_2(V, F)$ can be given a structure of a separable Hilbert space for the scalar product

$$((A,B)) = (\theta(A), \theta(B)) = \sum_{n,m=1}^{\infty} \langle f_m^*, Ae_n \rangle \langle f_m^*, Be_n \rangle.$$

This scalar product defines the norm

$$\|A\|_2 = \sqrt{((A,A))} = \left(\sum_{n,m=1}^{\infty} |\langle f_m^*, Ae_n \rangle|^2 \right)^{1/2} = \left(\sum_{n=1}^{\infty} \|Ae_n\|_F^2 \right)^{1/2}.$$

Consider the dual bases $e_n^* = Je_n$ and $f_m = K^{-1}f_m^*$ where $J \in \mathscr{L}(V, V^*)$ and $K \in \mathscr{L}(F, F^*)$ are duality operators. Since

$$\langle f_k^*, (e_n^* \otimes f_m)(e_j) \rangle = \langle e_n^*, e_j \rangle \langle f_k^*, f_m \rangle,$$

it follows that the family of operators $e_n^* \otimes f_m$ is orthonormal.

Moreover, the operator A can be written

(10) $$A = \sum_{n,m=1}^{\infty} \langle f_m^*, Ae_n \rangle (e_n^* \otimes f_m)$$

since for all $x \in V$,

$$Ax = A\left(\sum_{n=1}^{\infty} \langle e_n^*, x \rangle e_n \right) = \sum_{n=1}^{\infty} \langle e_n^*, x \rangle Ae_n$$

$$= \sum_{n=1}^{\infty} \langle e_n^*, x \rangle \sum_{m=1}^{\infty} \langle f_m^*, Ae_n \rangle f_m$$

$$= \sum_{n,m=1}^{\infty} \langle f_m^*, Ae_n \rangle (e_n^* \otimes f_m)(x).$$

Hence the sequence $\{e_n^* \otimes f_m\}$ forms an orthonormal base of $\mathscr{L}_2(V, F)$. ∎

Remark 1

If $\{e_n\}$ and $\{f_m^*\}$ are orthonormal bases of the Hilbert spaces V and F, then the double sequence of coefficients

$$((A, e_n^* \otimes f_m)) = \langle f_m^*, Ae_n \rangle = a_{m,n}$$

of A with respect to the orthonormal base $\{e_n^* \otimes f_m\}_{n,m}$ can be interpreted as a square summable "infinite matrix." ∎

Example 2. Integral Operators

Let Ω_1 and Ω_2 be two open sets in $\mathbb{R}^{n_1}$ and $\mathbb{R}^{n_2}$ and let $K \in L^2(\Omega_1 \times \Omega_2)$ be a square summable function on the product $\Omega_1 \times \Omega_2$. We associate to the function K (called the kernel) the integral operator $\tilde{K}$, which associates to every function x defined on Ω_1 the function $\tilde{K}x$ defined on Ω_2 by

(11) $$\tilde{K}x(\omega_2) = \int_{\Omega_1} K(\omega_1, \omega_2) x(\omega_1) \, d\omega_1.$$ ▲

PROPOSITION 1
If $K \in L^2(\Omega_1 \times \Omega_2)$, then $\tilde{K}$ is a Hilbert-Schmidt operator from $L^2(\Omega_1)$ to $L^2(\Omega_2)$ such that

(12) $$\|\tilde{K}\|_2 = \|K\|_{L^2(\Omega_1 \times \Omega_2)}.$$ ■

We show in Section 6 that, conversely, every Hilbert-Schmidt operator from $L^2(\Omega_1)$ to $L^2(\Omega_2)$ is associated with a kernel $K \in L^2(\Omega_1 \times \Omega_2)$ (the kernel theorem, Theorem 6.2).

Proof. To this end we consider the orthonormal bases $\{e_n(\cdot)\}_n$ and $\{f_m(\cdot)\}_m$ of $L^2(\Omega_1)$ and of $L^2(\Omega_2)$. Then

$$\|\tilde{K}\|_2^2 = \sum_{n,m=1}^{\infty} |\langle \tilde{K}e_n, f_m \rangle|^2$$

$$= \sum_{n,m=1}^{\infty} |\int_{\Omega_2} \int_{\Omega_1} K(\omega_1, \omega_2) e_n(\omega_1) f_m(\omega_2) \, d\omega_1 \, d\omega_2|^2.$$

Since the double sequence of functions $\{\omega_1, \omega_2\} \mapsto e_n(\omega_1) f_m(\omega_2)$ is clearly orthonormal in $L^2(\Omega_1 \times \Omega_2)$, the Bessel inequality (Theorem 1.7.1) implies that this series is convergent and that

$$\|\tilde{K}\|_2^2 \leq \|K\|_{L^2(\Omega_1 \times \Omega_2)}^2.$$

Hence $\tilde{K}$ is a Hilbert-Schmidt operator. In order to show that $\|\tilde{K}\|_2 = \|K\|$, it suffices to verify that the double sequence $\{e_n, f_m\}_{n,m}$ forms a base, that is, that it generates a dense subspace. According to the density criterion (Theorem 2.2.1), it is enough to verify that if $K \in L^2(\Omega_1 \times \Omega_2)$ is such that

$$\int_{\Omega_1} \int_{\Omega_2} K(\omega_1, \omega_2) e_n(\omega_1) f_m(\omega_2) \, d\omega_1 \, d\omega_2 = 0 \quad \text{for all} \quad n, m,$$

then $K = 0$. Since $\int_{\Omega_2} (\tilde{K}e_n)(\omega_2) f_m(\omega_2) \, d\omega_2 = 0$ for all m, it follows that $\tilde{K}e_n(\omega_2)$ is identically zero for all $\omega_2 \in \Omega_2$ except in a set of measure zero Γ_n. If ω_2 does not belong to the set of measure zero $\Gamma = \bigcup_{n=1}^{\infty} \Gamma_n$, it follows that $\int_{\Omega_1} K(\omega_1, \omega_2) e_n(\omega_1) d\omega_1 = 0$ for all n and that, therefore, $K(\omega_1, \omega_2) = 0$ for all $\omega_1 \in \Omega_1$ (except on a set of measure zero). Thus $K = 0$. ■

We are now going to establish the elementary properties of Hilbert-Schmidt operators.

PROPOSITION 2
Let $A \in \mathscr{L}_2(V, F)$ be a Hilbert-Schmidt operator. Then

(13) $$\|A\|_{\mathscr{L}(V,F)} \leq \|A\|_2,$$

and the transpose $A^* \in \mathscr{L}_2(F^*, V^*)$ is a Hilbert-Schmidt operator satisfying

(14) $$\|A^*\|_2 = \|A\|_2.$$

Finally, if $A \in \mathscr{L}_2(V, F)$ and $B \in \mathscr{L}(F, G)$, then $BA \in \mathscr{L}_2(V, G)$ and

(15) $$\|BA\|_2 \leq \|A\|_2 \|B\|_{\mathscr{L}(F,G)}.$$

If $A \in \mathscr{L}_2(V, F)$ and $B \in \mathscr{L}(U, V)$ then $AB \in \mathscr{L}_2(U, F)$ and

(16) $$\|AB\|_2 \leq \|A\|_2 \|B\|_{\mathscr{L}(U,V)}. \qquad \blacktriangle$$

Proof. Let $\{f_m^*\}$ be an orthonormal base of F^*. Then

$$\|Ax\|^2 = \sum_{m=1}^{\infty} |\langle f_m^*, Ax \rangle|^2 = \sum_{m=1}^{\infty} |\langle A^* f_m^*, x \rangle|^2$$
$$\leq \sum_{m=1}^{\infty} \|A^* f_m^*\|^2 \|x\|^2 \leq \|x\|^2 \|A\|_2^2$$

according to inequality (1). Hence $\|A\|_{\mathscr{L}(V,F)} \leq \|A\|_2$. Moreover, inequality (1) implies that $\|A^*\|_2^2 = \sum_{m=1}^{\infty} \|A^* f_m^*\|^2 = \|A\|_2^2$ and, therefore, that A^* is a Hilbert-Schmidt operator. Finally, if $\{e_n\}$ is an orthonormal base of V,

$$\|BA\|_2 = \sum_{n=1}^{\infty} \|BAe_n\|_G^2 \leq \|B\|_{\mathscr{L}(F,G)}^2 \sum_{n=1}^{\infty} \|Ae_n\|_F^2 = \|B\|_{\mathscr{L}(F,G)}^2 \|A\|_2.$$

Moreover,

$$\|AB\|_2 = \|(AB)^*\|_2 = \|B^* A^*\|_2 \leq \|A^*\|_2 \|B^*\|_{\mathscr{L}(V^*, U^*)} = \|A\|_2 \|B\|_{\mathscr{L}(U,V)}. \qquad \blacksquare$$

The Hilbert-Schmidt operators are compact and hence possess the properties of these operators that we have established in Chapter 11.

PROPOSITION 3
The space of operators of finite rank is dense in $\mathscr{L}_2(V, F)$. The Hilbert-Schmidt operators are compact. $\qquad \blacktriangle$

Proof. Let $A \in \mathscr{L}_2(V, F)$ and let $\{e_n\}$ be an orthonormal base of V. We denote by A the operator of finite rank defined by

$$A_k e_n = \begin{cases} Ae_n & \text{if} \quad n \leq k \\ 0 & \text{if} \quad n \geq k+1; \end{cases}$$

then

$$\|A - A_k\|_2^2 = \sum_{n=1}^{\infty} \|(A - A_k)e_n\|_F^2 = \sum_{n=k+1}^{\infty} \|Ae_n\|_F^2.$$

Since the series $\sum \|Ae_n\|_F^2$ converges, it follows that $\|A - A_k\|_2^2$ converges to zero and hence that A is the limit in $\mathscr{L}_2(V, F)$ of a sequence of operators of finite rank. Since

$$\|A - A_k\|_{\mathscr{L}(V,F)} \leq \|A - A_k\|_2 \mapsto 0.$$

we find that A is compact. (See Proposition 11.1.1.) ∎

The following result is then a consequence of Theorem 11.3.2.

THEOREM 2
A compact operator $A \in \mathscr{L}(U, F)$ is a Hilbert-Schmidt operator if and only if there exist orthonormal bases $\{e_n^*\}_n$ and $\{f_n\}_n$ of U^* and of F and a sequence $\lambda = \{\lambda_n\}_n \in l^2(\mathbb{N})$ of positive scalars such that

(17) $$Ax = \sum_{n=1}^{\infty} \lambda_n \langle e_n^*, x \rangle f_n.$$ ▲

Proof. Indeed, when A is written in the form (17), we have

(18) $$\|A\|_2 = \left(\sum_{n=1}^{\infty} |\lambda_n|^2 \right)^{1/2}.$$

If A is compact, Theorem 11.3.1 implies (17) and, consequently, that A is a Hilbert-Schmidt operator when $\{\lambda_n\} \in l^2$.

The converse is true, since every Hilbert-Schmidt operator is compact and therefore can be written in the form (17), from which it follows, according to (18), that $\{\lambda_n\} \in l^2$. ∎

Remark 2. Realization of the Dual of $\mathscr{L}_2(V, F)$

We could choose $\mathscr{L}_2(F^*, V^*)$ as a realization of the dual of $\mathscr{L}_2(V, F)$ for the duality pairing

$$\langle\langle M, A \rangle\rangle = \sum_{m,n=1}^{\infty} \langle Mf_m^*, e_n \rangle \langle f_m^*, Ae_n \rangle$$

$$= \sum_{n,m=1}^{\infty} \langle f_m^*, M^* e_n \rangle \langle f_m^*, Ae_n \rangle = ((M^*, A)).$$

However, in many problems it is convenient to use another realization of the dual of $\mathscr{L}_2(V, F)$.

PROPOSITION 4

The bilinear form on $\mathscr{L}_2(V^*, F^*) \times \mathscr{L}_2(V, F)$ defined by

(19) $$\langle M, A \rangle = \sum_{m,n=1}^{\infty} \langle Me_n^*, f_m \rangle \langle f_m^*, Ae_n \rangle$$

is a duality pairing on $\mathscr{L}_2(V^*, F^*)$ and $\mathscr{L}_2(V, F)$. It can also be written as

$$\langle M, A \rangle = \sum_{n=1}^{\infty} \langle Me_n^*, Ae_n \rangle_{F^* \times F} = \sum_{m=1}^{\infty} \langle A^* f_m^*, M^* f_m \rangle_{V^* \times V}.$$

If K is the duality operator from V onto V^* and J the duality operator from F onto F^*, the duality operator from $\mathscr{L}_2(V, F)$ onto $\mathscr{L}_2(V^*, F^*)$ is the operator $A \mapsto JAK^{-1}$. ▲

Proof. If $e^* \in V^*$, $e \in V$, $Me^* \in F^*$, and $Ae \in F$, we know that

$$\langle Me^*, Ae \rangle_{F^* \times F} = \sum_{m=1}^{\infty} \langle Me^*, f_m \rangle \langle f_m^*, Ae \rangle,$$

since $\{f_m\}$ is an orthonormal base of F and $\{f_m^*\}$ (where $f_m^* = Jf_m$) is its dual base. It follows then that

$$\langle M, A \rangle = \sum_{n=1}^{\infty} \langle Me_n^*, Ae_n \rangle_{F^* \times F} = \sum_{m,n=1}^{\infty} \langle Me_n^*, f_m \rangle \langle f_m^*, Ae_n \rangle$$

$$= \sum_{n,m=1}^{\infty} \langle e_n^*, M^* f_m \rangle \langle A^* f_m^*, e_n \rangle = \sum_{m=1}^{\infty} \langle A^* f_m^*, M^* f_m \rangle_{V^* \times V}$$

is a bilinear form on $\mathscr{L}_2(V^*, F^*) \times \mathscr{L}_2(V, F)$ that *does not depend* on the choice of the orthonormal bases $\{e_n\}$ and $\{f_m\}$.

It is clear that the isomorphism

$$A \in \mathscr{L}(V, F) \mapsto JAK^{-1} \in \mathscr{L}(V^*, F^*)$$

satisfies

$$((A, B)) = \sum_{m,n=1}^{\infty} \langle f_m^*, Ae_n \rangle \langle f_m^*, Be_n \rangle$$

$$= \sum_{m,n=1}^{\infty} \langle Jf_m, AK^{-1} e_n^* \rangle \langle f_m^*, Be_n \rangle$$

$$= \sum_{m,n=1}^{\infty} \langle JAK^{-1} e_n^*, f_m \rangle \langle f_m^*, Be_n \rangle$$

$$= \langle JAK^{-1}, B \rangle.$$

Hence $\mathscr{L}_2(V^*, F^*)$ is a realization of the dual and the bilinear form $\langle M, A \rangle = ((J^{-1} MK, A))$ is the duality pairing on $\mathscr{L}_2(V^*, F^*) \times \mathscr{L}_2(V, F)$. ■

2. THE FUNDAMENTAL ISOMORPHISM THEOREM

Let us consider three *separable* Hilbert spaces V, F, G, and the Hilbert spaces $\mathscr{L}_2(V,F)$ and $\mathscr{L}_2(F,G)$ of Hilbert-Schmidt operators.

THEOREM 1
The space $\mathscr{L}_2(\mathscr{L}_2(V,F);G)$ of Hilbert-Schmidt operators from $\mathscr{L}_2(V,F)$ to G is isometric to the space $\mathscr{L}_2(V^*;\mathscr{L}_2(F,G))$ of Hilbert-Schmidt operators from V^* to $\mathscr{L}_2(F,G)$. ▲

Proof. Let $A \in \mathscr{L}_2(V^*;\mathscr{L}_2(F,G))$ be the operator that to every $e^* \in V^*$ associates the operator $A(e^*) \in \mathscr{L}_2(F,G)$. This Hilbert-Schmidt operator $A(e^*)$ has its norm equal to

$$\|A(e^*)\|^2_{\mathscr{L}_2(F,G)} = \sum_{j,k=1}^{\infty} |\langle A(e^*)f_j, g_k^* \rangle|^2 \tag{1}$$

where $\{f_j\}$ and $\{g_k^*\}$ are orthonormal bases of F and of G^*, respectively. The norm of A is, therefore, equal to

$$\|A\|^2_2 = \sum_{m=1}^{\infty} \|A(e_m^*)\|^2_{\mathscr{L}_2(F,G)} = \sum_{m,j,k=1}^{\infty} |\langle A(e_m^*)f_j, g_k^* \rangle|^2 \tag{2}$$

where $\{e_m^*\}$ is an orthonormal base of V^*.

Let $M \in \mathscr{L}_2(V,F)$ be a Hilbert-Schmidt operator defined with respect to the orthonormal base $\{e_m^* \otimes f_j\}$ by

$$M = \sum_{m,j=1}^{\infty} a_{mj} e_m^* \otimes f_j; \quad \{a_{mj}\} \in l^2(\mathbb{N} \times \mathbb{N}). \tag{3}$$

We associate to A the operator $\tilde{A} \in \mathscr{L}(\mathscr{L}_2(V,F);G)$ defined by

$$\tilde{A}(M) = \sum_{m,j=1}^{\infty} a_{mj} A(e_m^*) f_j \in G. \tag{4}$$

It is clear that $\tilde{A}$ is a linear operator satisfying

$$\tilde{A}(e_m^* \otimes f_j) = A(e_m^*) f_j. \tag{5}$$

Hence

$$\|\tilde{A}\|^2_{\mathscr{L}_2(\mathscr{L}_2(V,F),G)} = \sum_{m,j=1}^{\infty} \|A(e_m^* \otimes f_j)\|^2_G \tag{6}$$
$$= \sum_{m,j=1}^{\infty} \|A(e_m^*)f_j\|^2_G = \sum_{m,j,k=1}^{\infty} |\langle A(e_m^*)f_j, x_k \rangle|^2 = \|A\|^2_2,$$

which implies that $\tilde{A}$ is a Hilbert-Schmidt operator. The mapping

$$A \in \mathscr{L}_2(V^*, \mathscr{L}_2(F, G)) \mapsto \tilde{A} \in \mathscr{L}_2(\mathscr{L}_2(V, F), G)$$

is, therefore, an isometry. It is clearly surjective; let $\tilde{A} \in \mathscr{L}_2(\mathscr{L}_2(V, F); G)$. Then for all fixed $e^* \in V^*$, $e^* \otimes f \in \mathscr{L}_2(V, F)$. The operator $A(e^*)$ defined by

(7) $$A(e^*): f \in F \mapsto \tilde{A}(e^* \otimes f) \in G$$

is a Hilbert-Schmidt operator related to $\tilde{A}$ by formula (4). Hence we verify that

$$A: e^* \mapsto A(e^*) \in \mathscr{L}_2(F, G)$$

is a Hilbert-Schmidt operator from V^* to $\mathscr{L}_2(F, G)$. ∎

3. HILBERT TENSOR PRODUCTS

Theorem 2.1 allows us to construct a Hilbert space $E \hat{\otimes} F$, which is the completion of the tensor product $E \otimes F$ of the two separable Hilbert spaces E and F.

First of all we observe the following lemma (analogous to Lemma 1.1).

LEMMA 1
Let E, F, and G be three separable Hilbert spaces and $a: E \times F \mapsto G$ be a bilinear mapping from $E \times F$ to G. If the series

(1) $$\sum_{m,j=1}^{\infty} \|a(\bar{e}_m, \bar{f}_j)\|_G^2 < +\infty$$

converges for the orthonormal bases $\{\bar{e}_m\}$ and $\{\bar{f}_j\}$ of E and of F, then

(2) $$\sum_{m,j=1}^{\infty} \|a(e_m, f_j)\|_G^2 = \sum_{m,j,k=1}^{\infty} |\langle a(e_m, f_j), g_k^* \rangle|^2 = \sum_{m=1}^{\infty} \|A e_m\|_{\mathscr{L}_2(F,G)}^2,$$

no matter which orthonormal bases $\{e_m\}, \{f_j\}$, and (g_k^*) of E, F, and G^* are chosen, where $A \in \mathscr{L}_2(E, \mathscr{L}_2(F, G))$ is the operator associated to a by

(3) $$A(e)f = a(e, f) \quad \text{for all} \quad e \in E, \, f \in F. \quad \blacktriangle$$

Proof. Indeed, for every orthonormal base $\{g_k^*\}$ of G^* we have

$$\sum_{m,j=1}^{\infty} \|a(e_m, f_j)\|_G^2 = \sum_{m,j,k=1}^{\infty} |\langle a(e_m, f_j), g_k^* \rangle|^2$$
$$= \sum_{m=1}^{\infty} \sum_{j,k=1}^{\infty} |\langle A(e_m)f_j, g_k^* \rangle|^2 = \sum_{m=1}^{\infty} \|A(e_m)\|_{\mathscr{L}_2(F,G)}^2$$
$$= \|A\|_{\mathscr{L}_2(E, \mathscr{L}_2(F,G))}^2. \quad \blacksquare$$

DEFINITION 1

We say that a bilinear mapping $a: E \times F \mapsto G$ is Hilbert-Schmidt if the series

$$(4) \qquad \|a\|_2 = \left(\sum_{m,k,j=1}^{\infty} |\langle a(e_m, f_j), g_k^* \rangle|^2 \right)^{1/2} < +\infty$$

is convergent for arbitrary orthonormal bases $\{e_m\}$, $\{f_j\}$, and $\{g_k^*\}$ of E, F, and G^*.

We denote by $\mathcal{B}_2(E \times F; G)$ the space of Hilbert-Schmidt bilinear mappings with the norm $\|a\|_2$.

It is clear that the mapping.

$$a \in \mathcal{B}_2(E \times F; G) \mapsto A \in \mathcal{L}_2(E, \mathcal{L}_2(F, G))$$

is a surjective isometry. ▲

Taking $V = E^*$, Theorem 2.1 can be reformulated as follows:

THEOREM 1

Let E and F be two separable Hilbert spaces. The Hilbert space $\mathcal{L}_2(E^*, F)$ has the following property: for every Hilbert space G and every Hilbert-Schmidt bilinear mapping $a \in \mathcal{B}_2(E \times F; G)$, there exists a unique Hilbert-Schmidt operator $A \in \mathcal{L}_2(\mathcal{L}_2(E^*, F), G)$ such that the diagram

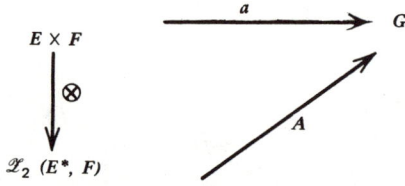

is commutative, where $\otimes$ is the Hilbert-Schmidt bilinear mapping from $E \times F$ to $\mathcal{L}_2(E^*, F)$ defined by

$$e \otimes f : y \in E^* \mapsto \langle e, y \rangle f \in F.$$ ▲

Remark 1

It is obvious that the space $\mathcal{L}_2(E^*, F)$ possessing the preceding property is *unique up to an isomorphism*: If there exists a Hilbert space X_1 and a Hilbert-Schmidt bilinear mapping $\otimes_1$ from $E \times F$ to X_1 possessing the preceding property, then there exists a Hilbert-Schmidt isometry j from

$\mathscr{L}_2(E^*, F)$ onto X_1 such that the diagram

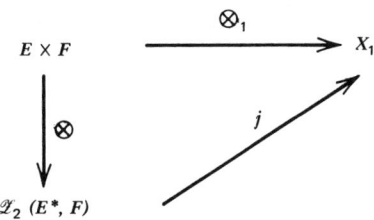

is commutative. Indeed, the mapping j is none other than the mapping A of diagram (5) where we take $G = X_1$ and $a = \otimes_1$. This same diagram where $\mathscr{L}_2(E^*, F)$ is replaced by X_1, where $\otimes$ is replaced by $\otimes_1$, and where we take $G = \mathscr{L}_2(E^*, F)$ and $a = \otimes$ implies the existence of an isometry that is precisely the inverse of j. ■

Remark 2

The property stated in the preceding theorem is called the *universal property* defining the Hilbert tensor product for which the space $\mathscr{L}_2(E^*, F)$ is the solution (unique up to an isomorphism). ■

DEFINITION 2
We say the space $\mathscr{L}_2(E^*, F)$ is the Hilbert tensor product *of the separable Hilbert spaces E and F*. We denote this space by

(6) $$E \hat{\otimes} F = \mathscr{L}_2(E^*; F)$$

and its elements by

(7) $$M = \sum_{m,j=1}^{\infty} a_{mj} e_m \otimes f_j = \sum_{m=1}^{\infty} e_m \otimes y_m$$

where

$$\{a_{mj}\}_{m,j} \in l^2(\mathbb{N} \times \mathbb{N})$$

and where

$$y_m = \sum_{j=1}^{\infty} a_{mj} y_m,$$

when the orthonormal bases $\{e_m\}$ and $\{f_j\}$ of E and of F have been chosen. ▲

We remark that the scalar product of $E \hat{\otimes} F$ satisfies

$$((x_1 \otimes y_1, x_2 \otimes y_2)) = ((x_1, x_2))_E ((y_1, y_2))_F.$$

Proposition 1.4 shows that the bilinear form

$$\langle e \otimes f, x \otimes y \rangle = \langle e, x \rangle_{E^* \times E} \langle f, y \rangle_{F^* \times F}$$

has an extension to a bilinear form on $E^* \hat{\otimes} F^* \times E \hat{\otimes} F$ that is a duality pairing on $E^* \hat{\otimes} F^*$ and $E \hat{\otimes} F$. If $K \in \mathscr{L}(E, E^*)$ and $J \in \mathscr{L}(F, F^*)$ denote the duality operators from E onto E^* and from F onto F^*, then the mapping $K \otimes J \in \mathscr{L}(E \hat{\otimes} F, E^* \hat{\otimes} F^*)$ defined by

(8) $\quad M \in E \hat{\otimes} F = \mathscr{L}_2(E^*, F) \mapsto (K \otimes J)M = JMK \in \mathscr{L}_2(E, F^*) = E^* \otimes F^*$

is the duality operator from $E \hat{\otimes} F$ onto $E^* \hat{\otimes} F^*$.

Then Theorem 2.1 can again be reformulated as

THEOREM 2
The space $\mathscr{L}_2(E \hat{\otimes} F, G)$ of Hilbert-Schmidt operators from $E \hat{\otimes} F$ to G is isometric to the Hilbert tensor product $E^* \hat{\otimes} \mathscr{L}_2(F, G)$ of E^* and the space of Hilbert-Schmidt operators from F to G. In particular, $(E \hat{\otimes} F)^*$ is isometric to $E^* \hat{\otimes} F^*$. ▲

Proof. Indeed, we replace V by E^* in Theorem 2.1 and use the notation of the tensor product. ∎

It is convenient to make explicit the isometry j. Let $A \in \mathscr{L}_2(E \hat{\otimes} F; G)$. We associate to A the operators $\hat{A}(x) \in \mathscr{L}_2(F, G)$ defined by

$$\hat{A}(x) : y \in F \mapsto \hat{A}(x)y = A(x \otimes y).$$

If $\{e_m\}$ is an orthonormal base of E and $\{e_m^*\}$ is its dual base, the isometry is defined by

(9) $\quad A \in \mathscr{L}_2(E \hat{\otimes} F; G) \mapsto jA = \sum_{m=1}^{\infty} e_m^* \otimes \hat{A}(e_m) \in E^* \hat{\otimes} \mathscr{L}_2(F, G).$

Indeed, since

$$x = \sum_{m=1}^{\infty} \langle e_m^*, x \rangle e_m,$$

then

$$A(x \otimes y) = \sum_{m=1}^{\infty} \langle e_m^*, x \rangle \hat{A}(e_m)y$$

$$= \sum_{m=1}^{\infty} (e_m^* \otimes \hat{A}(e_m))(x \otimes y). \quad \blacksquare$$

We consider the Hilbert tensor product as an associative and commutative law of composition for which $\mathbb{R}$ is the neutral element.

PROPOSITION 1
Let E, F, and G be three separable Hilbert spaces. Then

(10) $\quad\quad\quad (E \hat{\otimes} F) \hat{\otimes} G \quad \text{is isometric to } E \hat{\otimes} (F \hat{\otimes} G)$.

(11) $\quad\quad\quad\quad\quad E \hat{\otimes} \mathbb{R} \quad \text{is isometric to } E$.

(12) $\quad\quad\quad\quad\quad E \hat{\otimes} F \quad \text{is isometric to } F \hat{\otimes} E$.

(13) $\quad\quad (E_1 \times E_2) \hat{\otimes} F \quad \text{is isometric to } (E_1 \hat{\otimes} F) \times (E_2 \hat{\otimes} F)$. ▲

Proof. The first assertion is a consequence of Theorem 2: Since $(E \hat{\otimes} F)^*$ is identified with $E^* \hat{\otimes} F^*$, it follows that

$$(E \hat{\otimes} F) \hat{\otimes} G = \mathscr{L}_2((E \hat{\otimes} F)^*; G) = \mathscr{L}_2(E^* \hat{\otimes} F^*, G)$$

is isomorphic to

$$\mathscr{L}_2(E^*, \mathscr{L}_2(F^* \otimes G)) = \mathscr{L}_2(E^*, F \hat{\otimes} G) = E \hat{\otimes} (F \hat{\otimes} G)$$

according to Theorem 2.

The second assertion follows from Proposition 1.2, which implies that every continuous linear form $e \in E = \mathscr{L}(E^*, \mathbb{R})$ on E^* is Hilbert-Schmidt, that is, that

$$e \in E = \mathscr{L}_2(E^*, \mathbb{R}) = E \hat{\otimes} \mathbb{R}.$$

The third statement is a consequence of Proposition 1.2, which implies that the transposition

$$M \in E \hat{\otimes} F = \mathscr{L}_2(E^*, F) \mapsto M^* \in \mathscr{L}_2(F^*, E) = F \hat{\otimes} E$$

is a surjective isometry.

Finally, the last assertion follows from the following isometry relations:

$$(E_1 \times E_2) \hat{\otimes} F = \mathscr{L}_2(E_1^* \times E_2^*, F) = \mathscr{L}_2(E_1^*, F) \times \mathscr{L}_2(E_2^*, F)$$
$$= (E_1 \hat{\otimes} F) \times (E_2 \hat{\otimes} F),$$

since every linear operator M from $E_1^* \times E_2^*$ to F can be written in a unique fashion as

$$M(x_1, x_2) = M_1 x_1 + M_2 x_2 \quad \text{where} \quad M_i \in \mathscr{L}(E_i^*, F) \quad (i = 1, 2). \blacksquare$$

4. THE TENSOR PRODUCT OF CONTINUOUS LINEAR OPERATORS

Let us consider two pairs of Hilbert spaces E_i and F_i $(i = 1, 2)$ and the operators

$$A \in \mathscr{L}(E_1, E_2), B \in \mathscr{L}(F_1, F_2).$$

We are going to associate to them a linear operator denoted by

$$A \otimes B \quad \text{from} \quad E_1 \hat{\otimes} F_1 \quad \text{to} \quad E_2 \hat{\otimes} F_2.$$

To do this we recall that

$$E_i \hat{\otimes} F_i = \mathscr{L}_2(E_i^*, F_i).$$

If $M \in \mathscr{L}_2(E_1^*, F_1)$, we know that the operator BMA^* from E_2^* to F_2 is also a Hilbert-Schmidt operator and that

(1) $$\begin{aligned} \|BMA^*\|_2 &\leq \|B\|_{\mathscr{L}(F_1, F_2)} \|M\|_2 \cdot \|A^*\|_{\mathscr{L}(E_2^*, E_1^*)} \\ &= \|B\|_{\mathscr{L}(F_1 F_2)} \|A\|_{\mathscr{L}(E_1, E_2)} \|M\|_2. \end{aligned}$$

(See proposition 1.2.)

Hence the mapping

$$M \in \mathscr{L}_2(E_1^*, F_1) = E_1 \hat{\otimes} F_1 \mapsto BMA^* \in \mathscr{L}_2(E_2^*, F_2) = E_2 \hat{\otimes} F_2$$

is continuous and linear.

DEFINITION 1

We denote by $A \otimes B$ the continuous linear operator from $E_1 \hat{\otimes} F_1$ to $E_2 \hat{\otimes} F_2$, which is defined by $(A \otimes B)M = BMA^*$. We say that $A \otimes B$ is the tensor product of A and B. ▲

PROPOSITION 1

Let $A \in \mathscr{L}(E_1, E_2)$ and $B \in \mathscr{L}(F_1, F_2)$. Then $A \otimes B \in \mathscr{L}(E_1 \hat{\otimes} E_2, F_1 \hat{\otimes} F_2)$ has the following properties:

(2) $$\begin{cases} \text{i.} & \|A \otimes B\| \leq \|A\| \|B\|. \\ \text{ii.} & (A \otimes B)(x \otimes y) = Ax \otimes By \quad \forall x \in E, \ \forall y \in F. \\ \text{iii.} & (A \otimes B)^* = A^* \otimes B^*. \end{cases}$$

If $A_0 \in \mathscr{L}(E_0, E_1)$ and $B_0 \in \mathscr{L}(F_0, F_1)$, then

(3) $$(A \otimes B)(A_0 \otimes B_0) = AA_0 \otimes BB_0.$$

In particular, if A and B are both left invertible (respectively, right invertible), then $A \hat{\otimes} B$ is left invertible (respectively, right invertible). If A and B are isomorphisms then $A \otimes B$ is an isomorphism. The tensor product of two projectors is a projector. ▲

Proof. It is clear that (2)i follows from (1). Let us establish (2)ii; since $M = x \otimes y$ is the linear operator $e \mapsto \langle e, x \rangle y$ from E^* to F, the operator $(A \otimes B)M = BMA^*$ is defined by

$$e \to \langle A^*e, x \rangle By = \langle e, Ax \rangle By;$$

it is the operator $Ax \otimes By$. It is also easy to show that $(A \otimes B)^* = A^* \otimes B^*$ by verifying this equality on elements of the form $e_2 \otimes f_2$ of $E_2^* \otimes F_2^*$:

$$\begin{aligned}\langle (A \otimes B)^*(e_2 \otimes f_2), x_1 \otimes y_1 \rangle &= \langle e_2 \otimes f_2, (A \otimes B)(x_1 \otimes y_1) \rangle \\ &= \langle e_2 \otimes f_2, Ax_1 \otimes By_1 \rangle = \langle e_2, Ax_1 \rangle \langle f_2, By_1 \rangle \\ &= \langle A^* e_2, x_1 \rangle \langle B^* f_2, y_1 \rangle \\ &= \langle A^* e_2 \otimes B^* f_2, x_1 \otimes y_1 \rangle \\ &= \langle (A^* \otimes B^*)(e_2 \otimes f_2), x_1 \otimes y_1 \rangle.\end{aligned}$$

Finally we verify (3) as follows:

$$\begin{aligned}(A \otimes B)(A_0 \otimes B_0)M &= (A \otimes B)(B_0 M A_0^*) = BB_0 M A_0^* A^* \\ &= BB_0 M(AA_0)^* = ((AA_0) \otimes BB_0)M,\end{aligned}$$

where $M \in \mathcal{L}(E_0^*, F_0) = E_0 \hat{\otimes} F_0$. ∎

In particular, if $A \in \mathcal{L}(E_1, E_2)$, we denote by $\vec{A}_F$ (or by $\vec{A}$, if no confusion is possible) the operator $\vec{A}_F = (A \otimes 1_F)$ from $E_1 \hat{\otimes} F$ to $E_2 \hat{\otimes} F$.

Let us recall the following definition:

If $M \in E_1 \hat{\otimes} F = \mathcal{L}_2(E_1^*, F)$ is defined by $\varphi \in E_1^* \mapsto M(\varphi) \in F$, then $\vec{A}M \in E_2 \hat{\otimes} F$ is defined by

(4) $$\forall \varphi \in E_2^*, \quad \vec{A}M(\varphi) = M(A^* \varphi).$$ ∎

Suppose that the space E is the domain of an operator A in the following sense:

Let $A \in \mathcal{L}(H, V)$ be a linear operator from a Hilbert space H to V, U a subspace of V, and E the domain of A defined by

(5) $$D(A) = \{x \in H \quad \text{such that} \quad Ax \in U\}$$

with the "norm of the graph"

(6) $$\|x\|_{D(A)} = (\|x\|_H^2 + \|Ax\|_U^2)^{1/2},$$

which makes continuous the injection from $D(A)$ to H as well as the operator A from $D(A)$ to U. The space $D(A)$ is a Hilbert space if the graph of A is closed in $H \times U$.

For every Hilbert space F the injection 1_F from $D(A) \hat{\otimes} F$ to $H \hat{\otimes} F$ and the operator $\vec{A}_F$ from $D(A) \hat{\otimes} F$ (respectively, from $H \hat{\otimes} F$) to $U \hat{\otimes} F$ (respectively, to $V \hat{\otimes} F$) are continuous.

The question arises as to whether $D(A) \hat{\otimes} F$ is equal to the domain $D(\vec{A}, F)$ defined by

(7) $$D(\vec{A}, F) = \{M \in H \hat{\otimes} F \quad \text{such that} \quad \vec{A}_F M \in U \hat{\otimes} F\}$$

with the norm of the graph

(8) $$\|M\|_{D(\vec{A},F)} = (\|M\|^2_{H\hat{\otimes}F} + \|\vec{A}_F M\|^2_{U\hat{\otimes}F})^{1/2}.$$

THEOREM 1
The Hilbert tensor product $D(A)\hat{\otimes}F$ of the domain $D(A)$ and a Hilbert space F is isometric to the domain $D(\vec{A},F)$. Moreover, $\mathscr{L}_2(D(\vec{A},F^*)^*,G)$ is isometric to $D(\vec{A},\mathscr{L}_2(F,G))$ for every Hilbert space G. ▲

Proof. Indeed, let $M \in D(\vec{A},F)$; that is,
$$M \in H \hat{\otimes} F = \mathscr{L}_2(H^*,F)$$
satisfying
$$\vec{A}_F M = MA^* \in U \hat{\otimes} F = \mathscr{L}_2(U^*,F).$$
Then
$$M^* \in \mathscr{L}_2(F^*,H) \quad \text{and} \quad AM^* \in \mathscr{L}_2(F^*,U).$$
Hence, for all $f \in F^*$, $M^*f \in H$, and $AM^*f \in U$. Consequently, $M^*f \in D(A)$. Therefore M^* is a linear operator from F^* to $D(A)$. It is a Hilbert-Schmidt operator. Indeed, for every orthonormal base $\{f_m^*\}_m$ of F^* we obtain

$$\|M^*\|^2_{\mathscr{L}_2(F^*,D(A))} = \sum_{m=1}^{\infty} \|M^*f_m\|^2_{D(A)}$$
$$= \sum_{m=1}^{\infty} \|M^*f_m\|^2_H + \sum_{m=1}^{\infty} \|AM^*f_m\|^2_U$$
$$= \|M^*\|^2_{\mathscr{L}_2(F^*,H)} + \|AM^*\|^2_{\mathscr{L}_2(F^*,U)}$$
$$= \|M\|^2_{\mathscr{L}_2(H^*,F)} + \|MA^*\|^2_{\mathscr{L}_2(U^*,F)}$$
$$= \|M\|^2_{H\hat{\otimes}F} + \|\vec{A}_F M\|^2_{U\hat{\otimes}F}.$$

We now apply Theorem 3.2 with $E = D(A)^*$, after remarking that $D(A)^* \hat{\otimes} F = (D(A) \hat{\otimes} F^*)^* = D(\vec{A},F^*)^*$. ∎

Remark 1. Tensor Product by the Domain of an Unbounded Operator

Let $(D(A), A)$ be an unbounded operator from U to E that is closed and has dense domain. (See Chapter 5, Section 5.) We consider its adjoint $(D(A^*), A^*)$, an unbounded operator from E^* to U^*, which is also closed and has dense domain. Then $\vec{A}^* = A^* \otimes 1_F$ is a continuous linear operator from $D(A^*)\hat{\otimes}F$ to $V^* \hat{\otimes} F$ and from $E^* \hat{\otimes} F$ to $D(A)^* \hat{\otimes} F$, since A^* is a continuous operator from $D(A^*)$ to V^* and from E^* to $D(A)^*$. (We recall that V^* is a dense subspace of $D(A)^*$ and that $D(A^*) = \{e \in E^* \text{ such that } A^*e \in V^*\}$.) ∎

THEOREM 2

The Hilbert tensor product $D(A^*)\hat{\otimes} F$ is the domain $D(\vec{A^*}, F)$ of the operator $\vec{A^*}$ in the sense that

(9) $\quad D(\vec{A^*}, F) = \{M \in E^* \hat{\otimes} F \quad \text{such that} \quad \vec{A^*} M \in V^* \hat{\otimes} F\}.$

The unbounded operator $(D(\vec{A^*}, F), \vec{A^*})$ from $E^* \hat{\otimes} F$ to $V^* \hat{\otimes} F$ is closed and has dense domain. ▲

Proof. **a.** The first part of the statement of the theorem is established as in Theorem 1: If $M \in E^* \hat{\otimes} F = \mathscr{L}_2(E, F)$ is such that $\vec{A^*} M = MA \in V^* \hat{\otimes} F = \mathscr{L}_2(V, F)$, we must show that $M \in D(A^*) \hat{\otimes} F = \mathscr{L}_2(D(A^*)^*, F)$. But by transposition, $M^* \in \mathscr{L}_2(F^*, E^*)$ and $A^* M^* \in \mathscr{L}_2(F^*, V^*)$; consequently, for all $p \in F^*$, $M^* p \in E^*$ and $A^* M^* p \in V^*$ and, therefore, $M^* p \in D(A^*)$.

This shows that M^* is a linear operator from F^* to $D(A^*)$; it is in fact a Hilbert-Schmidt operator, since it is easy to verify that

$$\|M^*\|^2_{\mathscr{L}_2(F^*, D(A^*))} = \|M^*\|^2_{\mathscr{L}_2(F^*, E^*)} + \|A^* M^*\|^2_{\mathscr{L}_2(F^*, V^*)}$$
$$= \|M\|^2_{\mathscr{L}_2(E, F)} + \|MA\|^2_{\mathscr{L}_2(V, F)}.$$

Consequently, $M \in \mathscr{L}_2(D(A^*)^*, F) = D(A^*) \hat{\otimes} F$.

b. Let us show that the operator is closed, that is, that if a sequence $M_n \in D(\vec{A^*})$ converges to zero in $E^* \hat{\otimes} F$ and if the sequence $\vec{A^*} M_n$ converges to M in $V^* \hat{\otimes} F$, then $M = 0$. But, by transposition, we know that M_n^* converges to zero and $A^* M_n^*$ converges to M^* in $\mathscr{L}_2(F^*, E^*)$ and $\mathscr{L}_2(F^*, V^*)$, respectively. Hence for all $p \in F^*$, $M_n^* p$ converges to zero in F^* and $A^* M_n^* p$ converges to $M^* p$ in V^*. Since $D(\vec{A^*})$ is closed (see Chapter 5, Section 5), $M^* p = 0$. Hence $M^* = 0$ and, consequently, $M = 0$.

c. $D(\vec{A^*}, F)$ is dense in $E^* \hat{\otimes} F$. We use the density criterion (Theorem 2.2.1): if $T \in (E^* \hat{\otimes} F)^* = E \hat{\otimes} F^*$ satisfies $\langle T, M \rangle = 0$ for all $M \in D(\vec{A^*}, F) = D(A^*) \hat{\otimes} F$, then $T = 0$. We take $M = u \otimes f$ where $u \in D(A^*)$ and $f \in F$. Hence

$$\langle T, u \otimes f \rangle = \langle Tu, f \rangle = 0, \quad \forall u \in D(A^*), \quad \forall f \in F.$$

This shows that $Tu = 0, \forall u \in D(A^*)$, and, consequently, that $Tu = 0, \forall u \in E^*$, since $D(A^*)$ is dense in E^*. (See Chapter 5, Section 5.) Therefore $T = 0$. ∎

5. THE HILBERT TENSOR PRODUCT BY l^2

We have defined the space $l^2(\mathbb{N}, F)$ of square summable sequences $y = \{y_m\}_n$ with values in a Hilbert space F, which is a Hilbert space of the scalar product

(1) $\quad\quad\quad\quad ((y, z)) = \sum_{n=1}^{\infty} ((y_n, z_n)).$

We are going to show that

(2) $$l^2(\mathbb{N}, F) = l^2(\mathbb{N}) \hat{\otimes} F.$$

PROPOSITION 1
The space $l^2(\mathbb{N}, F)$ is isometric to the Hilbert tensor product of $l^2(\mathbb{N})$ and F. Every Hilbert-Schmidt operator K from $l^2(\mathbb{N}, F)$ to G is defined by a square summable sequence $\{K_n\}_n$ of Hilbert-Schmidt operators $K_n \in \mathscr{L}_2(F, G)$:

(3) $$K(\{y_n\}_n) = \sum_{n=1}^{\infty} K_n y_n \in G.$$ ▲

Proof. We consider the orthonormal base $\{e_n\}$ of $l^2(\mathbb{N})$ defined by

$$e_n = \{0, \ldots, 0, 1, 0, \ldots, 0, \ldots\}$$

and an orthonormal base $\{f_j\}_j$ of F. The mapping θ from $l^2(\mathbb{N}) \hat{\otimes} F$ to $l^2(\mathbb{N}, F)$ defined by

(4) $$\theta\left(\sum_{m=1}^{\infty} e_m \otimes f_m\right) = \{f_m\}_m$$

is an isometry, since

$$\left(\left(\sum_{m=1}^{\infty} e_m \otimes f_m, \sum_{n=1}^{\infty} e_n \otimes g_n\right)\right) = \sum_{m=1}^{\infty} ((f_m, g_m)).$$

It is clearly surjective.

Theorem 3.2 then implies that $\mathscr{L}_2(l^2(\mathbb{N}, F), G)$ is isometric to $l^2(\mathbb{N}, \mathscr{L}_2(F, G))$ (taking $E = l^2(\mathbb{N})$). Hence $K \in \mathscr{L}_2(l^2(\mathbb{N}, F), G)$ is identified to the square summable sequence of operators $K_n \in \mathscr{L}_2(F, G)$ defined by $K_n f = K(e_n \otimes f)$. ∎

6. THE HILBERT TENSOR PRODUCT BY L^2

Let F be a Hilbert space, Ω an open set in $\mathbb{R}^n$ with the Lebesgue measure $d\omega$. (What follows is valid for any other measure space $(\Omega, d\mu(\omega))$.)

We define $L^2(\Omega, F)$ as the space of classes (for the equivalence relation; y is equivalent to z if $y(\omega) - z(\omega) = 0$ almost everywhere) of measurable functions $\omega \to y(\omega)$ with values in F such that

(1) $$\|y\| = \left(\int_\Omega \|y(\omega)\|_F^2 \, d\omega\right)^{1/2} < +\infty.$$

This is a Hilbert space for the scalar product

(2) $$((y, z)) = \int_\Omega ((y(\omega), z(\omega))) \, d\omega.$$

The bilinear form

(3) $$\langle f, z \rangle = \int_\Omega \langle f(\omega), z(\omega) \rangle \, d\omega$$

is a duality pairing on the spaces $L^2(\Omega, F^*)$ and $L^2(\Omega, F)$.

We are going to show that $L^2(\Omega, F)$ is the Hilbert tensor product of $L^2(\Omega)$ and F and use Theorem 3.2.

THEOREM 1

Let F be a Hilbert space. Then the space $L^2(\Omega, F)$ is isometric to the Hilbert tensor product $L^2(\Omega) \hat{\otimes} F$ of $L^2(\Omega)$ by F.

Every Hilbert-Schmidt operator A from $L^2(\Omega, F)$ to G is defined by a square summable function $\omega \mapsto A(\omega) \in \mathscr{L}_2(F, G)$ of Hilbert-Schmidt operators $A(\omega)$ from F to G:

(4) $$A\varphi = \int_\Omega A(\omega) \varphi(\omega) \, d\omega \qquad \forall \varphi \in L^2(\Omega, F).$$ ▲

Proof. We consider an orthonormal base $\{\varphi_n\}$ of $L^2(\Omega)$ and an orthonormal base $\{f_j\}$ of F. The mapping θ from $L^2(\Omega) \hat{\otimes} F$ to $L^2(\Omega, F)$ defined by

(5) $$\theta\left(\sum_{m,j=1}^\infty a_{mj} \varphi_m \otimes f_j\right) = \omega \to \sum_{m,j=1}^\infty a_m^j \varphi_m(\omega) f_j$$

is an isometry, since

$$\int_\Omega \left(\left(\sum_{m,j} a_m^j \varphi_m(\omega) f_j, \sum_{n,k} b_n^k \varphi_n(\omega) f_k\right)\right) d\omega$$
$$\sum_{m,j,n,k} \int_\Omega a_m^j b_n^k ((f_j, f_k)) \varphi_m(\omega) \varphi_n(\omega) \, d\omega$$
$$= \sum_{m,j} a_m^j b_m^j.$$

Hence $\theta(L^2(\Omega) \hat{\otimes} F)$ is closed in $L^2(\Omega, F)$. It suffices for us to show that it is dense in $L^2(\Omega, F)$. Therefore let $\psi \in L^2(\Omega, F^*) = L^2(\Omega, F)^*$ be a square summable function with values in the dual space F^* of F that is identically zero on $\theta(L^2(\Omega) \hat{\otimes} F)$:

(6) $$\forall m, \forall j, \quad \int_\Omega \langle \psi(\omega), \theta(\varphi_m \otimes f_j) \rangle \, d\omega = \int \varphi_m(\omega) \langle \psi(\omega), f_j \rangle \, d\omega = 0$$

According to the density criterion (Theorem 2.2.1), we must show that $\psi(\omega)$ is zero (almost everywhere). But for every j the function $\omega \mapsto \langle \psi(\omega), f_j \rangle$ is zero almost everywhere; hence $\langle \psi(\omega), f_j \rangle = 0$, except possibly on a subset $\Omega_j \subset \Omega$ of measure zero.

Let $\Omega_\infty = \bigcup_{j=1}^\infty \Omega_j$ be the countable union of the sets Ω_j of measure zero, which is of measure zero. Then for every ω which does not belong to Ω_∞,

(7) $$\psi(\omega) = \sum_{j=1}^{\infty} \langle \psi(\omega), f_j \rangle f_j = 0.$$

Therefore $\psi(\omega) = 0$ almost everywhere, and we have shown that θ is surjective.

Applying Theorem 3.2 with $E = L^2(\Omega)$ and using the preceding results, we find that the spaces $\mathscr{L}_2(L^2(\Omega, F), G)$ and $L^2(\Omega, \mathscr{L}_2(F, G))$ are isomorphic. We are going to give explicitly the isomorphism that associates to the operator $A \in \mathscr{L}_2(L^2(\Omega, F), G)$ the square summable function $\omega \mapsto A(\omega) \in \mathscr{L}_2(F, G)$.

First of all we associate to $A \in \mathscr{L}_2(L^2(\Omega) \hat{\otimes} F, G)$ the operators $\hat{A}(\varphi) \in \mathscr{L}_2(F, G)$ defined by

(8) $$\hat{A}(\varphi) y = A(\varphi \otimes y) \qquad \forall \varphi \in L^2(\Omega), \quad \forall y \in F.$$

Let $\{f_j\}$ and $\{g_k\}$ be orthonormal bases of F and G. Then the components of the Hilbert-Schmidt operator $\hat{A}(\varphi)$ in the base $\{f_j^* \otimes g_k\}_{j,k}$ are defined by the $u_{jk}(\varphi) = \langle \hat{A}(\varphi) f_j, g_k^* \rangle$:

(9) $$\hat{A}(\varphi) y = \sum_{j,k=1}^{\infty} u_{jk}(\varphi) \langle f_j^*, y \rangle g_k.$$

But $\varphi \mapsto u_{jk}(\varphi)$ is a continuous linear form on $L^2(\Omega)$; hence there exists a function $a_{jk} \in L^2(\Omega)$ such that

$$u_{jk}(\varphi) = \int_\Omega a_{jk}(\omega) \varphi(\omega) \, d\omega.$$

Consequently, $\hat{A}(\varphi) y$ can be written

(10) $$\hat{A}(\varphi) y = \sum_{j,k=1}^{\infty} \int_\Omega a_{jk}(\omega) \langle f_j^*, y \rangle g_k \varphi(\omega) \, d\omega.$$

We associate to almost every ω the Hilbert-Schmidt operator $A(\omega)$ with components $a_{jk}(\omega)$:

(11) $$A(\omega) y = \sum_{j,k=1}^{\infty} a_{jk}(\omega) \langle f_j^*, y \rangle g_k.$$

Hence we can write

(12) $$\hat{A}(\varphi) y = \int_\Omega A(\omega) y \varphi(\omega) \, d\omega = A(\varphi \otimes y) \in F. \qquad \blacksquare$$

The Kernel Theorem

Theorem 1 is going to imply the kernel theorem, the converse of Proposition 1.

THEOREM 2

An operator $\tilde{K}$ from $L^2(\Omega_1)$ to $L^2(\Omega_2)$ is a Hilbert-Schmidt operator if and only if it is associated with a kernel $K \in L^2(\Omega_1 \times \Omega_2)$. ▲

Proof. Indeed, the space $\mathscr{L}_2(L^2(\Omega_1), L^2(\Omega_2))$ of Hilbert-Schmidt operators from $L^2(\Omega_1)$ to $L^2(\Omega_2)$ is equal to $L^2(\Omega_1) \hat{\otimes} L^2(\Omega_2)$ (since $L^2(\Omega_1)$ is identified with its dual space). According to Theorem 1, it is isomorphic to the space $L^2(\Omega_1, L^2(\Omega_2))$ (or to the space $L^2(\Omega_2, L^2(\Omega_1))$). But by the Fubini theorem the space $L^2(\Omega_1, L^2(\Omega_2))$ is isomorphic to the space $L^2(\Omega_1 \times \Omega_2)$ of square summable functions on $\Omega = \Omega_1 \times \Omega_2$. ∎

7. THE TENSOR PRODUCT BY THE SOBOLEV SPACE H^m

We are going to use the results of the preceding paragraphs to characterize the Hilbert tensor product $H^m(\Omega) \hat{\otimes} F$ of $H^m(\Omega)$ and F as a Sobolev space of vector-valued functions.

Thus let Ω be an open set in $\mathbb{R}^n$. Let us recall (Theorem 6.1) that every vector-valued function $M \in L^2(\Omega, F) = \mathscr{L}_2(L^2(\Omega), F)$ can be considered as a Hilbert-Schmidt operator:

(1) $$M : \varphi \in L^2(\Omega) \mapsto M(\varphi) \in F,$$

where the elements of the domain of M are called *test functions*.

We are going to define the *derivatives* D^k of $M \in L^2(\Omega, F)$ if $|k| \leq m$ as follows: $D^k M$ is the Hilbert-Schmidt operator

(2) $$D^k M : \varphi \in H_0^m(\Omega) \mapsto M((-1)^{|k|} D^k \varphi) \in F.$$

It is clear that formula (2) is meaningful. We find again the derivative in the sense of distributions in the case where $F = \mathbb{R}$, since $M \in L^2(\Omega, \mathbb{R})$ is indeed a linear form $\varphi \to M(\varphi) = \int_\Omega M(\omega) \varphi(\omega) \, d\omega$ and since its derivative $D^k M$ is indeed defined by

$$\varphi \in H_0^m(\Omega) \to M((-1)^{|k|} D^k \varphi) = (-1)^{|k|} \int_\Omega M(\omega) D^k \varphi(\omega) \, d\omega.$$

[See Chapter 7, Section 3, formula (1) if Ω is an interval of $\mathbb{R}$; chapter 9, section 1, formula (8) if Ω is an open set in $\mathbb{R}^n$.] Hence $D^k M \in \mathscr{L}_2(H_0^m(\Omega), F) = H^{-m}(\Omega) \hat{\otimes} F$.

It is then natural to define the *Sobolev space $H^m(\Omega, F)$ as the subspace of vector-valued functions $M \in L^2(\Omega, F)$ all of whose derivatives $D^k M \in L^2(\Omega, F)$ for $|k| \leq m$*, with the norm of the graph:

(3) $$\|M\|_{H^m(\Omega, F)} = \left(\sum_{|k| \leq m} \|D^k M\|_{L^2(\Omega, F)}^2 \right)^{1/2}.$$

We then deduce from Theorem 6.1 the following characterization of Sobolev spaces of vector-valued functions.

THEOREM 1
The Sobolev space $H^m(\Omega, F)$ is isometric to the Hilbert tensor product $H^m(\Omega) \hat{\otimes} F$ of $H^m(\Omega)$ and F. For every pair of Hilbert spaces F and G, we have

(4) $$\mathscr{L}_2(H^m(\Omega, F^*)^* ; G) = H^m(\Omega, \mathscr{L}_2(F, G)).$$ ▲

Proof. We are going to apply Theorem 4.1 with $H = L^2(\Omega)$, $U = L^2(\Omega)^{\|m\|}$ and $V = H^{-m}(\Omega)^{\|m\|}$ where $\|m\|$ is the number of n-tuples k of integers such that $|k| \leq m$.

Consider the operator $A \in \mathscr{L}(L^2(\Omega), H^{-m}(\Omega)^{\|m\|})$ defined by

(5) $$\forall \varphi \in L^2(\Omega), \quad A\varphi = \{D^k \varphi\}_{|k| \leq m} \in H^{-m}(\Omega)^{\|m\|}.$$

Then $H^m(\Omega) = D(A)$ is indeed the domain of A.

Let us recall that the transpose $A^* \in \mathscr{L}(H_0^m(\Omega)^{\|m\|}, L^2(\Omega))$ is defined by

(6) $$\vec{\varphi} = \{\varphi_k\}_{|k| \leq m} \mapsto A^* \varphi = \sum_{|k| \leq m} (-1)^{|k|} D^k \varphi_k \in L^2(\Omega).$$

Then consider the operator
$$\vec{A} = A \otimes 1_F \in \mathscr{L}(L^2(\Omega) \hat{\otimes} F, H^{-m}(\Omega)^{\|m\|} \hat{\otimes} F).$$

If $M \in L^2(\Omega, F) = L^2(\Omega) \otimes F$, $\vec{A}M$ is defined by

(7) $$\vec{\varphi} = \{\varphi_k\}_{|k| \leq m} \in H_0^m(\Omega)^{\|m\|} \to \vec{A}M(\vec{\varphi})$$
$$= \sum_{|k| \leq m} (-1)^{|k|} M(D^k \varphi_k) = \sum_{|k| \leq m} (D^k M)(\varphi_k) \in F$$

[according to definition (2) of $D^k M$].

In other words, $\vec{A}$ is the linear operator that associates to every $M \in L^2(\Omega, F)$ the sequence

$$\vec{A}M = \{D^k M\}_{|k| \leq m} \in (H^{-m}(\Omega) \hat{\otimes} F)^{\|m\|} = H^{-m}(\Omega)^{\|m\|} \hat{\otimes} F.$$

Hence $H^m(\Omega, F) = D(\vec{A}, F)$ is indeed the domain of $\vec{A}$. Thus Theorem 1 follows from Theorem 6.1. ∎

From this result we deduce a kernel theorem characterizing the Hilbert-Schmidt operators from the dual space $H^m(\Omega_1)^*$ of a Sobolev space to a Sobolev space $H^p(\Omega_2)$.

THEOREM 2
Let Ω_1 and Ω_2 be two open sets in $\mathbb{R}^{n_1}$ and $\mathbb{R}^{n_2}$, respectively. Then $H^m(\Omega_1, H^p(\Omega_2))$ is isometric to the subspace $H^{m,p}(\Omega_1 \times \Omega_2)$ of functions

$M(.,.) \in L^2(\Omega_1 \times \Omega_2)$ whose derivatives $D_1^k D_2^l M(.,.)$ (in the sense of distributions) belong to $L^2(\Omega_1 \times \Omega_2)$ for $|k| \leqq m$ and $|l| \leqq p$.

Every Hilbert-Schmidt operator $\tilde{K}$ from $H^m(\Omega_1)^*$ to $H^p(\Omega_2)$ is associated with a kernel

$$K \in H^{m,p}(\Omega_1 \times \Omega_2) : (\tilde{K}f)(\omega_2) = (f, K(.,\omega_2)).$$

for all $f \in H^m(\Omega_1)^*$. ▲

Proof. Let $M \in H^m(\Omega_1, H^p(\Omega_2))$. This is the space of functions $M \in L^2(\Omega_1, H^p(\Omega_2))$ such that $D_1^k M \in L^2(\Omega, H^p(\Omega_2))$. But to say that $D_1^k M \in L^2(\Omega_1, H^p(\Omega_2))$ amounts to saying that for every $\varphi \in H_0^m(\Omega_1)$,

$$(D_1^k M)(\varphi) = (-1)^{|k|} \int_{\Omega_1} M(\omega_1,.) D^k \varphi(\omega_1) \, d\omega_1 = \int_{\Omega_1} D_1^k M(\omega_1,.) \varphi(\omega_1) \, d\omega_1$$

belongs to $H^p(\Omega_2)$. Consequently, $H^m(\Omega_1, H^p(\Omega_2))$ is indeed isometric to $H^{m,p}(\Omega_1 \times \Omega_2)$.

Theorem 1, with $F = \mathbb{R}$ and $G = H^p(\Omega)$, implies that

$$\mathscr{L}_2(H^m(\Omega_1)^*, H^p(\Omega_2)) = H^m(\Omega_1, \mathscr{L}_2(\mathbb{R}, H^p(\Omega_2))) = H^m(\Omega_1, H^p(\Omega_2))$$
$$= H^{m,p}(\Omega_1 \times \Omega_2). \blacksquare$$

CHAPTER 13

Boundary Value Problems

We devote this chapter to an elementary study of boundary value problems for second order elliptic differential equations, to a brief introduction to the calculus of variations (which motivates the study of boundary value problems), and to optimal control theory for linear differential systems. In fact each of these subjects requires one or several books in order to develop its theory. Nevertheless, by adopting an abstract approach and limiting our objectives to the essential theorems, we can use the results we established previously to give the reader a precise idea of these theories.

Our essential tool is the *abstract form of Green's formula*. We have already observed the following situation: The closure $V_0 = H_0^1(\Omega)$ of $\mathscr{D}(\Omega)$ in the Sobolev space $V = H^1(\Omega)$ is distinct from V and when the open set Ω with boundary Γ is "regular," $V_0 = H_0^1(\Omega)$ is the kernel of the trace operator γ_0 from $V = H^1(\Omega)$ onto $T = H^{1/2}(\Gamma)$ (the trace theorem). Finally, this kernel is sufficiently large in the sense that $V = H^1(\Omega)$ and $V_0 = H_0^1(\Omega)$ are both contained and dense in $H = L^2(\Omega)$.

In other words, we are in the situation where once and for all the spaces V, H, and T and $\gamma \in \mathscr{L}(V, T)$ are given such that

   i. $V \subset H$ with a finer topology.
  ii. $V_0 = \mathrm{Ker}\, \gamma$ is dense in H.
 iii. $\gamma \in \mathscr{L}(V, T)$ is surjective.

Then we associate to every operator $A \in \mathscr{L}(V, E^*)$ its restriction $A_0 \in \mathscr{L}(V_0, E^*)$ and their transposes $A^* \in \mathscr{L}(E, V^*)$ and $A_0^* \in \mathscr{L}(E, V_0^*)$. As we have already seen for differential operators on Sobolev spaces, it is easier to use A_0^* than A^*. We say then that A_0^* *is the formal adjoint of A* and that

$$E(A_0^*) = \{e \in E \quad \text{such that} \quad A_0^* e \in H\}$$

is its domain. In using A^* we pass from the properties of A to those of A^* by the formula $[e, Ax] - (A^*x, e) = 0$ if $x \in V$ and $e \in E$. In using A_0^* we need an analogous formula; hence we prove *Green's formula*. We can associate to A (and to γ) a *unique* operator $\beta^* \in \mathscr{L}(E(A_0^*), T^*)$ such that

(1) $$[e, Ax] - (A_0^* e, x) = \langle \beta^* e, \gamma x \rangle$$

holds for all $x \in V$ and $e \in E(A_0^*)$.

Thus it is clear that this formula will play a crucial role every time that we use not the transpose A^* of A, but its formal adjoint A_0^* and the operator β^*.

We prove in Section 2 a form of Green's formula for a continuous bilinear form a on $V \times V$; we associate to a the formal operator $\Lambda \in \mathscr{L}(V, V_0^*)$ defined by

$$(\Lambda x, y) = a(x, y) \quad \text{for all} \quad y \in V_0$$

and its domain $V(\Lambda) = \{x \in V \text{ such that } \Lambda x \in H\}$. Then we prove that there exists a unique operator $\delta \in \mathscr{L}(V(\Lambda), T^*)$ such that

$$a(x, y) = (\Lambda x, y) + \langle \delta x, \gamma y \rangle \quad \text{if} \quad x \in V(\Lambda), \quad y \in V.$$

In Section 3 we deduce that when the bilinear form a is V-elliptic, the operators

$$\Lambda \times \delta : V(\Lambda) \mapsto H \times T^* \quad \text{and} \quad \Lambda \times \gamma : V(\Lambda) \mapsto H \times T$$

are isomorphisms. In other words, the Neumann problem

$$\Lambda x = f, \quad \delta x = \varphi$$

and the Dirichlet problem

$$\Lambda x = f, \quad \gamma x = \psi$$

(as well as many other boundary value problems) have unique solutions.

When the form a is symmetric, we can even use the Fredholm alternative of Chapter 11, Section 4.

We shall apply these results in order to study some concrete boundary value problems for concrete differential operators.

Although we are assured of the existence and uniqueness of the solution of an elliptic Neumann problem, we have nevertheless developed no methods that would enable us to calculate it. Hence we present in Section 5 a brief introduction to methods that allow us to approximate the solution of a Neumann problem by solutions of systems of a finite (though large) number of linear equations that can be solved using computers.

Section 6 contains some complementary results on the properties of the operators $\Lambda \times \gamma$ and $\Lambda \times \delta$ that may be omitted during a first reading.

We take up in Section 7 the study of unilateral boundary value problems:

$$\Lambda x = f, \quad \gamma x \leq \varphi, \quad \delta x \leq \psi, \quad \text{and} \quad \langle \gamma x - \varphi, \delta x - \psi \rangle = 0,$$

showing that they are equivalent to certain variational inequalities and using the results of Chapter 3, Section 7.

The problems of *calculus of variations* are of the following form: to find

$\bar{x} \in V$ such that

$$v = [\mathscr{F}(\bar{x}, A\bar{x}) + f(\gamma\bar{x})] = \min_{x \in V}[\mathscr{F}(x, Ax) + f(\gamma x)]$$

where $\mathscr{F}$ is a convex lower semicontinuous function from $H \times E^*$ to $]-\infty, +\infty]$. Hence these are problems analogous to the minimization problems treated in Chapter 10, Sections 4 and 5. We associate with them a dual problem; to find $\bar{p} \in E(A_0^*)$ such that

$$-v = \mathscr{F}^*(-A_0^*\bar{p}, \bar{p}) + f^*(-\beta^*\bar{p}) = \min_{p \in E(A_0^*)}[\mathscr{F}^*(-A_0^*p, p) + f^*(-\beta^*p)].$$

Then we show that $\bar{x}$ and $\bar{p}$ are optimal solutions if and only if

$$\{-A_0^*\bar{p}, \bar{p}\} \in \partial\mathscr{F}(\bar{x}, A\bar{x}) \quad \text{and} \quad -\beta^*\bar{p} \in \partial f(\gamma\bar{x})$$

or, indeed, in the Hamiltonian form, if and only if

(2) $\begin{cases} \text{i.} & A\bar{x} \in \partial_p \mathscr{H}(\bar{x}, \bar{p}). \\ \text{ii.} & A_0^*\bar{p} \in -\partial_x(-\mathscr{H}(\bar{x}, \bar{p})). \\ \text{iii.} & -\beta^*\bar{p} \in \partial f(\gamma\bar{x}). \end{cases}$

It is in this way that *we obtain most of the boundary value problems*. We take $V = H^1(\Omega)$, $E = L^2(\Omega)^n$ and as the operator, the operator $Ax = \text{grad}\, x$, whose formal adjoint is $A_0^*p = -\text{div}\, p$. We set

$$\mathscr{F}(x, p) = \frac{1}{2}\int_\Omega \left(|x(\omega)|^2 + \sum_{i=1}^n (|p_i(\omega)|^2)\right) d\omega.$$

Equations (2) become

$$\text{grad}\, \bar{x} = \bar{p}, \quad \text{div}\, \bar{p} = \bar{x}, \quad \text{and} \quad -\beta^*\bar{p} \in \partial f(\gamma\bar{x}),$$

that is, the boundary value problem

$$-\text{div}\cdot\text{grad}\, \bar{x} + \bar{x} = -\Delta\bar{x} + \bar{x} = 0 \quad \text{and} \quad -\beta^* \text{grad}\, \bar{x} \in \partial f(\gamma_0 \bar{x})$$

for the operator $-\Delta + 1$.

The problems of optimal control, to which we shall give a brief look, are analogous; they consist in seeking the pair $\{\bar{x}, \bar{u}\}$ (state-control) minimizing a cost function $\{x, u\} \mapsto \mathscr{F}(x, u) + f(\gamma x)$ under the constraints $Ax = Bu$ describing a system where the state x is determined by the control u.

1. THE FORMAL ADJOINT OF AN OPERATOR AND GREEN'S FORMULA

Suppose we are given

(1) \qquad a pivot space H, $\mathscr{D} \subset H$ dense in H

and

(2) $\begin{cases} \text{i.} & \text{a vector space } V \text{ embedded in } H. \\ \text{ii.} & \mathcal{D} \subset V \subset H. \end{cases}$

Let V_0 be the closure of $\mathcal{D}$ in V, that is the associated normal space. (See Definition 5.3.1.)

If $A \in \mathcal{L}(V, E^*)$ is a continuous linear operator from V to the dual E^* of a Hilbert space E, we denote by $A_0 = A|_{V_0} \in \mathcal{L}(V_0, E^*)$ the restriction of A to V_0. We obtain by transposition

(3) $\qquad A^* \in \mathcal{L}(E, V^*), \qquad A_0^* \in \mathcal{L}(E, V_0^*).$

We are going to "compare" these two operators.

Since V_0 is a normal space, we identify H with a dense subspace of V_0^*:

(4) $\qquad \mathcal{D} \subset V_0 \subset H \subset V_0^*.$

(See Chapter 5, Section 3.)

In order not to confuse A^* and A_0^* we introduce the following definition.

DEFINITION 1

Let V be a Hilbert space satisfying (2), V_0 *the associated normal space,* $A \in \mathcal{L}(V, E^*)$ *and* $A_0 \in \mathcal{L}(V_0, E^*)$. *We say that*

(5) $\qquad A_0^* \in \mathcal{L}(E, V_0^*)$

defined by

(6) $\qquad (A_0^* e, \varphi) = [e, A\varphi] \qquad \forall e \in E, \quad \forall \varphi \in V_0$

is the formal adjoint *of A.* ▲

To compare $A^* \in \mathcal{L}(E, V^*)$ and $A_0^* \in \mathcal{L}(E, V_0^*)$, we are going to "shift" the operator A^* so that it has its values in H rather than in V_0^* and "direct" it to V^*. More precisely, if j and j_0 denote the canonical injections from V and V_0 to H, their transposes $j^* \in \mathcal{L}(H, V^*)$ and $j_0^* \in \mathcal{L}(H, V_0^*)$ are both embeddings, where j_0^* only is identified with the canonical injection. Hence we are in the following situation:

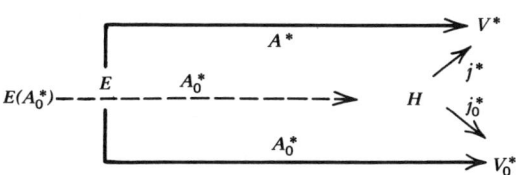

DEFINITION 2
We say that the subspace $E(A_0^*)$ defined by

(7) $$E(A_0^*) = \{e \in E \text{ such that } A_0^* e \in H\}$$

with the scalar product

(8) $$((e,f))_{E(A_0^*)} = ((e,f))_E + (A_0^* e, A_0^* f)$$

is the domain of the operator A_0^*. (See Chapter 5, Section 5, if necessary.) ▲

We show that $E(A_0^*)$ is a Hilbert space at the end of this section (Theorem 4, following).

First of all, we are going to *compare* the operators A^* and A_0^*, which is possible since the operators $A_0^* j^*$ and A^* are both defined on $E(A_0^*)$ and have their values in V^*.

The following lemma gives us a first method for comparing A_0^* and A^*.

LEMMA 1
Let j be the injection from V to H. Then the image of $E(A_0^*)$ under the operator

$$A^* - j^* A_0^* \in \mathscr{L}(E(A_0^*), V^*)$$

is contained in $V_0^\perp$. ▲

Proof. Indeed, we take $e \in E(A_0^*)$ and $x \in V_0$. Then

$$(j^* A_0^* e - A^* e, x) = (A_0^* e, jx) - [Ax, e]$$
$$= (A_0^* e, x) - [Ax, e] = [A_0 x - Ax, e] = 0,$$

since $jx = x$ and $Ax = A_0 x$ when $x \in V_0$. ∎

We are going to complete this formula when V_0 is considered as the kernel of a surjective operator from V onto a Hilbert space T.

Hence suppose that

(9) $\begin{cases} \text{i.} & \gamma \in \mathscr{L}(V, T) \text{ where } T \text{ is a Hilbert space and where } \gamma \text{ is surjective.} \\ \text{ii.} & V_0 = \operatorname{Ker} \gamma. \end{cases}$

THEOREM 1
Suppose that the normal space V_0 associated to V is the kernel of a surjective operator γ from V onto a Hilbert space T. Then there exists a unique operator $\beta^* \in \mathscr{L}(E(A_0^*), T^*)$ (depending on A and on γ) such that the abstract form of Green's formula

(10) $$[e, Ax] - (A_0^* e, x) = \langle \beta^* e, \gamma x \rangle_{T^* \times T}$$

holds for all $e \in E(A_0^*)$, for all $x \in V$. ▲

CH. 13, SEC. 1   THE FORMAL ADJOINT OF AN OPERATOR   285

Proof. Let $\sigma \in \mathscr{L}(T, V)$ be a right inverse of γ (which exists according to Proposition 4.6.1). Then $\sigma\gamma$ is a projector whose kernel is V_0. Its transpose $\gamma^*\sigma^*$ is a projector from V^* onto $V_0^\perp = (\operatorname{Ker} \sigma\gamma)^\perp$. We set

(11) $$\beta^* = \sigma^*(A^* - j^*A_0^*) \in \mathscr{L}(E(A_0^*), T^*)$$

because $\sigma^* \in \mathscr{L}(V^*, T^*)$. Since $(j^*A_0^* - A^*)e \in V_0^\perp$ when $e \in E(A_0^*)$, according to Lemma 1, it follows that

$$(A^* - j^*A_0^*)e = \gamma^*\sigma^*(A^* - j^*A_0^*)e = \gamma^*\beta^*e,$$

and, consequently, that for all $x \in V$,

$$[e, Ax] - (A_0^*e, x) = \langle A^*e - j^*A_0^*e, x \rangle_{V^* \times V}$$
$$= \langle \gamma^*\beta^*e, x \rangle_{V^* \times V} = \langle \beta^*e, \gamma x \rangle_{T^* \times T}.$$

It remains to be proved that the operator β^* does not depend on the choice of the right inverse σ of γ.

Indeed, if $\beta^* \in \mathscr{L}(E(A^*), T^*)$ satisfies Green's formula (10), we can write for all $e \in E(A_0^*)$

(12) $$A^*e - j^*A_0^*e = \gamma^*\beta^*e.$$

If σ is an arbitrary right inverse of γ, σ^* is a left inverse of γ^* and, consequently, (12) implies that

$$\beta^*e = \sigma^*\gamma^*e = \sigma^*(A^* - j^*A_0^*)e$$

for any σ.   ∎

Remark 1

Since V_0 is a closed subspace of the Hilbert space V, we can always take as the space T the orthogonal complement of V_0 and for γ the orthogonal projector from V onto $V_0^\oplus = T$, whose kernel is V_0.   ∎

If we take $V = H^m(\Omega)$ and $V_0 = H_0^m(\Omega)$, the closure of $\mathscr{D}(\Omega)$ in $H^m(\Omega)$, the trace theorems of Chapters 7 and 9 imply that hypothesis (9) is satisfied and that we can prove Green's formula.

Example 1.   Formula for Integration by Parts

We are going to prove that we can "integrate by parts" in the Sobolev spaces of functions of one variable thanks to the abstract form of Green's formula of Theorem 1.

This is possible since we have a trace theorem. Let $\Omega = \,]a, b[$ be an interval of $\mathbb{R}$. Consider the operator $G^m \in \mathscr{L}(H^m(\Omega), L^2(\Omega)^{m+1})$ defined by

(13) $$G^m x = \{x, Dx, \ldots, D^p x, \ldots, D^m x\}.$$

Since $\mathscr{D} = \mathscr{D}(\Omega)$ is dense in $H_0^m(\Omega)$, $V_0 = H_0^m(\Omega)$ is a normal space that is the kernel of the operator $\gamma \in \mathscr{L}(H^m(\Omega), \mathbb{R}^{2m})$ defined by

(14) $$\gamma x = \{\gamma_0 x, \ldots, \gamma_{m-1} x\} \in \mathbb{R}^{2m}$$

where

(15) $$\gamma_k x = \{D^k x(a), D^k(b)\} \in \mathbb{R}^2.$$

If G_0^m denotes the restriction to $H_0^m(\Omega)$ of G^m, we have seen that its transpose $G_0^{m*} \in \mathscr{L}(L^2(\Omega)^{m+1}, H^{-m}(\Omega))$ is defined by

(16) $$G_0^{m*} e = \sum_{k=0}^{m} (-1)^k D^k e^k \quad \text{where} \quad e = \{e^k\}_{0 \leq k \leq m} \in L^2(\Omega)^{m+1}.$$

We denote by

(17) $$H(\Omega, G_0^{m*}) = \{e \in L^2(\Omega)^{m+1} \text{ such that } G_0^{m*} e \in L^2(\Omega)\}$$

the domain of G_0^{m*}, with the norm of the graph. ∎

THEOREM 2
There exist unique operators $\beta_j^* \in \mathscr{L}(H(\Omega, G_0^{*m}), \mathbb{R}^2)$ $(0 \leq j \leq m-1)$ such that

(18) $$\sum_{k=0}^{m} \int_a^b D^k x(\omega) e^k(\omega) \, d\omega = \int_a^b \left(\sum_{k=0}^{m} (-1)^k D^k e^k(\omega) \right) x(\omega) \, d\omega + \sum_{j=0}^{m-1} \langle \beta_{m-j-1}^*, \gamma_j x \rangle$$

for all $x \in H^m(\Omega)$, for all $e \in H(\Omega, G_0^{m*})$. ▲

Proof. This result is a consequence of Theorem 1 with
$$V = H^m(\Omega), \quad V_0 = H_0^m(\Omega), \quad H = L^2(\Omega), \quad T = \mathbb{R}^{2m},$$
$$\gamma = (\gamma_0, \ldots, \gamma_{m-1}), \quad E = L^2(\Omega)^{m+1}, \quad A = G^m, \quad A_0^* = G_0^{m*};$$
there exists a unique operator $\beta^* \in \mathscr{L}(H(\Omega, G_0^{m*}), \mathbb{R}^{2m})$, which we write as
$$\beta^* = \{\beta_{m-1}^*, \ldots, \beta_{m-j-1}^*, \ldots, \beta_0^*\} \quad \text{where} \quad \beta_j^* \in \mathscr{L}(H(\Omega, G_0^{m*}); \mathbb{R}^2). \blacksquare$$

If we restrict the functions x and the components e^k of e to be infinitely differentiable functions, the usual formulas for integration by parts of order k imply that

(19) $$\beta_j^* e = \left\{ -\sum_{l=0}^{j} (-1)^{m-j} D^{j-l} e^{m-l}(a), \sum_{l=0}^{j} (-1)^{m-j} D^{j-l} e^{m-l}(b) \right\}.$$

Indeed, the formula for integration by parts of order k implies that if $k \geq 1$,

$$\int_a^b e^k \cdot D^k x \, d\omega = (-1)^k \int_a^b D^k e^k \cdot x \, dx + \sum_{j=0}^{k-1} (-1)^j D^{k-1-j} e^k \cdot D^j x \Big]_a^b.$$

Adding these equalities from $k = 0$ to m, we obtain

$$\sum_{k=0}^m \int_a^b D^k x \cdot e^k \, d\omega - \int_a^b \left(\sum_{k=0}^m (-1)^k D^k e^k \right) x \, d\omega$$

$$= \sum_{k=0}^m \sum_{j=0}^{k-1} (-1)^j D^{k-1-j} e^k \cdot D^j x \Big]_a^b$$

$$= \sum_{j=0}^{m-1} D^j x \sum_{k=j+1}^m (-1)^j D^{k-1-j} e^k \Big]_a^b$$

$$= \sum_{j=0}^{m-1} \left(\sum_{l=0}^{m-j-1} (-1)^j D^{m-j-1-l} e^{m-l}(b) \right) D^j x(b)$$

$$- \sum_{j=0}^{m-1} \left(\sum_{l=0}^{m-j-1} (-1)^j D^{m-j-1-l} e^{m-l}(a) \right) D^j x(a). \quad \blacksquare$$

Remark 2

Since the operators β_j^* satisfying the formula for integration by parts (18) are *unique*, we write the formulas (19) even when $e \in H(\Omega, G_0^{m*})$. $\blacksquare$

Example 2. (The Formula of Ostrogradski)

We consider the Sobolev space $V = H^1(\Omega)$ as the domain of the operator $G = \text{grad}$ (gradient):

(20) $\quad H^1(\Omega) = \{x \in L^2(\Omega) \text{ such that } \text{grad } x = \{D_1 x, \ldots, D_n x\} \in L^2(\Omega)^n\}.$

We denote by $G_0 = G|_{H_0^1(\Omega)}$ the restriction of the gradient G to $H_0^1(\Omega)$. We verify without difficulty that the formal adjoint G_0^* of G is the operator $-\text{div}$ (divergence):

(21) $\quad G_0^* e = - \sum_{i=1}^n D_i e_i = - \text{div } e \in H^{-1}(\Omega),$

since by definition of the derivatives in the sense of distributions of the functions of $L^2(\Omega)$, we have

$$(G_0^* e, x) = [e, G_0 x] = \sum_{i=1}^n \int_\Omega e_i(\omega) D_i x(\omega) \, d\omega$$

$$= - \sum_{i=1}^n \int_\Omega (D_i e_i)(\omega) x(\omega) \, d\omega = - (\text{div } e, x)$$

for all $x \in H_0^1(\Omega)$.

We denote by

(22) $\quad H(\Omega, \text{div}) = \{e \in L^2(\Omega)^n \quad \text{such that} \quad \text{div } e \in L^2(\Omega)\}$

the domain of the operator div.

Suppose now that Ω is "regular." Theorem 9.5.1 implies that

(23) $\quad H^1_0(\Omega) = \text{Ker } \gamma$ where $\gamma = \gamma_0 \in \mathscr{L}(H^1(\Omega), H^{1/2}(\Gamma))$ is surjective. ∎

THEOREM 3

Suppose that Ω is regular. Then there exists a unique operator $\beta^* = \beta_1^* \in \mathscr{L}(H^1(\Omega, \text{div}), H^{-1/2}(\Gamma))$ such that

(24) $\qquad [\text{grad } x, e] + (\text{div } e, x) = \langle \beta_1^* e, \gamma_0 x \rangle$

for all $x \in H^1(\Omega)$ and all $e \in H^1(\Omega, \text{div})$. ▲

Proof. We apply Theorem 1 with

$$V = H^1(\Omega), \quad V_0 = H^1_0(\Omega), \quad E = L^2(\Omega)^n, \quad G = \text{grad},$$
$$G_0^* = -\text{div}, \quad E(A_0^*) = H(\Omega, \text{div}), \quad T = H^{1/2}(\Gamma),$$
$$T^* = H^{-1/2}(\Gamma), \quad \text{and } \gamma = \gamma_0.$$
∎

If we suppose in addition that the functions x and e_j are infinitely differentiable on an open set containing $\bar{\Omega}$, then we know that

(25) $\quad \int_\Omega \langle \text{grad } x, e \rangle \, d\omega = -\int_\Omega (\text{div } e) x \, d\omega + \int_\Gamma \langle e, v \rangle x \, d\sigma$

where $d\sigma$ is the *superficial measure* on Γ and $\langle e, v(\omega) \rangle = \sum_{j=1}^n e_j(\omega) \cdot \cos(v(\omega), \omega_j)$, $v(\omega)$ being the *normal vector to* Γ *at* ω, exterior to Ω. We also say that $\cos(v(\omega), \omega_j)$ is the *j*th *director cosine* of the normal vector v.

Comparing formulas (24) and (25) and agreeing to denote by $\int_\Gamma y(\omega) x(\omega) \, d\sigma(\omega)$ the duality pairing on $H^{-1/2}(\Gamma)$ and $H^{1/2}(\Gamma)$ [although we do not at all have the right to write $y(\omega)$ when $y \in H^{-1/2}(\Gamma)$], we obtain

(26) $\qquad \beta_1^* e = \sum_{j=1}^n e_j \cos(v, \omega_j)$

where v is the exterior normal vector to Γ. Since β_1^* is the unique operator of $\mathscr{L}(H^1(\Omega, \text{div}), H^{-1/2}(\Gamma))$ satisfying formula (24), we write (26) even when e runs over $H^1(\Omega, \text{div})$.

Remark 3

Taking for x the constant function equal to one, we obtain the formula of

Ostrogradski:

(27) $$\int_\Omega \operatorname{div} e \, d\omega = \int_\Gamma \langle e, v \rangle \, d\sigma \qquad \forall e \in H^1(\Omega, \operatorname{div}),$$

which can be interpreted by saying that the flow of a vector field on the boundary Γ of a domain Ω is equal to the integral on Ω of its divergence. ∎

Remark 4

The problem arises of studying the dependence of β^* with respect to γ when the operator $A \in \mathscr{L}(V, E^*)$ is fixed.

Let γ_1 be another right invertible operator from V onto T_1 whose kernel is equal to V_0. Then there exists an isomorphism θ from T_1 onto T such that

(28) $$\gamma = \theta \gamma_1.$$

Indeed, the operators $\tilde{\gamma} \in \mathscr{L}(V/V_0, T)$ and $\tilde{\gamma}_1 \in \mathscr{L}(V/V_0, T_1)$, derived from γ and γ_1 by taking the quotient, are isomorphisms. Hence $\theta = \tilde{\gamma} \tilde{\gamma}_1^{-1}$ is an isomorphism from T_1 onto T satisfying (28). ∎

PROPOSITION 1
Suppose that $\gamma \in \mathscr{L}(V, T)$ and $\gamma_1 \in \mathscr{L}(V, T_1)$ are two surjective operators with kernel V_0 and let $\theta \in \mathscr{L}(T_1, T)$ be the isomorphism such that $\gamma = \theta \gamma_1$. Then

(29) $$\beta_1^* = \theta^* \beta^*. \qquad \blacktriangle$$

Proof. Theorem 1 shows that for all $e \in E(A_0^*)$ and $x \in V$

$$\langle \beta_1^* e, \gamma_1 x \rangle = \langle \beta^* e, \gamma x \rangle = \langle \beta^* e, \theta \gamma_1 x \rangle = \langle \theta^* \beta^* e, \gamma_1 x \rangle.$$

Since γ_1 is surjective from V onto T_1, it follows that $\beta_1^* e = \theta^* \beta^* e$. ∎

Now we are going to show that $E(A_0^*)$ is a Hilbert space. For this we must show that its graph

(30) $$G(A_0^*) = \{e, A_0^* e\}_{e \in E(A_0^*)} \subset E \times H$$

is closed in $E \times H$.

By symmetry, we introduce the subspace

(31) $$G(A_0) = \{-A_0 x, x\}_{x \in V_0} \subset E^* \times H,$$

which is closed only if V_0 is the domain of A_0 in the sense that V_0 is complete for the norm of the graph, that is, if there exist constants c_0 and c_1 such that

(32) $$\forall x \in V_0; \quad c_1 \|x\|_{V_0}^2 \leq \|x\|_H^2 + \|Ax\|_{E^*}^2 \leq c_0 \|x\|_{V_0}^2.$$

LEMMA 2
The graph $G(A_0^*) \subset E \times H$ is the orthogonal complement of $G(A_0) \subset E^* \times H$. ▲

Proof. Indeed, since the bilinear form $[g, e] + (x, h)$ is a duality pairing on $E^* \times H$ and $E \times H$, an element $\{e, h\} \in E \times H$ is orthogonal to $G(A_0)$ if and only if

(33) $\qquad [-A_0 x, e] + (x, h) = 0 \qquad$ for all $\qquad x \in V_0$.

This implies that $h = A_0^* e$ according to the very definition of the formal adjoint. Since $h = A_0^* e \in H$, it follows that $e \in E(A^*)$ and, consequently, that $\{e, h\} = \{e, A_0^* e\} \in G(A_0^*)$. ∎

This lemma allows us to prove the following theorem.

THEOREM 4
The domain $E(A_0^*)$ of the formal adjoint A_0^* of A is a Hilbert space for the scalar product (8).

The domain $E(A_0^*)$ is dense in E if the normal space V_0 is complete for the norm of the graph:

$$(\|x\|_H^2 + \|Ax\|_{E^*}^2)^{1/2}.$$ ▲

Proof. Lemma 2 shows that $G(A_0^*) = G(A_0)^\perp$ is closed in $E \times H$ and therefore that $E(A_0^*)$ is complete. Moreover, V_0 is complete for the norm of the graph $(\|x\|_H^2 + \|Ax\|_{E^*}^2)^{1/2}$ if and only if $G(A_0)$ is closed in $E^* \times H$, that is, if and only if

(34) $\qquad G(A_0) = G(A_0)^{\perp\perp} = G(A_0^*)^\perp$

(according to Lemma 2).

Let us show then that $E(A_0^*)$ is dense: According to the density criterion (Theorem 2.2.1), we must prove that $g \in E^*$ is identically zero when g vanishes on $E(A_0^*)$. But to say that g vanishes on $E(A_0^*)$ means that $[g, e] = [g, e] + (-0, A_0^* e) = 0$ for all $e \in E(A_0^*)$, that is, that

$$\{g, 0\} \in G(A_0) = G(A_0^*)^\perp$$

and hence that $g = A_0 0 = 0$. ∎

THEOREM 5
The space $H^1(\Omega, \text{div})$ is a dense Hilbert space in $L^2(\Omega)^n$. If the open set Ω is regular, the operator β_1^* is surjective from $H^1(\Omega, \text{div})$ onto $H^{-1/2}(\Gamma)$ and its kernel $H_0^1(\Omega, \text{div})$ is the closure of $\mathcal{D}(\Omega)^n$. ▲

Proof. **a.** The first statement follows from Theorem 4.
b. Let us show that the operator β_1^* is surjective. Let $\varphi \in H^{-1/2}(\Gamma)$. Consider the form $\gamma_0^* \varphi : y \mapsto \langle \varphi, \gamma_0 y \rangle$, which is continuous and linear on $H^1(\Omega)$. Since the scalar product of $H^1(\Omega)$ is written $((x, y))_1 = [\operatorname{grad} x, \operatorname{grad} y] + (x, y)$, there exists a unique solution $x \in H^1(\Omega)$ such that

(35) $\quad [\operatorname{grad} x, \operatorname{grad} y] + (x, y) = \langle \varphi, \gamma_0 y \rangle \quad$ for all $\quad y \in H^1(\Omega)$.

Taking $y \in H_0^1(\Omega)$, we deduce that $-\operatorname{div} \operatorname{grad} x + x = 0$. Hence $e = \operatorname{grand} x$ is a solution of $\operatorname{div} e = x \in L^2(\Omega)$. Then $e \in H^1(\Omega, \operatorname{div})$. We write Green's formula (24) and comparing it with (35), we obtain $\langle \beta_1^* e, \gamma_0 y \rangle = \langle \varphi, \gamma_0 y \rangle$ for all $y \in H^1(\Omega)$. Since γ_0 is surjective, it follows that $\beta_1^* e = \varphi$. Hence β_1^* is indeed surjective.

c. Let us show that $\mathscr{D}(\Omega)^n$ is dense in $H_0^1(\Omega, \operatorname{div}) = \operatorname{Ker} \beta_1^*$. Since $\mathscr{D}(\Omega)^n$ is dense in $H_0^1(\Omega)^n$, it suffices for us to show that $H_0^1(\Omega)^n$ is dense in $H_0^1(\Omega, \operatorname{div})$. It is clear that $H_0^1(\Omega)^n$ is contained in $\operatorname{Ker} \beta_1^*$. Hence it suffices for us to show that $[H_0^1(\Omega)^n]^\perp \subset [\operatorname{Ker} \beta_1^*]^\perp = \operatorname{Im} \beta_1^{**}$ (which is closed since β_1^* is surjective). (See Theorem 4.4.2.) Hence let $l \in H^1(\Omega, \operatorname{div})^*$ vanish on $H_0^1(\Omega)^n$. It can be written

$$l(e) = [p, e] + (x, \operatorname{div} e) \quad \text{where} \quad p \in L^2(\Omega)^n \quad \text{and} \quad x \in L^2(\Omega).$$

Since $l(e) = 0$ for all $e \in H_0^1(\Omega)^n$, we have

$$0 = l(e) = [p, e] - [\operatorname{grad} x, e] \quad (\text{where } \operatorname{grad} x \in H^{-1}(\Omega)^n).$$

This implies that $\operatorname{grad} x = p \in L^2(\Omega)^n$ and, therefore, that $x \in H^1(\Omega)$. We can then apply Green's formula (24): for all $e \in H^1(\Omega, \operatorname{div})$

$$l(e) = [p - \operatorname{grad} x, e] + \langle \beta_1^* e, \gamma_0 x \rangle$$
$$= \langle \beta_1^* e, \gamma_0 x \rangle = (\beta_1^{**} \gamma_0 x, e).$$

Hence $l = \beta_1^{**} \gamma_0 x \in \operatorname{Im} \beta_1^{**} = (\operatorname{Ker} \beta_1^*)^\perp$. ∎

2. GREEN'S FORMULA FOR BILINEAR FORMS

We are going to consider the case where $E = V$ and where we are given

(1) $\quad A \in \mathscr{L}(V, V^*) \quad$ (or the continuous bilinear form $a(x, y) = \langle Ax, y \rangle$).

Let us recall that

(2) $\quad \begin{cases} \text{i.} & H \text{ is a pivot space} \\ \text{ii.} & \mathscr{D} \subset H \text{ is dense} \\ \text{iii.} & \mathscr{D} \subset V \subset H \end{cases}$

and that we suppose that

(3) $\begin{cases} \text{i.} & \text{the associated normal space } V_0 \text{ is equal to Ker } \gamma \text{ where} \\ \text{ii.} & \gamma \in \mathscr{L}(V, T) \text{ is surjective.} \end{cases}$

We apply the preceding results to

(4) $$A^* \in \mathscr{L}(V, V^*).$$

DEFINITION 1
We say that the formal adjoint of A^*:

$$\Lambda = (A^*|_{V_0})^* \in \mathscr{L}(V, V_0^*)$$

is the formal "operator" associated with $A \in \mathscr{L}(V, V^*)$ or with the bilinear form $a(x, y) = \langle Ax, y \rangle$. We denote by

(5) $$V(\Lambda) = \{x \in V \quad \text{such that} \quad \Lambda x \in H\}$$

its domain with the scalar product

(6) $$((x, y))_{V(\Lambda)} = ((x, y))_V + (\Lambda x, \Lambda y). \qquad \blacktriangle$$

We are going to reformulate the preceding results in the following theorem.

THEOREM 1
The formal operator $\Lambda \in \mathscr{L}(V, V_0^*)$ is associated with the bilinear form $a(x, y)$ by

(7) $$\forall x \in V, \quad \forall y \in V_0, \quad a(x, y) = (\Lambda x, y).$$

Its domain $V(\Lambda)$ is a Hilbert space. If

(8) $$V_0 \text{ is complete for the norm } (\|x\|_H^2 + \|A^*x\|_*^2)^{1/2}.$$

then $V(\Lambda)$ is dense in V. Finally, if $V_0 = \text{Ker } \gamma$ where $\gamma \in \mathscr{L}(V, T)$ is surjective, there exists a unique operator $\delta \in \mathscr{L}(V(\Lambda), T^*)$ for which Green's formula

(9) $$\forall x \in V(\Lambda), \quad \forall y \in V, \quad a(x, y) = (\Lambda x, y) + \langle \delta x, \gamma y \rangle$$

is satisfied. $\qquad \blacktriangle$

Proof. To write that $\Lambda = (A^*|_{V_0})^*$ is the formal adjoint of A^* is equivalent to writing that if $x \in V, y \in V_0$

$$(\Lambda x, y) = ((A^*|_{V_0})^* x, y) = (x, (A^*|_{V_0})y) = \langle x, A^*y \rangle_{V \times V^*}$$
$$= \langle Ax, y \rangle_{V^* \times V} = a(x, y).$$

Theorem 1.4 shows that $V(\Lambda) = E((A^*|_{V_0})^*)$ is a Hilbert space that is dense in V when (8) is satisfied.

Finally, Theorem 1.1 applied to A^* where we have set $\delta = \beta^*$ implies that Green's formula (9) holds: if $x \in V(\Lambda)$, $y \in V$,

$$a(x, y) = [x, A^*y] = (\Lambda x, y) + \langle \beta^* x, \gamma y \rangle$$
$$= (\Lambda x, y) + \langle \delta x, \gamma y \rangle. \qquad \blacksquare$$

Remark 1

If Λ is the formal operator associated with the bilinear form $a(x, y)$, then if $\lambda \in \mathbb{R}$, the operator $\Lambda + \lambda$ is associated with the bilinear form $a(x, y) + \lambda(x, y)$. It is clear that the domain of $\Lambda + \lambda$

(10) $\qquad V(\Lambda + \lambda) = V(\Lambda)$ does not depend on λ

and that the operator δ associated with $\Lambda + \lambda$ and with γ by Green's formula does not depend on λ:

(11) $\qquad \delta(\Lambda + \lambda; \gamma) = \delta(\Lambda, \gamma). \qquad \blacksquare$

DEFINITION 2
We agree to denote by $\Lambda^* \in \mathscr{L}(V, V_0^*) \cap \mathscr{L}(V(\Lambda^*), H)$ the operator associated with the bilinear form $a_*(x, y) = a(y, x) = \langle A^*x, y \rangle$ and by $\delta^* \in \mathscr{L}(V(\Lambda^*), T^*)$ the operator defined by Green's formula

(12) $\qquad \forall x \in V(\Lambda^*), \quad \forall y \in V, \quad a_*(x, y) = (\Lambda^* x, y) + \langle \delta^* x, \gamma y \rangle. \qquad \blacktriangle$

We point out immediately that despite the notation, Λ^* *is not the transpose of* Λ. (It is the formal adjoint of A!.)

PROPOSITION 1
The operators $\Lambda, \Lambda^*, \delta, \delta^*$ satisfy the following formula

(13) $\quad \forall x \in V(\Lambda^*), \quad Vy \in V(\Lambda), \quad (x, \Lambda y) - (\Lambda^* x, y) = \langle \delta y, \gamma x \rangle - \langle \delta^* x, \gamma y \rangle.$
$\qquad \blacktriangle$

Proof. Formula (13) follows from the Green formulas (9) and (12) and the equality:

$$a_*(x, y) = a(y, x). \qquad \blacksquare$$

Example 1

Suppose we are given

(14) $\qquad$ piecewise continuous functions v_{pq} on $\Omega = \,]a, b[$ for $p, q \leq m$

and let us denote by

(15) $$\mathscr{V}(\omega) = (v_{pq}(\omega))_{0 \leq p,q \leq m}$$

the matrix of coefficients v_{pq}. Consider the bilinear form a defined on $H^m(\Omega) \times H^m(\Omega)$ by

(16) $$a(x, y) = \sum_{p,q=0}^{m} \int_a^b v_{pq}(\omega) D^p x(\omega) D^q y(\omega) \, d\omega$$

as x and y run over $H^m(\Omega)$. When y runs only over $H_0^m(\Omega)$, it follows that

(17) $$a(x, y) = (\Lambda x, y)$$

where Λ is the differential operator of order $2m$ of the form:

(18) $$\Lambda x = \sum_{0 \leq p,q \leq m} (-1)^q D^q (v_{pq}(\omega) D^p x(\omega)) \in H^{-m}(\Omega).$$

We consider its domain:

(19) $$H^m(\Omega, \Lambda) = \{x \in H^m(\Omega) \quad \text{such that} \quad \Lambda x \in L^2(\Omega)\}. \quad \blacksquare$$

THEOREM 2
There exists a unique operator $\delta = (\delta_m, \ldots, \delta_j, \ldots, \delta_{2m-1})$ where $\delta_j \in \mathscr{L}(H^m(\Omega, \Lambda), \mathbb{R}^2)$ satisfying

(20) $$a(x, y) = (\Lambda x, y) + \sum_{j=0}^{m-1} \langle \delta_{2m-j-1} x, \gamma_j y \rangle$$

for all $x \in H^m(\Omega, \Lambda)$, $y \in H^m(\Omega)$. The operator δ_{m+j} can be written as follows:

(21) $$\begin{cases} \delta_{m+j} x = \left\{ (-1)^{m-j-1} \sum_{l=0}^{j} \sum_{p=0}^{m} D^{j-l}(v_{p,m-l} D^p x)(a), \right. \\ \left. (-1)^{m-j} \sum_{l=0}^{j} \sum_{p=0}^{m} D^{j-l}(v_{p,m-l} D^p x)(b) \right\} \end{cases}$$

when the function x and the coefficients v_{pq} are sufficiently differentiable. ▲

Proof. The existence and uniqueness of δ are a consequence of Theorem 1 with $V = H^m(\Omega)$, $V_0 = H_0^m(\Omega)$, $T = \mathbb{R}^{2m}$, a defined by (16) and Λ defined by (18).

To calculate δ, we note that

(22) $$\Lambda = G_0^{m*} \mathscr{V} G^m$$

where $\mathscr{V} \in \mathscr{L}(L^2(\Omega)^{m+1}, L^2(\Omega)^{m+1})$ is the operator defined by

(23) $$(\mathscr{V} e)^q(\omega) = \sum_{p=0}^{m} v_{pq}(\omega) e^p(\omega).$$

We also note that

(24) $\quad H^m(\Omega, \Lambda) = \{x \in H^m(\Omega) \quad \text{such that} \quad \mathscr{V} G^m x \in H(\Omega, G_0^{m*})\}$

Hence the formula for integration by parts (Theorem 1.2) with $e = \mathscr{V} G^m x$ shows that

$$a(x, y) = [\mathscr{V} G^m x, G^m y]$$
$$= (\Lambda x, y) + \sum_{j=0}^{m-1} \beta^*_{m-j-1}(\mathscr{V} G^m x), \gamma_j y \rangle.$$

Thus it follows that if $0 \leq j \leq m - 1$, $\delta_{m+j} = \beta^*_j(\mathscr{V} G^m x)$, which implies (21). ∎

In particular consider the case where

(25) $\begin{cases} \text{i.} & a(x, y) = \int_a^b v(\omega) Dx(\omega) Dy(\omega) \, d\omega \\ \text{ii.} & \Lambda x(\omega) = - D(v(\omega) Dx(\omega)). \end{cases}$

Then Green's formula (20) becomes

$$a(x, y) = \int_a^b v(\omega) Dx(\omega) Dy(\omega) \, d\omega$$
$$= \int_a^b \Lambda x(\omega) y(\omega) \, d\omega + \langle \delta x, \gamma y \rangle$$

for all $\quad x \in H^1(\Omega, \Lambda) \quad$ and all $\quad y \in H^1(\Omega)$

where

(26) $\quad\quad\quad\quad\quad\quad \gamma y = \{y(a), y(b)\}$

and where δ is the extension of the operator defined by

(27) $\quad\quad\quad\quad\quad \delta x = \{- v(a) Dx(a), v(b) Dx(b)\}.$

when v and x are smooth. ∎

Example 2. Green's Formula

Let us consider a *regular* open set with boundary Γ and $n^2 + 1$ functions $v_{ij}(i, j = 1, \ldots, n)$ and v_0 satisfying

(28) $\quad\quad$ the functions v_0 and v_{ij} are continuous on $\bar{\Omega}$.

We denote by $\mathscr{V}(\omega) = \{v_{ij}(\omega)\}_{i,j=1,\ldots,n}$ the matrix of coefficients $v_{ij}(\omega)$.
Consider the bilinear form a defined on $H^1(\Omega) \times H^1(\Omega)$ by

(29) $\quad\quad a(x, y) = \sum_{i,j=1}^n \int_\Omega v_{ij}(\omega) D_i x D_j y \, d\omega + \int_\Omega v_0(\omega) x(\omega) y(\omega) \, d\omega$
$\quad\quad\quad\quad = [\mathscr{V} \operatorname{grad} x, \operatorname{grad} y] + (v_0 x, y).$

It is clear that the formal operator associated with it is defined by

$$\Lambda x = - \sum_{i,j=1}^{n} D_j(v_{ij}(\omega)D_i x(\omega)) + v_0(\omega)x(\omega) \tag{30}$$

and that

$$\begin{aligned}H^1(\Omega), \Lambda) &= \{x \in H^1(\Omega) \quad \text{such that} \quad \Lambda x \in L^2(\Omega)\} \\ &= \{x \in H^1(\Omega) \quad \text{such that} \quad \mathscr{V} \cdot \text{grad } x \in H^1(\Omega, \text{div})\}\end{aligned} \tag{31}$$

is the domain of Λ in $H^1(\Omega)$. In fact we can also write that

$$\Lambda x = - \text{div } \mathscr{V} \cdot \text{grad } x + v_0 \cdot x. \tag{32}$$

▲

THEOREM 3
The operator

$$\frac{\partial}{\partial v_\Lambda} = \beta_1^* \mathscr{V} \text{ grad} \in \mathscr{L}(H^1(\Omega, \Lambda), H^{-1/2}(\Gamma)) \tag{33}$$

is the unique operator such that Green's formula,

$$a(x, y) = \int_\Omega (\Lambda x) y \, d\omega + \int_\Gamma \frac{\partial x}{\partial v_\Lambda} y \, d\sigma, \tag{34}$$

holds for all $x \in H^1(\Omega, \Lambda)$ and $y \in H^1(\Omega)$.

▲

Proof. Theorem 3 follows from Theorem 1 and Theorem 1.3. ∎

Let us recall that the Laplacian Δ is the composition of the gradient and the divergence

$$\Delta = \text{div grad} = -G_0^* G. \tag{35}$$

We introduce its domain in $H^1(\Omega)$ defined by

$$\begin{aligned}H^1(\Omega, \Delta) &= \{x \in H^1(\Omega) \quad \text{such that} \quad \Delta x \in L^2(\Omega)\} \\ &= \{x \in H^1(\Omega) \quad \text{such that grad } x \in H^1(\Omega, \text{div})\}.\end{aligned} \tag{36}$$

Green's formula (34) becomes

$$\begin{cases} \int_\Omega \text{grad } x \cdot \text{grad } y \, d\omega = \int_\Omega -\Delta y \cdot x \, d\omega + \int_\Gamma \frac{\partial y}{\partial v} \cdot x \, d\sigma \\ \text{if} \quad x \in H^1(\Omega) \quad \text{and} \quad y \in H^1(\Omega, \Delta), \end{cases} \tag{37}$$

where we denote by

$$\frac{\partial}{\partial v} = \beta_1^* \text{ grad} \in \mathscr{L}(H^1(\Omega, \Delta), H^{-1/2}(\Gamma)) \tag{38}$$

the *normal derivative operator.*

Formula (37) *is the "Green's formula" that is meaningful when x runs over* $H^1(\Omega)$ *and y runs overs* $H^1(\Omega, \Delta)$.

3. ABSTRACT VARIATIONAL BOUNDARY VALUE PROBLEMS

Abstract Neumann Problems

We suppose that we are still in the situation of the hypotheses (2) and (3) of Chapter 13, Section 2. We are going to begin by studying the operator

$$\Lambda \times \delta : V(\Lambda) \mapsto H \times T^*.$$

PROPOSITION 1
Suppose we are given

(1) $$f \in H, \quad \varphi \in T^*.$$

The following problems are equivalent:
a. *to find* $x \in V(\Lambda)$ *satisfying*

(2) $$\begin{cases} \text{i.} & \Lambda x + \lambda x = f \\ \text{ii.} & \delta x = \varphi; \end{cases}$$

b. *to find* $x \in V$ *satisfying*

(3) $$\forall y \in V, \quad a(x, y) + \lambda(x, y) = (f, y) + \langle \varphi, \gamma y \rangle. \quad \blacktriangle$$

Proof. Suppose that $x \in V(\Lambda)$ is a solution of problem **a**. Then $x \in V$. Moreover, using Green's formula we obtain

$$a(x, y) + \lambda(x, y) = (\Lambda x + \lambda x, y) + \langle \delta x, \gamma y \rangle = (f, y) + \langle \varphi, \gamma y \rangle \quad \text{for all} \quad y \in V.$$

Conversely, suppose that $x \in V$ is a solution of problem **b**. As y runs over V_0, formula (3) becomes

(4) $$a(x, y) + \lambda(x, y) = (\Lambda x + \lambda x, y) = (f, y) \quad \forall y \in V_0.$$

Hence $\Lambda x = f - \lambda x \in V_0^*$. Since $f - \lambda x$ belongs to H, this implies that $x \in V(\Lambda)$. Thus we can apply Green's formula; we deduce from (3) that for all $y \in V$,

$$\langle \delta x, \gamma y \rangle = a(x, y) + \lambda(x, y) - (\Lambda x + \lambda x, y)$$
$$= a(x, y) + \lambda(x, y) - (f, y) = \langle \varphi, \gamma y \rangle.$$

Since γ is a surjective operator from V onto T, we obtain the equality $\delta x = \varphi$. ∎

Abstract Dirichlet Problems

We are now going to study the operator

$$\Lambda \times \gamma : V(\Lambda) \to H \times T.$$

PROPOSITION 2
Suppose we are given

(5) $$f \in H, \quad \psi \in T.$$

The following problems are equivalent:
a. *to find $x \in V(\Lambda)$ satisfying*

(6) $$\begin{cases} \text{i.} & \Lambda x + \lambda x = f \\ \text{ii.} & \gamma x = \varphi; \end{cases}$$

b. *to find $x = z + x_0$ where $x_0 \in V$ satisfies $\gamma x_0 = \varphi$ and where $z \in V_0$ satisfies*

(7) $$a(z, y) + \lambda(z, y) = (f - \Lambda x_0 - \lambda x_0, y) \quad \forall y \in V_0. \quad \blacktriangle$$

Proof. If $x \in V(\Lambda)$ is a solution of problem **a** and if $x_0 \in V$ satisfies $\gamma x_0 = \varphi$, then $z = x - x_0 \in V_0$, since

$$\gamma z = \gamma x - \gamma x_0 = \varphi - \varphi = 0$$

and satisfies when y ranges over V_0,

$$\begin{aligned} a(z, y) + \lambda(z, y) &= a(x - x_0, y) + \lambda(x - x_0, y) \\ &= (\Lambda x + \lambda x - \Lambda x_0 - \lambda x_0, y) \\ &= (f - \Lambda x_0 - \lambda x_0, y) \quad \forall y \in V_0. \end{aligned}$$

Conversely, let $z \in V_0$ be a solution of the variational equation (7). Then z satisfies

(8) $$\Lambda z + \lambda z = f - \Lambda x_0 - \lambda x_0.$$

This implies that if we set $x = z + x_0$,

$$\Lambda x = f - \lambda x.$$

Since $f - \lambda x \in H$, it follows that $x \in V(\Lambda)$. Moreover,

$$\gamma x = \gamma z + \gamma x_0 = \gamma x_0 = \varphi \in T. \quad \blacksquare$$

DEFINITION 1
Let us say that T is a **trace space,** *that the operator $\gamma \in \mathcal{L}(V, T)$ such that $V_0 = \operatorname{Ker} \gamma$ is the* **Dirichlet operator,** *and that the operator δ is the* **Neumann operator** *of Λ. Problems (2) and (6) are called the* **Neumann variational problem** *and the* **Dirichlet variational problem,** *respectively.* $\blacktriangle$

Abstract Boundary Value Problems

The Neumann variational problem and the Dirichlet variational problem are equivalent to the *variational equations* on the spaces V and V_0, respectively. We are going to characterize the equivalent problems to the variational equations on a closed subspace W of V such that

(9) $$V_0 \subset W \subset V.$$

If

(10) $\quad \omega_1 \in \mathscr{L}(T, T)$ is a projector from T onto $T_1 = \omega_1(T)$

and if

(11) $$\gamma_1 = \omega_1 \gamma \in \mathscr{L}(V, T_1),$$

then

(12) $\quad\quad W = \operatorname{Ker} \gamma_1 \quad$ is such that $\quad V_0 \subset W \subset V.$

Remark 1

Conversely, we can associate to every closed subspace W of V satisfying (1) a projector $\omega_1 \in \mathscr{L}(T, T)$ such that

$$W = \operatorname{Ker} \gamma_1 \quad \text{where} \quad \gamma_1 = \omega_1 \gamma.$$

We are going to set

(13) $$\omega_2 = 1 - \omega_1$$

and for $i = 1, 2$

(14) $\quad T_i = \omega_i T, \quad T_i^* = \omega_i^* T^*, \quad \gamma_i = \omega_i \gamma \quad \text{and} \quad \delta_i = \omega_i^* \delta.$

We are going to study the operators of the form

(15) $$\Lambda \times \gamma_1 \times \delta_2 : V(\Lambda) \to H \times T_1 \times T_2^*. \quad\blacksquare$$

Remark 2

We find again the Dirichlet problem with $\omega_1 = 1$ and the Neumann problem with $\omega_1 = 0$. $\quad\blacksquare$

PROPOSITION 3

Suppose we are given

(16) $$f \in H, \quad \varphi_1 \in T_1, \quad \psi_2 \in T_2^*.$$

The following problems are equivalent:
a. *to find $x \in V(\Lambda)$ satisfying*

(17) $\quad \begin{cases} \text{i.} & \Lambda x + \lambda x = f \\ \text{ii.} & \gamma_1 x = \varphi_1 \\ \text{iii.} & \delta_2 x = \psi_2; \end{cases}$

b. *to find $x = z + x_1$ where $x_1 \in V$ satisfies $\gamma_1 x_1 = \varphi_1$ and where $z \in W = \operatorname{Ker} \gamma_1$ satisfies*

(18) $\quad a(z, y) + \lambda(z, y) = (f - \lambda x_1, y) - a(x_1, y) + \langle \psi_2, \gamma_2 y \rangle \quad \text{for all} \quad y \in W.$ ▲

If we take x_1 in $V(\Lambda)$, then (18) can be written

$$\forall y \in W, \quad a(z, y) + \lambda(z, y) = (f - \Lambda x_1 - \lambda x_1, y) + \langle \psi_2 - \delta_2 x_1, \gamma_2 y \rangle$$

DEFINITION 2
The boundary value problems of the form (17) *are called* **variational boundary value problems,** *and problem* (18) *is called the* **variational form.** *These problems are said to be homogeneous if $\varphi_1 = 0$ and $\psi_2 = 0$.* ▲

Proof of Proposition 3. Suppose that $x \in V(\Lambda)$ is a solution of problem **a**. Then $x \in V$ and $z = x - x_1 \in W$ since $\gamma_1 z = \gamma_1 x - \gamma_1 x_1 = 0$. Moreover, it follows from Green's formula that if $y \in W$

$$\begin{aligned} a(z, y) + \lambda(z, y) &= a(x - x_1, y) + \lambda(x - x_1, y) \\ &= (\Lambda x - \lambda x - \lambda x_1, y) - a(x_1, y) \\ &\quad + \langle \delta_2 x, \gamma_2 y \rangle + \langle \delta_1 x, \gamma_1 y \rangle \\ &= (f - \lambda x_1, y) - a(x_1, y) + \langle \psi_2, \gamma_2 y \rangle \\ &= (f - \Lambda x_1 - \lambda x_1, y) + \langle \psi_2 - \delta_2 x_1, \gamma_2 y \rangle \\ &\text{when } x_1 \in V(\Lambda). \end{aligned}$$

Conversely, let $x_1 \in V$ be a solution of the equation $\gamma_1 x_1 = \varphi_1$ and $z \in W$ be a solution of the variational equation (18). Taking $y \in V_0$, we deduce from (18) that

$$((\Lambda + \lambda)z, y) = a(z, y) + \lambda(z, y) = (f - (\Lambda + \lambda)x_1, y)$$

for all $y \in V_0$. This implies that if we set $x = z + x_1$, then $\Lambda x + \lambda x = f$. Since $f \in H$, it follows that $x \in V(\Lambda)$. Moreover, $\gamma_1 x = \gamma_1 z + \gamma_1 x_1 = \gamma_1 x_1 = \varphi_1$.

It remains to show that $\delta_2 x = \psi_2$. To this end we use Green's formula: If $y \in W$, then

$$\begin{aligned} a(x, y) + \lambda(x, y) &= ((\Lambda + \lambda)x, y) + \langle \delta_2 x, \gamma_2 y \rangle \\ &= (f, y) + \langle \delta_2 x, \gamma_2 y \rangle. \end{aligned}$$

By comparing with (18), this implies that

$$\langle \delta_2 x, \gamma_2 y \rangle = \langle \psi_2, \gamma_2 y \rangle \quad \text{for all} \quad y \in W.$$

Since γ_2 is *surjective* from W onto T_2, it follows that $\psi_2 = \delta_2 x$. ∎

Existence and Uniqueness of Solutions to Boundary Value Problems

We are able to deduce an existence theorem.

THEOREM 1
Suppose that hypotheses (2) and (3) in Section 2 are satisfied. Let $W = \mathrm{Ker}\, \gamma_1$ be a closed subspace of V. Suppose that the restriction of A to W is (W, H)-coercive, that is, that there exists μ_W and $c > 0$ such that

(19) $$\forall x \in W, \quad a(x, x) + \mu_W \|x\|_H^2 \geq c \|x\|_V^2.$$

Then for all $\lambda \geq \mu_W$, $(\Lambda + \lambda) \times \gamma_1 \times \delta_2$ is an isomorphism from $V(\Lambda)$ onto $H \times T_1 \times T_2^*$. ▲

Proof. This theorem is a consequence of the Lax-Milgram theorem, Theorem 3.6.1, which implies the existence and uniqueness of a solution to problem (18), and of Proposition 3 above, which implies that this solution is the unique solution to problem (17). Hence $(\Lambda + \lambda) \times \gamma_1 \times \delta_2$ is a bijective operator from $V(\Lambda)$ onto $H \times T_1 \times T_2^*$, which is an isomorphism according to the Banach theorem, Theorem 4.3.3. ∎

THEOREM 2
Suppose that hypotheses (2) and (3) in Section 2 are satisfied, that A is W-elliptic, that

(20) $$\text{the injection from } V \text{ to } H \text{ is compact,}$$

and that

(21) $$\text{the bilinear form } a \text{ is symmetric.}$$

Then there exist a sequence $\{e_n\}_n$ of elements e_n of W and a countable sequence of numbers μ_n approaching $-\infty$ such that

(22) $$(\Lambda + \mu_n) e_n = 0, \quad \gamma_1 e_n = 0, \quad \delta_2 e_n = 0.$$

Consider the subspace $D(A_W)$ of $V(\Lambda)$ defined by

(23) $$D(A_W) = \{x \in V(\Lambda) \quad \text{such that} \quad \gamma_1 x = 0 \text{ and } \delta_2 x = 0\}.$$

The following alternative holds: If $\forall n, \lambda \neq \mu_n$, then $(\Lambda + \lambda)$ is an isomorphism from $D(A_W)$ onto H and if $\lambda = \mu_{n_0}$, then $(\Lambda + \lambda)$ is an isomorphism from

$D(A_W)/W_{n_0}$ onto $W_{n_0}^\perp$ where W_{n_0} is the eigensubspace of μ_{n_0} (which is finite dimensional).

If $\lambda \neq \mu_n$ for all $n \in \mathbb{N}$, the solution $x \in D(A_W)$ of the boundary value problem

(24) $\qquad (\Lambda + \lambda)x = f, \qquad \gamma_1 x = 0, \qquad \delta_2 x = 0$

can be written

(25) $$x = \sum_{n=1}^{\infty} \frac{\mu_n}{\mu_n - \lambda}(f, e_n)e_n,$$

and if $\lambda = \mu_{n_0}$, a solution x of (24) can be written

(26) $$x = \sum_{n \notin N(n_0)} \frac{\mu_n}{\mu_n - \lambda}(f, e_n)e_n + \sum_{n \in N(n_0)} a_n e_n$$

where $N(n_0) = \{n \text{ such that } \mu_n = \mu_{n_0}\}$ and the a_n's are arbitrary scalars. ▲

Proof. We consider W with the scalar product $((x, y))_W = a(x, y)$ [equivalent to the initial scalar product since A is W-elliptic]. From Theorem 11.4.1 we know that there exist an orthonormal base $\{e_n\}_n$ of W and a countable sequence of scalars μ_n approaching $-\infty$ such that

(27) $\qquad a(e_n, y) + \mu_n(e_n, y) = 0 \qquad$ for all $\qquad y \in W$.

Proposition 3 then implies that e_n is the solution of the boundary value problem (22). Hence every element $x \in D(A_W)$ can be written

(28) $$x = \sum_{n=1}^{\infty} ((x, e_n))e_n = \sum_{n=1}^{\infty} (\Lambda x, e_n)e_n;$$

consequently,

(29) $$(\Lambda + \lambda)x = \sum_{n=1}^{\infty} ((x, e_n))(\lambda - \mu_n)e_n = \sum_{n=1}^{\infty} (\Lambda x, e_n)(\lambda - \mu_n)e_n.$$

Since Λ is an isomorphism from $D(A_W)$ onto H, according to Theorem 1, it follows from (29) with $\lambda = 0$ that every element $f = \Lambda x$ of H can be written

(30) $$f = \sum_{n=1}^{\infty} -\mu_n(f, e_n)e_n.$$

Hence a solution $x = \sum_{n=1}^{\infty} c_n e_n$ of the boundary value problem (24) is a solution of

(31) $$\sum_{n=1}^{\infty} (\lambda - \mu_n)c_n e_n = \sum_{n=1}^{\infty} -\mu_n(f, e_n)e_n.$$

Thus we deduce formulas (25) and (26) from (31). ∎

Characterization of $D(A_W)$

It can be useful to characterize the set $D(A_W)$ [defined by (23)] of elements of $V(\Lambda)$ satisfying the homogeneous boundary conditions $\gamma_1 x = 0$ and $\delta_2 x = 0$ by means of the theory of unbounded operators. (See Chapter 5, Section 5.)

Since the restriction $a_W(x, y)$ of the form $a(x, y)$ to $W \times W$ is continuous, it defines an operator A_W from W to W^* by

(32) $\qquad \langle A_W x, y \rangle = a(x, y) \qquad \forall x, y \in W.$

Since the transpose j_W^* of the injection j_W from W to H is an embedding from H into W^*, we are going to introduce the domain $D(A_W)$ of A_W defined by

(33) $\qquad D(A_W) = \{x \in W \text{ such that } A_W x \in H\}.$

We are going to characterize this domain as follows:

THEOREM 3
The domain $D(A_W)$ defined by (33) is also equal to

(34) $\qquad D(A_W) = \{x \in V(\Lambda) \quad \text{such that} \quad \gamma_1 x = 0 \text{ and } \delta_2 x = 0\},$

and the restriction of A_W to W is equal to Λ. If $A_W \in \mathscr{L}(W, W^*)$ is an isomorphism, the domain $D(A_W) = \operatorname{Ker}(\gamma_1 \times \delta_2)$ is dense in W and in H, the operator $\gamma_1 \times \delta_2 \in \mathscr{L}(V(\Lambda), T_1 \times T_2^*)$ is surjective, and the image of $D(A_W)$ under $\gamma_2 \times \delta_1$ is dense in $T_2 \times T_1^*$. ▲

Proof. **a.** Indeed, to say that $x \in D(A_W)$ defined by (33) is to say that $A_W x \in H$, or that $x \in W$ is a solution of the equation

$$\forall y \in W, \quad a(x, y) = (A_W x, y).$$

According to Proposition 3, this amounts to saying that x is a solution of

(35) $\qquad x \in V(\Lambda), \quad \Lambda x = A_W x, \quad \gamma_1 x = 0, \quad \text{and} \quad \delta_2 x = 0.$

b. Since $A_W \in \mathscr{L}(W, W^*)$ is an isomorphism, it follows from Proposition 3 (as for Theorem 3) that $\Lambda \times (\gamma_1 \times \delta_2)$ is an isomorphism from $V(\Lambda)$ onto $H \times (T_1 \times T_2^*)$; in particular, it follows that

(36) $\begin{cases} \text{i.} & \Lambda \text{ is an isomorphism } D(A_W) \text{ onto } H. \\ \text{ii.} & (\gamma_1 \times \delta_2) \text{ is an isomorphism from the kernel } N(\Lambda) \text{ of } \Lambda \\ & \text{onto } T_1 \times T_2^*. \end{cases}$

and that

(37) $\qquad V(\Lambda)$ is the direct sum of $N(\Lambda)$ and $D(A_W)$.

c. Since A_W is an isomorphism from W onto W^* and from $D(A_W)$ onto H

and since H is identified with a dense subspace of W^*, it follows that $D(A_W)$ is dense in W. Since W is dense in H, it follows that $D(A_W)$ *is dense in H*.

d. Finally let us show that $(\gamma_2 \times \delta_1)D(A_W)$ is dense in $T_2 \times T_1^*$, using the density criterion (Theorem 2.2.1). Let $\{\psi_2, \varphi_1\} \in T_2^* \times T_1$ be a linear form that vanishes on $(\gamma_2 \times \delta_1)D(A_W)$ and let us show that it is identically zero. Let $\bar{x} \in V(\Lambda^*)$ be the unique solution of the problem

(38) $$\Lambda^*\bar{x} = 0, \quad \gamma_1\bar{x} = -\varphi_1, \quad \delta_2^*\bar{x} = \psi_2.$$

Consequently, for all $y \in D(A_W) = \text{Ker}(\gamma_1 \times \delta_2)$,

$$\begin{aligned}
0 &= \langle \psi_2, \gamma_2 y \rangle + \langle \delta_1 y, \varphi_1 \rangle \\
&= \langle \delta_2^*\bar{x}, \gamma_2 y \rangle - \langle \delta_1 y, \gamma_1 \bar{x} \rangle \\
&= \langle \delta_2^*\bar{x}, \gamma_2 y \rangle + \langle \delta_1^*\bar{x}, \gamma_1 y \rangle - \langle \delta_1 y, \gamma_1 \bar{x} \rangle - \langle \delta_2 y, \gamma_2 \bar{x} \rangle \\
&= (\Lambda^*\bar{x}, y) - (\bar{x}, \Lambda y) = -(\bar{x}, \Lambda y)
\end{aligned}$$

Since $\Lambda = A_W$ is an isomorphism from $D(A_W)$ onto H, it follows that $\bar{x} = 0$; consequently, (38) implies that

$$\varphi_1 = -\gamma_1\bar{x} = 0 \quad \text{and} \quad \psi_2 = \delta_2^*\bar{x} = 0.$$

This completes the proof of the theorem. ∎

4. EXAMPLES OF BOUNDARY VALUE PROBLEMS

The Sturm-Liouville Problem

We begin by giving some examples of boundary value problems for second order differential equations (Sturm-Liouville problems). Let us consider the operator Λ defined by

(1) $$\Lambda x(\omega) = -D(v(\omega)Dx(\omega))$$

where v is assumed to piecewise continuous on an interval $]a, b[$. Its domain is

(2) $$H^1(\Omega, \Lambda) = \{x \in H^1(\Omega) \text{ such that } D(v(\omega)Dx(\omega)) \in L^2(\Omega)\}.$$

It is the formal operator associated with the bilinear form

(3) $$a(x, y) = \int_a^b v(\omega)Dx(\omega)Dy(\omega)\, d\omega.$$

We set

(4) $$c = \min_{\omega \in \Omega} v(\omega).$$

If $c > 0$, the bilinear form $a(x, y) + \lambda(x, y)$ is $H^1(\Omega)$-elliptic if $\lambda > 0$, since

(5) $\qquad a(x, x) + \lambda(x, x) \geq c|Dx|^2 + \lambda|x|^2 \geq \min(c, \lambda) \|x\|_1^2.$

We are going to be able to apply the results on boundary value problems from the preceding section. Suppose we are given

(6) $\qquad f \in L^2(\Omega), \qquad \varphi = \{\varphi^a, \varphi^b\} \in \mathbb{R}^2.$

We are going to look for a solution $x \in H^1(\Omega, \Lambda)$ of various boundary value problems for the differential equation

(7) $\qquad -D(v(\omega)Dx(\omega)) + \lambda x(\omega) = f(\omega) \qquad \text{if} \qquad \omega \in \,]a, b[$

(the equality is considered in the sense of $L^2(\Omega)$). Propositions 3.1, 3.2, and 3.3 imply the following results.

The *Neumann problem* (7) with

(8) $\qquad v(a)Dx(a) = \varphi^a, \qquad v(b)Dx(b) = \varphi^b$

is equivalent to looking for $x \in H^1(\Omega)$, which is a solution of the variational equations

(9) $\qquad a(x, y) + \lambda(x, y) = (f, y) + \varphi^b y(b) - \varphi^a y(a) \qquad \forall y \in H^1(\Omega).$

The *Dirichlet problem* (7) with

(10) $\qquad x(a) = \varphi^a, \qquad x(b) = \varphi^b$

is equivalent to looking for $x = z + x_0$ where $x_0 \in H^1(\Omega)$ satisfies $\gamma x_0 = \varphi$ and where $z \in H_0^1(\Omega)$ is a solution of the variational equations

(11) $\qquad a(z, y) + \lambda(z, y) = (f - \Lambda x_0 - \lambda x_0, y) \qquad \forall y \in H_0^1(\Omega).$

The *mixed boundary value problem* (7) with

(12) $\qquad x(a) = \varphi^a, \qquad v(b)Dx(b) = \varphi^b$

is equivalent to looking for $x = z + x_0$ where $x_0 \in H^1(\Omega)$ satisfies $\gamma_1 x_0 = x_0(a) = \varphi^a$ and where $z \in H_a^1(\Omega) = \{z \in H^1(\Omega)$ such that $x(a) = 0\}$ is a solution of the variational equations

(13) $\quad a(x, y) + \lambda(z, y) = (f - \lambda x_0, y) - a(x_0, y) + \varphi^b y(b) \qquad \forall y \in H_a^1(\Omega).$

THEOREM 1
If $c = \min_{\omega \in \Omega} v(\omega)$ and λ are strictly positive, there exist unique solutions to the Neumann problem, the Dirichlet problem, and the mixed boundary value problem. For each of these problems, there exist a base of functions $\{e_n\}_n$ and a sequence of scalars $\mu_n \leq 0$ approaching $-\infty$ satisfying $D(v(\omega)De_n(\omega)) = \mu_n e_n(\omega)$ and the homogeneous boundary conditions.

The following alternative holds: If $\lambda \neq \mu_n$ for all n, there exists a unique solution to the homogeneous boundary value problem defined by

$$(14) \qquad x = \sum_{n=1}^{\infty} \left(\frac{\mu_n}{\mu_n - \lambda} \int_\Omega f(\omega) e_n(\omega)\, d\omega \right) e_n(\omega).$$

If $\lambda = \mu_{n_0}$, there exist solutions to the homogeneous boundary value problem of the form

$$(15) \qquad x = \sum_{n \notin N(n_0)} \left(\frac{\mu_n}{\mu_n - \lambda} \int_\Omega f(\omega) e_n(\omega)\, d\omega \right) e_n(\omega) + \sum_{n \in N(n_0)} a_n e_n(\omega). \qquad \blacktriangle$$

Proof. We use the existence Theorems 3.1 and 3.2 and the fact that the injection from $H^1(\Omega)$ to $L^2(\Omega)$ is compact when $\Omega =]a, b[$ is a bounded interval. ∎

Boundary Value Problems for Differential Equations of Order $2m$.

Consider piecewise continuous functions v_{pq} on $\Omega =]a, b[$ for $p, q \leq m$ and the operator Λ defined by

$$(16) \qquad \Lambda x = \sum_{p,q=0}^{m} (-1)^q D^q(v_{pq}(\omega) D^p x(\omega))$$

with domain $H^m(\Omega, \Lambda) = \{x \in H^m(\Omega) \text{ such that } \Lambda x \in L^2(\Omega)\}$. It is associated with the bilinear form

$$(17) \qquad a(x, y) = \sum_{p,q=0}^{m} \int_a^b v_{pq}(\omega) D^p x(\omega) D^q y(\omega)\, d\omega$$

which is continuous on $H^m(\Omega) \times H^m(\Omega)$ and is $H^m(\Omega)$-elliptic if the functions v_{pq} are continuous and if

$$(18) \qquad \begin{cases} \exists c > 0 \text{ such that } \sum_{p,q=0}^{m} v_{pq}(\omega) a^p a^q > 0 \text{ for all} \\ a \in \mathbb{R}^{m+1} \text{ and all } \omega. \end{cases}$$

THEOREM 2

Suppose that hypothesis (18) is satisfied. Let $I \cup J$ be a partition of the set $\{0, 1, \ldots, m-1\}$. Consider the operators γ_j and δ_{m+j} defined by (21) in Section 2. Then the operator $\Lambda \times \{\gamma_i\}_{i \in I} \times \{\delta_{2m-j-1}\}_{j \in J}$ is an isomorphism from $H^m(\Omega, \Lambda)$ onto $L^2(\Omega) \times \mathbb{R}^{2m}$. $\blacktriangle$

Proof. The proof is an immediate consequence of Proposition 2.3, where we associate to the partition $I \cup J$ the projector ω_1 from $\mathbb{R}^{2m} = \prod_{j=0}^{m-1} \mathbb{R}^2$ onto $\prod_{i \in I} \mathbb{R}^2$, in such a way that

$$\gamma_1 x = \omega_1 \gamma x = \{\gamma_i x\}_{i \in I} \text{ and } \delta_2 x = (1 - \omega_1^*)\delta x = \{\delta_{2m-j-1} x\}_{j \in J}. \qquad \blacksquare$$

Remark 1

Taking $I = \{0, 1, \ldots, m\}$, we obtain the isomorphism theorem that allows us to solve the Dirichlet problem and taking $I = \emptyset$ the isomorphism theorem that allows us to solve the Neumann problem. $\qquad \blacksquare$

Boundary Value Problems for Second Order Partial Differential Equations

Consider a "regular" open set Ω of $\mathbb{R}^n$ with boundary Γ and $n^2 + 1$ functions $v_{ij}(i, j = 1, \ldots, n)$ and v_0 that are *continuous* on Ω. We introduce the bilinear form

$$(19) \qquad a(x, y) = \sum_{i,j=1}^{n} \int_{\Omega} v_{ij}(\omega) D_i x \cdot D_j y \, d\omega + \int_{\Omega} v_0(\omega) x \cdot y \, d\omega,$$

which is continuous on $H^1(\Omega) \times H^1(\Omega)$, $H^1(\Omega)$-elliptic if

$$(20) \qquad \inf_{\omega \in \Omega} v_0(\omega) = c_0 > 0 \text{ and } \inf_{\omega \in \Omega} \inf_{a \in \mathbb{R}^n} \frac{\sum_{i,j=1}^{n} v_{ij}(\omega) a_i a_j}{\|a\|^2} = c_1 > 0.$$

(The second of these conditions is the *ellipticity condition* in the usual sense of second order partial differential equations.) We consider the associated formal operator

$$(21) \qquad \Lambda x(\omega) = - \sum_{i,j=1}^{n} D_j(v_{ij}(\omega) D_i x(\omega)) + v_0(\omega) x(\omega)$$

as well as the trace operators defined by

$$(22) \qquad \gamma x(\omega) = \gamma_0 x(\omega) = x(\omega)|_\Gamma \quad \text{and} \quad \delta x = \frac{\partial}{\partial \nu_\Lambda} x,$$

which are continuous from $H^1(\Omega)$ onto $H^{1/2}(\Gamma)$ and from $H^1(\Omega, \Lambda) = \{x \in H^1(\Omega) \text{ such that } \Lambda x \in L^2(\Omega)\}$ to $H^{-1/2}(\Gamma)$, respectively (Theorem 2.3).

Suppose we are given $f \in L^2(\Omega)$. We are going to look for a solution $x \in H^1(\Omega, \Lambda)$ of the partial differential equation

$$(23) \qquad - \sum_{i,j=1}^{n} D_j(v_{ij}(\omega) D_i x(\omega)) + v_0(\omega) x(\omega) = f(\omega)$$

[the equality considered in the sense of $L^2(\Omega)$] satisfying various boundary conditions.

The *Dirichlet problem* (23) with

(24) $$\gamma_0 x = 0$$

is equivalent to looking for

(25) $x \in H_0^1(\Omega)$ satisfying $a(x,y) = (f,y)$ for all $y \in H_0^1(\Omega)$

and has a unique solution if the ellipticity conditions (20) are satisfied.

The *Neumann problem* (23) with

(26) $$\frac{\partial x}{\partial v_\Lambda} = \varphi \quad \text{where } \varphi \text{ is given in } H^{-1/2}(\Gamma)$$

is equivalent to looking for

(27) $x \in H^1(\Omega)$ satisfying $a(x,y) = (f,y) + \langle \varphi, \gamma_0 y \rangle$ for all $y \in H^1(\Omega)$

and has a unique solution if the ellipticity conditions (20) are satisfied.

The *oblique problem* (23) with

(28) $$\frac{\partial x}{\partial v_\Lambda} + \alpha(\omega) x = \varphi \quad \text{where } \alpha \text{ is continuous on } \Gamma \text{ and } \varphi \in H^{-1/2}(\Gamma)$$

is equivalent to looking for

(29) $$\begin{cases} x \in H^1(\Omega) \quad \text{satisfying} \quad a(x,y) + \int_\Gamma \alpha(\omega) x(\omega) y(\omega) \, d\sigma \\ = (f,y) + \langle \varphi, \gamma_0 y \rangle \quad \text{for all} \quad y \in H^1(\Omega) \end{cases}$$

and has a unique solution if $\alpha(\omega) \geq 0$ for all $\omega \in \Gamma$ and if the ellipticity conditions (20) are satisfied.

The Interface Problem

Suppose that Ω is divided into two open sets Ω_1 and Ω_2 where $\bar{\Omega}_1 \subset \Omega$ and

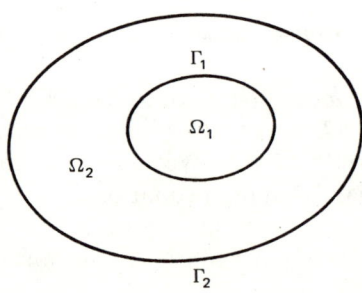

$\Omega_2 = \text{Int}(\Omega - \Omega_1)$. We denote by Γ_1 the boundary of Ω_1 and by Γ_2 the boundary of Ω. The boundary of Ω_2 is, therefore, the union of Γ_1 and Γ_2. We suppose that Ω_1 and Ω_2 are regular.

We assume we are given on each open set $\Omega_k(k=1,2)$ functions $v_{ij}^k(.)$ and $v_0^k(.)(k=1,2)$ and the differential operators Λ^k defined by

$$(30) \quad \Lambda^k x(\omega) = -\sum_{i,j=1}^n D_j(v_{ij}^k(\omega)D_i x)(\omega) + v_0^k(\omega)x(\omega) \quad \text{if} \quad \omega \in \Omega^k.$$

We consider two functions $f^k \in L^2(\Omega_k)(k=1,2)$, and we attempt to solve the following *interface problem* to find $x^k \in L^2(\Omega_k)$ satisfying

$$(31) \quad \begin{cases} \text{i.} & x^1 \in H^1(\Omega_1, \Lambda^1), \quad x^2 \in H^2(\Omega_2, \Lambda^2) \\ \text{ii.} & \Lambda^1 x^1 = f^1 \quad \text{on} \quad \Omega_1; \quad \Lambda^2 x^2 = f^2 \quad \text{on} \quad \Omega^2 \\ \text{iii.} & \dfrac{\partial x^2}{\partial v_{\Lambda^2}^2} = 0 \quad \text{on} \quad \Gamma_2 \\ \text{iv.} & \gamma_0^1 x^1 = \gamma_0^1 x^2 \quad \text{on} \quad \Gamma_1; \quad \dfrac{\partial x^1}{\partial v_{\Lambda^1}^1} = \dfrac{\partial x^2}{\partial v_{\Lambda^2}^1} \quad \text{on} \quad \Gamma_1 \end{cases}$$

where γ_0^1 is the trace operator of order zero on Γ_1 and where $\partial/\partial v_{\Lambda^j}^k$ denotes the normal derivative with respect to $\Gamma_k(k=1,2)$. We consider the following spaces:

$$(32) \quad \begin{cases} \text{i.} & H = L^2(\Omega_1) \times L^2(\Omega_2) = L^2(\Omega). \\ \text{ii.} & T = H^{1/2}(\Gamma_1) \times H^{1/2}(\Gamma_2). \\ \text{iii.} & V = \{x = \{x^1, x^2\} \in H^1(\Omega_1) \times H^1(\Omega_2) \text{ such that } \gamma_0^1 x^1 = \gamma_0^1 x^2\}. \end{cases}$$

It is clear that

$$(33) \quad V_0 = H_0^1(\Omega_1) \times H_0^1(\Omega_2) = \text{Ker } \gamma$$

is the kernel of the trace operator $\gamma \in \mathscr{L}(V, T)$ defined by $\gamma x = \{\gamma_0^1 x^2, \gamma_0^2 x^2\}$ where γ_0^2 is the trace operator on Γ_2.

We consider the bilinear form defined on $V \times V$ by

$$(34) \quad a(x,y) = a^1(x^1, y^1) + a^2(x^2, y^2)$$

where

$$(35) \quad a^k(x^k, y^k) = \sum_{i,j=1}^n \int_{\Omega_k} v_{ij}^k(\omega) D_i x^k D_j y^k \, d\omega + \int_{\Omega_k} v_0^k(\omega) xy \, d\omega.$$

The formal operator associated with the bilinear form a is the operator Λ associating to every $x = \{x^1, x^2\} \in V$ the distribution $\Lambda x = \{\Lambda^1 x^1, \Lambda^2 x^2\} \in H^{-1}(\Omega_1) \times H^{-1}(\Omega_2)$. Its domain $V(\Lambda)$ is then the space $H^1(\Omega_1, \Lambda^1) \times H^1(\Omega_2, \Lambda^2)$.

We can write Green's formulas

$$(36) \qquad a^1(x^1, y^1) = (\Lambda^1 x^1, y^1) + \int_{\Gamma_1} \frac{\partial x^1}{\partial v_{\Lambda^1}^1} \cdot y^1 \, d\sigma_1$$

and

$$(37) \qquad a^2(x^2, y^2) = (\Lambda^2 x^2, y^2) + \int_{\Gamma_2} \frac{\partial x^2}{\partial v_{\Lambda^2}^2} \cdot y^2 \, d\sigma_2 - \int_{\Gamma_1} \frac{\partial x^2}{\partial v_{\Lambda^2}^1} \cdot y^2 \, d\sigma_1$$

when $x^k \in H^1(\Omega_k, \Lambda^k)$ and $y^k \in H^1(\Omega_k)$ (for $k = 1, 2$). Consequently, if $y \in V$, we deduce from the equality $\gamma_0^1 y^1 = \gamma_0^1 y^2$ that

$$(38) \qquad a(x, y) = (\Lambda x, y) + \int_{\Gamma_2} \frac{\partial x^2}{\partial v_{\Lambda^2}^2} \cdot y^2 \, d\sigma_2 + \int_{\Gamma_1} \left(\frac{\partial x^1}{\partial v_{\Lambda^1}^1} - \frac{\partial x^2}{\partial v_{\Lambda^2}^1} \right) \cdot y^1 \, d\sigma_1$$

where we have set $(\Lambda x, y) = (\Lambda^1 x^1, y^1) + (\Lambda^2 x^2, y^2)$.

Summing up, we have

$$(39) \qquad a(x, y) = (\Lambda x, y) + \langle \delta x, \gamma y \rangle \quad \text{if} \quad x \in V(\Lambda), \quad y \in V,$$

where $\delta \in \mathscr{L}(V(\Lambda), H^{-1/2}(\Gamma_1) \times H^{-1/2}(\Gamma_2))$ is defined by

$$(40) \qquad \delta x = \left\{ \frac{\partial x^1}{\partial v_{\Lambda^1}^1} - \frac{\partial x^2}{\partial v_{\Lambda^2}^1}, \frac{\partial x^2}{\partial v_{\Lambda^2}^2} \right\}.$$

Consequently, the interface problem (31) is equivalent to looking for

$$(41) \qquad \begin{cases} x = \{x^1, x^2\} \in V \quad \text{satisfying} \\ a(x, y) = \int_{\Omega_1} f^1 y^1 \, d\omega + \int_{\Omega_2} f^2 y^2 \, d\omega \quad \text{for all} \quad y \in \{y^1, y^2\} \in V. \end{cases}$$

If the coefficients v_{ij}^k and v_0 satisfy the ellipticity conditions (20) for $k = 1, 2$, this problem has a unique solution.

5. APPROXIMATION OF SOLUTIONS TO NEUMANN PROBLEMS

We suppose that we are still under the hypotheses (2) and (3) of Section 2. Consider the case where we are given $f \in H$ and $\varphi \in T^*$ and suppose that the bilinear form $a(u, v)$ is V-elliptic:

$$(1) \qquad \exists c > 0 \quad \text{such that} \quad a(x, x) \geq c \|x\|^2 \quad \text{for all} \quad x \in V.$$

Then there exists a unique solution $x \in V$ to the equation

$$(2) \qquad \forall y \in V, \quad a(x, y) = (f, y) + \langle \varphi, \gamma y \rangle,$$

that is, to the Neumann problem

(3) $\quad x \in V(\Lambda), \quad \Lambda x = f \quad \text{and} \quad \delta x = \varphi.$

We are going to approximate the solution x of (2) by solutions x_h of analogous problems set in spaces V_h [in general finite dimensional $V_h = \mathbb{R}^{n(h)}$, where the dimension $n(h)$ increases as the parameter h approaches zero]. We pass from V_h to V by means of injective operators $p_h \in \mathscr{L}(V_h, V)$ with closed images (called *extension* operators). We denote by $\hat{r}_h = p_h^+ \in \mathscr{L}(V, V_h)$ their orthogonal right inverses (see Definition 4.6.1); then $t_h = p_h \hat{r}_h$ is the orthogonal projector onto $p_h V_h$. A family $\{V_h, p_h, \hat{r}_h\}_h$ is called a *family of approximations* of V. We say that it is *convergent* if

(4) $\quad \forall x \in V, \quad p_h \hat{r}_h x \to x \quad \text{as } h \text{ approaches zero}.$

If $U \subset V$ with a stronger topology, we measure the *speed of the convergence* by means of the error functions:

(5) $\quad e_U^V(p_h) = \| 1 - p_h \hat{r}_h \|_{\mathscr{L}(U,V)} = \sup_{x \in U} \frac{\| x - p_h \hat{r}_h x \|_V}{\| x \|_U}.$

(See Chapter 11, Section 6.)

Remark 1

If $V_h = \mathbb{R}^{n(h)}$ is a finite dimensional space, being given p_h amounts to being given $n(h)$ linearly independent vectors e_h^i related to p_h by the formula

(6) $\quad p_h x_h = \sum_i x_h^i e_h^i.$ ∎

Example 1

We can associate with every orthonormal base $\{e_n\}_{n \geq 1}$ of V the extension operators $p_n x_n = \sum_{i=1}^n x_n^i e_i$. Then $\hat{r}_n$ is defined by $(\hat{r}_n x)^i = ((x, e_i))$. The approximations thus constructed are called *Galerkin approximations*. We have seen examples of such approximations in Chapter 8, Sections 1, 2, and 3. Theorem 11.6.1 shows that *if the injection from U to V is compact*, the Galerkin approximation associated with the base of eigenvectors of $K^{-1}J$ is that which minimizes the error function. ∎

Example 2

We find in Chapter 8, Section 6 examples of approximations of Sobolev spaces. ∎

Hence consider a family $\{V_h, p_h, \hat{r}_h\}_h$ of convergent approximations of V. We can associate to problem (2) the following approximating problems: to find $x_h \in V_h$ that is a solution of

(7) $\quad\quad \forall y_h \in V_h, \quad a(p_h x_h, p_h y_h) = (f, p_h y_h) + \langle \varphi, \gamma p_h y_h \rangle.$

Since the bilinear form $a(p_h x_h, p_h y_h)$ is V_h-elliptic:

(8) $\quad\quad \forall y_h \in V_h, \quad a(p_h y_h, p_h y_h) \geq c \| p_h y_h \|^2,$

we deduce from the Lax-Milgram theorem, Theorem 3.6.1, the existence of a unique solution $x_h \in V_h$ of (7). We are going to show that it converges to the solution x of the Neumann problem (3).

THEOREM 1
Suppose that hypotheses (1) and, in Section 2, hypotheses (2) and (3) are satisfied and that we are given convergent approximations $\{V_h, p_h, \hat{r}_h\}$ of V. Then the solution x of the Neumann problem (3) is the limit of the family of the $p_h x_h$'s where the x_h's are the solutions of the problems (7). The error in V is bounded above in the following fashion when $x \in U \subset V$:

(9) $\quad\quad \| x - p_h x_h \| \leq Mc^{-1} \| x - p_h \hat{r}_h x \|_V \leq Mc^{-1} \| x \|_U e_U^V(p_h)$

(where M is the norm of the bilinear form a) and the error in H is bounded above as follows:

(10) $\quad\quad \| x - p_h x_h \|_H \leq M^2 c^{-1} \| x \|_U e_{D^*}^V(p_h) e_U^V(p_h)$

where $D^* = \{ y \in V(\Lambda^*) \text{ such that } \delta^* = 0 \}$ is supplied with the norm $\| \Lambda^* x \|_H$. ▲

Proof. First of all, it follows from (2) with $y = p_h y_h$ and from (7) that

(11) $\quad\quad a(x - p_h x_h, p_h y_h) = 0 \quad \text{for all} \quad y_h \in V_h.$

Consequently, we deduce from the V-ellipticity of a and from (11) the following upper estimates:

$$c \| x - p_h x_h \|^2 \leq a(x - p_h x_h, x - p_h x_h)$$
$$= a(x - p_h x_h, x - p_h \hat{r}_h x) \leq M \| x - p_h x_h \| \| x - p_h \hat{r}_h x \|$$

and hence inequality (9) (using the definition of $e_U^V(p_h)$). We can write

(12) $\quad\quad a(x - p_h x_h, y) = a_*(y, x - p_h x_h) = (\Lambda^* y, x - p_h x_h)$

when y belongs to the set $D^* = \{ y \in V(\Lambda^*) \text{ such that } \delta^* y = 0 \}$. Since Λ^* is an isomorphism from D^* onto H (see Theorem 3.1), it follows from (12) that

(13) $\quad\quad \| x - p_h x_h \|_H = \sup_{y \in D^*} \frac{|(\Lambda^* y, x - p_h x_h)|}{\| \Lambda^* y \|_H} = \sup_{y \in D^*} \frac{|a(x - p_h x_h, y)|}{\| \Lambda^* y \|_H}.$

Moreover,

(14)
$$|a(x - p_h x_h, y)| = |a(x - p_h x_h, y - p_h \hat{r}_h y)|$$
$$\leq M \|x - p_h x_h\| \|y - p_h \hat{r}_h y\|$$
$$\leq M \|x - p_h x_h\| e_{D^*}^V(p_h) \|y\|_{D^*}$$

where D^* is the closed subspace of $V(\Lambda^*)$ with the norm $\|y\|_{D^*} = \|\Lambda^* y\|_H$ [equivalent to that of $V(\Lambda^*)$]. Inequality (10) then results from inequalities (13), (14), and (9). ∎

Remark 2

We can prove an analogous approximation theorem for the Dirichlet problems and other variational boundary value problems. ∎

Example 3

We are only going to apply the preceding results to the Sturm-Liouville problem: to find $x \in H^1(\Omega, \Lambda)$ such that

(15) $$\begin{cases} \text{i.} & -D(v(\omega)Dx(\omega)) + \lambda x(\omega) = f(\omega) \quad \text{(in the sense of } L^2(\Omega)) \\ \text{ii.} & v(a)Dx(a) = \varphi^a \quad \text{and} \quad v(b)Dx(b) = \varphi^b \end{cases}$$

by using the approximations (V^h, p_h^1, r_h) defined in Chapter 8, Section 6. We set $\Omega =]0, 1[$ and $h = 1/n$. The knots of the grid are, therefore, $jh (0 \leq j \leq n)$. The restriction of $p_h^1 x_h$ to Ω can thus be written

(16) $$p_h^1 x_h(\omega) = \sum_{j=0}^{n-1} \left(x_h^j + (x_h^{j+1} - x_h^j)\left(\frac{\omega}{h} - j\right) \right) \theta_h^j(\omega).$$

(This is in fact the piecewise linear function that interpolates the values x_h^j at the knots jh of the grid.)

Since problem (15) is equivalent to the variational equation

(17) $$\begin{cases} \int_0^1 v(\omega)Dx(\omega)Dy(\omega)d\omega + \lambda \int_0^1 x(\omega)y(\omega)d\omega \\ = \int_0^1 f(\omega)y(\omega)d\omega + \varphi^1 y(1) - \varphi^0 y(0) \quad \forall y \in H^1(0, 1), \end{cases}$$

we can approximate the solution x of (15) by the solution $x_h \in \mathbb{R}^{n+1}$ of the variational problem

(18) $$\begin{cases} \int_0^1 v(\omega)Dp_h^1 x_h \cdot Dp_h^1 y_h \, d\omega + \lambda \int_0^1 p_h^1 x_h p_h^1 y_h \, d\omega \\ = \int_0^1 f(\omega) p_h^1 y_h \, d\omega + \varphi^b p_h^1 y_n(1) - \varphi^a p_h^1 y_n(0). \end{cases}$$

This problem is of the form

(19) $$A_h x_h = f_h$$

where x_h and f_h belong to $\mathbb{R}^{n+1}$ and A_h is a matrix of $\mathbb{R}^{n+1}$. We are going to calculate the components of f_h and the coefficients of A_h. To do this we set

(20) $$\begin{cases} \text{i.} & b_j = -\dfrac{1}{h^2} \int\limits_{jh}^{(j+1)h} v(\omega)\, d\omega - \dfrac{\lambda h}{6} & \text{if} & j = 0, \ldots, n \\[1em] \text{ii.} & a_0 = \dfrac{1}{h^2} \int\limits_0^h v(\omega)\, d\omega + \dfrac{\lambda h}{3} & \text{if} & j = 0 \\[1em] \text{iii.} & a_j = \dfrac{1}{h^2} \int\limits_{(j-1)h}^{(j+1)h} v(\omega)\, d\omega + \dfrac{8\lambda h}{3} & \text{if} & j = 1, \ldots, n-1 \\[1em] \text{iv.} & a_n = \dfrac{1}{h^2} \int\limits_{(n-1)h}^{1} v(\omega)\, d\omega + \dfrac{7\lambda h}{3} & \text{if} & j = n \end{cases}$$

and

(21) $$f_h^j = \begin{cases} \int\limits_0^h f(\omega)\left(-\dfrac{\omega}{h} + 1\right) d\omega - \varphi^0 & \text{if} \quad j = 0 \\[1em] \int\limits_{(j-1)h}^{jh} f(\omega)\left(\dfrac{\omega}{h} - j + 1\right) d\omega \\[1em] \quad + \int\limits_{jh}^{(j+1)h} f(\omega)\left(-\dfrac{\omega}{h} + j + 1\right) d\omega & \text{if} \quad j = 1, \ldots, n-1 \\[1em] \int\limits_{(n-1)h}^{1} f(\omega)\left(\dfrac{\omega}{h} - n + 1\right) d\omega + \varphi^1 & \text{if} \quad j = n. \end{cases}$$

Then the components f_h^j of f_h are defined by (21) and the matrix A_h is of the form

$$\begin{vmatrix} a_0 & b_0 & 0 & 0 & 0 \ldots\ldots\ldots\ldots\ldots & 0 & 0 & 0 \\ 0 & b_0 & a_1 & b_1 & 0 \ldots\ldots\ldots\ldots\ldots & 0 & 0 & 0 \\ \vdots & & & & & & & \\ 0 & 0 & 0 & 0 & 0 \ldots b_{j-1} \quad a_j \quad b_j \ldots 0 & 0 & 0 \\ \vdots & & & & & & & \\ 0 & 0 & 0 & 0 & 0 \ldots\ldots\ldots\ldots\ldots & 0 & b_{n-1} & a_n \end{vmatrix}$$

Indeed, we can write

$$a(p_h^1 x_h, p_h^1 y_h) = \sum_{j=0}^{n-1} \int\limits_{jh}^{(j+1)h} g_j(\omega)\, d\omega$$

where, according to formula (16), we obtain

$$g_j(\omega) = v(\omega)(x_h^{j+1} - x_h^j)(y_h^{j+1} - y_h^j)$$
$$+ \lambda\left(x_h^j + (x_h^{j+1} - x_h^j)\left(\frac{\omega}{h} - j\right)\right)\left(y_h^j + (y_h^{j+1} - y_h^j)\left(\frac{\omega}{h} - j\right)\right).$$

Consequently, integrating and simplifying the terms, we obtain

$$a(p_h^1 x_h, p_h^1 y_h) = (h_0 x_h^1 + a_0 x_h^0)y_h^0$$
$$+ \sum_{j=1}^{n-1}(b_j x_h^{j+1} + a_j x_h^j + b_{j-1} x_h^{j-1})y_h^j + (a_n x_h^n + b_{n-1} x_h^{n-1})y_h^n$$
$$= \sum_{j=0}^{n}(A_h x_h)_j y_h^j.$$

The calculation of the components f_h^j of f_h is easy. ∎

Theorem 1 and the results of Chapter 8, Section 6 imply the following result.

PROPOSITION 1
Suppose that $\inf_{\omega \in \Omega} v(\omega) > 0$ *and that* $\lambda > 0$. *Then problems (18) have unique solutions* x_h *such that* $p_h^1 x_h$ *converges to the solution* x *of the Neumann problem (15). If the solution* x *belongs to* $H^2(\Omega)$, *there exists a constant* $c > 0$ *such that*

(22) $$\|x - p_h^1 x_h\|_{H^1(\Omega)} \leq ch\|x\|_{H^2(\Omega)}.$$

If, in addition, $D^* \subset H^2(\Omega)$ [*i.e., every solution of the homogeneous Neumann problem belongs to* $H^2(\Omega)$], *then there exists a constant* $d > 0$ *such that*

(23) $$\|x - p_h^1 x_h\|_{L^2(\Omega)} \leq dh^2\|x\|_{H^2(\Omega)}.$$ ▲

6. RESTRICTION AND EXTENSION OF THE FORMAL ADJOINT

We have shown that the operator $\Lambda \times \gamma \times \delta$ is defined from $V(\Lambda)$ onto $H \times T \times T^*$.

Suppose that there exist spaces U, R, and T satisfying

(1)
$\begin{cases} \text{i.} & U \subset V(\Lambda) \cap V(\Lambda^*). \\ \text{ii.} & S \subset T \subset R, \text{ each space being dense in the following ones.} \\ \text{iii.} & \gamma \times \delta \text{ and } \gamma \times \delta^* \text{ are surjective operators from } U \text{ onto } S \times R^*. \\ \text{iv.} & U_0 = \text{Ker}(\gamma \times \delta) = \text{Ker}(\gamma \times \delta^*) \text{ is dense in } H. \end{cases}$

We can then apply Theorem 1.1 with V replaced by U, V_0 by U_0, T by $S \times R^*$, γ by $\gamma \times \delta$, and A by $\Lambda \in \mathcal{L}(U, H)$.

First of all, the *formal adjoint* $\Lambda_0^* \in \mathscr{L}(H, U_0^*)$ of Λ is the *unique extension by density of the formal operator* $\Lambda^* \in \mathscr{L}(V(\Lambda^*), H)$ associated with the form a_* defined by $a_*(x, y) = a(y, x)$, since if $x \in V(\Lambda^*)$ and if $y \in U_0$,

$$(\Lambda_0^* x, y) = (x, \Lambda_0 y) = (x, \Lambda y) = a(y, x) = a_*(x, y) = (\Lambda^* x, y).$$

Hence we agree to set $\Lambda_0^* = \Lambda^*$.

Consequently, the domain of the formal adjoint Λ^* of Λ is the space

(2) $$H(\Lambda^*) = \{x \in H \quad \text{such that} \quad \Lambda^* x \in H\}$$

with the scalar product

(3) $$((x, y))_{H(\Lambda^*)} = (x, y)_H + (\Lambda^* x, \Lambda^* y)_H.$$

Theorem 1.1 implies the existence of a unique operator denoted $\{\alpha_1^*, \alpha_2^*\} \in \mathscr{L}(H(\Lambda), S^* \times R)$ satisfying

$$\forall x \in H(\Lambda), \quad \forall y \in U \quad (x, \Lambda y) - (\Lambda^* x, y) = \langle \alpha_1^* x, \gamma y \rangle + \langle \delta y, \alpha_2^* x \rangle.$$

But Proposition 13.2.1 implies that if $x \in V(\Lambda^*) \subset H(\Lambda^*)$ and $y \in U \subset V$, we have

(4) $$(x, \Lambda y) - \langle \Lambda^* x, y \rangle = -\langle \delta^* x, \gamma y \rangle + \langle \delta y, \gamma x \rangle.$$

From this it follows that $\alpha_1^* \in \mathscr{L}(H(\Lambda), S^*)$ is an extension of $-\delta^* \in \mathscr{L}(V(\Lambda^*), T^*)$ and that $\alpha_2^* \in \mathscr{L}(H(\Lambda^*), R)$ is an extension of $\gamma \in \mathscr{L}(V(\Lambda^*), T)$. This motivates us to set, henceforth, $\alpha_1^* = -\delta^*$ and $\alpha_2^* = \gamma$.

We obtain analogous results for Λ, γ, δ by interchanging the roles of $a(x, y)$ and $a_*(x, y)$.

Summing up, we have established the following results.

PROPOSITION 1
Suppose that the hypotheses (1) *hold. Then we can extend*

(5) $$\begin{cases} \text{i.} & \Lambda \in \mathscr{L}(U, H) \cap \mathscr{L}(V(\Lambda), H) \quad \text{to} \quad \Lambda \in \mathscr{L}(H(\Lambda), H) \\ \text{ii.} & \Lambda^* \in \mathscr{L}(U, H) \cap \mathscr{L}(V(\Lambda^*), H) \quad \text{to} \quad \Lambda^* \in (H(\Lambda), H) \end{cases}$$

and extend

(6) $$\begin{cases} \text{i.} & \gamma \in \mathscr{L}(U, S) \cap \mathscr{L}(V(\Lambda), T) \quad \text{to} \quad \gamma \in \mathscr{L}(H(\Lambda), R) \\ \text{ii.} & \delta \in \mathscr{L}(U, R^*) \cap \mathscr{L}(V(\Lambda), T^*) \quad \text{to} \quad \delta \in \mathscr{L}(H(\Lambda), S^*) \\ \text{iii.} & \delta^* \in \mathscr{L}(U, R^*) \cap \mathscr{L}(V(\Lambda^*), T^*) \quad \text{to} \quad \delta^* \in \mathscr{L}(H(\Lambda^*), S^*) \end{cases}$$

in such a way that Green's formula

(7) $$(x, \Lambda y) - (\Lambda^* x, y) = \langle \delta y, \gamma x \rangle - \langle \delta^* x, \gamma y \rangle$$

is valid for all $x \in H(\Lambda^*)$ *and for all* $y \in U$ [*or for all* $x \in U$ *and for all* $y \in H(\Lambda)$]. ▲

THEOREM 1

Suppose that hypotheses (1) are satisfied and that A is V-elliptic. Then if $D(A^*) \subset U$, the operator $\Lambda \times \delta$ is an isomorphism from $V(\Lambda)$ onto $H \times T^*$, from U onto $H \times R^*$, and from $H(\Lambda)$ onto $H \times S^*$. If A is only V_0-elliptic and if $D(A_{V_0}^*) \subset U$, then the operator $\Lambda \times \gamma$ is an isomorphism from $V(\Lambda)$ onto $H \times T$, from U onto $H \times S$, and from $H(\Lambda)$ onto $H \times R$. Analogous assertions are true when Λ is replaced by Λ^* and δ by δ^*. ▲

Proof. Consider the case of the operator $\Lambda \times \delta$, which is an isomorphism from $V(\Lambda)$ onto $H \times T^*$ according to Theorem 3.1.

a. Let us show that $\Lambda \times \delta$ is an isomorphism from $H(\Lambda)$ onto $H \times S^*$.

Since it is clear that $\Lambda \times \delta$ is continuous and injective, it is sufficient to prove that it is surjective, that is, that there exists $x \in H(\Lambda)$ such that $\Lambda x = f$, $\delta x = \varphi$ when $f \in H$, $\varphi \in S^*$. We know that Λ^* is an isomorphism from $D(A^*) = \text{Ker } \delta^*$ onto H (according to Theorem 3.3 with $W = V$ and $A = A^*$). Its transpose is an isomorphism from H onto $D(A^*)^*$. Since the linear form

$$y \mapsto (f, y) + \langle \varphi, \gamma y \rangle$$

is continuous on $D(A^*)$, it follows that there exists a unique solution $x \in H$ satisfying

(8) $\qquad \forall y \in D(A^*), \qquad (x, \Lambda^* y) = (f, y) - \langle \varphi, \gamma y \rangle.$

As y runs over U_0, we obtain that

$$(x, \Lambda^* y) = (x, (\Lambda^*|_{U_0}) y) = (\Lambda x, y) = (f, y)$$

for all $y \in U_0$. Hence $\Lambda x = f$, and since $f \in H$, x belongs to $H(\Lambda)$. Since $D(A^*) \subset U$, we can then apply Green's formula (7) where Λ is replaced by Λ^*. We deduce from (8) that if $y \in D(A^*) = \text{Ker } \delta^*$,

$$(x, \Lambda^* y) = (\Lambda x, y) + \langle \delta^* y, \gamma x \rangle - \langle \delta x, \gamma y \rangle$$
$$= (f, y) - \langle \varphi, \gamma y \rangle$$

and, consequently, that $\langle \delta x - \varphi, \gamma y \rangle = 0$ for all $y \in D(A^*)$. Since $\gamma D(A^*)$ is dense in T according to Theorem 3.3, it follows that $\delta x = \varphi$.

We have, therefore, proved that $\Lambda \times \delta$ is a surjective operator from $H(\Lambda)$ onto $H \times S^*$. It is then a continuous bijective operator that, according to the Banach theorem, Theorem 4.3.3, is an isomorphism.

b. Let us show that $\Lambda \times \delta$ is an isomorphism from U onto $H \times R^*$, that is, in fact, that it is surjective. To this end we shall prove that its image is closed and dense.

The image is dense: According to the density criterion (Theorem 2.2.1), we must show that if $\{x, \psi\} \in H \times S$ satisfies

$$(x, \Lambda y) + \langle \psi, \delta y \rangle = 0$$

for all $y \in U$, then $x = 0$ and $\psi = 0$. But, taking $y \in U_0$, we obtain $\Lambda^* x = 0$. Hence $x \in H(\Lambda^*)$. Green's formula then tells us that

$$\delta^* x = 0 \quad \text{and} \quad \psi = -\gamma x.$$

Since $(\Lambda^* \times \delta^*)$ is an isomorphism from $H(\Lambda^*)$ onto $H \times S^*$ (according to the first part of the proof where Λ and δ are replaced by Λ^* and δ^*), it follows that

$$x = 0 \quad \text{and} \quad \psi = -\gamma x = 0.$$

The image is closed: Suppose that a sequence $x_n \in U$ is such that Λx_n converges to f in H and δx_n converges to φ in R^*. Since $\delta \in \mathscr{L}(U, R^*)$ is surjective, there exists a right inverse $\rho \in \mathscr{L}(R^*, U)$. Hence $y_n = x_n - \rho \delta x_n$ belongs to $D(A) = \operatorname{Ker} \delta$ and $\Lambda y_n = \Lambda x_n - \Lambda \rho \delta x_n$ converges to $f - \Lambda \rho \varphi$ in H. Since $\Lambda = A$ is an isomorphism from $D(A)$ onto H, according to Theorem 3.3, it follows that the sequence y_n converges to an element y satisfying $\Lambda y = f - \Lambda \rho \varphi$. Hence $x = y + \rho \varphi$ belongs to U and satisfies

$$\Lambda x = f \quad \text{and} \quad \delta x = \delta y + \varphi = \varphi \quad \text{since} \quad \delta y = 0.$$

The theorem is therefore established for $\Lambda \times \delta$.

The proof for $\Lambda \times \gamma$ is analogous, since the roles played by γ and δ are symmetric. ∎

Example 1

Consider the Neumann problem

$$(9) \quad \begin{cases} \text{i.} & \Lambda x = -\sum_{i,j=1}^{n} D_j(v_{ij}(\omega) D_i x(\omega)) + v_0(\omega) x(\omega) = f(\omega). \\ \text{ii.} & \delta x = \dfrac{\partial}{\partial v_\Lambda} x = \psi. \end{cases}$$

We know that under the ellipticity hypotheses (20) of Section 4 $\Lambda \times \delta$ is an isomorphism from $H^1(\Omega, \Lambda)$ onto $L^2(\Omega) \times H^{-1/2}(\Gamma)$.

Let us *accept* that if Ω is "regular" and if the functions v_{ij} and v_0 are sufficiently differentiable, then the solutions x of the homogeneous Neumann problem

$$(10) \quad \begin{cases} \Lambda^* x = -\sum_{i,j=1}^{n} D_i(v_{ij} D_j x(\omega)) + v_0(\omega) x(\omega) = f(\omega) \\ \delta^* x = \dfrac{\partial}{\partial v_{\Lambda^*}} x = 0 \end{cases}$$

belong to $H^2(\Omega)$ as f runs over $L^2(\Omega)$ and that $H^2(\Omega)$ is contained in $H^1(\Omega, \Lambda)$

and $H^1(\Omega, \Lambda^*)$. Suppose as well that

$$\gamma_0 \times \frac{\partial}{\partial v_\Lambda} \text{ and } \gamma_0 \times \frac{\partial}{\partial v_{\Lambda^*}}$$

are surjective operators from $H^2(\Omega)$ onto $H^{3/2}(\Gamma) \times H^{1/2}(\Gamma)$ whose kernels coincide with $H_0^2(\Omega)$. Then the hypotheses (1) are satisfied with

$$U = H^2(\Omega), \quad S = H^{3/2}(\Gamma), \quad T = H^{1/2}(\Gamma), \quad R = H^{-1/2}(\Gamma), \quad \text{and}$$

$$D^* = \left\{ x \in H^1(\Omega, \Lambda^*) \text{ such that } \frac{\partial}{\partial v_{\Lambda^*}} x = 0 \right\} \subset H^2(\Omega).$$

Theorem 1 then implies that $\Lambda \times \frac{\partial}{\partial v_\Lambda}$ is an isomorphism from $H^2(\Omega)$ onto $L^2(\Omega) \times H^{1/2}(\Gamma)$, from $H^1(\Omega, \Lambda)$ onto $L^2(\Omega) \times H^{-1/2}(\Gamma)$ and from $H^0(\Omega, \Lambda)$ onto $L^2(\Omega) \times H^{-3/2}(\Gamma)$ where $H^0(\Omega, \Lambda) = \{x \in L^2(\Omega) \text{ such that } \Lambda x \in L^2(\Omega)\}$. ∎

7. UNILATERAL BOUNDARY VALUE PROBLEMS

Suppose that $V, H,$ and T are Hilbert spaces such that

(1) $\begin{cases} \text{i.} & V \subset H \text{ with a continuous injection.} \\ \text{ii.} & V_0 = \text{Ker } \gamma \text{ is dense in } H \text{ where} \\ \text{iii.} & \gamma \in \mathscr{L}(V, T) \text{ is surjective.} \end{cases}$

Suppose we are given

(2) a closed convex cone $P \subset T$, and set $P^+ \subset T^*$ as its positive polar cone.

Let $a(x, y)$ be a continuous bilinear form on $V \times V, \Lambda \in \mathscr{L}(V, V_0^*) \cap \mathscr{L}(V(\Lambda), H)$ its associated formal operator and $\delta = \delta(\Lambda, \delta)$ the associated Neumann operator.

PROPOSITION 1
Suppose we are given

(3) $\qquad f \in H, \qquad \varphi \in T, \qquad \text{and} \qquad \psi \in T^*.$

Then the following problems are equivalent:
a. *To find $x \in V(\Lambda)$ satisfying*

(4) $\begin{cases} \text{i.} & \Lambda x = f. \\ \text{ii.} & \gamma x - \varphi \in P. \\ \text{iii.} & \delta x - \psi \in P^+. \\ \text{iv.} & \langle \delta x - \psi, \gamma x - \varphi \rangle = 0. \end{cases}$

b. To find $x \in V$ satisfying $x = z + x_1$ where x_1 satisfies $\gamma x_1 = \varphi$ and where z is a solution of the variational inequalities

(5) $\quad \begin{cases} \text{i.} & z \in Q = \gamma^{-1}(P). \\ \text{ii.} & a(z, z - y) \leq (f, z - y) - a(x_1, z - y) + \langle \psi, \gamma(z - y) \rangle \text{ for all } y \in Q. \end{cases}$ ▲

DEFINITION 1
Problem (3) is called a unilateral boundary value problem. ▲

Proof of Proposition 1. Let $x \in V(\Lambda)$ be a solution to problem **a** and x_1 a solution of $\gamma x_1 = \varphi$. Then $\gamma z = \gamma x - \varphi \in P$. Moreover, if $y \in Q$, that is, if $\gamma y \in P$, we obtain

$$a(z, y) = a(x, y) - a(x_1, y) = (\Lambda x, y) + \langle \delta x, \gamma y \rangle - a(x_1, y)$$
$$\geq (f, y) - a(x_1, y) + \langle \psi, \gamma y \rangle$$

since $\delta x - \psi \in P^+$. Moreover, we deduce from the inequality

$$\langle \delta x - \psi, \gamma x - \varphi \rangle = \langle \delta x - \psi, \gamma z \rangle = 0$$

that

$$a(z, z) = (\Lambda x, z) - a(x_1, z) + \langle \delta x, \gamma z \rangle$$
$$= (f, z) - a(x_1, z) + \langle \psi, \gamma z \rangle.$$

Hence z is indeed a solution to problem **b**.

Conversely, let $x = z + x_1$ where x_1 is a solution of the equation $\gamma x_1 = \varphi$ and where z is a solution of the inequalities (5), that is, of

(6) $\quad a(z + x_1, y) = a(x, y) \geq (f, y) + \langle \psi, \gamma y \rangle \geq 0 \quad \forall y \in Q$

and

(7) $\quad a(z + x_1, z) = a(x, z) = (f, z) + \langle \psi, \gamma z \rangle.$

First of all, the fact that γz belongs to $Q = \gamma^{-1} P$ implies that

(8) $\quad \gamma x - \varphi = \gamma(x - x_1) = \gamma z \in P.$

Then since V_0 is a vector subspace of the cone Q, the inequalities (5) imply the equalities

$$(\Lambda x, y) = (f, y) \quad \text{for all} \quad y \in V_0$$

and, consequently, that

(9) $\quad \Lambda x = f \quad \text{and} \quad x \in V(\Lambda).$

Hence we can apply Green's formula and obtain

$$(\Lambda x, y) + \langle \delta x, \gamma y \rangle \geq (f, y) + \langle \psi, \gamma y \rangle \quad \forall y \in Q,$$

and, therefore, that $\langle \delta x - \psi, \theta \rangle \geq 0$ for all $\theta \in P$. Thus $\delta x - \psi \in P^+$.

Finally we deduce from equality (9) that
$$a(x,z) = (\Lambda x, z) + \langle \delta x, \gamma z \rangle = (f, z) + \langle \psi, \gamma z \rangle$$
and, consequently, that
(10) $$\langle \delta x - \psi, \gamma z \rangle = \langle \delta x - \psi, \gamma x - \varphi \rangle = 0. \qquad \blacksquare$$

Proposition 1 and Theorem 3.7.1 imply the existence and uniqueness of a solution to problem (4).

THEOREM 1
Suppose that hypotheses (1) are satisfied and that the bilinear form a is V-elliptic. Then there exists a unique solution to the unilateral boundary value problem (4).
$\blacktriangle$

Examples

Consider a regular open set $\Omega \subset \mathbb{R}^n$ with boundary Γ, a second order operator Λ defined by

(11) $$\Lambda x(\omega) = - \sum_{i,j=1}^{n} D_j(v_{ij}(\omega) D_i x(\omega)) + v_0(\omega) x(\omega)$$

as well as the associated trace operators $\gamma_0 \in \mathscr{L}(H^1(\Omega), H^{1/2}(\Gamma))$ and $\dfrac{\partial}{\partial v_\Lambda} \in \mathscr{L}(H^1(\Omega, \Lambda), H^{-1/2}(\Gamma))$.

Consider the following case:

(12) $P = \{\varphi \in H^{1/2}(\Gamma) \quad \text{such that} \quad \varphi(\omega) \geqq 0 \quad \text{almost everywhere on } \Gamma\}$

of "positive" functions on the boundary Γ. We say that the elements of the positive polar cone P^+ are "positive" distributions $\psi \in H^{-1/2}(\Gamma)$.

Suppose we are given

(13) $$f \in L^2(\Omega), \qquad \varphi \in H^{1/2}(\Gamma) \quad \text{and} \quad \psi \in H^{-1/2}(\Gamma).$$

We then deduce from Theorem 1 that under the ellipticity hypotheses (20) of Section 4 there exists a unique solution to the unilateral boundary value problem:

(14) $\begin{cases} \text{i.} \quad -\sum_{i,j=1}^{n} D_j(v_{ij}(\omega) D_i x(\omega)) + v_0(\omega) x(\omega) = f(\omega). \\ \text{ii.} \quad \gamma_0 x \geqq \varphi \quad \text{and} \quad \dfrac{\partial}{\partial v_\Lambda} x \geqq \psi. \\ \text{iii.} \quad \int_\Gamma \left(\dfrac{\partial}{\partial v_\Lambda} x - \psi \right)(\gamma_0 x - \varphi) \, d\sigma = 0. \end{cases}$ $\qquad \blacksquare$

8. INTRODUCTION TO CALCULUS OF VARIATIONS

Consider the spaces $V, H,$ and T and $\gamma \in \mathscr{L}(V, T)$ such that

(1) $\begin{cases} \text{i.} & V \subset H \quad \text{with a continuous injection.} \\ \text{ii.} & \gamma \quad \text{is surjective.} \\ \text{iii.} & V \text{ and } V_0 = \operatorname{Ker} \gamma \quad \text{are dense in } H. \end{cases}$

Let $A \in \mathscr{L}(V, E)$ be a continuous linear operator. We suppose that H and E are pivot spaces.

Suppose that we are given a function

$$\mathscr{F} : H \times E \times T \mapsto]-\infty, +\infty],$$

which we shall, in what follows, suppose to satisfy

(2) $\mathscr{F}$ is a convex lower semicontinuous function with nonempty domain.

The problems of *calculus of variations* present themselves in the following abstract form:

(3) $$v = \inf_{x \in V} \mathscr{F}(x, Ax, \gamma x).$$

We associate to A its formal adjoint $A_0^* \in \mathscr{L}(E(A_0^*), H)$ and the operator $\beta^* \in \mathscr{L}(E(A_0^*), T^*)$. We then consider the problem

(4) $$v^* = \inf_{p \in E(A_0^*)} \mathscr{F}^*(-A_0^* p, p, -\beta^* p)$$

where $\mathscr{F}^* : H \times E \times T^* \mapsto]-\infty, +\infty]$ is the conjugate function of $\mathscr{F}$. Green's formula (Theorem 1.1) implies that

(5) $$0 \leq v + v^*$$

since

$$0 = [p, Ax] - (A_0^* p, x) - \langle \beta^* p, \gamma x \rangle \leq \mathscr{F}(x, Ax, \gamma x) + \mathscr{F}^*(-A_0^* p, p, -\beta^* p)$$

for all $x \in V, p \in E(A_0^*)$.

DEFINITION 1
We say that $\bar{p} \in E(A_0^*)$ is a Lagrange multiplier if and only if

(6) $\begin{cases} \text{i.} & 0 = v + v^*. \\ \text{ii.} & v^* = \mathscr{F}^*(-A_0^* \bar{p}, \bar{p}, -\beta^* \bar{p}). \end{cases}$ ▲

PROPOSITION 1
A necessary and sufficient condition for $\bar{x}$ to minimize $x \mapsto \mathscr{F}(x, Ax, \gamma x)$ and

for $\bar{p}$ to be a Lagrange multiplier is that

(7) $$\{-A_0^*\bar{p}, \bar{p}, -\beta^*\bar{p}\} \in \partial \mathscr{F}(\bar{x}, A\bar{x}, \gamma\bar{x}). \qquad \blacktriangle$$

Proof. The proof is analogous to that of Proposition 10.4.2: The triple $\{-A_0^*\bar{p}, \bar{p}, -\beta^*\bar{p}\}$ belongs to $\partial \mathscr{F}(\bar{x}, A\bar{x}, \gamma\bar{x})$ if and only if

$$\mathscr{F}(\bar{x}, A\bar{x}, \gamma\bar{x}) + \mathscr{F}^*(-A_0^*\bar{p}, \bar{p}, -\beta^*\bar{p})$$
$$= (-A_0^*\bar{p}, \bar{x}) + [\bar{p}, A\bar{x}] + \langle -\beta^*\bar{p}, \bar{x} \rangle = 0$$

according to Green's formula. It follows that

$$0 = v + v^* = \mathscr{F}(\bar{x}, A\bar{x}, \gamma\bar{x}) + \mathscr{F}^*(-A_0^*\bar{p}, \bar{p}, -\beta^*\bar{p}). \qquad \blacksquare$$

We are now going to express the extremality relations (7) in the Hamiltonian form.

DEFINITION 2

We say that the function $\mathscr{H}: H \times E \times T \mapsto]-\infty, +\infty]$ defined by

(8) $$\mathscr{H}(x, p, \psi) = \sup_{e \in E, \varphi \in T} \{[p, e] + \langle \psi, \varphi \rangle - F(x, e, \varphi)\}$$

is the Hamiltonian associated with $\mathscr{F}$. $\qquad \blacktriangle$

We can then express the extremality conditions (7) in the form of a Hamiltonian system (see Chapter 10, Section 5).

PROPOSITION 2

A necessary and sufficient condition for $\bar{x}$ to minimize $x \mapsto \mathscr{H}(x, Ax, \gamma x)$ and for $\bar{p}$ to be a Lagrange multiplier is

(9) $$\begin{cases} \text{i.} & \{A\bar{x}, \gamma\bar{x}\} \in \partial_{p,\varphi} \mathscr{H}(\bar{x}, \bar{p}, -\beta^*\bar{p}) \\ \text{ii.} & A_0^*\bar{p} \in \bar{\partial}_x \mathscr{H}(\bar{x}, \bar{p}, -\beta^*\bar{p}) \end{cases}$$

where $\bar{\partial}_x \mathscr{H}(\bar{x}, \bar{p}, -\beta^*\bar{p}) = -\partial_x(-\mathscr{H}(\bar{x}, \bar{p}, -\beta^*\bar{p}))$. $\qquad \blacktriangle$

Proof. The proof is analogous to that of Proposition 10.5.1. Let $\{-A_0^*\bar{p}, \bar{p}, -\beta^*\bar{p}\} \in \partial \mathscr{F}(\bar{x}, A\bar{x}, \gamma\bar{x})$ (according to Proposition 1). Then

(10) $$\begin{cases} \mathscr{F}(\bar{x}, A\bar{x}, \gamma\bar{x}) - \mathscr{F}(x, e, \varphi) \le (-A_0^*\bar{p}, \bar{x} - x) + [\bar{p}, A\bar{x} - e] \\ \qquad - \langle \beta^*\bar{p}, \gamma\bar{x} - \varphi \rangle. \end{cases}$$

Taking $x = \bar{x}$, we deduce that $\{A\bar{x}, \gamma\bar{x}\}$ belongs to $\partial_{p,\varphi} \mathscr{H}(\bar{x}, \bar{p}, -\beta^*\bar{p})$ since $\mathscr{H}(\bar{x}, \cdot, \cdot)$ is the conjugate function of $\mathscr{F}(\bar{x}, \cdot, \cdot)$. This amounts to saying that

$$[\bar{p}, A\bar{x}] - \langle \beta^*\bar{p}, \gamma\bar{x} \rangle = \mathscr{F}(\bar{x}, A\bar{x}, \gamma\bar{x}) + \mathscr{H}(\bar{x}, \bar{p}, -\beta^*\bar{p}).$$

We can then write (10) in the form
$$[\bar{p}, e] + \langle -\beta^*\bar{p}, \varphi \rangle - \mathcal{F}(x, e, \varphi) \leq (-A_0^*\bar{p}, \bar{x} - x) + \mathcal{H}(\bar{x}, \bar{p}, -\beta^*\bar{p}).$$
Taking the supremum as e runs over E and φ runs over T, we obtain $A_0^*\bar{p} \in \bar{\partial}_x \mathcal{H}(\bar{x}, \bar{p}, -\beta^*\bar{p})$.

Conversely, suppose that $\{\bar{x}, \bar{p}\}$ is a solution of the Hamiltonian system (9). It follows from (9)i that

(11) $\qquad [\bar{p}, A\bar{x}] + \langle -\beta^*\bar{p}, \gamma\bar{x} \rangle = \mathcal{F}(\bar{x}, A\bar{x}, \gamma\bar{x}) + \mathcal{H}(\bar{x}, \bar{p}, -\beta^*\bar{p}).$

We obtain the following by the definition of $\mathcal{H}$:

(12) $\qquad [\bar{p}, e] - \langle \beta^*\bar{p}, \varphi \rangle \leq \mathcal{F}(x, e, \varphi) + \mathcal{H}(x, \bar{p}, -\beta^*\bar{p}).$

Since $A_0^*\bar{p} \in \bar{\partial}_x \mathcal{H}(\bar{x}, \bar{p}, -\beta^*\bar{p})$, we have

(13) $\qquad \mathcal{H}(x, \bar{p}, -\beta^*\bar{p}) - \mathcal{H}(\bar{x}, \bar{p}, -\beta^*\bar{p}) \leq (-A_0^*\bar{p}, \bar{x} - x).$

These last three inequalities imply that
$$\mathcal{F}(\bar{x}, A\bar{x}, \gamma\bar{x}) - \mathcal{F}(x, e, \varphi) \leq (-A_0^*\bar{p}, \bar{x} - x) + [\bar{p}, A\bar{x} - e]$$
$$+ \langle -\beta^*\bar{p}, \gamma\bar{x} - \varphi \rangle,$$
which expresses that $\{-A_0^*\bar{p}, \bar{p}, -\beta^*\bar{p}\} \in \partial \mathcal{F}(\bar{x}, A\bar{x}, \gamma\bar{x})$. ∎

Remark 1

In the case where

(14) $\qquad\qquad \mathcal{F}(x, e, \varphi) = \mathcal{F}_0(x, e) + f(\varphi)$

the Hamiltonian is written as follows:

(15) $\qquad\qquad \mathcal{H}(x, p, \psi) = \mathcal{H}_0(x, p) - f^*(\psi)$

where $\mathcal{H}_0(x, p) = \sup_{e \in E} \{[p, e] - \mathcal{F}(x, e)\}$. The extremality conditions of Proposition 1 are written as follows:

(16) $\begin{cases} \text{i.} & \{-A_0^*\bar{p}, \bar{p}\} \in \partial \mathcal{F}_0(\bar{x}, A\bar{x}) \\ \text{ii.} & -\beta^*\bar{p} \in \partial f(\gamma\bar{x}) \end{cases}$

and those of Proposition 2:

(17) $\begin{cases} \text{i.} & A\bar{x} \in \partial_p \mathcal{H}_0(\bar{x}, \bar{p}) \quad\text{and}\quad A_0^*\bar{p} \in \bar{\partial}_x \mathcal{H}_0(\bar{x}, \bar{p}). \\ \text{ii.} & -\beta^*\bar{p} \in \partial f(\gamma\bar{x}). \end{cases}$

Conditions (16)i are the *Euler-Lagrange conditions*; conditions (17)i are the *Hamiltonian conditions*; and condition (17)ii is the *transversality condition*. If $\mathcal{F}_0(x, e) = \frac{1}{2}(\|x\|_H^2 + \|e\|_E^2)$, equations (16)i become $\{-A_0^*\bar{p}, \bar{p}\} = \{\bar{x}, \mathcal{V} A\bar{x}\}$ if $\mathcal{V}$ is the duality operator $\langle \mathcal{V}x, x \rangle = \|x\|_H^2$. Eliminating $\bar{p}$, we obtain the system

$$\begin{cases} \text{i.} & A_0^* \mathscr{V} A\bar{x} + \bar{x} = 0 \\ \text{ii.} & -\beta^* \mathscr{V} A\bar{x} \in \partial f(\gamma \bar{x}), \end{cases}$$

which is a type of *multivalued boundary value problem*. ■

Example 1. The Variational Dirichlet Problems

Consider a regular open set $\Omega \subset \mathbb{R}^n$ with boundary Γ. We set $V = H^1(\Omega)$, $H = L^2(\Omega)$, $T = H^{1/2}(\Gamma)$, $E = L^2(\Omega)^n$, $\gamma = \gamma_0 \in \mathscr{L}(H^1(\Omega), H^{1/2}(\Gamma))$ the trace operator on Γ and $A = G = \text{grad}$. Then $A_0^* = -\text{div}$ and $\beta^* = \langle ., v \rangle$. (See Theorem 1.3.)

Consider a function $\mathscr{F}_0 : L^2(\Omega) \times L^2(\Omega)^n \mapsto]-\infty, +\infty]$ that is convex and lower semicontinuous, and of nonempty domain. We then define the variational Dirichlet problem as follows:

$$(18) \quad v = \inf_{\substack{x \in H^1(\Omega) \\ \gamma_0 x = \varphi}} \mathscr{F}_0(x, \text{grad } x) = \inf_{x \in H^1(\Omega)} [\mathscr{F}_0(x, \text{grad } x) + \Psi_{\{\varphi\}}(\gamma_0 x)]$$

where φ is given in $H^{1/2}(\Gamma)$ and $\Psi_{\{\varphi\}}$ is the indicator function of $\{\varphi\}$.

We then obtain

$$(19) \quad v^* = \inf_{p \in H^1(\Omega,, \text{div})} \{\mathscr{F}_0^*(\text{div } p, p) - \int_\Gamma \langle p, v \rangle \varphi \, d\sigma\}.$$

This is a *Neumann variational problem*.

The Euler-Lagrange system then becomes

$$(20) \quad \begin{cases} \text{i.} & \{\text{div } \bar{p}, \bar{p}\} \in \partial \mathscr{F}_0(\bar{x}, \text{grad } \bar{x}), \\ \text{ii.} & \gamma \bar{x} = \varphi. \end{cases}$$

Let $\mathscr{H}_0$ be the Hamiltoniam defined by

$$(21) \quad \mathscr{H}_0(x, p) = \sup_{e \in L^2(\Omega)^n} \left\{ \sum_{i=1}^n \int_\Omega p_i(\omega) e_i(\omega) \, d\omega - \mathscr{F}_0(x, e) \right\}.$$

The Hamiltonian system can be written as follows:

$$(22) \quad \begin{cases} \text{i.} & \text{grad } \bar{x} \in \partial_p \mathscr{H}_0(\bar{x}, \bar{p}). \\ \text{ii.} & -\text{div } \bar{p} \in \bar{\partial}_x \mathscr{H}_0(\bar{x}, \bar{p}). \\ \text{iii.} & \gamma \bar{x} = \xi. \end{cases}$$

▲

Example 2. The Case of Quadratic Functionals

Let us take the following as an example:

$$(23) \quad \mathscr{F}_0(x, e) = \frac{1}{2} \left\{ \int_\Omega |x(\omega)|^2 \, d\omega + \sum_{i,j=1}^n \int_\Omega v_{ij}(\omega) x_i(\omega) x_j(\omega) \, d\omega \right\}$$

where *the functions* v_{ij} *are continuous on* Ω *and satisfy* $v_{ij} = v_{ji}$ for all i,j. Suppose that we also have the *ellipticity condition*:

(24) $\exists c > 0$ such that $\sum_{i,j=1}^{n} v_{ij}(\omega) a_i a_j \geq c \sum_{i=1}^{n} a_i^2$ $\forall \omega \in \Omega$, $\forall a \in \mathbb{R}^n$.

Then
$$\partial \mathscr{F}_0(\bar{x}, \bar{p}) = \{x, \mathscr{V}\bar{p}\}$$

where $\mathscr{V} p(\omega) = \{\sum_{i=1}^{n} v_{ij}(\omega) p(\omega)\}_j$. The Euler-Lagrange system becomes $\operatorname{div} \bar{p} = \bar{x}$ and $\mathscr{V} \operatorname{grad} \bar{x} = \bar{p}, \gamma \bar{x} = \varphi$; that is,

(25) $\begin{cases} \text{i.} & -\operatorname{div} \mathscr{V} \operatorname{grad} \bar{x} + \bar{x} = 0. \\ \text{ii.} & \gamma \bar{x} = \varphi. \end{cases}$

This is the Dirichlet problem for the elliptic operator $\Lambda x = -\operatorname{div} \mathscr{V} \operatorname{grad} x + x$ (See Chapter 13, Section 4.)   ▲

9. INTRODUCTION TO OPTIMAL CONTROL

Let us now consider a Hilbert space U representing the space of "controls" of a system described abstractly by

(1) $$Ax - Bu = 0$$

[where $A \in \mathscr{L}(V, E)$ and $B \in \mathscr{L}(U, E)$] relating the "control" u to the "state" x of the system.

We suppose that we are given a loss (or cost) function $\mathscr{F} : H \times U \times T \mapsto]-\infty, +\infty]$ satisfying

(2) $\mathscr{F}$ is convex, lower semicontinuous, and has a nonempty domain.

Optimal control problems are formulated abstractly as follows:

(3) $$v = \inf_{Ax - Bu = 0} \mathscr{F}(x, u, \gamma x).$$

We associate with $\mathscr{F}$ the Hamiltonian $\mathscr{H} : H \times U^* \times T^* \mapsto]-\infty, +\infty]$ defined by

(4) $$\mathscr{H}(x, e, \psi) = \sup_{\substack{u \in U \\ \varphi \in T}} \{[e, u] + \langle \psi, \varphi \rangle - \mathscr{F}(x, u, \varphi)\}.$$

We associate with the optimal control problem the following problem:

(5) $$v^* = \inf_{p \in E(A_0^*)} \mathscr{F}^*(-A_0^* p, B^* p, -\beta^* p).$$

We say that $\bar{p} \in E(A_0^*)$ is an *optimal co-state* if

(6) $\begin{cases} \text{i.} & v + v^* = 0. \\ \text{ii.} & \bar{p} \quad \text{minimizes} \quad p \mapsto \mathscr{F}^*(-A_0^* p, B^* p, -\beta^* p). \end{cases}$

Since Green's formula is written

(7) $\begin{aligned} 0 &= [p, Ax] - (A_0^* p, x) - \langle \beta^* p, \gamma x \rangle \\ &= [B^* p, u] - (A_0^* p, x) - \langle \beta^* p, \gamma x \rangle \end{aligned}$

when $Ax = Bu$, it follows from the definition of the conjugate functions that

$$0 \leq \mathscr{F}(x, u, \gamma x) + \mathscr{F}^*(-A_0^* p, B^* p, -\beta^* p)$$

and from the definition of the subdifferential that $\{\bar{x}, \bar{u}\}$ is a *solution of the optimal control problem and that $\bar{p}$ is an optimal co-state if and only if*

(8) $\begin{cases} \text{i.} & A\bar{x} = B\bar{u}. \\ \text{ii.} & \{-A_0^* \bar{p}, B^* \bar{p}, -\beta^* \bar{p}\} \in \partial \mathscr{F}(\bar{x}, \bar{u}, \gamma \bar{x}). \end{cases}$

An argument analogous to the proof of Proposition 8.2 implies that the *extremality conditions* (8) are equivalent to

(9) $\begin{cases} \text{i.} & A\bar{x} = B\bar{u}. \\ \text{ii.} & A_0^* \bar{p} \in \bar{\partial}_x \mathscr{H}(\bar{x}, B^* \bar{p}, -\beta^* \bar{p}). \\ \text{iii.} & \{\bar{u}, \gamma \bar{x}\} \in \partial_{e, \psi} \mathscr{H}(\bar{x}, B^* \bar{p}, -\beta^* \bar{p}). \end{cases}$

Let us consider in particular the case where

(10) $\mathscr{F}(x, u, \varphi) = \mathscr{F}_0(x, u) + f(\varphi).$

It follows that

$$\mathscr{H}(x, e, \psi) = \mathscr{H}_0(x, e) - f^*(\psi)$$

where $\mathscr{H}_0(x, e) = \sup_u \{[e, u] - \mathscr{F}_0(x, u)\}.$

Relations (8) and (9) then become

(11) $\begin{cases} \text{i.} & A\bar{x} = B\bar{u}. \\ \text{ii.} & \{-A_0^* \bar{p}, B^* \bar{p}\} \in \partial \mathscr{F}_0(\bar{x}, \bar{u}). \\ \text{iii.} & -\beta^* \bar{p} \in \partial f(\gamma \bar{x}). \end{cases}$

and

(12) $\begin{cases} \text{i.} & A\bar{x} = B\bar{u}. \\ \text{ii.} & A_0^* \bar{p} \in \bar{\partial}_x \mathscr{H}_0(\bar{x}, B^* \bar{p}). \\ \text{iii.} & \bar{u} \in \partial_e \mathscr{H}_0(\bar{x}, B^* \bar{p}). \\ \text{iv.} & -\beta^* \bar{p} \in \partial f(\gamma \bar{x}). \end{cases}$

The relation (12)ii is called the *adjoint equation* of the system and the condition (12)iv the *transversality condition*. Condition (12)iii is equivalent to

(13) $\bar{u}$ maximizes $\langle \bar{p}, Bu \rangle - \mathscr{F}_0(\bar{x}, u)$.

It is known as the *Pontryagin maximum principle*. Eliminating $\bar{u}$ in equations (12), we obtain

(14) $\begin{cases} \text{i.} & A\bar{x} \in B\partial_e \mathscr{H}_0(\bar{x}, \beta^*\bar{p}). \\ \text{ii.} & A_0^*\bar{p} \in \bar{\partial}_x \mathscr{H}_0(\bar{x}, B^*\bar{p}). \\ \text{iii.} & -\beta^*\bar{p} \in \partial f(\gamma \bar{x}). \end{cases}$

This is type of boundary value problem for a system of multivated equations.

Example 1

Consider the case where $\Omega =]t, T[$ and where we take

$$V = H^1(\Omega, \mathbb{R}^n), \quad H = L^2(\Omega, \mathbb{R}^n), \quad U = L^2(\Omega, \mathbb{R}^p),$$
$$\gamma x = \{x(t), x(T)\}, \quad \beta^* p = \{-p(t), p(T)\}.$$

Suppose that we are given the families of matrices $M(\cdot)$ and $B(\cdot)$ belonging to $\mathscr{C}(\Omega, \mathscr{L}(\mathbb{R}^n, \mathbb{R}^n))$ and $\mathscr{C}(\Omega, \mathscr{L}(\mathbb{R}^p, \mathbb{R}^n))$ and

$\begin{cases} \text{a convex lower semicontinuous function } \mathscr{F}_0 \text{ from} \\ L^2(\Omega, \mathbb{R}^n) \times L^2(\Omega, \mathbb{R}^p) \text{ to }]-\infty, +\infty] \text{ with nonempty domain.} \end{cases}$

Let G be a closed convex subset of $\mathbb{R}^n \times \mathbb{R}^n$. The optimal control problem becomes

(15) $v = \inf \mathscr{F}_0(x, u)$

where the following conditions are required to be satisfied:

(16) $\begin{cases} \text{i.} & \dot{x}(\omega) - M(\omega)x(\omega) = B(\omega)u(\omega) \\ \text{ii.} & \{x(t), x(T)\} \in G \end{cases}$

where $\dot{x} = (d/d\omega)x = Dx$.

System (16)i describes the evolution of the system, conditions (16)ii describe constraints on the initial state $x(t)$ and the final state $x(T)$ of the system.

We denote by $M \in \mathscr{L}(L^2(\Omega, \mathbb{R}^n), L^2(\Omega, \mathbb{R}^n))$, $B \in \mathscr{L}(L^2(\Omega, \mathbb{R}^p), L^2(\Omega, \mathbb{R}^n))$ and $A \in \mathscr{L}(H^1(\Omega, \mathbb{R}^n), L^2(\Omega, \mathbb{R}^n))$ the operators defined by $(Mx)(\omega) = M(\omega)x(\omega)$, $Bx(\omega) = B(\omega)x(\omega)$ and $Ax = \dot{x} - Mx$. We take for $f = \Psi_G$ the indicator function of G and for $f^* = \sigma_G$ its conjugate function, the support function of G.

Since the optimal control problem is written as follows:

(17) $$v = \inf_{\dot{x} - Mx = Bu} [\mathscr{F}_0(x, u) + \Psi_G(\{x(t), x(T)\})],$$

and since the formal adjoint A_0^* of A is defined by $A_0^* p = -\dot{p} - M^* p$ (where $M^* p(\omega) = M^*(\omega) p(\omega)$), it follows that

(18) $$v^* = \inf_{p \in H^1(\Omega, \mathbb{R}^n)} [\mathscr{F}_0(\dot{p} + M^* p, B^* p) + \sigma_G(\{p(t), -p(T)\})].$$

The Euler-Lagrange conditions then become

(19) $\begin{cases} \text{i.} & \dot{\bar{x}} - M\bar{x} = B\bar{u}. \\ \text{ii.} & \{\dot{\bar{p}} + M^* \bar{p}, B^* \bar{p}\} \in \partial \mathscr{F}_0(\bar{x}, \bar{u}). \\ \text{iii.} & \{\bar{p}(t), -\bar{p}(T)\} \text{ is normal to } G \text{ at the point } \{\bar{x}(t), \bar{x}(T)\}. \end{cases}$

Conditions (19)iii are the *transversality conditions*. We can write conditions (19)ii in the Hamiltonian form as follows:

(20) $\begin{cases} \text{i.} & \dot{\bar{x}} - M\bar{x} = B\bar{u}. \\ \text{ii.} & \dot{\bar{p}} + M^* \bar{p} \in -\bar{\partial}_x \mathscr{H}_0(\bar{x}, B^* \bar{p}). \\ \text{iii.} & \bar{u} \in \partial_e \mathscr{H}_0(\bar{x}, B^* \bar{p}). \\ \text{iv.} & \{-\bar{p}(t), \bar{p}(T)\} \text{ is normal to } G \text{ at the point } \{\bar{x}(t), \bar{x}(T)\}. \end{cases}$

Condition (20)ii is the *adjoint differential equation*, whereas (20)iii expresses the *Pontryagin maximum principle*, namely, that the control $\bar{u}$ maximizes the function

$$u \mapsto \int_t^T \langle \bar{p}(\omega), B(\omega) u(\omega) \rangle \, d\omega - \mathscr{F}_0(\bar{x}, u) \quad \text{on} \quad L^2(\Omega, \mathbb{R}^p). \quad \blacksquare$$

Example 2. **The Case of Quadratic Functionals**

Suppose that

(21) $\begin{cases} \text{i.} & \mathscr{F}_0(x, u) = \dfrac{1}{2} \int_t^T [j(\omega, x(\omega) - x_0(\omega))^2 + k(\omega, u(\omega) - u_0(\omega))^2] \, d\omega \\ \text{ii.} & f(\varphi, \psi) = \Psi_\xi(\varphi) + \tfrac{1}{2} g(\psi - \psi_T)^2 \end{cases}$

where we have set

(22) $\begin{cases} \text{i.} & j(\omega, x)^2 = \langle J(\omega) x, x \rangle, \; g(\psi)^2 = \langle G\psi, \psi \rangle \\ \text{ii.} & k(\omega, u)^2 = \langle K(\omega) u, u \rangle \end{cases}$

where $J(.), K(.)$ belong to $\mathscr{C}(\Omega, \mathscr{L}(\mathbb{R}^n, \mathbb{R}^{n*}))$ and $\mathscr{C}(\Omega, \mathscr{L}(\mathbb{R}^p, \mathbb{R}^{p*}))$, respectively, and where, for all ω, $J(\omega)$, $K(\omega)$, and G are duality operators.

The differential system describing the evolution of the state is

(23) $\begin{cases} \text{i.} & \dot{x} = Mx + Bu. \\ \text{ii.} & x(t) = \xi \text{ where } \xi \text{ is given.} \end{cases}$

The adjoint system then becomes

(24) $\begin{cases} \text{i.} & -\dot{p} = M^*p - J(x - x_0) \\ \text{ii.} & p(T) = G(x(T) - \psi_T), \end{cases}$

since the Hamiltonian is written as follows:

$$\mathcal{H}(x, e, \psi) = -\frac{1}{2}\int_t^T j(\omega), x(\omega) - x_0(\omega))^2 \, d\omega$$
$$+ \frac{1}{2}\int_t^T [k_*(\omega, e(\omega))^2 + \langle e(\omega), u_0(\omega) \rangle] \, d\omega + \tfrac{1}{2}g_*(\psi)^2 + \langle \psi_T, \psi \rangle$$

where $k_*(\omega, e) = \langle K(\omega)^{-1} e, e \rangle, g_*(\psi)^2 = \langle G^{-1}\psi, \psi \rangle$.

The maximum principle affirms that the optimal control $\bar{u}$ is the differential with respect to e of $\mathcal{H}_0$ at the point $\{\bar{x}, B^*p\}$, that is, that

(25) $\qquad \bar{u}(\omega) = u_0(\omega) + K(\omega)^{-1} B^*(\omega) \bar{p}(\omega).$

Putting this value of $\bar{u}$ in the initial system and setting:

(26) $\qquad N(\omega) = B(\omega)K(\omega)^{-1}B^*(\omega),$

we characterize the optimal state $\bar{x}$ and co-state $\bar{p}$ as solutions of

(27) $\begin{cases} \text{i.} & \dot{\bar{x}} = M\bar{x} + N\bar{p} + Bu_0. \\ \text{ii.} & -\dot{\bar{p}} = M^*\bar{p} - J(\bar{x} - x_0). \\ \text{iii.} & \bar{x}(t) = \xi, \quad \bar{p}(T) = G(\bar{x}(T) - \psi_T). \end{cases}$

In other words, the evolution of the *optimal state* $x(\omega)$ *and of the adjoint optimal state* $p(\omega)$ *is the solution of a boundary value problem for a differential system.* ∎

CHAPTER 14

Differential-Operational Equations and Semigroups of Operators

We devote this short chapter to an introduction to differential-operational equations: Consider an unbounded operator $(D(A), A)$. (See Chapter 5, Section 5.) We propose to look for a solution $t \mapsto x(t) \in D(A)$ to the differential-operational equation

$$\frac{dx}{dt} + Ax(t) = f(t); \qquad x(0) = x_0$$

where $f: [0, \infty] \mapsto H$ is continuously differentiable at $x_0 \in D(A)$.

Let us recall that if $A \in \mathscr{L}(H, H)$ is a bounded operator (i.e., $D(A) = H$), the solution of this equation is written as follows:

$$x(t) = G(t)x_0 + \int_0^t G(t-s)f(s)\,ds$$

where, for all t, $G(t) = \exp(-tA) \in \mathscr{L}(H, H)$. These operators $G(t)$ have the properties

(1) $\begin{cases} \text{i.} & G(0) = 1; \quad G(t+s) = G(t)G(s) \\ \text{ii.} & t \mapsto G(t)x \text{ is continuous from } [0, \infty[\text{ to } H \text{ for all } x \end{cases}$

as well as the property

(2) $\qquad \forall x \in H, \quad \lim_{h \to 0} \frac{G(h) - G(0)}{h} x = -Ax.$

We are going to show that these properties can be extended to certain unbounded operators; this will allow us to solve boundary value problems (Dirichlet's problem, for example) for parabolic problems: to find $t \mapsto x(t) \in V(\Lambda)$ satisfying

$\begin{cases} \text{i.} & \dfrac{dx}{dt} + Ax(t) = f(t) \quad \text{for all} \quad t > 0. \\ \text{ii.} & x(0) = x_0 \quad \text{where} \quad x_0 \in V(\Lambda) \quad \text{satisfies} \quad \gamma x_0 = 0. \\ \text{iii.} & \gamma x(t) = 0 \quad \text{for all} \quad t \geq 0. \end{cases}$

To this end we call a semigroup of operators $G(t)$ any family of operators $G(t)$ satisfying (1). On the other hand, the relation (2) does not always hold; we are led to introduce the subspace

$$D(A) = \left\{ x \in H \text{ such that } \lim_{h \to 0} \frac{G(h) - G(0)}{h} x = -Ax \text{ exists} \right\}.$$

The unbounded operator (see Chapter 5, Section 5) $(D(A), A)$ is called the *infinitesimal generator* of the semigroup. It is a closed operator with dense domain such that $A + \lambda$ is invertible when $\lambda > \beta$ and satisfies

$$\|(A + \lambda)^{-n}\|_{\mathscr{L}(H,H)} \leq \frac{M}{(\lambda - \beta)^n} \text{ for all } n \in \mathbb{N}.$$

We show in Section 2 that these conditions characterize the infinitesimal generators of semigroups of operators (theorem of Hille-Philips).

In particular, if V is embedded in a pivot space H, every V-elliptic operator $A \in \mathscr{L}(V, V^*)$ is the infinitesimal generator of a semigroup.

We conclude this chapter by a brief introduction to the theory of the representation of *systems*. We suppose we are given a Hilbert space U, the space of "inputs" of the *system*, and a Hilbert space Y, the space of "outputs."

We fix an input $u \in U$, and we suppose that if we stimulate the system at the instant $-t \leq 0$ by an (instantaneous) *impulse u*, we observe at the instant $t = 0$ the state $F(u, t) = y(t)$. The problem of the representation of this system is to find a *model* by which one can describe the output $y(t)$ as a function of the input u, when the system is autonomous (known as the "black box" problem). In fact, we shall construct a Hilbert space X (the space of states), an operator $G \in \mathscr{L}(U, X)$, an operator $H \in \mathscr{L}(X, Y)$, and an unbounded operator $(D(A), A)$ from X to X such that $y(t)$ is obtained from u in the following fashion:

$$\begin{cases} \text{i.} & y(t) = Hx(t) \\ \text{ii.} & \dfrac{dx}{dt} + Ax = 0 \\ \text{iii.} & x(0) = Gu \end{cases}$$

if $Gu \in D(A)$.

1. SEMIGROUPS OF OPERATORS

Let H be a Hilbert space and

(1) $A \in \mathscr{L}(H, H)$ a continuous linear operator from H to itself.

We can then define the exponential

(2) $$G(t) = \exp(-tA)$$

by the series

(3) $$\exp(-tA)x = \sum_{n=0}^{\infty} \frac{(-1)^n}{n!} t^n A^n x \quad \text{for all} \quad x \in H$$

since this series in convergent in H.

These operators $G(t)$ clearly satisfy

(4) $$G(t+s) = G(t)G(s)$$

and

(5) $$\frac{d}{dt} G(t) = -AG(t) = -G(t)A$$

in the sense that for all $x \in H$,

$$\lim_{h \to 0} \frac{G(t+h) - G(t)}{h} x = -AG(t)x = -G(t)Ax.$$

Consequently, $x(t) = G(t)x_0$ is the unique solution of the linear differential equation

(6) $$\begin{cases} \text{i.} & \dfrac{dx}{dt} + Ax(t) = 0 \\ \text{ii.} & x(0) = x_0 \end{cases}$$

where x_0 is given in H.

We set ourselves the problem of finding out whether we can generalize these properties to the case of unbounded operators $(D(A), A)$ of H (see Chapter 5, Section 5), and conversely, if all the operators $G(t) \in \mathscr{L}(H, H)$ satisfying (4) can be considered as *exponential of unbounded operators*. This will allow us to solve differential equations analogous to (6).

DEFINITION 1

We say that a family G of operators

$$G(t) \in \mathscr{L}(H, H) (t \in [0, \infty[)$$

is a continuous semigroup of operators if

(7) $$\begin{cases} \text{i.} & G(t)G(s) = G(t+s) \quad \text{for all} \quad t, s \geq 0. \\ \text{ii.} & G(0) = 1. \\ \text{iii.} & \text{for all} \quad x \in H, t \in [0, \infty[\to G(t)x \quad \text{is continuous.} \end{cases}$$ ▲

First of all we verify the following result.

PROPOSITION 1

There exist constants $M > 0$ and β such that

(8) $$\|G(t)\| \leq Me^{\beta t}.$$

Proof. This amounts to showing that

(9) $$\lim_{t \to \infty} \frac{\log \|G(t)\|}{t} = \beta < +\infty.$$

We observe that

$$\log \|G(t+s)\| \leq \log \|G(t)\| \|G(s)\|$$
$$= \log \|G(t)\| + \log \|G(s)\|,$$

according to (1)i. Set

$$\beta = \inf_{t \geq 0} \frac{\log \|G(t)\|}{t} < +\infty$$

and let us establish (9). We fix $\varepsilon > 0$. Then there exists t_0 such that if $t \leq t_0$,

$$\log \|G(t)\| \leq (\beta + \varepsilon)t \leq (\beta + \varepsilon)t_0.$$

Moreover, for every $x \in X$,

$$\sup_{0 \leq t \leq t_0} \|G(t)x\| < +\infty$$

since $t \mapsto G(t)x$ is continuous on the compact set $[0, t_0]$. From Theorem 4.1.1 it follows that

$$\sup_{0 \leq t \leq t_0} \|G(t)\| = \alpha < +\infty.$$

Let t and n be such that $t > t_0$ and $nt_0 \leq t \leq (n+1)t_0$. Then

$$\beta \leq \frac{\log \|G(t)\|}{t} \leq \frac{\log \|G(nt_0)\|}{t} + \frac{\log \|G(t - nt_0)\|}{t}$$
$$= n \frac{\log \|G(t_0)\|}{t} + \frac{\log \|G(t - nt_0)\|}{t}$$
$$\leq \frac{n}{t}(\beta + \varepsilon)t_0 + \frac{\log \|G(t - nt_0)\|}{t} \leq (\beta + \varepsilon) + \frac{\log \|G(t - nt_0)\|}{t}.$$

Since $t - nt_0 \leq t_0$, we know that $\|G(t - nt_0)\| \leq \alpha$ for any value of t. Consequently, we have

$$\beta \leq \frac{\log \|G(t)\|}{t} \leq \beta + \varepsilon + \frac{\log \alpha}{t} \leq \beta + 2\varepsilon$$

for t sufficiently large. This proves that

$$\beta = \lim_{t \to \infty} \frac{\log \|G(t)\|}{t}.$$ ■

DEFINITION 2
We denote by $\mathscr{C}(M,\beta)$ the set of semigroups such that $\|G(t)\| \leq Me^{\beta t}$.
If $\beta = 0$, we say that the semigroup G is equibounded. If $\beta = 0$ and $M = 1$, we say that G is a semigroup of contractions. ▲

Replacing, if necessary, $G(t)$ by $e^{-\beta t}G(t)$, we can always *assume that G is an equibounded semigroup* (and this we shall do henceforth).

DEFINITION 3
If G is an equibounded semigroup and $\varphi \in L^1(0, \infty)$, then $G(\varphi) \in \mathscr{L}(H, H)$ denotes the operator defined by

(10) $$\forall x \in H, \quad G(\varphi)x = \int_0^\infty G(t)x\varphi(t)\,dt.$$ ▲

We are now going to associate with a semigroup G an unbounded operator $(D(A), A)$.

DEFINITION 4
We denote by $D(A)$ the vector subspace of elements $x \in H$ satisfying

(11) $$\lim_{h \to \infty} \frac{(G(h) - G(0))x}{h} \text{ exists.}$$

If $x \in D(A)$, we set

(12) $$-Ax = \lim_{h \to 0} \frac{G(h)x - x}{h}$$

We say that $(D(A), A)$ is the infinitesimal generator *of G*. ▲

Remark 1

It is clear that if $(D(A), A)$ is the infinitesimal generator of G, $(D(A), A + \beta)$ is the infinitesimal generator of the semigroup $\{e^{-\beta t}G(t)\}$, in such a way that we replace A by $A + \beta$ every time that we replace $\{G(t)\}$ by the equibounded semigroup $\{e^{-\beta t}G(t)\}$. ■

Before stating the characteristic properties of the semigroup G, we prove the following lemma.

LEMMA 1
If $x \in D(A)$, then $x(t) = G(t)x$ belongs to $D(A)$ and $t \mapsto x(t)$ is differentiable and satisfies

(13) $$\frac{dx(t)}{dt} = -AG(t)x = -G(t)Ax \quad \text{if} \quad t > 0.$$

If φ is a continuously differentiable function on $[0, \infty[$ and if φ and $\varphi' = \dfrac{d\varphi}{dt}$ are integrable on $]0, \infty[$, then $G(\varphi)x$ belongs to $D(A)$ and

(14) $$AG(\varphi) = G(\varphi)A = G(\varphi') + \varphi(0)G(0). \quad \blacktriangle$$

Proof. Taking limits in the following equalities yields formula (13):

$$\frac{x(t+h) - x(t)}{h} = \frac{G(h) - G(0)}{h} G(t)x = G(t) \frac{G(h)x - x}{h}$$

and

$$\frac{x(t) - x(t-h)}{h} = G(t-h) \frac{G(h)x - x}{h} \quad \text{if} \quad 0 < h < t.$$

Now let us establish (14). We can extend φ to a continuous and differentiable function $\tilde{\varphi}$ on $[-1, \infty]$. Hence we can write

$$\frac{G(0) - G(h)}{h} \int_0^\infty G(t)x\varphi(t)\,dt$$
$$= \int_0^\infty G(t)x\left(\frac{\varphi(t) - \varphi(t-h)}{h}\right) dt + \frac{1}{h}\int_0^h G(t)x\tilde{\varphi}(t-h)\,dt$$
$$= \int_0^\infty G(t)x\left(\frac{\varphi(t) - \varphi(t-h)}{h}\right) dt + \int_0^1 G(th)x\tilde{\varphi}((t-1)h)\,dt.$$

Letting h approach zero, we obtain

(15) $$AG(\varphi)x = \int_0^\infty G(t)x\varphi'(t)\,dt + \int_0^1 G(0)x\tilde{\varphi}(0)\,dt$$
$$= G(\varphi')x + \varphi(0)x. \quad \blacksquare$$

We are now going to establish the characteristic properties of the infinitesimal generator of a semigroup.

THEOREM 1
Let G be an equibounded semigroup of operators. Its infinitestimal generator

$(D(A), A)$ possesses the following properties:

(16) $\begin{cases} \text{i.} & D(A) \text{ is dense in } H. \\ \text{ii.} & (D(A), A) \text{ is closed.} \\ \text{iii.} & \text{for every } \lambda > 0, A + \lambda \in \mathscr{L}(D(A), H) \text{ is invertible and} \\ & \|(A + \lambda)^{-n}\|_{\mathscr{L}(H,H)} \leq \dfrac{M}{\lambda^n}. \end{cases}$

Moreover, we can write

(17) $$(A + \lambda)^{-1} = G(e^{-\lambda t}).$$

▲

Proof. **a.** We begin by showing (17). We take

$$\varphi(t) = e^{-\lambda t} \in L^1(0, \infty).$$

Then $\varphi'(t) = -\lambda e^{-\lambda t}$, and formula (8) of Lemma 1 leads us to

$$AG(e^{-\lambda t})x = G(e^{-\lambda t})Ax = -\lambda G(e^{-\lambda t})x + x,$$

that is, to

$$(A + \lambda)[G(e^{-\lambda t})]x = G(e^{-\lambda t})(A + \lambda)x = x.$$

Hence $G(e^{-\lambda t})$ is an inverse of $A + \lambda$ when $\lambda > 0$.

b. We next establish (16)iii, where $M = \sup_t \|G(t)\|$. Indeed,

$$(A + \lambda)^{-n}x = G(e^{-\lambda t})^n x = \int_0^\infty e^{-\lambda t_1} G(t_1)\, dt_1 \ldots \int_0^\infty e^{-\lambda t_n} G(t_n) x\, dt_n$$

$$= \int_0^\infty \ldots \int_0^\infty e^{-\lambda(t_1 + \ldots + t_n)} G(t_1 + \ldots + t_n) x\, dt_1 \ldots dt_n.$$

It follows that

$$\|(A + \lambda)^{-n}x\|_H \leq \int_0^\infty \ldots \int_0^\infty e^{-\lambda(t_1 + \ldots + t_n)} \|G(t_1 + \ldots + t_n)x\|\, dt_1 \ldots dt_n$$

$$\leq M\|x\| \int_0^\infty \ldots \int_0^\infty e^{-\lambda t_1} \ldots e^{-\lambda t_n}\, dt_1 \ldots dt_n \leq \frac{M}{\lambda^n} \|x\|_H.$$

c. We show that $(D(A), A)$ is closed: Let $\{x_n\}_n$ be a sequence of elements $x_n \in D(A)$ such that $x_n \to x$ in H and $Ax_n = f_n$ converges to f in H. Then

$$x_n = (A + 1)^{-1}(f_n + x_n).$$

Since $f_n + x_n$ converges to $f + x$ in H, $(A + 1)^{-1}(f_n + x_n)$ converges to $(A + 1)^{-1}(f + x)$ in H; consequently, we obtain $x = (A + 1)^{-1}(f + x)$. This implies that $f = Ax$ and, therefore, that $x \in D(A)$.

d. Finally we show that $D(A)$ is dense in H. Let φ be a continuously differentiable function with compact support in $]0, \infty[$ such that $\int_0^\infty \varphi(s)\,ds = 1$. Let $\varphi_h(s) = \varphi(s/h)/h$. Then $G(\varphi_h)x$ belongs to $D(A)$ according to Lemma 1, and $G(\varphi_h)x$ converges to x in H, since we can write

$$G(\varphi_h)x = \frac{1}{h}\int_0^\infty G(t)x\varphi\left(\frac{t}{h}\right)dt = \int_0^\infty G(th)x\varphi(t)\,dt$$

from which $G(\varphi_h)x$ converges to $\int_0^\infty G(0)x\varphi(t)\,dt = x$. ∎

It is remarkable that conditions (16) are also sufficient. The proof of this important theorem is the subject of the following section.

2. CHARACTERIZATION OF INFINITESIMAL GENERATORS OF SEMIGROUPS

THEOREM 1 (HILLE-PHILIPS)
Suppose that an unbounded operator $(D(A), A)$ satisfies

(1) $\begin{cases} \text{i.} & D(A) \text{ is dense in } H. \\ \text{ii.} & (D(A), A) \text{ is closed.} \\ \text{iii.} & \forall \lambda > 0, \quad \forall n \geq 1, \quad (A + \lambda)^{-n} \in \mathscr{L}(H, H) \text{ and satisfies} \\ & \|(A + \lambda)^{-n}\|_{\mathscr{L}(H,H)} \leq \dfrac{M}{\lambda^n}. \end{cases}$

Then $(D(A), A)$ is the infinitesimal generator of an equibounded semigroup $G = \{G(t)\}$ that satisfies

(2) $$\sup_t \|G(t)\| \leq M.$$ ▲

Before proving this theorem we give some of its important consequences.

COROLLARY 1
Suppose that an unbounded operator $(D(A), A)$ satisfies

(3) $\begin{cases} \text{i.} & D(A) \text{ is dense in } H. \\ \text{ii.} & (D(A), A) \text{ is closed.} \\ \text{iii.} & \forall \lambda > 0, \quad (A + \lambda)^{-1} \in \mathscr{L}(H, H) \quad \text{and} \quad \|(A + \lambda)^{-1}\| \leq \dfrac{1}{\lambda}. \end{cases}$

Then $(D(A), A)$ is the infinitesimal generator of a semigroup of contractions. ▲

Proof of Corollary 1. Indeed condition (3)iii implies hypothesis (1)iii of the theorem with $M = 1$. ∎

COROLLARY 2
Suppose that

(4) $\quad \begin{cases} \text{i.} & H \text{ is a pivot space.} \\ \text{ii.} & V \text{ is a Hilbert space embedded in } H. \end{cases}$

Let $A \in \mathscr{L}(V, V^*)$ *be an operator with domain*

(5) $\qquad\qquad D(A) = \{x \in V \text{ such that } Ax \in H\}.$

If

(6) $\qquad\qquad A \text{ is } V\text{-elliptic},$

then $(D(A), A)$ *is the infinitesimal generator of a semigroup of contractions of* H. ▲

Proof of Corollary 2. Indeed, if A is V-elliptic, it is clear that $(D(A), A)$ is a closed operator with dense domain (see Chapter 5, Section 5). Moreover, if $\lambda > 0$ and $x \in D(A)$,

$$\lambda \|x\|_H^2 \leq \lambda \|x\|_H^2 + c\|x\|_V^2 \leq \lambda \|x\|_H^2 + (Ax, x)$$
$$= ((A + \lambda)x, x) \leq \|(A + \lambda)x\|_H \|x\|_H.$$

Hence if $x = (A + \lambda)^{-1} y \in D(A)$ where $y \in H$, we obtain

$$\lambda \|(A + \lambda)^{-1} y\|_H \leq \|y\|_H,$$

which implies that condition (3)iii of Corollary 1 is satisfied. Hence $(D(A), A)$ is the infinitesimal generator of a semigroup of contractions G. ∎

Proof of Theorem 1. We set, if $n \in \mathbb{N}$,

(7) $\qquad\qquad R_n = \left(1 + \frac{1}{n} A\right)^{-1} = n(A + n)^{-1} \in \mathscr{L}(H, D(A)).$

We obtain the identities $n(1 - R_n) = R_n A = AR_n$ and $R_n R_m = R_m R_n$. Hypothesis (3)iii can then be written as follows:

(8) $\qquad\qquad \|R_n^k\| = n^k \|(A + n)^{-k}\| \leq M.$

It follows, for $k = 1$, that if $x \in D(A)$, $\|x - R_n x\| = \|R_n Ax\|/n \leq M\|Ax\|/n$. Hence $R_n x$ converges to x for all $x \in D(A)$. Since $\|R_n\| \leq M$ and since $D(A)$ is dense in H, it follows from Theorem 4.1.2 that

(9) $\qquad\qquad \forall x \in H, \quad R_n x \text{ converges to } x.$

We then introduce the operator $A_n \in \mathcal{L}(H, H)$ defined by

(10) $$A_n = R_n A x = A R_n x = n(1 - R_n)$$

and called the *Yosida approximation of A*, since, according to (9),

(11) $$\forall x \in D(A), \quad A_n x = R_n A x \text{ converges to } A x.$$

We then consider the semigroup $G_n(t)$ defined by

(12) $$G_n(t) = \exp(-A_n t) = e^{-nt} \exp(n R_n t)$$

of which A_n is the infinitesimal generator. We are going to prove that it converges to a semigroup whose infinitesimal generator is $(D(A), A)$. We remark that the semigroups are equibounded:

(13) $$\|G_n\| \leq M$$

since

$$\|G_n(t)\| = e^{-nt} \|\exp(n R_n(t))\| \leq e^{-nt} \sum_{k=0}^{\infty} \frac{(nt)^k}{k!} \|R_n^k x\|$$

$$\leq e^{-nt} M \|x\| \sum_{k=0}^{\infty} \frac{(nt)^k}{k!} = M \|x\|$$

according to (8). We shall show that the operators $G_n(t)$ converge pointwise to $G(t)$.

To this end we verify that for every $x \in D(A)$, $G_n(t)x$ is a Cauchy sequence of elements of H. Indeed,

$$\frac{d}{ds}(G_m(t-s)G_n(s)x) = G_m(t-s)A_n G_n(s)x - G_m(t-s)A_m G_n(s)x$$

if $0 < s < t$.

Moreover, $A_n G_n(s)x = G_n(s) A_n x$ and also $A_m G_n(s)x = G_n(s) A_m x$ (because $A_m = A R_m$ and $A_n = R_n A$ commute since R_n and R_m commute). Hence

$$\frac{d}{ds}(G_m(t-s)G_n(s)x) = G_m(t-s)G_n(s)(A_n x - A_m x).$$

Integrating from zero to t, we obtain

(14) $$\begin{cases} \|G_n(t)x - G_m(t)x\| \leq \int_0^t \|G_m(t-s)G_n(s)\| \, \|A_n x - A_m x\| \, ds \\ \leq M^2 t \|A_n x - A_m x\|. \end{cases}$$

Since $A_n x$ is a Cauchy sequence when $x \in D(A)$ [by (11)], it follows that $G_n x$ is a Cauchy sequence when $x \in D(A)$ and, consequently, that $G_n(t)x$ converges to an element $G(t)x$ for all $x \in D(A)$.

Since $D(A)$ is dense in H, Theorem 4.1.2 and (13) imply that $G_n(t)$ converges (uniformly on every compact set of H) to a continuous linear operator $G(t) \in \mathscr{L}(H, H)$ of norm at most equal to M. It is clear that the operators $G(t)$ satisfy the semigroup condition $G(t + s)x = G(t)G(s)x$.

Let us verify that for all x, $t \to G(t)x$ is continuous. The functions $t \to G_n(t)x$ are continuous. Moreover, for every $\varepsilon > 0$, there exist $y \in D(A)$ and $N(\varepsilon)$ such that $\|x - y\| \leq \varepsilon$ and $\|G_n(t)y - G_m(t)y\| \leq M^2 t\varepsilon \leq M^2 T\varepsilon$ [according to (14)]. Hence for every $t \in [0, T]$, we obtain $\|G_n(t)x - G_m(t)x\| \leq \|(G_n(t) - G_m(t))(x - y)\| + \|G_n(t)y - G_m(t)y\| \leq 2M\varepsilon + M^2 T\varepsilon$. Letting m approach infinity, we deduce that $\|G_n(t)x - G(t)x\| \leq \varepsilon(2M + M^2 T)$ for all $n \geq N(\varepsilon)$ independent of $t \in [0, T]$. Hence $t \mapsto G(t)x$ is a uniform limit of continuous functions and is, consequently, continuous.

Thus $G(t)$ is a continuous equibounded semigroup of operators that possesses an infinitesimal generator $(D(B), B)$. In order to finish let us show that $D(B) = D(A)$ and that $B = A$. If $x \in D(A)$, $G_n(t)A_n x$ converges to $G(t)Ax$ uniformly on every compact set $[0, T]$ according to (11) and (13). Integrating $(d/dt)G_n(t)x = -G_n(t)A_n x$ from zero to h, we find that

$$\frac{G_n(h)x - x}{h} = -\frac{1}{h}\int_0^h G_n(s)A_n x\, ds$$

converges to

$$\frac{G(h)x - x}{h} = -\frac{1}{h}\int_0^h G(s)Ax\, ds = -\int_0^1 G(th)Ax\, dx.$$

Therefore, letting h approach zero, we obtain $x \in D(B)$ and $Bx = Ax$. Since $(D(B), B)$ is the infinitesimal generator of $G(t)$, $B + 1$ is a bijection from $D(B)$ onto H. Moreover, $A + 1$ is a bijection from $D(A)$ onto H by hypothesis (3)iii. Hence $D(A) = D(B)$. ∎

Let F be a Hilbert space. We associate with $(D(A), A)$ the tensor product $D(A^*) \hat{\otimes} F$, the domain of the unbounded operator $\overrightarrow{A^*} = A^* \otimes 1_F$ on $H \hat{\otimes} F$. Theorem 12.4.2 implies that $(D(A^*) \hat{\otimes} F, \overrightarrow{A^*})$ is a closed operator with dense domain when $(D(A), A)$ is closed and has dense domain.

THEOREM 2
Suppose that $(D(A), A)$ is the infinitesimal generator of an equibounded semigroup $G(t)$. Then $(D(A^*) \hat{\otimes} F, \overrightarrow{A^*})$ is the infinitesimal generator of a semigroup of operators of $H \hat{\otimes} F$. ▲

Proof. We already know that $(D(A^*) \hat{\otimes} F, \overrightarrow{A^*})$ is closed and has dense domain. Since $A + \lambda$ is an isomorphism from $D(A)$ onto H if $\lambda > 0$, it follows that $\overrightarrow{A^*} + \lambda$ is an isomorphism from $D(A^*) \hat{\otimes} F$ onto $H \hat{\otimes} F$. Moreover, if

$\|(A+\lambda)^{-m}\|_{\mathscr{L}(H,H)} \leq M/\lambda^m$, it follows that for all $N \in H \hat{\otimes} F = \mathscr{L}_2(H,F)$,

$$\|(\vec{A}^* + \lambda)^{-m} N\|_{\mathscr{L}_2(H,F)} = \|N(A+\lambda)^{-m}\|_{\mathscr{L}_2(H,F)}$$

$$\leq \|N\|_{\mathscr{L}_2(H,F)} \|(A+\lambda)^{-m}\|_{\mathscr{L}(H,H)} \leq \frac{M}{\lambda^m} \|N\|_{\mathscr{L}_2(H,F)}.$$

Hence the norm of $(\vec{A}^* + \lambda)^{-m}$ in the space of continuous linear operator from $H \hat{\otimes} F$ to $H \hat{\otimes} F$ is bounded by M/λ^m. It is, therefore, the infinitesimal generator of a semigroup of operators. ∎

3. DIFFERENTIAL-OPERATIONAL EQUATIONS

THEOREM 1
Suppose we are given

(1) $\begin{cases} \text{an unbounded operator } (D(A), A) \text{ that is the infinitesimal} \\ \text{generator of a continuous semigroup } G(t) \end{cases}$

(2) *a function* $f : [0, \infty[\mapsto H$, *continuously differentiable*

and

(3) $\qquad\qquad\qquad x_0 \in D(A).$

Then there exists a unique solution to the differential-operational equation

(4) $\begin{cases} \text{i.} \quad \dfrac{dx}{dt} + Ax(t) = f(t) \qquad \forall t > 0 \\ \text{ii.} \quad x(0) = x_0 \end{cases}$

which is continuous from $[0, \infty]$ *to* $D(A)$ *and differentiable from* $]0, \infty[$ *to* H. *This solution is defined by*

(5) $\qquad\qquad x(t) = G(t)x_0 + \int_0^t G(t-s)f(s)\,ds.$ ▲

Proof. We set

(6) $\begin{cases} \text{i.} \quad y(t) = G(t)x_0 \\ \text{ii.} \quad z(t) = \int_0^t G(t-s)f(s)\,ds \\ \qquad = \int_0^t G(t-s)\left[f(0) + \int_0^s f'(r)\,dr\right]ds \\ \qquad = \int_0^t G(t-s)f(0)\,ds + \int_0^t \left[\int_r^t G(t-s)\,ds\right]f'(r)\,dr. \end{cases}$

It is clear that, by definition of the infinitesimal generator,

(7) $$y(0) = x_0 \quad \text{and} \quad \frac{dy}{dt} = -Ay(t).$$

Moreover,

(8) $$z(0) = 0$$

For proving that $x(t)$ is a solution of (4), it remains to verify that

(9) $$\frac{dz}{dt} = -Az(t) + f(t)$$

To this end we remark that

(10) $$A \int_r^t G(s) \, ds = G(r) - G(t) \quad \text{if} \quad 0 \leq r \leq t,$$

since

$$\frac{1 - G(h)}{h} \int_r^t G(s)x \, ds = \frac{1}{h} \left[\int_r^t G(s)x \, ds - \int_{r+h}^{t+h} G(s)x \, ds \right]$$

$$= \frac{1}{h} \int_r^{r+h} G(s)x \, ds - \frac{1}{h} \int_t^{t+h} G(s)x \, ds.$$

We deduce from (10) the formula

(11) $$A \int_r^t G(t - s) \, ds = A \int_0^{t-r} G(s) \, ds = 1 - G(t - r)$$

if $0 \leq r \leq t$. Consequently, it follows from the definition (6)ii of $z(t)$ and from (11) that

(12) $$\begin{cases} Az(t) = A \int_0^t G(t - s) f(0) \, ds + \int_0^t \left[A \int_r^t G(t - s) \, ds \right] f'(r) \, dr \\ = (1 - G(t)) f(0) + \int_0^t [1 - G(t - r)] f'(r) \, dr \\ = f(t) - G(t) f(0) - \int_0^t G(t - r) f'(r) \, dr. \end{cases}$$

Moreover, since $z(t) = \int_0^t G(t - s) f(s) \, ds$, we deduce that

(13) $$\frac{dz}{dt} = G(t) f(0) + \int_0^t G(t - r) f'(r) \, dr.$$

Indeed,

$$\frac{z(t+h)-z(t)}{h} = \frac{1}{h}\left[\int_0^{t+h} G(t+h-s)f(s)\,ds - \int_0^t G(t-s)f(s)\,ds\right]$$

$$= G(t)\frac{1}{h}\int_0^h G(-s)f(s+h)\,ds + \int_0^t G(t-s)\frac{f(s+h)-f(s)}{h}.$$

Equalities (12) and (13) then imply that (9) holds. Let us show that the solution is unique. If $x_1(t)$ and $x_2(t)$ are two solutions, then $x(t) = x_1(t) - x_2(t)$ is a solution of the differential equation.

(14) $\begin{cases} \text{i.} & \dfrac{dx}{dt} + Ax = 0. \\ \text{ii.} & x(0) = 0. \end{cases}$

Since $x(t) = x_1(t) - x_2(t)$ is a continuous function from $[0, \infty[$ to $D(A)$, we obtain

$$\frac{d}{ds}[G(t-s)x(s)] = G(t-s)Ax(s) + G(t-s)\frac{dx}{ds}(s)$$

because

$$\frac{G(t-s-h)x(s+h) - G(t-s)x(s)}{h} = G(t-s)\frac{G(0)-G(h)}{h}x(s+h)$$

$$+ G(t-s)\frac{x(s+h)-x(s)}{h}.$$

Then since $(d/ds)x(s) + Ax(s) = 0$, it follows that $(d/ds)[G(t-s)x(s)] = 0$ and, consequently, that the function $s \mapsto G(t-s)x(s)$ is constant. In particular, for $s = 0$ and for $s = t$ we obtain $x(t) = G(t-t)x(t) = G(t)x(0) = 0$. Hence $x(t) = x_1(t) - x_2(t) = 0$, which implies that the solution is unique. ∎

Application. Observation of Solutions

In many problems we are not particularly interested in the solution to the equation (4) (for $f = 0$) but only in the evolution of the image $Nx(t)$ under a Hilbert-Schmidt operator N from H to a Hilbert space F. In this case it is useful to know once and for all a family of operators $N(t) \in \mathscr{L}_2(H, F)$ such that

$$\forall x_0 \in D(A), \qquad Nx(t) = N(t)x_0,$$

that is, such that the image $Nx(t)$ of the solution $x(t)$ for a given initial state x_0 is the image of x_0 under the operator $N(t)$.

THEOREM 2
Suppose that $(D(A), A)$ is the infinitesimal generator of a semigroup $G(t)$. Consider the solution of the differential-operational equation

(15) $$\frac{dx(t)}{dt} + Ax(t) = 0; \quad x(0) = x_0$$

is given in $D(A)$.
  Let $N \in \mathscr{L}_2(D(A^*)^*, F)$ be a Hilbert-Schmidt operator. Then

(16) $$\forall x_0 \in D(A), \quad Nx(t) = N(t)x.$$

where $N(t) \in \mathscr{L}_2(D(A^*)^*, F)$ is the solution of the differential equation

(17) $$\frac{dN(t)}{dt} + \vec{A^*}N(t) = 0; \quad N(0) = N$$

where $\vec{A^*} = A^* \hat{\otimes} 1_F$ is an unbounded operator of domain $D(A^*) \hat{\otimes} F = \mathscr{L}_2(D(A^*)^*, F)$. ▲

Proof. Theorem 2.2 states that $(D(A^*) \hat{\otimes} F, \vec{A^*})$ is the infinitesimal generator of a semigroup and Theorem 1 that the differential-operational equation has a *unique solution* $N(t) \in D(A^*) \hat{\otimes} F = \mathscr{L}_2(D(A^*)^*, F)$. (See Theorem 12.4.2.) Let us check that this solution is equal to $N(t) = NG(t)$. Indeed,

$$N(0) = NG(0) = N \quad \text{and} \quad \frac{d}{dt}N(t) = N\frac{d}{dt}G(t)$$
$$= -NG(t)A = -\vec{A^*}(N(t)).$$

Hence

$$N(t)x_0 = NG(t)x = Nx(t) \text{ for all } x_0 \in D(A). \quad \blacksquare$$

4. BOUNDARY VALUE PROBLEM FOR PARABOLIC EQUATIONS

Theorem 3.1, Corollary 2.2, and Theorem 13.3.3 imply the existence and uniqueness of a solution to the boundary value problem for a differential equation of the following type:
  Consider a pivot space H, a Hilbert space $V \subset H$, and $\gamma \in \mathscr{L}(V, T)$ where

(1) $\begin{cases} \text{i.} & V \subset H \text{ with a continuous injection.} \\ \text{ii.} & V_0 = \operatorname{Ker} \gamma \text{ is dense in } H. \\ \text{iii.} & \gamma \text{ is surjective.} \end{cases}$

Consider a continuous bilinear form $a(x,y)$ on $V \times V$, its formal operator $\Lambda \in \mathscr{L}(V(\Lambda), H)$, and $\delta \in \mathscr{L}(V(\Lambda), T^*)$. We associate the following operators with a projector ω_1 from T to T the maps

(2) $\quad \gamma_1 = \omega_1 \gamma \in \mathscr{L}(V(\Lambda), T_1) \quad$ and $\quad \delta_2 = (1 - \omega_1^*)\delta \in \mathscr{L}(V(\Lambda), T_2^*)$.

THEOREM 1
Suppose that (1) is satisfied and that the bilinear form a is V-elliptic. If we are given $f \in \mathscr{C}^{(1)}(0, \infty ; H)$ and $x_0 \in V(\Lambda)$ satisfying $\gamma_1 x_0 = 0$ and $\delta_2 x_0 = 0$, then there exists a unique solution $x \in \mathscr{C}(0, \infty ; V(\Lambda))$ satisfying:

(3) $\quad \begin{cases} \text{i.} & \dfrac{dx}{dt} + \Lambda x(t) = f(t) \quad \text{for all} \quad t > 0. \\ \text{ii.} & \gamma_1 x(t) = 0 \quad \text{and} \quad \delta_2 x(t) = 0 \quad \text{for all} \quad t \geq 0. \\ \text{iii.} & x(0) = x_0. \end{cases}$ $\quad\blacktriangle$

Proof. We take $W = \operatorname{Ker} \gamma_1$ and $A_W \in \mathscr{L}(W, W^*)$ to be the operator associated with the bilinear form a restricted to W. (See Chapter 13, Section 3.) According to Theorem 13.3.3,

(4) $\quad D(A_W) = \{x \in V(\Lambda) \quad \text{such that} \quad \gamma_1 x = 0 \quad \text{and} \quad \delta_2 x = 0\}$

and $A_W x = \Lambda x$ if $x \in D(A_W)$. Corollary 2.2 implies that $(D(A_W), A_W)$ is the infinitesimal generator of a semigroup of contractions when the bilinear form a is V-elliptic. Theorem 3.1 implies the existence of a unique solution of (3)i and (3)iii, which satisfies (3)ii since $x(t) \in D(A_W)$ for all $t \geq 0$. $\quad\blacksquare$

Let $\Omega \subset \mathbb{R}^n$ be a regular open set with boundary Γ. Consider the bilinear form defined on $H^1(\Omega) \times H^1(\Omega)$ by

(5) $\quad a(x,y) = \sum_{i,j=1}^{n} \int_\Omega v_{ij}(\omega) D_i x D_j y \, d\omega + \int_\Omega v_0(\omega) xy \, d\omega$

where the functions v_{ij} and v_0 are continuous.
We take $\omega_1 = 0$, $W = V = H^1(\Omega)$. Then

(6) $\quad D(A) = \left\{ x \in H^1(\Omega, \Lambda) \quad \text{such that} \quad \dfrac{\partial x}{\partial v_\Lambda} = 0 \right\}$

and on $D(A)$, A coincides with the operator Λ defined by

(7) $\quad \Lambda x = -\sum_{i,j=1}^{n} D_j(v_{ij}(\omega) D_i x(\omega)) + v_0(\omega) x(\omega)$.

Suppose that the ellipticity conditions (20) of Chapter 13, Section 4, are satisfied. Then there exists a unique solution $x(t, \omega)$ of the Neumann problem for the parabolic partial differential equation

(8)
$$\begin{cases} \text{i.} & x \in \mathscr{C}(0, \infty; H^1(\Omega, \Lambda)) \\ \text{ii.} & \dfrac{\partial x}{\partial t} - \sum_{i,j=1}^{n} D_j(v_{ij}(\omega)D_i x(t, \omega)) + v_0(\omega) x(t, \omega) = f(t, \omega) \\ \text{iii.} & x(0, \omega) = x_0(\omega) \quad \text{where} \quad x_0 \in H^1(\Omega, \Lambda) \quad \text{satisfies} \quad \dfrac{\partial x_0}{\partial v_\Lambda} = 0 \\ \text{iv.} & \text{for all} \quad t \geq 0, \dfrac{\partial x}{\partial v_\Lambda}(t, \omega) = 0 \quad \text{on} \quad \Gamma \end{cases}$$

when $f \in \mathscr{C}^{(1)}(0, \infty; L^2(\Omega))$.

Taking $\omega_1 = 1$, $W = H_0^1(\Omega)$, Then

(9) $\qquad D(A_0) = \{x \in H^1(\Omega, \Lambda) \quad \text{such that} \quad \gamma_0 x = 0\}$.

If the ellipticity conditions (20) of Chapter 13, Section 4, are satisfied, there exists a unique solution of the Dirichlet problem for the parabolic partial differential equation

(10)
$$\begin{cases} \text{i.} & x \in \mathscr{C}(0, \infty; H^1(\Omega, \Lambda)). \\ \text{ii.} & \dfrac{\partial x}{\partial t} - \sum_{i,j=1}^{n} D_j(v_{ij}(\omega)D_i x(t, \omega)) + v_0(\omega) x(t, \omega) = f(t, \omega). \\ & x(0, \omega) = x_0(\omega) \quad \text{where} \quad x_0 \in H^1(\Omega, \Lambda) \quad \text{satisfies} \quad \gamma_0 x_0 = 0. \\ \text{iv.} & \text{for all} \quad t \geq 0, \quad \gamma_0 x(t, \omega) = 0 \quad \text{on} \quad \Gamma. \end{cases}$$ ∎

5. SYSTEMS THEORY: INTERNAL AND EXTERNAL REPRESENTATIONS

We describe a "system" as a mapping that associates to every input law (representing the excitation of the system during the past $]-\infty, 0]$) an output observed at time 0. We are going to try to *represent* such a system in a *more explicit* fashion. To this end we specify the spaces of input and output laws that we shall use as well as the nature of the systems we shall study.

We denote by U and Y the spaces of inputs and outputs that we assume to be separable Hilbert spaces.

Let $m \geq 1$. We take for the space of input laws the space:

(1) $\qquad\qquad\qquad H^m(-\infty, 0; U^*)^*,$

which is the dual of the Sobolev space of functions with values in U^*. We denote these input laws by f and the duality product on $H^m(-\infty, 0; U^*)^* \times$

$H^m(-\infty, 0; U^*)$ by

(2) $$[f, \varphi] = \int_{-\infty}^{0} \langle \varphi(t), df(t) \rangle$$

where $\langle .,. \rangle$ is the duality product on $U^* \times U$.

We recall that according to Theorem 12.7.1,

(3) $$\begin{cases} H^m(-\infty, 0; U^*)^* = (H^m(-\infty, 0) \hat{\otimes} U^*)^* \\ \quad = H^m(-\infty, 0)^* \hat{\otimes} U = \mathscr{L}_2(H^m(-\infty, 0), U). \end{cases}$$

In other words, we can also represent the input laws f as Hilbert-Schmidt operators:

(4) $$f : \psi \in H^m(-\infty, 0) \mapsto f(\psi) = \int_{-\infty}^{0} \psi(s) \, df(s) \in U.$$

These considerations imply that $H^m(-\infty, 0; U^*)^*$ contains the functions $f \in L^2(-\infty, 0; U)$ with values in U in the sense where

(5) $$[f, \varphi] = \int_{-\infty}^{0} \langle \varphi(t), f(t) \rangle \, dt$$

and the vector-valued Dirac measures $u \otimes \delta(s)$ defined by

(6) $$[u \otimes \delta(s), \varphi] = \langle \varphi(s), u \rangle,$$

which represent the *impulses* that instantaneously activate the input

$$u \in U \quad \text{at time} \quad s \leq 0.$$

DEFINITION 1

We say that every mapping $F \in \mathscr{L}_2(H^m(-\infty, 0; U^*)^*, Y)$ that associates to an input law f the output $F(f)$ observed at the instant 0 is a Hilbert-Schmidt system. ▲

The theory of Hilbert tensor products is going to allow us to represent F using a *kernel* $K \in H^m(0, \infty; \mathscr{L}_2(U, Y))$. In autonomous systems we make a more detailed study of the structure of these kernels by introducing a state space.

THEOREM 1

Let us consider a Hilbert-Schmidt system F. There exist Hilbert-Schmidt operators $K(s) \in \mathscr{L}_2(U, Y)$ such that $K(.) \in H^m(0, \infty; \mathscr{L}_2(U, Y))$ and such that

(7) $$F(f) = \int_{-\infty}^{0} K(-s) \, df(s) \in Y$$

for every input law f. In particular if $f \in L^2(-\infty, 0; U)$, we obtain

$$(8) \qquad F(f) = \int_{-\infty}^{0} K(-s) f(s) \, ds \in Y$$

and if $f = u \otimes \delta(-s)$ is an impulse, we obtain

$$(9) \qquad F(u \otimes \delta(-s)) = K(s)u \in Y. \qquad \blacktriangle$$

Proof. We know that $\mathscr{L}_2(H^m(-\infty, 0)^* \hat{\otimes} U; Y)$ is isometric to $H^m(-\infty, 0) \hat{\otimes} \mathscr{L}_2(U, Y)$ according to Theorem 12.3.3, that is, to $H^m(-\infty, 0; \mathscr{L}_2(U, Y))$ according to Theorem 12.7.1. Hence every Hilbert-Schmidt system is associated in a one-to-one fashion with a kernel $K \in H^m(0, \infty; \mathscr{L}_2(U, Y))$ where the $K(s) \in \mathscr{L}_2(U, Y)$ are Hilbert-Schmidt operators.

Let us recall this isometry. We associated to $\mu \in H^m(-\infty, 0)^*$ and to $F \in \mathscr{L}_2(H^m(-\infty, 0)^* \otimes U; Y)$ the operator $\hat{F}(\mu) \in \mathscr{L}_2(U, Y)$ defined by

$$(10) \qquad \forall u \in U, \qquad \hat{F}(\mu) u = F(\mu \otimes u) \in Y.$$

Let $\{u_j\}_j$ and $\{y_k\}_k$ be orthonormal bases of U and of Y, $\{u_j^*\}_j$ and $\{y_k^*\}_k$ their dual bases. Then we can write

$$\hat{F}(\mu) u = \sum_{j,k=1}^{\infty} t_{jk}(\mu) \langle u_j^*, u \rangle y_k$$

where $t_{jk}(\mu) = \langle \hat{F}(\mu) u_j, y_k^* \rangle$ are the components of $\hat{F}(\mu)$ in the orthonormal base $\{u_j^* \otimes y_k\}_{j,k}$. But since the forms $\mu \mapsto t_{jk}(\mu)$ are continuous and linear on $H^m(-\infty, 0)^*$, they are represented by the functions $\varphi_{jk}(.) \in H^m(0, \infty)$ as $t_{jk}(\mu) = \int_{-\infty}^{0} \varphi_{jk}(-s) \, d\mu(s)$. Hence we can write

$$\hat{F}(\mu) u = \sum_{j,k=1}^{\infty} \int_{-\infty}^{0} \varphi_{jk}(-s) \, d\mu(s) \langle u_j^*, u \rangle y_k.$$

We then define $K(s)$ by

$$(11) \qquad K(s) u = \sum_{j,k=1}^{\infty} \varphi_{jk}(s) \langle u_j^*, u \rangle y_k.$$

It follows that

$$(12) \qquad F(u \otimes \mu) = \hat{F}(\mu) u = \int_{-\infty}^{0} K(-s) u \, d\mu(s).$$

Consequently, the transpose $F^* \in \mathscr{L}(Y^*, H^m(-\infty, 0; U^*))$ is defined by

$$\langle F^* y^*, u \otimes \mu \rangle = \langle y^*, F(u \otimes \mu) \rangle$$
$$= \int_{-\infty}^{0} \langle y^*, K(-s) u \rangle \, d\mu(s)$$
$$= \int_{-\infty}^{0} \langle K^*(-s) y^*, u \rangle \, d\mu(s).$$

Thus

(13) $\quad F^*y^* : s \mapsto K^*(-s)y^* \quad$ belongs to $\quad H^m(-\infty, 0; U^*)$.

Therefore, if $f \in H^m(-\infty, 0; U^*)^*$, we obtain

$$\langle y^*, F(f) \rangle = \langle F^*y^*, f \rangle$$
$$= \int_{-\infty}^{0} \langle K^*(-s)y^*, df(s) \rangle = \int_{-\infty}^{0} \langle y^*, K(-s) df(s) \rangle$$
$$= \langle y^*, \int_{-\infty}^{0} K(-s) df(s) \rangle.$$

Hence this implies formula (7), which we were trying to establish. ∎

DEFINITION 2
We say that the kernel $K \in H^m(0, \infty; \mathscr{L}_2(U, Y))$ is an **external representation** of the system F. ▲

We now introduce the operators

$$T(t) \in \mathscr{L}(H^m(-\infty, 0; U^*)^*, H^m(-\infty, 0; U^*)^*)$$

defined by

(14) $\qquad \langle T(t)f, \varphi \rangle = \int_{-\infty}^{0} \langle \varphi(s-t), df(s) \rangle$

for all $\varphi \in H^m(-\infty, 0; U^*)$.

The operators $T(t)$ are translation operators by t. For example, if $f \in L^2(-\infty, 0; U)$, we obtain

(15) $\qquad T(t)f(s) = \begin{cases} f(s+t) & \text{if} \quad s \in]-\infty, -t] \\ 0 & \text{if} \quad s \in]-t, 0] \end{cases}$

and if $f = u \otimes \delta(s)$,

(16) $\qquad T(t)(u \otimes \delta(s)) = u \otimes \delta_{s-t}$.

The operators $T(t)$ form a semigroup of contractions. Indeed, it is clear that $T(0) = 1$ and $T(t)T(s) = T(t+s)$ if $t, s \geq 0$. Also $\|T(t)\| \leq 1$ for all $t \geq 0$: first of all,

$$|\langle T(t)f, \varphi \rangle| = \left| \int_{-\infty}^{0} \langle \varphi(s-t), df(s) \rangle \right|$$
$$\leq \|\varphi(.-t)\|_{H^m(-\infty, 0, U^*)} \|f\|^2_{H^m(-\infty, 0, U^*)}$$

Moreover, if $t \geq 0$, we have

$$\|\varphi(.-t)\|^2_{H^m(-\infty,0,U^*)} = \sum_{k=0}^{m} \int_{-\infty}^{0} \|D^k\varphi(s-t)\|^2_{U^*}\,ds$$
$$= \sum_{k=0}^{m} \int_{-\infty}^{-t} \|D^k\varphi(s)\|^2_{U^*}\,ds \leq \|\varphi\|^2_{H^m(-\infty,0,U^*)}.$$

Hence it follows that

$$\|T(t)\| \leq 1 \qquad \text{for all} \qquad t \geq 0.$$

Let us show that the semigroup is continuous. Let t_n be a sequence that converges to t. To show that $T(t_n)f$ converges to $T(t)f$ for all $f \in H^m(-\infty, 0; U^*)^*$, it suffices to show that $T(t_n)f$ converges to $T(t)f$ for all $f \in L^2(-\infty, 0; U)$ (which is dense in $H^m(-\infty, 0; U^*)^*$) according to Theorem 4.1.2. But if $f \in L^2(-\infty, 0; U)$, it is clear that $T(t_n)f$ converges to $T(t)f$ in $L^2(-\infty, 0; U)$, hence in $H^m(-\infty, 0; U^*)^*$.

DEFINITION 3
Let F be a system whose external representation is defined by the kernels $K(t)$. If we are given a Hilbert space X (called the state space), operators $H \in \mathscr{L}(X, Y)$ and $G \in \mathscr{L}(U, X)$, and a semigroup of contraction operators $\Gamma(t) \in \mathscr{L}(X, X)$ such that

(17) $\qquad\qquad$ for all $\quad t \geq 0, \quad K(t) = H\Gamma(t)G,$

then we say that $\{X, G, H, \Gamma(t)\}$ is an internal representation of a system F. We say that a representation is completely accessible if the mapping $f \mapsto \int_{-\infty}^{0} \Gamma(-s)G\,df(s)$ from $H^m(-\infty, 0; U^*)^*$ to X is surjective and that it is completely observable if the mapping that associates to every $x \in X$ the function $t \mapsto H\Gamma(t)x$ is injective. ▲

THEOREM 2
Let $F \in \mathscr{L}_2(H^m(-\infty, 0; U^*)^*; Y)$ be a Hilbert-Schmidt system. There exists an internal representation $\{X_0, G_0, H_0, \Gamma_0(t)\}$ that is both completely observable and completely accessible. ▲

Proof. We denote by $\mathscr{H}$ the linear operator from $H^m(-\infty, 0; U^*)^*$ to $\mathscr{C}(0, \infty; Y)$ defined by

(18) $\qquad\qquad \mathscr{H}f(t) = F(T(t)f).$

(This operator is called the *Hankel operator*.)

Let M be the kernel of $\mathscr{H}$. We take as the state space the quotient space

(19) $\qquad\qquad X_0 = H^m(-\infty, 0; U^*)^*/M$

Then $\mathscr{H} = Q_0 L_0$ where $L_0 \in \mathscr{L}(H^m(-\infty,0;U^*)^*;X_0)$ is the canonical *surjection* from the space of input laws onto the *state space* X_0 and Q_0 is an *injective* mapping from X_0 to $\mathscr{C}(0,\infty;Y)$.

Moreover, $T(t)$ maps M to itself; indeed, if $\mathscr{H}f(s) = 0$ for all $\delta \geq 0$, then $\mathscr{H}(T(t)f)(r) = F(T(r)T(t)f) = \mathscr{H}f(r+t) = 0$ for all $r \geq 0$. Hence the operator $L_0 T(t)$ from $H^m(-\infty,0;U^*)^*$ to X vanishes on M and, consequently, is well defined on the quotient space X_0 by writing

(20) $$L_0 T(t) = \Gamma_0(t) L_0$$

where $\Gamma_0(t) \in \mathscr{L}(X,X)$. Thus we can write

(21) $$\mathscr{H}(T(t)f) = Q_0 \Gamma_0(t) L_0(f).$$

Since L_0 is surjective, $\Gamma_0(0) = 1$ and $\Gamma_0(t+s) = \Gamma_0(t)\Gamma_0(s)$. For the same reason it is clear that $t \mapsto \Gamma_0(t)x = \Gamma_0(t)L_0 f = L_0(T(t)f)$ is continuous for all $x = L_0 f \in X_0$. Moreover, if $x = L_0 f \in X_0$, we have

$$\|\Gamma_0(t)x\| = \|\Gamma_0(t)L_0 f\| = \|L_0 T(t)f\| \leq \|T(t)f\| \leq \|f\|.$$

Hence $\|\Gamma_0(t)x\| \leq \inf_{L_0 f = x} \|f\| = \|x\|$ when X_0 has the scalar product of a quotient space. Therefore the operators $\Gamma_0(t)$ form a semigroup of contractions.

Finally we introduce the operators $G_0 \in \mathscr{L}(U,X)$ and $H_0 \in \mathscr{L}(X,Y)$ defined by

(22) $$\begin{cases} \text{i.} & G_0(u) = L_0(u \otimes \delta), \\ \text{ii.} & H_0 x = Q_0 x(0). \end{cases}$$

Consequently, we deduce from (9), (18), (21), and (22) that

(23) $$\begin{cases} K(t)u = F(u \otimes \delta(-t)) = F(T(t)(u \otimes \delta)) = \mathscr{H}(u \otimes \delta)(t). \\ \mathscr{H}(T(t)(u \otimes \delta))(0) = (Q_0 \Gamma_0(t) L_0(u \otimes \delta))(0). \\ \qquad = H_0 \Gamma_0(t) G_0(u). \end{cases}$$

Hence $\{X_0, G_0, H_0, \Gamma_0(t)\}$ defined by (19), (20), and (22) is an internal representation of F. Moreover, we have the following for all $t \geq 0$:

(24) $$[Q_0(L_0 f)](t) = (\mathscr{H}f)(t) = \mathscr{H}(T(t)f)(0)$$
$$= [Q_0 \Gamma_0(t) L_0(f)](0) = H_0 \Gamma_0(t) L_0 f.$$

Since L_0 is surjective, it follows that for all $x \in X$, $(Q_0 x)(t) = H_0 \Gamma_0(t)x$, that is, that $Q_0 x$ is the function $t \mapsto H_0 \Gamma_0(t)x$. Since Q_0 is injective, this amounts to saying that the representation is completely observable. Formulas (8), (14), and (23) show that

$$Q_0(L_0 f)(t) = F(T(t)f) = \int_{-\infty}^{0} H_0 \Gamma_0(-s) G_0 T(t) \, df(s)$$

$$= \int_{-\infty}^{0} H_0 \Gamma_0(t-s) G_0 \, df(s) = H_0 \Gamma_0(t) \int_{-\infty}^{0} \Gamma_0(-s) G_0 \, df(s)$$

$$= Q_0 \left[\int_{-\infty}^{0} \Gamma_0(-s) G_0 \, df(s) \right](t).$$

Hence $Q_0 L_0 f = Q_0 \int_{-\infty}^{0} \Gamma_0(-s) G_0 \, df(s)$. Since Q_0 is injective, this implies that $L_0 f = \int_{-\infty}^{0} \Gamma_0(-s) G_0 \, df(s)$. Since L_0 is surjective, the representation is completely accessible. ∎

In general if $\{X, G, H, \Gamma(t)\}$ is an internal representation, we set

(25) $\begin{cases} \text{i.} & Qx : t \mapsto H\Gamma(t)x. \\ \text{ii.} & Lf = \int_{-\infty}^{0} \Gamma(-s) G \, df(s). \end{cases}$

The internal representations that are both completely accessible and completely observable are unique up to an isomorphism and "minimal" in the following sense.

THEOREM 3
Let $\{X_0, G_0, H_0, \Gamma_0(t)\}$ be an internal representation that is both completely accessible and completely observable and $\{X, G, H, \Gamma(t)\}$ another internal representation of a Hilbert-Schmidt system F. Then there exists an injective mapping j from X_0 to X. If $\{X, G, H, \Gamma(t)\}$ is also completely accessible and completely observable, there exists an isomorphism j from X_0 onto X such that

(26) $\qquad H_0 = Hj, \qquad G = jG_0, \qquad \text{and} \qquad \Gamma(t) = j\Gamma_0(t)j^{-1}.$ ▲

Proof. **a.** Let L_0 and L be mappings from $H^m(-\infty, 0; U^*)^*$ to X associated with the internal representations by formula (25). Since L_0 is surjective, it has a right inverse B_0. Let $j = LB_0 \in \mathscr{L}(X_0, X)$. Let us show that j is injective. Suppose that $jx_0 = 0$. Then $LB_0 x_0 = 0$ and, consequently, $\mathscr{H}(B_0 x_0) = QLB_0 x_0 = Qj(x_0) = 0$. Hence $B_0 x_0 \in M$ and, therefore, $x_0 = L_0 B_0 x_0 = 0$. Thus j is injective.
b. Now suppose that $\{X, G, H, \Gamma(t)\}$ is *also* both completely observable and completely accessible. Then $\operatorname{Ker} Q = \operatorname{Ker} Q_0 = M$ and $L = jL_0$ where j is an isomorphism from X_0 onto X. It follows that $QjL_0 = Q_0 L_0 = \mathscr{H}$, that is, since L_0 is surjective, that $Qj = Q_0$. Consequently, $G = jG_0$ and

$Hj = H_0$. Since

$$\mathcal{H}(T(t)f) = Q\Gamma(t)Lf = Q_0\Gamma_0(t)L_0f = Qj\Gamma_0(t)j^{-1}Lf,$$

it follows that $\Gamma(t) = j\Gamma_0(t)j^{-1}$ because Q is injective and L is surjective. ■

Remark

This theorem implies that if the dimension of the state space of a completely observable and completely assessible representation is finite, it is minimum and is a characteristic of the system. ■

We denote by $(D(A), A)$ the infinitesimal generator of the semigroup of contractions $\Gamma(t)$ of completely accessible and completely observable internal representation $\{X, G, H, \Gamma(t)\}$. Consider an impulse $u \otimes \delta$ that activates u at the initial time and the outputs $y(t) = \mathcal{H}(u \otimes \delta)(t) = H\Gamma(t)G(u)$. If $G(u) \in D(A)$, Theorem 3.1 implies that the output $y(t)$ can be written as follows:

(27) $$y(t) = Hx(t),$$

where the evolution of the state $x(t)$ is governed by the differential-operational equation

(28) $$\begin{cases} \text{i.} & x(t) \in D(A). \\ \text{ii.} & \dfrac{dx}{dt} + Ax(t) = 0. \\ \text{iii.} & x(0) = G(u). \end{cases}$$

This differential-operational equation serves, therefore, as a model of the Hilbert-Schmidt system F.

CHAPTER 15

Introduction to Nonlinear Analysis

The aim of this final chapter is to give a precise presentation of the principal results from nonlinear analysis concerning the existence of fixed points of a correspondence and theorems of surjectivity. To give additional interest to this subject we mention some application to game theory and to economic equilibrium.

We have tried to unify the proofs by basing them all on the *Ky Fan inequality* and the use of continuous partitions of unity. These two techniques are, therefore, reviewed in the first section. (See [AAA], Chapter 5, Sections 5 and 6.)

We have not tried to give the most general theorems. Nevertheless we have given proofs that easily generalize to more general spaces.

In Section 1 we define the notion of upper hemicontinuity of a correspondence. We take up in Section 2 the fundamental *theorem of Browder–Ky Fan* on the existence of a *critical point* $\bar{x}$ for a correspondence S [i.e., the solution of the multivalued equation $0 \in S(\bar{x})$]. The proof we give is due to B. Cornet. We also prove the *Leray-Schauder theorem* by a method due to Granas. In Section 3, we derive from these proofs the *fixed-point theorems* (in particular, the important *theorem of Kakutani*) and the *theorems of surjectivity*. These theorems use the concept of *tangent cones*; we devote Section 4 to their properties.

In Section 5 we prove a theorem on the existence of solutions $\bar{x}$ to the *variational inequalities* for a correspondence S on a closed convex set X. We establish in Section 6 the existence of a solution to the *quasi-variational inequalities* by which we find again in Section 7 a theorem on the existence of a noncooperative equilibrium due to Arrow-Debreu-Nash. Section 8 is devoted to the proof of the existence theorem of Debreu-Gale-Nikaido from which we obtain the existence of an economic equilibrium. Finally, we adapt the famous Perron-Frobenius theorem on positive matrices to correspondences in Section 9.

1. UPPER HEMICONTINUOUS CORRESPONDENCES

We are going to study a whole class of nonlinear problems that reduce to a problem of the following form:

(1) $\qquad$ to find $\quad \bar{x} \in X \quad$ such that $\quad 0 \in S(\bar{x})$,

where S is a correspondence from X to a Hilbert space V that associates to every $x \in X$ a subset $S(x)$ of V that will always be nonempty, closed, and convex. If S is a mapping, problem (1) can be written in the more familiar form

(2) $\qquad$ to find $\quad \bar{x} \in X \quad$ such that $\quad S(\bar{x}) = 0$

of solving an equation. A solution $\bar{x}$ to (1) is called a *critical point* of S.

The use of correspondences is motivated by problems arising in particular in optimization theory, game theory, and mathematical economics.

In fact, we only use a few of the elements of the general theory of correspondences. We take advantage of the fact that the images $S(x)$ are *closed convex sets* to represent them by their support functions:

(3) $\qquad \forall p \in V^*, \quad$ we set $\quad \sigma(S(x), p) = \sup_{y \in S(x)} \langle p, y \rangle$

[since $y \in S(x)$ if and only if $\langle p, y \rangle \leqq \sigma(S(x), p)$ for all $p \in V^*$ by Theorem 2.5.1).]

DEFINITION 1
We say that a correspondence S is upper hemicontinuous at $x_0 \in X$ if and only if $\forall p \in V^*$, the function $x \mapsto \sigma(S(x), p)$ is upper semicontinuous at x_0. ▲

Every continuous mapping S from X to V clearly defines an upper hemicontinuous correspondence. (It is even sufficient that $\forall p \in V^*$, the functions $x \mapsto \langle p, S(x) \rangle$ be continuous.)

Let B be the unit ball of V. Let us recall that a correspondence S from X to V is *upper semicontinuous at* x_0 if for all $\varepsilon > 0$, there exists a neighborhood $N(x_0)$ of x_0 such that $S(x) \subset S(x_0) + \varepsilon B$ for all $x \in N(x_0)$. (See [AAA], Chapter 3, Section 7, p. 105). This implies that for all $p \in V^*, p_0 \in V^*$,

$$\sigma(S(x), p) \leqq \sigma(S(x_0), p) + \sigma(\varepsilon B, p) = \sigma(S(x_0), p) + \varepsilon \|p\|_* \leqq \sigma(S(x_0), p_0)$$
$$+ \|p - p_0\|_* \sup_{v \in S(x_0)} \|v\| + \varepsilon \|p\|_*.$$

PROPOSITION 1
Every upper semicontinuous correspondence is upper hemicontinuous. More-

over, if the images of S are bounded, the functions $\{x,p\} \mapsto \sigma(S(x),p)$ are upper semicontinuous. ▲

We give another example of an upper hemicontinuous correspondence.

PROPOSITION 2
Let f be a continuous convex function on the interior X (assumed to be nonempty) of its domain. Then the correspondence $x \mapsto \partial f(x)$ is upper hemicontinuous from X to U^*; moreover, the function $\{x,y\} \mapsto \sigma(\partial f(x), y)$ is upper semicontinuous on $X \times U$. ▲

Proof. At every point x where f is continuous, we know that $\partial f(x)$ is a nonempty closed convex subset whose support function satisfies

$$(4) \qquad \forall x \in X, \qquad \sigma(\partial f(x), y) = Df(x)(y) = \inf_{\theta > 0} \frac{f(x + \theta y) - f(x)}{\theta}$$

(Theorem 10.3.1). But since the functions $f(x + \theta y) - f(x)$ are continuous at every $\{x,y\}$ for θ sufficiently small, it follows that $\sigma(\partial f(x), y)$ is upper semicontinuous at $\{x,y\}$ (as the pointwise infimum of continuous functions. See [AAA], Theorem 3.7.1, p. 107.) ■

We recall the statement of the theorem on the existence of continuous partitions of unity that we are going to use.

THEOREM 1 (CONTINUOUS PARTITION OF UNITY)
For every covering of a metric space X by a finite sequence of n nonempty open sets V_i there exists a continuous partition of unity, *that is, n continuous functions β_i from X to $[0,1]$ such that*

$$(5) \quad \begin{cases} \text{i.} & \text{support} \quad \beta_i \subset V_i \quad \text{for} \quad i = 1, \ldots, n. \\ \text{ii.} & \sum_{i=1}^{n} \beta_i(x) = 1 \quad \text{for all} \quad x \in X. \end{cases}$$
▲

Proof. See Theorem 5.5.1 of [AAA], p. 198. ■

We are also going to make essential use of the Ky Fan inequality.

THEOREM 2 (KY FAN INEQUALITY)
Suppose that X is a compact convex subset and that $\varphi : X \times X \mapsto \mathbb{R}$ is a function satisfying

$$(6) \quad \begin{cases} \text{i.} & \forall y \in X, \quad x \mapsto \varphi(x,y) \text{ is lower semicontinuous.} \\ \text{ii.} & \forall x \in X, \quad y \mapsto \varphi(x,y) \text{ is concave.} \end{cases}$$

Then there exists $\bar{x} \in X$ such that

(7) $$\sup_{y \in X} \varphi(\bar{x}, y) \leq \sup_{y \in X} \varphi(y, y). \quad \blacktriangle$$

Proof. See Theorem 5.6.3 of [AAA], p. 203. ∎

Remark 1

We deduced the Ky Fan inequality from the Brouwer fixed point theorem. In fact, these two results are equivalent; we can deduce the Brouwer theorem from the Ky Fan inequality.

Indeed, let D be a continuous mapping from a compact convex set K of a finite dimensional vector space $\mathbb{R}^n$ to itself. We set

(8) $$\varphi(x, y) = \langle x - D(x), x - y \rangle$$

where $\langle \cdot, \cdot \rangle$ is the Euclidean scalar product on $\mathbb{R}^n$.

This function φ clearly satisfies the hypotheses of Theorem 2 (the Ky Fan inequality). Hence there exists an element $\bar{x} \in K$ such that

(9) $$\langle \bar{x} - D(\bar{x}), \bar{x} - y \rangle \leq 0 \quad \text{for all } y \in K.$$

By taking $y = D(\bar{x}) \in K$, we infer that $\bar{x} = D(\bar{x})$. ∎

2. EXISTENCE THEOREMS FOR A CRITICAL POINT OF A CORRESPONDENCE

Suppose we are given two Hilbert spaces U and V,

(1) $\begin{cases} \text{i.} & \text{a closed convex subset } X \text{ of } U \\ \text{ii.} & \text{a correspondence } S \text{ from } X \text{ to } V \end{cases}$

and a point $x \in X$.

We recall that the closed convex cone

(2) $$N_X(x) = \left\{ p \in U^* \text{ such that } \langle p, x \rangle = \max_{y \in X} \langle p, y \rangle \right\}$$

is the *normal cone to X at x*. (See Example 10.3.1.) The closed convex cone

$$T_X(x) = N_X(x)^- \subset U$$

is the *tangent cone to X at x*.

LEMMA 1

Let $L \in \mathcal{L}(U, V)$. The condition

(3) $$\forall x \in X, \quad -[L(T_X(x))] \cap S(x) \neq \varnothing$$

implies the property

(4) $$\forall x \in X, \quad \inf_{L^*p \in N_X(x)} \sigma(S(x), p) \geq 0.$$

The converse is true if the images $S(x)$ are nonempty convex and compact. ▲

Proof. **a.** We show that (3) implies (4). If $y = -Lz = -\lim Lz_n$ belongs to $S(x)$ where z_n belongs to $T_X(x)$ for all $n \in \mathbb{N}$ and if $p \in V^*$ satisfies $L^*p \in N_X(x) = T_X(x)^-$, we obtain that $\sigma(S(x), p) \geq \langle p, -Lz \rangle = -\lim_{n \to \infty} \langle L^*p, z_n \rangle \geq 0$, since $\langle L^*p, z_n \rangle \leq 0$ for all $n \in \mathbb{N}$.
b. Conversely, assume that $S(x)$ is convex and compact and that $-S(x) \cap cl(LT_X(x)) \neq \emptyset$. By the separation theorem for disjoint compact convex sets and closed convex sets, Theorem 2.3.3, we deduce the existence of $p_0 \in V^*$ satisfying

$$\sigma(S(x), p_0) < \inf_{z \in T_X(x)} \langle p_0, -Lz \rangle = \inf_{z \in T_X(x)} \langle -L^*p_0, z \rangle.$$

Since $T_X(x)$ is a cone, we infer that $L^*p_0 \in T_X(x)^- = N_X(x)$ and that $\sigma(X(x), p_0) < 0$. Hence (4) is negated. ∎

Let us point out the following obvious but important lemma.

LEMMA 2
If two correspondences S_1 and S_2 satisfy condition (3) [respectively (4)] and if λ_1, λ_2 are positive, then the correspondence $S = \lambda_1 S_1 + \lambda_2 S_2$ also satisfies condition (3) [respectively (4)].

In particular, for any $y \in X$, the correspondence $x \to S(x) + L(x - y)$ satisfies condition (3) [respectively (4)] whenever S does. ▲

Proof. The first statement is obvious. The second statement follows from the first and from the fact that the mapping $x \to L(x - y)$ satisfies conditions (3) [and (4)] when $y \in X$. ∎

DEFINITION 1
We say that S satisfies the **boundary condition** *on X with respect to $L \in \mathscr{L}(U, V)$ if condition (4) is satisfied.* ▲

We are going to use Theorem 1.2 (the Ky Fan inequality) to establish the existence of a critical point of S and the surjectivity of the correspondence $L + S$ from X onto $L(X)$.

THEOREM 1 (BROWDER–KY FAN)
Suppose that

(5) $\begin{cases} \text{i.} & X \text{ is compact and convex} \\ \text{ii.} & S \text{ is an upper hemicontinuous correspondence with nonempty closed convex values} \end{cases}$

and that

(6)   *S satisfies the boundary condition for an operator $L \in \mathscr{L}(U, V)$.*

Then there exists $\bar{x} \in X$ such that $0 \in S(\bar{x})$. Furthermore, $L + S$ is surjective in the sense that for all $y \in L(X)$, there exists $x \in X$ such that $y \in Lx + S(x)$. ▲

Proof. **a.** Let us assume that no critical point exists: for every $x \in X$, $0 \notin S(x)$. Since $S(x)$ is closed and convex, Theorem 2.4.1 on separation implies that there exists $p \in V^*$ such that $\sigma(S(x), p) < 0$. Consequently, X is covered by the sets $N(p) = \{x \in X$ such that $\sigma(S(x), p) < 0\}$. Since S is upper hemicontinuous, these sets are open. Hence since X is compact, we can cover it with a finite number of open sets $N(p_i)(i = 1, \ldots, n)$ and associate with this finite open covering a continuous partition of unity $\{\beta_i\}_{i=1,\ldots,n}$. (See Theorem 1.1.)

b. We introduce the real-valued function φ defined on $X \times X$ by

(7) $$\varphi(x, y) = -\sum_{i=1}^{n} \beta_i(x) \langle L^* p_i, x - y \rangle.$$

This function is clearly *lower semicontinuous* with respect to x (in fact, continuous) and *concave* with respect to y (in fact, affine). Thus the Ky Fan inequality implies the existence of $\bar{x} \in X$, which is a solution of

(8) $$\sup_{y \in X} \varphi(\bar{x}, y) \leq \sup_{y \in X} \varphi(y, y) = 0.$$

If we set $\bar{p} = \sum_{i=1}^{n} \beta_i(\bar{x}) p_i$, this amounts to saying that $\langle L^* \bar{p}, \bar{x} \rangle = \max_{y \in X} \langle L^* \bar{p}, y \rangle$; in other words, that $L^* \bar{p}$ belongs to the normal cone to X at x. Moreover, since $\sum_{i=1}^{n} \beta_i(\bar{x}) = 1$, the set I of the indices i such that $\beta_i(\bar{x}) > 0$ is nonempty. This means that for these $i \in I$, $\bar{x} \in \text{supp}(\beta_i) \subset N(\bar{p}_i)$ that is, that $\sigma(S(\bar{x}) p_i) < 0$. Hence

(9) $$\sigma(S(\bar{x}), \bar{p}) = \sigma(S(\bar{x}), \sum_{i \in I} \beta_i(\bar{x}), p_i) \leq \sum_{i \in I} \beta_i(\bar{x}) \sigma(S(\bar{x}), p_i) < 0$$

Consequently,

(10) $$\inf_{L^* p \in N_X(\bar{x})} \sigma(S(\bar{x}), p) \leq \sigma(S(\bar{x}), \bar{p}) < 0.$$

This is a contradiction to the boundary condition (4) with respect to L.

c. The second statement follows from by applying the first one to the correspondence $x \mapsto S(x) + Lx - y$, which satisfies the assumptions (5) and (6) of Theorem 1 (See Lemma 2.) ∎

Example 1

Consider the case where $U = V$ is a finite dimensional space. We take the case where $L = 1$. The simplest example of a compact convex set X is the unit ball $B = \{x \in U$ such that $\|x\| \leq 1\}$ of U, which we identify with its dual. ∎

THEOREM 2 (BROWDER)
Let B be the unit ball of a finite dimensional space U. Suppose that S is an upper hemicontinuous correspondence with closed convex values from B to U. If S satisfies the boundary condition

(11) $$\inf_{\|x\|=1} \sigma(S(x), x) \geq 0,$$

then there exists $\bar{x} \in B$ such that $0 \in S(\bar{x})$. ▲

Proof. Indeed, the unit ball is compact and convex. It suffices, therefore, to verify that S satisfies the boundary condition. But $N_B(x) = \{0\}$ if $\|x\| < 1$ and $N_B(x) = \{\lambda x\}_{\lambda > 0}$ if $\|x\| = 1$. Thus $\inf_{\lambda \geq 0} \sigma(S(x), \lambda x) \geq 0$ according to condition (11). Hence the hypotheses of Theorem 1 are satisfied. ∎

Example 2

We still consider the case where $U = V$ is a finite dimensional space. ∎

THEOREM 3 (ALTMAN)
Let B be the unit ball of a finite dimensional space U. Suppose that a correspondence S from B to U that is upper hemicontinuous with nonempty closed convex values satisfies

(12) $$\forall \|x\| = 1, \quad \forall u \in S(x), \quad \|u\|^2 \geq \|u - x\|^2 - \|x\|^2.$$

Then S has a critical point. ▲

Proof. Indeed, if $\|x\|$ is the Euclidean norm of U, then we obtain that for all $\|x\| = 1$,

$$\langle u, x \rangle = \tfrac{1}{2}(\|u\|^2 + \|x\|^2 - \|u - x\|^2) \geq 0 \quad \text{when} \quad u \in S(x)$$

and, consequently, that $\sigma(S(x), x) \geq 0$ if $\|x\| = 1$. Theorem 3 then follows from Theorem 2. ∎

Example 3. Critical Points of Coercive Correspondences

We are going to give an example of the theorem of the existence of a critical point when $X = U = V$ is a finite dimensional space. We introduce the following definitions. ∎

DEFINITION 2

We say that a correspondence S from U to U is **semicoercive** (respectively, coercive) *if*

(13) $$\lim_{\|x\| \to \infty} \frac{\sigma(S(x), x)}{\|x\|} > 0 \qquad (respectively, = \infty).$$ ▲

THEOREM 4

Suppose that S is a semicoercive upper hemicontinuous correspondence with closed values from a finite dimensional space U to itself. There exists a critical point x of S. ▲

Proof. Let
$$a = \lim_{\|x\| \to \infty} \frac{\sigma(S(x), x)}{\|x\|} > 0.$$

We take $\varepsilon = a/2$. Then there exists $A > 0$ such that
$$\frac{\sigma(S(x), x)}{\|x\|} \geq a - \varepsilon = \frac{a}{2}$$

when $\|x\| \geq A$. In particular, if $\|x\| = A$, $\sigma(S(x), x) \geq a/2 \, \|x\| = aA/2$.

Hence the restriction of S to the ball $B(A)$ of radius A satisfies the hypotheses of Theorem 2. Thus there exists a solution $\bar{x} \in B(A)$ to the equation $0 \in S(\bar{x})$. ∎

Perturbation of S: Leray-Schauder Theorem

We can deduce from Theorem 1 the Leray-Schauder theorem on the existence of critical points by the Poincaré method. To this end we adapt a proof due to Granas.

Consider the boundary $\partial X = X \cap \complement \text{Int } X$ of the compact convex set X (which is distinct from X if U is finite dimensional and X has a nonempty interior).

THEOREM 5 (LERAY-SCHAUDER)

Consider a compact convex set X with nonempty interior and a correspondence S from $X \times [0, 1]$ to V which is upperhemicontinuous with nonempty closed convex values.

Suppose that for $t = 0$ the correspondence $x \mapsto S(x, 0)$ satisfies the boundary condition for a suitable operator $L \in \mathscr{L}(U, V)$. Suppose also that

(14) $$\forall \lambda \in [0, 1[, \quad \forall x \in \partial X, \quad \text{then } 0 \notin S(x, \lambda).$$

Then there exists a critical point $\bar{x} \in X$ of the correspondence $x \to S(x, 1)$. ▲

Proof. We suppose that $x \mapsto S(x, 1)$ has no critical points in X and derive a contradiction. We set $A = \partial X$, a closed subset of X, and we introduce the subset

(15) $\qquad B = \{x \in X \text{ such that } \exists \lambda \in [0, 1] \text{ satisfying } 0 \in S(x, \lambda)\}$.

The set B is nonempty because it contains the critical points of $x \mapsto S(x, 0)$. It is closed (because S is upper hemicontinuous) and disjoint from A: If $x \in A$ and if $t \in [0, 1[$, assumption (14) implies that $x \notin B$. If $x \in A$ and if $t = 1$, $x \notin B$, since we supposed that $S(., 1)$ has no critical points.

We then introduce a continuous function φ from X to $[0, 1]$, that is, equal to zero on A and to one on B: for instance

$$\varphi(x) = \frac{d(x, A)}{d(x, A) + d(x, B)}$$

and the correspondence T defined by

(16) $\qquad T(x) = S(x, \varphi(x))$.

Then T is clearly upper hemicontinuous with nonempty closed convex values. It coincides with $S(x, 0)$ on A and, consequently, satisfies the hypotheses of Theorem 1. Thus the correspondence T has a critical point $\bar{x} \in X$ such that $0 \in T(\bar{x}) = S(\bar{x}, \varphi(\bar{x}))$. But this implies that $\bar{x} \in B$ and, therefore, that $\varphi(\bar{x}) = 1$, hence that $0 \in S(\bar{x}, 1)$. It follows that $S(., 1)$ has a critical point, which is a contradiction. ∎

In particular, we obtain the following result.

THEOREM 6
Let X be a compact convex subset with nonempty interior and let R and S be two correspondences from X to V that are upper hemicontinuous with nonempty closed convex values. Suppose that

(17) $\qquad R$ *satisfies the boundary condition for* $L \in \mathcal{L}(U, V)$

and that

(18) $\qquad \forall \mu \geq 0, \quad \forall x \in \partial X, \quad 0 \notin R(x) + \mu S(x)$.

Then the correspondence S has a critical point. ▲

Proof. We apply the preceding theorem with $S(x, t) = (1 - t)R(x) + tS(x)$. Condition (14) can be written $\forall \mu \geq 0, \forall x \in \partial X$, then $0 \notin R(x) + \mu S(x)$. It follows that $S(.) = S(., 1)$ has a critical point. ∎

We take $U = V$, a finite dimensional space, and $x_0 \in \text{Int } X$. Then the mapping $R(x) = x - x_0$ satisfies the boundary condition for $L = 1$. Hence we obtain the following theorem.

THEOREM 7
Let x_0 be an interior point of a compact convex set X of U and S a correspondence from X to U that is upper hemicontinuous with nonempty closed convex values. We suppose that

(19) $$\forall \mu > 0, \quad \forall x \in \partial X, \quad x_0 \notin x + \mu S(x).$$

Then S has a critical point. ▲

Generalization

We can replace the sufficient condition (4) by other sufficient conditions that we describe. Let f be a function from $X \times X$ to $\mathbb{R}$ satisfying the following hypotheses:

(20) $\begin{cases} \text{i.} & \forall x, \quad y \mapsto f(x, y) \text{ is concave} \\ \text{ii.} & \forall y, \quad x \mapsto f(x, y) \text{ is lower semicontinuous} \\ \text{iii.} & \forall y, \quad f(y, y) \leq 0. \end{cases}$

Let $x \in X \mapsto L(x) \in \mathscr{L}(U, V)$ be a *continuous* mapping associating to every $x \in X$ a continuous linear operator from U to V.

Theorem 3.3 (the Ky Fan inequality) implies that we can associate to every $p \in B^*$ a solution $x \in X$ of

(21) $$\sup_y [f(x, y) - \langle p, L(x)(x - y) \rangle] \leq 0.$$

The proof of Theorem 1 where the function φ defined by (7) is replaced by the function φ defined by

(22) $$\varphi(x, y) = f(x, y) - \sum_{i=1}^{n} \beta_i(x) \langle p_i, L(x)(x - y) \rangle$$

establishes the existence of a critical point $\bar{x}$ when condition (4) is replaced by

(23) $\begin{cases} \text{There exists a continuous mapping } x \in X \mapsto L(x) \in \mathscr{L}(U, V) \\ \text{and } f \text{ satisfying (20) such that for} \\ \text{all } p \in V^*, \text{ for every solution } x \text{ of (21), we have } \sigma(S(x), p) \geq 0. \end{cases}$

THEOREM 8
Suppose that the hypotheses (5) and (23) are satisfied. Then there exists a critical point $\bar{x}$ of S. ▲

Remark 1

We can also generalize Theorem 5 to this case by replacing the boundary $A = \partial X$ of X by the closed set A of solutions of (22) as p runs over B^*. ∎

3. FIXED POINT THEOREMS FOR A CORRESPONDENCE

Consider the case where $U = V$ is a Hilbert space (finite dimensional or not) and where R is a correspondence from X to U. We are going to give sufficient conditions for the existence of fixed points $\bar{x} \in R(\bar{x})$ of R. Since $\bar{x}$ is a fixed point of R if and only if $\bar{x}$ is a solution of $0 \in \bar{x} - R(\bar{x})$ [or $0 \in R(\bar{x}) - \bar{x}$], such results follow from Theorem 2.1.

THEOREM 1 (KAKUTANI)
Let X be a compact convex subset and R an upper hemicontinuous correspondence from X to itself with nonempty closed convex values. Then there exists a fixed point $\bar{x} \in R(\bar{x})$. ▲

Proof. We apply Theorem 2.1 to the correspondence $S = 1 - R$, which is also upper hemicontinuous with closed convex values. If x belongs to the boundary of X, then p is normal to X at x if and only if $\langle p, x \rangle \geq \langle p, y \rangle$ for all $y \in X$. In particular, this implies that $\sup_{y \in R(x)} \langle p, x - y \rangle \geq 0$. Hence $\inf_{p \in N_X(x)} \sup_{y \in R(x)} \langle p, x - y \rangle \geq 0$, which implies that the boundary condition is satisfied. ∎

More generally, we deduce from Theorem 2.1 the existence of a fixed point of R each time that the correspondence $1 - R$ (or $R - 1$) satisfies the boundary condition for $L = 1$.

DEFINITION 1
We say that R is "inward" (respectively, "outward") if $1 - R$ (respectively, $R - 1$) satisfies the boundary condition. ▲

In other words, R is "inward" if

(1) $$\forall x \in X, \quad \theta^+(x) = \inf_{p \in N_X(x)} \sup_{y \in R(x)} \langle p, x - y \rangle \geq 0$$

and R is "outward" if

(2) $$\forall x \in X, \quad \theta^-(x) = \inf_{p \in N_X(x)} \sup_{y \in R(x)} \langle p, y - x \rangle \geq 0.$$

If the images $R(x)$ are convex and compact, Lemma 2.1 implies that R is *inward if and only if*

(3) $$\forall x \in X, \quad R(x) \cap (x + T_X(x)) \neq \emptyset$$

and that R is *outward if and only if*

(4) $$\forall x \in X, \quad R(x) \cap (x - T_X(x)) \neq \emptyset.$$

We deduce from Theorem 2.1 the following result.

THEOREM 2
Let X be a nonempty compact convex subset and R an upper hemicontinuous correspondence with nonempty closed convex values.
a. *If R is inward, there exists a fixed point $\bar{x} \in X$ of R.*
b. *If R is outward, there exists a fixed point $x \in R(\bar{x})$ and, for all $y \in X$, there exists $x \in X$ satisfying $y \in R(x)$.* ▲

Proof. We apply Theorem 2.1
a. If R is inward there exists a critical point of $1 - R$, which is a fixed point of R.
b. If R is outward there exists a critical point of $R - 1$, which is a fixed point of R, and, for all $y \in X$, there exists a solution x to the multivalued equation $y \in x + R(x) - x = R(x)$. ■

We shall give a convenient sufficient condition implying that R is outward. We denote by

(5) $$\partial \sigma_X(p) = \{x \in X \text{ such that } \langle p, x \rangle = \sigma_X(p)\}.$$

[we remark that $x \in \partial \sigma_X(p)$ if and only if $p \in N_X(x)$.] Certain authors call the subsets $\partial \sigma_X(p)$ *support zones of X*. If the image under R of every support zone is contained in itself, then R is outward.

PROPOSITION 1
Let X be a compact convex subset and R a correspondence from X to X such that

(6) $$\forall p \in V^*, \quad R(\partial \sigma_X(p)) \subset \partial \sigma_X(p).$$

Then R is outward (and the conclusions of Theorem 2 hold). ▲

Proof. We take $x \in X$ and $p \in N_X(x)$. Then $x \in \partial \sigma_X(p)$. Hypothesis (6) implies the existence of $y_0 \in \partial \sigma_X(p) \cap R(x)$. Hence $\langle p, y_0 \rangle = \langle p, x \rangle = \sigma_X(p)$ and, consequently,

(7) $$\sup_{y \in R(x)} \langle p, y - x \rangle \geq \langle p, y_0 - x \rangle = 0.$$

Therefore $\theta^-(x) \geq 0$ for all $x \in X$, that is, R is outward. ■

Remark 1

In fact, we can replace hypothesis (6) by the weaker condition

(8) $\quad\quad \forall p \in U^*, \quad \forall x \in \partial \sigma_X(p), \quad R(x) \cap \partial \sigma_X(p) \neq \emptyset$ ∎

We can also use Theorem 2.3 to give a sufficient condition for the surjectivity of a correspondence from a finite dimensional space to itself.

THEOREM 3
Let R be a coercive upper hemicontinuous correspondence with nonempty closed convex values from a finite dimensional space U to itself. Then R is surjective. ▲

Proof. It suffices for us to show that for all $y \in U$ the correspondence $S = R - y$ is semicoercive. In this case Theorem 3 follows from Theorem 2.4. But it is clear that since R is coercive S is also. ∎

4. PROPERTIES OF NORMAL AND TANGENT CONES

The fundamental role played by the boundary condition in the preceding theorems on existence of critical and fixed points motivates a further study of normal and tangent cones to a closed convex subset.

We begin by a characterization of tangent cones.

PROPOSITION 1
Let X be a closed convex subset of U. Then

(1) $\quad\quad T_X(x) = cl\left(\bigcup_{\lambda > 0} \lambda(X - x) \right).$ ▲

Proof. **a.** First $T_X(x)^- \subset N_X(x)$; if $y \in X$, then $v = y - x \in T_X(x)$ and thus, if $p \in T_X(x)^-$, we obtain $\langle p, y - x \rangle \leq 0$. Hence p belongs to $N_X(x)$.
b. Conversely, $T_X(x) \subset N_X(x)^-$; if $p \in N_X(x)$ and if $v = \lim_{n \to \infty} \lambda_n(y_n - x) \in T_X(x)$ where $\lambda_n > 0$ and $y_n \in X$, we deduce that $\langle p, v \rangle \leq 0$, since $\langle p, \lambda_n(y_n - x) \rangle = \lambda_n \langle p, y_n - x \rangle \leq 0$ for all n. ∎

We also note that

PROPOSITION 2
If $\text{Int } X \neq \emptyset$ and if $x \in \text{Int } X$, then $T_X(x) = U$ and $N_X(x) = \{0\}$. If $X = \{x_0\}$, then $T_X(x_0) = \{0\}$ and $N_X(x_0) = U$. ▲

Proof. (Left as an exercise.) ∎

PROPOSITION 3
If X and Y are closed convex subsets satisfying $X \subset Y$, if $x \in X$, then

(2) $\qquad N_Y(x) \subset N_X(x) \qquad \text{and} \qquad T_X(x) \subset T_Y(x).$ ▲

Proof. If $p \in N_Y(x)$, then $\langle p, x \rangle = \max_{y \in Y} \langle p, y \rangle \geqq \max_{y \in X} \langle p, y \rangle \geqq \langle p, x \rangle$. Hence $p \in N_X(x)$. ∎

PROPOSITION 4
Let X^i be closed convex subsets of $U^i (i = 1, \ldots, n)$. If $X = \prod_{i=1}^n X^i$ and $x = (x^1, \ldots, x^n) \in X$, then

(3) $\qquad N_X(x) = \prod_{i=1}^n N_{X^i}(x^i) \qquad \text{and} \qquad T_X(x) = \prod_{i=1}^n T_{X^i}(x^i).$ ▲

Proof. Let $p = (p^1, \ldots, p^n) \in N_X(x)$. Then

$$\langle p, x \rangle = \sum_{i=1}^n \langle p^i, x^i \rangle = \max_{y^i \in X^i} \sum_{i=1}^n \langle p^i, y^i \rangle = \sum_{i=1}^n \max_{y^i \in X^i} \langle p^i, y^i \rangle.$$

This amounts to saying that for all $i = 1, \ldots, n$

$$\langle p^i, x^i \rangle = \max_{y^i \in X^i} \langle p^i, y^i \rangle, \qquad \text{that is, that} \qquad p^i \in N_{X^i}(x^i). \qquad \blacksquare$$

PROPOSITION 5
Let U and V be two Hilbert spaces, $X \subset U$ be a closed convex subset, and $A \in \mathcal{L}(U, V)$. Then, if $x \in X$,

(4) $\qquad \begin{cases} \textbf{i.} & N_{\overline{A(X)}}(Ax) = A^{*-1} N_X(x). \\ \textbf{ii.} & T_{\overline{A(X)}}(Ax) = cl(AT_X(x)). \end{cases}$ ▲

Proof. Since $\langle p, Ax \rangle = \max_{y \in X} \langle p, Ay \rangle = \max_{y \in X} \langle A^*p, y \rangle = \langle A^*p, x \rangle$, we obtain equality (4)i. By polarity, we deduce equality (4)ii due to Proposition 3.4.3. As a corollary we obtain ∎

PROPOSITION 6
Let X and Y be two closed convex subsets of U. Then

(5) $\qquad \begin{cases} \textbf{i.} & N_{\overline{X+Y}}(x+y) = N_X(x) \cap N_Y(x). \\ \textbf{ii.} & T_{\overline{X+Y}}(x+y) = cl(T_X(x) + T_X(y)). \end{cases}$ ▲

PROPOSITION 7
Let U and V be two Hilbert spaces, $X = U$ and $Y \subset V$ be two closed convex subsets, and $L \in \mathcal{L}(U, V)$.

… CH. 15, SEC. 4   PROPERTIES OF NORMAL AND TANGENT CONES

We set

(6) $$Z = \{x \in X \text{ such that } Lx \in Y\}.$$

We assume that

(7) $$0 \in \text{Int}(L(X) - Y).$$

Then, if $x \in Z$,

(8) $$\begin{cases} \text{i.} & N_Z(x) = N_X(x) + L^* N_Y(Lx). \\ \text{ii.} & T_Z(x) = T_X(x) \cap L^{-1} T_Y(Lx). \end{cases}$$ ▲

Proof. **a.** First, we prove that $N_X(x) + L^* N_Y(Lx) \subset N_Z(x)$. Indeed, if $r \in N_X(x)$ and $q \in N_Y(Lx)$, if we set $p = r + L^* q$, then $\langle p, x \rangle = \langle r, x \rangle + \langle q, Lx \rangle \geq \langle r, y \rangle + \langle q, Ly \rangle = \langle p, y \rangle$ for all $y \in Z$. Hence $p \in N_Z(x)$.
b. Conversely, we take $z \in Z$ and $p \in N_Z(z)$. We shall deduce from assumption (7) and the Fenchel theorem, Theorem 10.4.1, that $p = r + L^* q$ with $r \in N_X(z)$, $q \in N_Y(Lz)$. Indeed, we take $F(x, y) = \Psi_X(x) - \langle p, x \rangle + \Psi_Y(y)$ where Ψ_X and Ψ_Y are the indicator functions of X and Y; it is clear that $F(z, Lz) = \min_{x \in Z} F(x, Lx)$ since $p \in N_X(z)$. We can apply Theorem 10.4.1: There exists a Lagrange multiplier $q \in V^*$, which by Proposition 10.4.1, satisfies $-L^* q \in \partial \Psi_X(z) - p = N_X(z) - p$ and $q \in \partial \Psi_Y(Lz)$. Hence $r = p - L^* q \in N_X(z)$.
c. We deduce equality (8)ii from equality (8)i by Proposition 1.5.4 and 3.4. As a corollary, we obtain ■

PROPOSITION 8
Let X and Y be two closed convex subsets such that $0 \in \text{Int}(X - Y)$. Then, if $x \in X \cap Y$:

(9) $$N_{X \cap Y}(x) = N_X(x) + N_Y(x) \quad \text{and} \quad T_{X \cap Y}(x) = T_X(x) \cap T_Y(x).$$ ▲

We now characterize tangent cones to closed convex cones.

PROPOSITION 9
Let $P \subset V$ be a closed convex cone; then $N_P(x) = P^- \cap \{x\}^\perp$ and

(10) $v \in T_P(x)$ *if and only if* $\langle p, v \rangle \leq 0$ *for all $p \in P^-$ satisfying $\langle p, x \rangle = 0$.*
If P is a closed subspace, then $T_P(x) = P$ and $N_P(x) = P^\perp$. ▲

Proof. **a.** It is clear that $P^- \cap \{x\}^\perp$ is contained in $N_P(x)$.
b. Conversely, if $p \in N_P(x)$, then $\langle p, x \rangle = \max_{y \in P} \langle p, y \rangle = \sigma(P, p)$. Since P is a cone, we deduce that $\langle p, x \rangle = 0$ and $p \in P^-$. ■

PROPOSITION 10
Let $P \subset V$ be a closed convex cone, $p_0 \in P^-, p_0 \neq 0$; we set $X = \{x \in P$ such

that $\langle p_0, x \rangle = -1$. If X is not empty and if $x \in X$, then $N_X(x) = p_0 \mathbb{R}_+$ $(P^- \cap \{x\}^\perp)$ and

(11) $\begin{cases} v \in T_X(x) & \text{if and only if} \quad \langle p_0, v \rangle = 0 \quad \text{and} \quad \langle p, v \rangle \leq 0 \\ \text{for all} & p \in P^- \quad \text{satisfying} \quad \langle p, x \rangle = 0. \end{cases}$ ▲

Proof. Proposition 10 follows from Proposition 9 and Proposition 7 where X is replaced by P, $V = \mathbb{R}$ and $Y = \{-1\}$. ∎

As a corollary, we obtain the following useful result:

PROPOSITION 11
Let us consider $\mathbb{R}^n_+$ and $M^n = \{x \in \mathbb{R}^n_+ \text{ such that } \sum_{i=1}^n x_i = 1\}$. Let $I(x) = \{i = 1, \ldots, n \text{ such that } x_i = 0\}$. Then

(12) $v \in T_{\mathbb{R}^n_+}(x)$ if and only if $v_i \geq 0$ for all $i \in I(x)$

and

(13) $v \in T_{M^n}(x)$ if and only if $v_i \geq 0$ or all $i \in I(x)$ and $\sum_{i=1}^n v_i = 0$. ▲

5. VARIATIONAL INEQUALITIES

Let U be a Hilbert space, U^* its dual space, and X a *compact* convex subset of U. We consider a *correspondence* S from X to U^* with *nonempty compact convex values*.

We say that $\bar{x} \in X$ is a solution of the variational inequalities for the correspondence S on X if there exists $\bar{p} \in U^*$ such that

(1) $\begin{cases} \text{i.} & \bar{x} \in X, \quad \bar{p} \in S(\bar{x}). \\ \text{ii.} & \forall y \in X, \quad \langle \bar{p}, \bar{x} - y \rangle \leq 0. \end{cases}$

This is the same as saying

(2) $\qquad -N_X(\bar{x}) \cap S(\bar{x}) \neq \emptyset.$

Also, (2) amounts to saying that $\bar{x}$ is a critical point of the correspondence $x \to S(x) + N_X(x)$ (which is equal to S on Int X).

We point out that *any critical point $\bar{x}$ of S that belongs to X is a solution to* (1). The converse is true if we assume that

(3) $\qquad \forall x \in X, \quad -S(x) \subset JT_X(x)$

Indeed, if $\bar{x}$ is a solution to (1), then there exists $\bar{p}$ satisfying

$$-\bar{p} \in -S(\bar{x}) \cap N_X(\bar{x}) \subset JT_X(\bar{x}) \cap N_X(\bar{x}) = \{0\}$$

(See Theorem 1.5.1).

If S is a map $A \in \mathscr{L}(U, U^*)$, we already studied variational inequalities. (See Chapter 3, Section 7.)

THEOREM 1
Suppose that X is compact and convex and that S is an upper semicontinuous correspondence from X to U^* with nonempty compact convex values. There exists a solution $\bar{x} \in X$ to the variational inequalities on X. ▲

Proof. We set

$$\varphi(x, y) = -\sigma(S(x), y - x).$$

The function φ is concave with respect to y, lower semicontinuous with respect to x (by Proposition 1.1 since S is upper semicontinuous) and satisfies $\varphi(y, y) = 0$ for all $y \in X$. Since X is compact and convex, Theorem 1.2 (the Ky Fan inequality) implies the existence of $\bar{x} \in X$ such that

(4) $\qquad \forall y \in X, \qquad \varphi(\bar{x}, y) = -\sigma(S(\bar{x}), y - \bar{x}) \leq 0.$

Moreover, since $S(\bar{x})$ is compact and convex, since X is convex, and since the functions $y \mapsto \langle p, y - \bar{x} \rangle$ and $p \mapsto \langle p, y - \bar{x} \rangle$ are continuous and affine, (4) and the lop-sided minimax theorem, Theorem 2.7.1, imply that there exists $\bar{p} \in S(\bar{x})$ such that

(5) $\qquad \begin{cases} 0 \leq \inf_{y \in X} \sigma(S(\bar{x}), y - \bar{x}) = \inf_{y \in X} \sup_{p \in S(\bar{x})} \langle p, y - \bar{x} \rangle \\ = \inf_{y \in X} \langle \bar{p}, y - \bar{x} \rangle. \end{cases}$

(See Chapter 2, Section 7.) This shows that $\bar{x}$ is a solution of the variational inequalities. ∎

When U is finite dimensional, we can replace the hypothesis that X be compact by a hypothesis dealing with S.

DEFINITION 1
We say that S is strongly coercive on X if

(6) $\qquad \lim_{\|x\| \to \infty} \inf_{p \in S(x)} \frac{\langle p, x \rangle}{\|x\|} = +\infty$

and that S is bounded if

(7) $$\lim_{\|x\| \to \infty} \sup_{p \in S(x)} \frac{\|p\|}{\|x\|} < +\infty. \qquad \blacktriangle$$

THEOREM 2
Suppose that X is a closed convex subset of a finite dimensional space U and that S is an upper semicontinuous correspondence which is strongly coercive and bounded from X to U^* with nonempty compact convex values. Then there exists a solution $\bar{x} \in X$ to the variational inequalities (1). $\blacktriangle$

Proof. We introduce the convex compact subsets

$$X_n = \{x \in X \text{ such that } \|x\| \leq n\}.$$

For all $n \geq 1$, there exists, due to Theorem 1, a solution $x_n \in X_n$ to the variational inequalities:

(8) $$\forall y \in X_n, \quad 0 \leq \sigma(S(x_n), y - x_n).$$

For proving that the sequence of solutions x_n is bounded, we choose $y_0 \in X_1$ and we deduce from (8) that

(9) $$0 \leq \sigma(S(x_n), y_0 - x_n) = \sigma(S(x_n), y_0) - \inf_{p \in S(x_n)} \langle p, x_n \rangle.$$

Hypothesis (7) implies the existence of constants M and b such that if $\|x\| \geq b, \sigma(S(x), y_0) \leq M \|y_0\| \|x\|$. Hypothesis (6) implies the existence of c such that, if $\|x\| \geq c, \inf_{p \in S(x)} \langle p, x \rangle / \|x\| \geq 2M \|y_0\| \|x\|$. Consequently, if $a = \max(b, c)$, every nonzero solution x_n of (8) has norm inferior to a [otherwise (9) would imply that $0 \leq M \|y_0\| \|x_n\|$, which is impossible]. Hence the elements x_n belong to the compact convex set $X_a = \{x \in X \text{ such that } \|x\| \leq a\}$. Since $x \to \sigma(S(x), y - x)$ is upper semicontinuous, any cluster point $\bar{x}$, limit of a subsequence of solutions x_m to (8), satisfies

$$\forall y \in X, \quad 0 \leq \limsup_{m \geq \|y\|} \sigma(S(x_m), y - x_m) \leq \sigma(S(\bar{x}), y - \bar{x}).$$

We conclude that $\bar{x}$ is a solution to (1) as in the proof of Theorem 1. $\blacksquare$

6. QUASI-VARIATIONAL INEQUALITIES

In this section we are going to give a method for selecting a fixed point of a correspondence S.

THEOREM 1
We assume that

(1)   *X is a compact convex subset of a Hilbert space U*

and that

(2) $\begin{cases} \text{S is an upper hemicontinuous correspondence from} \\ \text{X to X with nonempty closed convex values.} \end{cases}$

We consider a function $f : X \times X \mapsto \mathbb{R}$ *satisfying*

(3) $\begin{cases} \text{i.} & \forall y \in X, x \mapsto f(x, y) \text{ is lower semicontinuous.} \\ \text{ii.} & \forall x \in X, y \mapsto f(x, y) \text{ is concave.} \\ \text{iii.} & \sup_{y \in X} f(y, y) \leq 0. \end{cases}$

Finally suppose that the correspondence S and the function f are related by the property

(4) $\qquad \left\{ x \in X \text{ such that } \alpha(x) = \sup_{y \in S(x)} f(x, y) \leq 0 \right\}$ *is closed.*

Then there exists $\bar{x} \in X$ *that is a solution of the quasi-variational inequalities*

(5) $\begin{cases} \text{i.} & \bar{x} \in S(\bar{x}). \\ \text{ii.} & \sup_{y \in S(\bar{x})} f(\bar{x}, y) \leq 0. \end{cases}$ ▲

Remark 1

Hypotheses (1) and (2) are those of the Kakutani theorem, Theorem 3.1, and hypotheses (1) and (3) are those of Theorem 1.2 (the Ky Fan inequality). Hypothesis (4) is a consistency hypothesis between S and f. ∎

Proof. We argue by contradiction: If the conclusion is false, then for all $x \in X$ either $\alpha(x) > 0$ or $x \notin S(x)$. Saying that $x \notin S(x)$ implies that there exists $p \in U^*$ such that $\langle p, x \rangle - \sigma(S(x), p) > 0$. We set

(6) $\begin{cases} \text{i.} & V_0 = \{x \in X \text{ such that } \alpha(x) > 0\}. \\ \text{ii.} & V(p) = \{x \in X \text{ such that } \langle p, x \rangle - \sigma(S(x), p) > 0\}. \end{cases}$

Then the negation of the conclusion can be expressed in the form

(7) $\qquad\qquad\qquad X \subset V_0 \cup \bigcup_{p \in U^*} V(p).$

Hypotheses (2) and (3)i imply that the sets V_0 and $V(p)$ are open. Since

X is compact, it follows that there exist $p_1, \ldots, p_n$ such that

$$(8) \qquad X \subset V_0 \cup \bigcup_{i=1}^{n} V(p_i)$$

and that there exists a *continuous partition of unity* $\{\beta_0, \beta_1, \ldots, \beta_n\}$ subordinate to this covering.

We then introduce the function $\varphi: X \times X \mapsto \mathbb{R}$ defined by

$$(9) \qquad \varphi(x, y) = \beta_0(x) f(x, y) + \sum_{i=1}^{n} \beta_i(x) \langle p_i, x - y \rangle.$$

This function φ is lower semicontinuous with respect to x (because of (3)i) and concave with respect to y [because of (3)ii]. Since X is compact and convex [from (1)] and since $\sup_y \varphi(y, y) = 0$ [from (3)iii], Theorem 1.2 (the Ky Fan inequality) implies the existence of $\bar{x}$ satisfying

$$(10) \qquad \sup_{y \in X} \varphi(\bar{x}, y) \leq 0.$$

We are going to contradict this inequality by proving that there exists $\bar{y} \in X$ such that

$$(11) \qquad \varphi(\bar{x}, \bar{y}) > 0.$$

We take

$$(12) \qquad \begin{cases} \bar{y} \in S(\bar{x}) \text{ arbitrary if } \alpha(\bar{x}) \leq 0 \text{ and satisfying} \\ f(\bar{x}, \bar{y}) \geq \dfrac{\alpha(\bar{x})}{2} \text{ if } \alpha(\bar{x}) > 0 \text{ (which is possible).} \end{cases}$$

Since $\{\beta_0, \beta_1, \ldots, \beta_n\}$ is a partition of unity, $\beta_i(\bar{x}) > 0$ for at least one index $i = 0, 1, \ldots, n$. Inequality (11) will then follow from the following statements:

$$(13) \qquad \begin{cases} \text{i.} & \beta_0(\bar{x}) > 0 \quad \text{implies that} \quad f(\bar{x}, \bar{y}) > 0. \\ \text{ii.} & \beta_i(\bar{x}) > 0 \quad \text{implies that} \quad \langle p_i, \bar{x} - \bar{y} \rangle > 0. \end{cases}$$

Let us verify these statements. If $\beta_0(\bar{x}) > 0$, then $\bar{x} \in V_0$ and, consequently, $\alpha(\bar{x}) > 0$. Hence $f(\bar{x}, \bar{y}) \geq \alpha(\bar{x})/2 > 0$. If $\beta_i(\bar{x}) > 0$, then $\bar{x} \in V(p_i)$ and, consequently, $\langle p_i, \bar{x} \rangle > \sigma(S(\bar{x}), p_i) \geq \langle p_i, \bar{y} \rangle$ because $\bar{y} \in S(\bar{x})$. Hence $\langle p_i, \bar{x} - \bar{y} \rangle > 0$. ∎

Remark 2

It is useful to have sufficient conditions implying hypothesis (4). One such is to say that the function $\alpha: x \mapsto \alpha(x) = \sup_{y \in S(x)} f(x, y)$ is lower semicontinuous. But the *maximum theorem* implies that if the correspondence S is *lower semicontinuous*, then α is lower semicontinuous. ∎

DEFINITION 1
We say that a correspondence S from a metric space X to a metric space Y is lower semicontinuous at x_0 if for every sequence of elements $x_n \in X$ converging to x_0 and for every element $y \in S(x_0)$ we can find a sequence of elements $y_n \in S(x_n)$ converging to y. ▲

PROPOSITION 1
Suppose that

(14) $\begin{cases} \text{i.} & f: X \times X \mapsto \mathbb{R} \text{ is lower semicontinuous.} \\ \text{ii.} & \text{the correspondence } S \text{ is lower semicontinuous.} \end{cases}$

Then the function $\alpha: x \mapsto \alpha(x) = \sup_{y \in S(x)} f(x, y)$ is itself lower semicontinuous. ▲

Proof. We must show that if a sequence of elements $x_n \in X$ converges to x_0, then $\alpha(x_0) \leq \lim_{n \to \infty} \inf \alpha(x_n)$. We fix $\varepsilon > 0$. There exists $y \in S(x_0)$ such that, by definition of $\alpha(x_0)$, $\alpha(x_0) \leq f(x_0, y_0) + \varepsilon/2$. Since $y_0 = \lim y_n$ where $y_n \in S(x_n)$ by (14)ii, we deduce from (14)i that $f(x_0, y_0) \leq f(x_n, y_n) + \varepsilon/2 \leq \alpha(x_n) + \varepsilon/2$ when $n \geq N(\varepsilon)$. Hence $\alpha(x_0) \leq \alpha(x_n) + \varepsilon$ when $n \geq N(\varepsilon)$. ∎

*7. NONCOOPERATIVE EQUILIBRIA IN N-PERSON GAMES

We continue the study of noncooperative games begun in Chapter 3, Section 8. We denote by $N = \{1, 2, \ldots, n\}$ the set of n players, and we set $\hat{i} = \{j \in N \text{ such that } j \neq i\}$. A noncooperative game is described when we are given

(1) $\begin{cases} \text{i.} & n \text{ sets of strategies } E^i. \\ \text{ii.} & n \text{ loss function } f_i: \prod_{i=1}^{n} E^i \to \mathbb{R}. \\ \text{iii.} & n \text{ correspondences } S_i: \prod_{j \neq i} E^j \to E^i. \end{cases}$

The loss function f_i of the ith player associates with each multistrategy $x = \{x^1, \ldots, x^n\} \in \prod_{i=1}^{n} E^i$ the loss $f_i(x) = f_i(x^1, \ldots, x^n)$ that he sustains.

Each player i distinguishes the strategy $x^i \in E^i$ that he controls from the multistrategies $x^{\hat{i}} = \{x^1, \ldots, , \ldots x^n\} \in \prod_{j \neq i} E^j$ employed by the adverse coalition $\hat{i}$. From his point of view, the multistrategies decompose as follows:

$$x = \{x^i, x^{\hat{i}}\} \in E^i \times \prod_{j \neq i} E^j.$$

We describe the noncooperative behavior of the players in the following fashion: If the adverse coalition employs the multistrategy $x^{\hat{i}} \in \prod_{j \neq i} E^j$, we suppose that the ith player chooses a strategy $x^i \in S_i(x^{\hat{i}})$ that minimizes the loss function $y^i \to f_i(y^i, x^{\hat{i}})$ on the subset $S_i(x^{\hat{i}})$.

DEFINITION 1
We say that a multistrategy $x = \{x^1, \ldots, x^n\} \in \prod_{i=1}^n E^i$ is a constrained noncooperative equilibrium if for every player i, we have

(2) $\quad \begin{cases} \text{i.} & x^i \in S_i(x^{\hat{i}}). \\ \text{ii.} & f_i(x^i, x^{\hat{i}}) = \min_{y^i \in S_i(x^{\hat{i}})} f_i(y^i, x^{\hat{i}}). \end{cases}$ ▲

THEOREM 1 (ARROW-DEBREU-NASH)
We suppose that

(3) $\quad \forall i \in N$, *the sets E^i are compact and convex*

and that

(4) $\quad \begin{cases} \forall i \in N, \text{ the correspondences } S_i \text{ from } \prod_{j \neq i} E^j \text{ to} \\ E^i \text{ are upper hemicontinuous and lower semicontinuous} \\ \text{with nonempty closed convex values.} \end{cases}$

Finally we suppose that

(5) $\quad \begin{cases} \forall i \in N, \text{ the functions } f_i \text{ are continuous and the} \\ \text{functions } y^i \mapsto f_i(y^i, x^{\hat{i}}) \text{ are convex.} \end{cases}$

Then there exists a constrained noncooperative equilibrium. ▲

Proof. We set

(6) $\quad \begin{cases} \text{i.} & X = \prod E^i. \\ \text{ii.} & \varphi(x, y) = \sum_{i=1}^n (f_i(x^i, x^{\hat{i}}) - f_i(y^i, x^{\hat{i}})). \\ \text{iii.} & S(x) = \prod_{i=1}^n S_i(x^{\hat{i}}). \end{cases}$

The set X is compact and convex. The correspondence S clearly has nonempty closed convex values and is upper hemicontinuous and lower semicontinuous. The function φ is continuous, and for all $x \in X$, the function $y \mapsto \varphi(x, y)$ is concave. Moreover, $\varphi(y, y) = 0$ for all $y \in X$. Thus Theorem 6.1 and Proposition 6.1 imply the existence of $\bar{x} \in X$ such that

(7) $\quad \begin{cases} \text{i.} & \bar{x} = \{\bar{x}^1, \ldots, \bar{x}^n\} \in S(\bar{x}) = \prod_{i=1}^n S_i(\bar{x}^{\hat{i}}). \\ \text{ii.} & \varphi(\bar{x}, y) \leq 0 \quad \text{for all} \quad y \in S(\bar{x}). \end{cases}$

Condition (7)i clearly implies the relations (2)i. In particular, we take $y = \{y^i, \bar{x}^{\hat{i}}\}$ where $y^i \in S_i(\bar{x}^{\hat{i}})$. The inequality $\varphi(\bar{x}, y) \leq 0$ can be written

(8) $$f_i(\bar{x}^i, \bar{x}^{\hat{i}}) - f_i(y^i, \bar{x}^{\hat{i}}) + \sum_{j \neq i} (f_j(\bar{x}^j, \bar{x}^{\hat{j}}) - f_j(y^j, \bar{x}^{\hat{j}})) \leq 0.$$

But $\bar{x} = \{\bar{x}^j, \bar{x}^{\hat{j}}\} = \{y^j, \bar{x}^{\hat{j}}\}$ each time that $j \neq i$. Hence (8) implies that

(9) $$\forall y^i \in S_i(\bar{x}^{\hat{i}}), \qquad f_i(x^i, \bar{x}^{\hat{i}}) \leq f_i(y^i, \bar{x}^{\hat{i}}),$$

that is, that $\bar{x}$ is a noncooperative equilibrium. ∎

*8. WALRAS EQUILIBRIA

We take up again the study of the general model of equilibrium that we treated in the quadratic case in Chapter 4, Section 7. (See also [AAA], Chapter 5, Section 8, p. 206.)

The theorem we are going to establish is the basis of the existence theorems in the theory of economic equilibrium. It is convenient to state it in the case $U = \mathbb{R}^n$. We consider the simplex $M^n = \{x \in \mathbb{R}^n_+ \text{ such that } \sum_{i=1}^n x_i = 1\}$.

THEOREM 1 (DEBREU-GALE-NIKAIDO)
Let S be a correspondence from M^n to $\mathbb{R}^n$ with nonempty values. If

(1) $$\begin{cases} \text{i.} & S \text{ is upper hemicontinuous} \\ \text{ii.} & \forall x \in M^n, S(x) - \mathbb{R}^n_+ \text{ is closed and convex} \\ \text{iii.} & \forall x \in M^n, \sigma(S(x), x) \geq 0 \qquad \text{(the Walras law)}, \end{cases}$$

then there exists $\bar{x} \in M^n$ such that $S(\bar{x}) \cap \mathbb{R}^n_+ \neq \emptyset$. ▲

Proof. We introduce the function φ defined on $M^n \times M^n$ by

(2) $$\varphi(x, y) = -\sigma(S(x), y).$$

This function is concave with respect to y [since $y \mapsto \sigma(S(x), y)$ is convex] and lower semicontinuous with respect to x [since, because S is upper hemicontinuous, $x \mapsto \sigma(S(x), y)$ is upper semicontinuous]. Since M^n is compact and convex, the Ky Fan theorem implies the existence of $\bar{x} \in M^n$ such that $\sup_{y \in M^n} \varphi(\bar{x}, y) \leq \sup_{y \in M^n} \varphi(y, y) \leq 0$ [according to (1)iii,] that is, such that

(3) $$0 \leq \sigma(S(\bar{x}), y) \qquad \text{for all} \qquad y \in M^n.$$

This condition is equivalent to

(4) $$0 \leq \sigma(S(\bar{x}) - \mathbb{R}^n_+, y) \qquad \text{for all} \qquad y \in U$$

since $\sigma(-\mathbb{R}^n_+, y) = 0$ if $y \in \mathbb{R}^n_+$ and $\sigma(-\mathbb{R}^n_+, y) = +\infty$ if $y \notin \mathbb{R}^n_+$. (See Proposition 2.5.1.) Since $S(\bar{x}) - \mathbb{R}^n_+$ is closed and convex, (4) implies that $0 \in S(\bar{x}) - \mathbb{R}^n_+$, that is, that $S(\bar{x}) \cap \mathbb{R}^n_+ \neq \emptyset$. ∎

Example 1. Existence of a Walras Equilibrium

We consider a space $U = \mathbb{R}^n$ representing a *commodity space*. Then its dual space $\mathbb{R}^{n*}$ represents the price space and $M^n = \{p \in \mathbb{R}^{n*}_+ \text{ such that } \sum_{i=1}^n p^i = 1\}$ is the set of positive normalized prices.

Suppose we are given a subset $Y \subset \mathbb{R}^n$ of available commodities where

(5) $\qquad Y$ is a closed convex subset such that $Y = Y - \mathbb{R}^n_+$.

Then its support function $r(p) = \sigma_Y(p) = \sup_{y \in Y} \langle p, y \rangle$ can be interpreted as the *maximum income* function obtained by Y. Now consider I consumers $i = 1, \ldots, I$, whose problem is to find $x^i \in \mathbb{R}^n$ such that the sum of the consumptions x^i is available, that is,

(6) $$\sum_{i=1}^I x^i \in Y.$$

To choose such an allocation $x = \{x^1, \ldots, x^I\}$, we suppose that the behavior of the consumers is described as follows:

a. The income $r(p)$ is allotted to the I consumers, who then dispose of $r_i(p)$ where

(7) $$r(p) = \sum_{i=1}^I r_i(p).$$

b. Each consumer chooses a commodity x^i in the set $D_i(p)$ where D_i is the demand correspondence that associates to each $p \in M^n$ a *nonempty compact convex subset* contained in the budget set

(8) $\qquad B_i(p) = \{x^i \in \mathbb{R}^n \text{ such that } \langle p, x^i \rangle \leq r_i(p)\}$

of the consumer.

The problem then arises of verifying if it is possible to find a price $\bar{p}$ (*called the Walras equilibrium*) such that each consumer can find $\bar{x}^i \in D_i(\bar{p})$ satisfying $\sum_{i=1}^I \bar{x}^i \in Y$.

THEOREM 2
We assume (5), (7), *and*

(9) $\qquad$ *the demand correspondences D_i are upper hemicontinuous.*

Then there exists a Walras equilibrium. ▲

Proof. We apply Theorem 1 with $S(p) = -\sum_{i=1}^{I} D_i(p) + Y$ (where x is replaced by p). Then $S(p) = S(p) - \mathbb{R}_+^n$ is closed and convex, since the $D_i(p)$'s are compact and convex and $Y = Y - \mathbb{R}_+^n$ is closed and convex. The correspondence S is clearly upper hemicontinuous, being the sum of upper hemicontinuous correspondences. Finally since $D_i(p) \subset B_i(p)$, it follows that the Walras law is satisfied. Hence there exists $\bar{p}$ such that $0 \in S(\bar{p}) = -\sum_{i=1}^{I} D_i(\bar{p}) + Y$ and, consequently, there exist $\bar{x}_i$'s $\in D_i(\bar{p})$ such that $\sum_{i=1}^{I} \bar{x}_i \in Y$. ■

*9. THE PERRON-FROBENIUS THEOREM FOR CORRESPONDENCES

Let $M^n = \{x \in \mathbb{R}_+^n \text{ such that } \sum_{i=1}^{n} x_i = 1\}$. We consider

(1)
- **i.** a continuous map F from M^n to $\mathbb{R}^n$.
- **ii.** an upper hemicontinuous correspondence G from M^n to $\mathbb{R}_+^n$ with compact convex values such that $x \to \sigma(G(x), p_0)$ is continuous (where $p_0 = (1, \ldots, 1)$).

We posit the following positivity assumptions:

(2)
- **i.** $\forall x \in M^n$, $\langle p_0, F(x) \rangle > 0$.
- **ii.** $\forall x \in M^n$, $\sigma(G(x), p_0) > 0$.

We also assume that F satisfies the following boundary conditions:

(3)   If $x_i = 0$, then the ith component $F_i(x)$ of $F(x)$ is nonpositive.

Since M^n is compact, the following number δ is positive:

(4) $$\delta = \sup_{x \in M^n} \frac{\sigma(G(x), p_0)}{\langle p_0, F(x) \rangle}$$

THEOREM 1
We posit assumptions (1), (2), and (3).
a. There exist $\bar{x} \in M^n$ and $\bar{\lambda} \in \,]0, \delta]$ such that

(5) $\bar{\lambda} F(\bar{x}) \in G(\bar{x})$ (*Perron's property*).

b. If $\mu > \delta$ and if $y \in M^n$, there exist $x \in M^n$ and $\beta > 0$ such that

(6) $\mu F(x) \in G(x) + \beta y$ (*Frobenius's property*).

c. If we assume, moreover, that

(7) $\forall x \in M^n, \quad G(x) \subset \mathring{\mathbb{R}}_+^n$

then the solutions $\bar{x}$ of (5) and x of (6) belong to $\mathring{\mathbb{R}}_+^n$. ▲

Proof. **a.** For proving the Perron property, we introduce the correspondence S defined by

$$(8) \qquad S(x) = -G(x) + \frac{\sigma(G(x), p_0)}{\langle p_0, F(x) \rangle} F(x)$$

It is clearly upper hemicontinuous with closed convex values. Let us check that for all $x \in M^n$, $-S(x) \cap T_{M^n}(x) \neq \emptyset$. Let $w \in G(x)$ satisfying $\langle p_0, w \rangle = \sigma(G(x), p_0)$. Then

$$v = \frac{\sigma(G(x), p_0)}{\langle p_0, F(x) \rangle} F(x) - w$$

belongs to $S(x)$ and satisfies $\langle p_0, v \rangle = 0$; since $w \in G(x) \subset \mathbb{R}_+^n$, assumption (3) implies that $-v_i \geq 0$ whenever $x_i = 0$. Proposition 4.11 shows that the assumptions of the Browder–Ky Fan theorem hold. Then there exists a critical point $\bar{x}$ of S, that is, a solution $\bar{x}$ of (5) with

$$\bar{\lambda} = \frac{\sigma(G(\bar{x}), p_0)}{\langle p_0, F(\bar{x}) \rangle} > 0.$$

b. For proving the Frobenius property, we introduce the correspondence R defined by

$$(9) \qquad R(x) = -G(x) + \mu F(x) - (\mu \langle p_0, F(x) \rangle - \sigma(G(x), p_0)) y,$$

which is clearly upper hemicontinuous with closed convex values. Since $\mu > \delta$, we know that $\sigma(G(x), p_0) < \mu \langle p_0, F(x) \rangle$ by the very definition of δ. We choose $w \in G(x)$ such that $\langle p_0, w \rangle = \sigma(G(x), p_0)$ and we set $v = -w + \mu F(x) - (\mu \langle p_0, F(x) \rangle - \sigma(G(x), p_0)) y \in R(x)$. It is clear that $\langle p_0, v \rangle = 0$ and that assumption (3) implies that $-v_i \geq 0$ whenever $x_i = 0$. Then the Browder–Ky Fan theorem implies the existence of a critical point of R, which is a solution of (6) with $\beta = \mu \langle p_0, F(x) \rangle - \sigma(G(x), p_0) > 0$.

c. If $G(x) \subset \mathring{\mathbb{R}}_+^n$ for all $x \in M^n$, we deduce from (5) that $F(\bar{x}) \in \mathring{\mathbb{R}}_+^n$ (respectively, from (6), that $F(x) \in \mathring{\mathbb{R}}_+^n$). Then assumption (3) implies that $\bar{x}$ (respectively, x) belongs to $\mathring{\mathbb{R}}_+^n$. ∎

Remark 1

Relation $\bar{\lambda} F(\bar{x}) = (\sigma(G(\bar{x}), p_0) / \langle p_0, F(\bar{x}) \rangle) F(\bar{x}) \in G(\bar{x})$ implies that

$$(10) \qquad p_0 = (1, \ldots, 1) \text{ belongs to the normal cone to } G(\bar{x}) \text{ at } \bar{\lambda} F(\bar{x}). \qquad \blacksquare$$

Since the mapping $F = 1$ obviously satisfies assumptions (1)i, (2)i, and (3), we obtain the following corollary.

THEOREM 2
Let us assume that the correspondence G satisfies properties (1)ii and (2)ii. Let $\delta = \max_{x \in M^n} \sigma(G(x), p_0)$. Then there exist $\bar{\lambda} \in \,]0, \delta]$ and $\bar{x} \in M^n$ such that

(11) $\qquad\qquad \bar{\lambda}\bar{x} \in G(\bar{x}) \qquad$ *(Perron's property).*

If $\mu > \delta$ and $y \in M^n$, there exist $x \in M^n$ and $\beta > 0$ such that

(12) $\qquad\qquad \mu\bar{x} \in G(\bar{x}) + \beta y \qquad$ *(Frobeniens's property).* ▲

Selection of Results

For the convenience of the reader we group here the principal results in this book. Less important properties and more general statements have not been selected; so the reader in need of them should go back and find them in the text.

1. GENERAL PROPERTIES

1.1. Best Approximation (Chapter 1, Section 4)

Let V be a Hilbert space and M be a nonempty *closed convex* subset, $x \in V$ and $tx \in M$. The following properties are equivalent:

$$\begin{cases} \text{i.} & \|x - tx\| = \min_{y \in M} \|x - y\|. \\ \text{ii.} & \forall y \in M, \quad ((tx - x, tx - y)) \leq 0. \end{cases}$$

Moreover, we can associate to every $x \in V$ a *unique* element $tx \in M$ satisfying either property i or ii.

The map t (called *projector*) satisfies the following properties:

$$\begin{cases} \text{i.} & t^2 = t. \\ \text{ii.} & \|tx - ty\| \leq \|x - y\|. \\ \text{iii.} & ((tx - ty, x - y)) \geq 0. \end{cases}$$

1.2. Orthogonal Projectors (Chapter 1, Section 5)

If M is a *closed vector subspace*, the projector t (called *orthogonal projector*) is a *linear* operator whose *norm equals* 1, satisfying

$$\begin{cases} \text{i.} & M = \operatorname{Im} t, \quad M^{\oplus} = \operatorname{Ker} t. \\ \text{ii.} & ((tx, y)) = ((x, ty)) \quad \text{for all} \quad x, y \in V. \end{cases}$$

1.3. Extension (Chapter 1, Section 3 and Chapter 2, Section 1)

Let V_1 and V_2 be two Hilbert spaces, $M_1 \subset V_1$ and $M_2 \subset V_2$ vector subspaces,

$A \in \mathscr{L}(M_1, M_2; W)$ a bilinear map from $M_1 \times M_2$ to a Hilbert space W.
a. There exists a *unique* bilinear map $\bar{A} \in \mathscr{L}(\bar{M}_1, \bar{M}_2; W)$ from $\bar{M}_1 \times \bar{M}_2$ to W that *extends* A.
b. There exists a bilinear map $\tilde{A} \in \mathscr{L}(V_1, V_2; W)$ from $V_1 \times V_2$ to W that extends $\bar{A}$ and satisfies

$$\|\bar{A}\|_{\mathscr{L}(\bar{M}_1, \bar{M}_2; W)} = \|\tilde{A}\|_{\mathscr{L}(V_1, V_2; W)}.$$

1.4. Isometry Between a Hilbert Space and Its Dual (Chapter 3, Section 1)

Let V be a Hilbert space and V^* be its (topological) dual. The map $J \in \mathscr{L}(V, V^*)$ that associates with every $x \in V$ the differential of the function $y \to \frac{1}{2}\|y\|^2$ at x is a *linear isometry from V onto V^** satisfying

$$\forall x, y \in V, \quad ((x, y)) = \langle Jx, y \rangle,$$

where $\langle \cdot, \cdot \rangle$ is the *duality product* on $V^* \times V$. The isometry J is called a duality operator.

1.5. Realization of the Dual (Chapter 3, Section 2)

A Hilbert space F is isometric to the dual V^* of a Hilbert space V if and only if there exists a bilinear form $\langle \cdot, \cdot \rangle$ on $F \times V$ (called a *duality pairing*) satisfying
a. $\langle f, x \rangle$ is nondegenerate on $F \times V$.

b. $\|f\|_F = \sup_{x \in V} \dfrac{|\langle f, x \rangle|}{\|x\|_V}.$

In this case there exists an *isometry* $J \in \mathscr{L}(V, F)$ from V onto F such that

$$\forall x, y \in V, \quad ((x, y)) = \langle Jx, y \rangle.$$

1.6. Density Criterion (Chapter 2, Section 2)

A subset $D \subset V$ spans a *dense vector subspace* of V if and only if any continuous linear form $f \in V^*$ vanishing on D is equal to zero, that is, if and only if $D^\perp = \{0\}$.

2. PROPERTIES OF CONTINUOUS LINEAR OPERATORS

2.1. Transposition of Operators (Chapter 3, Sections 3 and 4 and Chapter 4, Section 4)

a. An operator $A \in \mathscr{L}(V, F)$ is *injective* if and only if its transpose $A^* \in \mathscr{L}(F^*, V^*)$ has a dense image.

b. An operator $A \in \mathscr{L}(V, F)$ is *surjective* if and only if its transpose $A^* \in \mathscr{L}(F^*, V^*)$ is an *isomorphism from F^** onto *its closed image*.
c. The *image* of $A \in \mathscr{L}(V, F)$ is *closed* if and only if the *image* of its transpose $A^* \in \mathscr{L}(F^*, V^*)$ *is closed* (closed range theorem).
d. Ker $A = (\text{Im } A^*)^\perp$ and Ker $A^* = (\text{Im } A)^\perp$.

2.2. Banach's Open Mapping Principle (Chapter 4, Section 3)

a. Let $A \in \mathscr{L}(V, F)$ be *surjective*. There exists a constant $c > 0$ such that $\forall f \in F$, there *exists a solution to the equation $Ax = f$* satisfying $\|x\| \leq c \|f\|$.
b. Let $A \in \mathscr{L}(V, F)$ be *bijective*. Then its *inverse* A^{-1} is continuous.
c. A linear map A from V to F is *continuous* if and only if its *graph is closed*.

2.3. Lax-Milgram Theorem (Chapter 3, Section 6)

An operator $A \in \mathscr{L}(V, V^*)$ is *V-elliptic* if there exists a constant $c > 0$ such that

$$\forall x \in V, \quad \langle Ax, x \rangle \geq c \|x\|^2.$$

Any V-elliptic operator is an isomorphism.

2.4. Banach-Steinhauss's Uniform Boundedness Principle (Chapter 4, Section 1)

a. A sequence of *continuous linear* operators $A_n \in \mathscr{L}(V, F)$ is *bounded* if and only if

$$\forall x \in V, \quad \sup_{n \geq 0} \|A_n x\| \quad \text{is finite}.$$

b. Let us consider a *bounded* sequence of operators $A_n \in \mathscr{L}(V, F)$. Let $D \subset V$ be a *dense* subspace. Then the property

    i. $\quad \forall x \in D, \quad \lim_{n \to \infty} \|A_n x - Ax\| = 0$

implies the property

    ii. $\quad \forall \text{ compact } K \subset V, \quad \lim_{n \to \infty} \sup_{x \in K} \|A_n x - Ax\| = 0.$

In this case A is *linear and continuous*. Moreover,
    iii. If $x_p \to x$, then $A_n x_p$ converges to Ax.
c. Let us consider a sequence of operators $A_n \in \mathscr{L}(V, F)$ such that, $\forall x \in V$, $A_n x$ converges to some limit. Then there exists $A \in \mathscr{L}(V, F)$ such that **ii** holds.

2.5. Mean Ergodic Theorem (Chapter 4, Section 2)

Let $A \in \mathscr{L}(V,V)$ such that $\|A\| \leq 1$. The sequence of operators $T_n = \frac{1}{n}\sum_{k=0}^{n-1} A^k$ converges pointwise to the projector T_∞ onto $\operatorname{Ker}(A-1)$ whose kernel is $\overline{\operatorname{Im}(A-1)}$.

2.6. Left and Right Inverses (Chapter 4, Sections 5 and 6)

a. $A \in \mathscr{L}(V,F)$ has a *continuous linear* left (respectively, right) inverse if and only if A is *injective* with a *closed image* (respectively, A is *surjective*).

b. If $A \in \mathscr{L}(V,F)$ is *surjective* and if J is the duality operator of V, then $A^+ = J^{-1}A^*(AJ^{-1}A^*)^{-1} \in \mathscr{L}(F,V)$ is a right inverse (called the *right orthogonal inverse*).

c. If $A \in \mathscr{L}(V,F)$ is *injective with a closed image* and if K is the duality operator of F, then $A^- = (A^*KA)^{-1}A^*K \in \mathscr{L}(F,V)$ is a left inverse (called the *left orthogonal inverse*).

3. SEPARATION THEOREMS AND POLARITY

3.1. Separation of Convex Subsets (Chapter 2, Sections 3 and 4)

Let M and N be nonempty *disjoint* subsets of a Hilbert space V.

a. If $M - N$ is *closed and convex*, then

i. $\exists p \in V^*$ such that $\sup_{x \in M} \langle p, x \rangle < \inf_{x \in N} \langle p, x \rangle$.

b. In particular, if M is *convex* and *compact* and if N is *convex* and *closed*, property i holds.

c. If $V = \mathbb{R}^n$ is *finite dimensional* and if $M - N$ is *convex*, then

ii. $\exists p \in V^*, p \neq 0$ such that $\sup_{x \in M} \langle p, x \rangle \leq \inf_{x \in N} \langle p, x \rangle$.

3.2. Support Functions (Chapter 2, Section 5)

Let $M \subset V$ be nonempty and $\sigma_M(p) = \sup_{x \in M} \langle p, x \rangle$ its support function.

a. $\overline{\operatorname{co}}(M) = \{x \in V \text{ such that } \forall p \in V^*, \langle p, x \rangle \leq \sigma_M(p)\}$.

b. $p \to \sigma_M(p)$ is *convex, positively homogeneous*, and *lower semicontinuous*. It is *positive* when $0 \in M$ and *finite* when M is bounded.

c. If M is a *cone*, $\sigma_M(p) = \begin{cases} 0 & \text{if} \quad p \in M^- \\ +\infty & \text{if} \quad p \notin M^- \end{cases}$

d. If $\alpha, \beta > 0$, $\sigma_{\alpha M + \beta N}(p) = \alpha \sigma_M(p) + \beta \sigma_N(p)$.
e. $\sigma_M(p) \leq \sigma_N(p)$ if and only if $\overline{co}(M) \subset \overline{co}(N)$.
f. If $M = \bigcup_{i \in I} M_i$, then $\sigma_M(p) = \sup_{i \in I} \sigma_{M_i}(p)$.
g. $\sigma_M = \Psi_M^*$ (Ψ_M is the indicator of M).

3.3. Polarity (Chapter 1, Section 5 and Chapter 3, Section 4)

a. Let M be a *cone* and N be any subset. Then
$$(M + N)^- = M^- \cap N^-.$$

b. If moreover $M + N$ is a *closed convex cone*, then
$$M + N = (M^- \cap N^-)^-.$$

c. If $A \in \mathscr{L}(V, F)$ and if $M \subset V$ is any subset, then
$$(A(M))^- = A^{*-1}(M^-).$$

d. If $A(M)$ is a *closed convex cone*, then
$$A(M) = [A^{*-1}(M^-)]^-.$$

4. CONSTRUCTION OF HILBERT SPACES

4.1. Elementary Operations (Chapter 3, Section 5)

a. A *closed* subspace M of a Hilbert space V is a Hilbert space whose dual is $V^*/M^\perp$.

b. If M is a *closed* subspace of a Hilbert space V, the *quotient space* V/M supplied with the norm $\|x\| = \inf_{y \in M} \|x - y\|$ is a Hilbert space whose dual is $M^\perp$.

c. Let V_i be n Hilbert space; then the product $V = \prod_{i=1}^n V_i$ is a Hilbert space for the scalar product $((x, y)) = \sum_{i=1}^n ((x_i, y_i))$, whose dual is $V^* = \prod_{i=1}^n V_i^*$.

4.2. Dual of a Dense Subspace (Chapter 3, Section 5)

Let V and H be two Hilbert spaces such that

    i. $V \subset H$, the injection is *continuous* and V is *dense* in H.

Then H^* can be identified with a dense subspace of V^*. If H is a pivot space (i.e., $H = H^*$), then

    ii. $V \subset H \subset V^*$, the injections are continuous, each space is dense in the larger spaces.

4.3. Initial Topology (Chapter 5, Section 1)

Assume that n linear operators A_i from a vector space V to Hilbert spaces F_i are *closed* and *collectively injective*. Then V is a Hilbert space for the scalar product

$$((x, y)) = \sum_{i=1}^{n} ((A_i x, A_i y))_{F_i}$$

If K_i is the duality operator of $F_i (i = 1, \ldots, n)$, then $K = \sum_{i=1}^{n} A_i^* K_i A_i$ is the duality operator of V.

4.4. Final Topology (Chapter 5, Section 2)

Let A be a *surjective* linear map from a Hilbert space V onto a vector space F. Then F is a Hilbert space for the scalar product $((f, g)) = ((A^+ f, A^+ g))_V$, whose duality operator is $(AJ^{-1}A^*)^{-1}$.

4.5. Minimal and Maximal Domains of a Family of Operators (Chapter 5, Section 4)

Let $\mathscr{D}$ be a vector space.

a. Let $\mathscr{A}$ be a *closed* family of n linear operators A_i from $\mathscr{D}$ to $\mathscr{D}$. Let H be a completion of $\mathscr{D}$ for a scalar product (x, y). We choose H as a pivot space. There exists a completion $H_0(\mathscr{A})$ of $\mathscr{D}$ for the scalar product $((x, y)) = (x, y) + \sum_{i=1}^{n}(A_i x, A_i y)$ that is *contained in* H. The operators A_i can be extended to continuous linear operators from $H_0(\mathscr{A})$ to H.

b. If we assume that there exist n operators A_i^* from $\mathscr{D}$ to $\mathscr{D}$ satisfying

$$\forall x, y \in \mathscr{D}, \quad (A_i x, y) = (x, A_i^* y)$$

and if $\mathscr{A}^*$ denotes the family of A_i^*, then A_i *can be extended* to a continuous linear operator from $H_0(\mathscr{A})$ to H *and* from H to $H_0(\mathscr{A}^*)^*$.

c. The vector space $H(\mathscr{A}) = \{x \in H \text{ such that } A_i x \in H \text{ for } i = 1, \ldots, n\}$ is a Hilbert space for the scalar product $((x, y))$. The closure of $\mathscr{D}$ in $H(\mathscr{A})$ is $H_0(\mathscr{A})$.

5. COMPACT OPERATORS

5.1. Spectral Decompositions (Chapter 11, Sections 1 and 2)

Let U and V be two Hilbert spaces such that

(1) U is *dense* in V and the injection from U to V is *compact*.

Let $K \in \mathscr{L}(U, U^*)$ and $J \in \mathscr{L}(V, V^*)$ be the duality operators of U and V. There exist an orthonormal base $\{e_n\}$ of vectors of U and a sequence λ_n of positive scalars satisfying

$$\begin{cases} \text{i.} & n \geq 0, \quad Je_n = \lambda_n Ke_n \\ \text{ii.} & \lambda_1 \geq \lambda_2 \geq \ldots \geq \lambda_n \ldots; \quad \text{either the sequence is finite or} \\ & \lim_{n \to \infty} \lambda_n = 0 \\ \text{iii.} & ((e_n, e_k))_U = ((e_n, e_k))_V = 0 \quad \text{if} \quad n \neq k \\ \text{iv.} & \|e_n\|_U = 1; \quad \|e_n\|_V = \sqrt{\lambda_n} \quad \forall n \geq 0. \end{cases}$$

5.2. Characterization of Compact Operators (Chapter 11, Section 3)

Let $A \in \mathscr{L}(U, F)$. It is a *compact* operator if and only if there exist orthonormal bases $\{e_n^*\}_n$ and $\{f_n\}_n$ of U^* and F and a decreasing sequence of positive scalar λ_n, either finite or converging to zero, such that

$$(2) \qquad \forall x \in U, \quad Ax = \sum_{n=1}^{\infty} \lambda_n \langle e_n^*, x \rangle f_n.$$

It is a *Hilbert-Schmidt* operator if and only if $\{\lambda_n\}_n \in l^2$.

5.3. The Fredholm Alternative (Chapter 11, Section 4)

Let V be a Hilbert space and H be a pivot space such that

(3) V is *dense* in H and the injection from V to H is *compact*.

Let $A \in \mathscr{L}(V, V^*)$ be a *self-transposed V-elliptic* operator. There exists a sequence of *negative* scalars $\mu_n < 0$ going to $-\infty$ such that, if $\lambda \neq \mu_n$ for all n, $A + \lambda$ is an isomorphism from V onto V^*. Furthermore, there exists an orthogonal base $\{e_n\}_n$ of V such that

$$(A + \lambda)^{-1} f = \sum_{n=1}^{\infty} \frac{\mu_n}{\mu_n - \lambda} \langle f, e_n \rangle.$$

If $\lambda = \mu_{n_0}$, if $N(n_0) = \{n \in N \text{ such that } \mu_n = \mu_{n_0}\}$, if F_{n_0} is the closed subspace spanned by e_n when $n \notin N(n_0)$ and if $f \in F_{n_0}$, then

$$(A + \lambda)^{-1} f = \sum_{n \in N(n_0)} \frac{\mu_n}{\mu_n - \lambda} \langle f, e_n \rangle + F_{n_0}^{\perp}.$$

5.4. Perturbation of an Isomorphism (Chapter 11, Section 7)

Let $K \in \mathscr{L}(U, F)$ be an *isomorphism* and $J \in \mathscr{L}(U, F)$ be *compact*. If $\lambda \neq 0$, we set $A_\lambda = \lambda K - J$, $N_\lambda = \text{Ker } A_\lambda$, and $F_\lambda = \text{Im } A_\lambda$. Then N_λ is *finite dimensional* and F_λ is *closed*. If λ is not an eigenvalue, A_λ is an isomorphism from U onto F. The sequence of distinct eigenvalues is either finite or converges to zero. Finally, $N_\lambda \neq \{0\}$ if and only if $N_\lambda^* = \text{Ker } A_\lambda^* \neq 0$.

5.5. Hilbert Tensor Products (Chapter 12, Section 3, 4, and 5)

a. The *Hilbert tensor product* of two Hilbert spaces E and F is, by definition, the Hilbert space $E \hat{\otimes} F = \mathscr{L}_2(E^*, F)$ of *Hilbert-Schmidt operators*.

b. The space $\mathscr{L}_2(E \hat{\otimes} F, G)$ is isometric to $E^* \hat{\otimes} \mathscr{L}_2(F, G)$.

c. If $A \in \mathscr{L}(H, V)$, if $U \subset V$, and if $D(A) = \{x \in H \text{ such that } Ax \in U\}$, then $D(A) \hat{\otimes} F = \{M \in H \hat{\otimes} F \text{ such that } \vec{A}M \in U \hat{\otimes} F\}$, where $\vec{A} = A \otimes 1_F$.

d. We have $l^2(N, F) = l^2(N) \hat{\otimes} F$, $L^2(\Omega, F) = L^2(\Omega) \hat{\otimes} F$ and $H^m(\Omega, F) = H^m(\Omega) \hat{\otimes} F$.

5.6. The Kernel Theorem (Chapter 12, Section 6 and 7)

a. An operator $A \in \mathscr{L}(L^2(\Omega_1), L^2(\Omega_2))$ is a *Hilbert-Schimdt operator* if and only if there exists $K \in L^2(\Omega_1 \times \Omega_2)$ such that

$$(Ax)(\omega_2) = \int_\Omega K(\omega_1, \omega_2) x(\omega_1) d\omega_1.$$

b. An operator $A \in \mathscr{L}_2(H^m(\Omega, F^*)^*, G)$ can be written in a unique way in the form $Af = \int_\Omega \langle K(\omega), df(\omega) \rangle$ where $K(.) \in H^m(\Omega, \mathscr{L}_2(F, G))$.

6. SEMIGROUP OPERATORS

6.1. The Hille-Philips Theorem (Chapter 14, Sections 1 and 2)

a. An unbounded operator $(D(A), A)$ of H is the *infinitesimal generator* of an equibounded semigroup of operators $G(t) \in \mathscr{L}(H, H)$ if and only if $(D(A), A)$ is *closed*, with *dense domain* and satisfies

(1) $\quad \forall \lambda > 0, \quad (A + \lambda)$ is invertible and $\|(A + \lambda)^{-n}\|_{\mathscr{L}(H,H)} \leq \dfrac{M}{\lambda^n} \forall n > 0$

for some constant M.

b. Let $V \subset H$, V dense in H, the injection from V to H is continuous. Let $A \in \mathscr{L}(V, V^*)$ be a *V-elliptic* operator and $D(A) = \{x \in V \text{ such that } Ax \in H\}$ the domain of the associated unbounded operator $(D(A), A)$. The latter is the infinitesimal generator of a semigroup of contractions.

6.2. Differential Equations (Chapter 14, Section 3)

Let $(D(A), A)$ be the *infinitesimal generator of* a continuous semigroup. If $f : [0, \infty] \to H$ is continuously differentiable and if $x_0 \in D(A)$, then $x(t) = G(t)x_0 + \int_0^t G(t-s)f(s)\,ds$ is the unique solution of the differential equation

$$\begin{cases} \text{i.} & x(.) \in \mathscr{C}(0, \infty; D(A)). \\ \text{ii.} & \dfrac{dx}{dt} + Ax = f. \\ \text{iii.} & x(0) = x_0. \end{cases}$$

7. THE GREEN'S FORMULA

7.1. The Trace Property (Chapter 13, Section 1)

We say that three Hilbert spaces V, H, T and $\gamma \in \mathscr{L}(V, T)$ satisfy the abstract trace property when

$$(1) \quad \begin{cases} \text{i.} & V \subset H, \text{ the injection from } V \text{ to } H \text{ is } continuous. \\ \text{ii.} & V_0 = \operatorname{Ker} \gamma \text{ is } dense \text{ in } H. \\ \text{iii.} & \gamma \text{ is } surjective. \end{cases}$$

7.2. The Green Formula for Operators (Chapter 13, Section 1)

If $A \in \mathscr{L}(V, E^*)$, we set $A_0 = A|_{V_0} \in \mathscr{L}(V_0, E^*)$ and $E(A_0^*) = \{e \in E \text{ such that } A_0^* e \in H\}$. We assume (1).

There exists a *unique* operator $\beta^* \in \mathscr{L}(E(A_0^*), T^*)$ such that

$$\forall x \in V, \quad \forall e \in E(A_0^*), \quad [e, Ax] - (A_0^* e, x) = \langle \beta^* e, \gamma x \rangle$$

7.3. The Green Formula for Bilinear Forms (Chapter 13, Section 2)

We assume (1).

If $a(x, y)$ is a *continous bilinear form* on $V \times V$, we define its *formal operator* $\Lambda \in \mathscr{L}(V, V_0^*)$ by

$$(\Lambda x, y) = a(x, y) \quad \forall x \in V, \quad \forall y \in V_0$$

and its *domain* by $V(\Lambda) = \{x \in V \text{ such that } \Lambda x \in H\}$. There exists a *unique* operator $\delta \in \mathscr{L}(V(\Lambda), T^*)$ such that

$$\forall x \in V(\Lambda), \quad \forall y \in V, \quad a(x, y) = (\Lambda x, y) + \langle \delta x, \gamma y \rangle.$$

7.4. Theorems of Existence and Uniqueness (Chapter 13, Sections 3 and 7; Chapter 14, Section 4)

We assume (1). Let $a(x, y)$ be a V-*elliptic* continuous bilinear form.
a. $\Lambda \times \gamma$ is an *isomorphism* from $V(\Lambda)$ onto $H \times T$
b. $\Lambda \times \delta$ is an *isomorphism* from $V(\Lambda)$ onto $H \times T^*$.
c. If $f \in \mathscr{C}^1(0, \infty, H)$ and $x_0 \in V(\Lambda)$ satisfies $\gamma x_0 = 0$, there exists a *unique* solution $x(.) \in \mathscr{C}(0, \infty; V(\Lambda))$ to the differential equation

(2) $\quad \dfrac{dx}{dt} + \Lambda x = f, \quad x(0) = x_0, \quad \text{and} \quad \gamma x(t) = 0 \quad \forall t > 0.$

d. If $f \in \mathscr{C}^1(0, \infty, H)$ and $x_0 \in V(\Lambda)$ satisfies $\delta x_0 = 0$, there exists a *unique* solution $x(.) \in \mathscr{C}(0, \infty; V(\Lambda))$ to the differential equation

(3) $\quad \dfrac{dx}{dt} + \Lambda x = f, \quad x(0) = x_0, \quad \text{and} \quad \delta x(t) = 0 \quad \forall t > 0$

e. If $P \subset T$ is a *closed convex cone*, if $f \in H, \varphi \in T$ and $\psi \in T^*$ are given, there exists a *unique* solution $x \in V(\Lambda)$ to

(4) $\quad \Lambda x = f, \quad \gamma x - \varphi \in P, \quad \delta x - \psi \in P^+, \quad \langle \delta x - \psi, \gamma x - \varphi \rangle = 0.$

8. CONVEX ANALYSIS AND OPTIMIZATION

8.1. Conjugate Functions (Chapter 10, Section 1)

A function $f: U \to \,]{-\infty}, +\infty]$ with nonempty domain is *convex and lower semicontinuous* if and only if $f = f^{**}$.

8.2. Subdifferentiability (Chapter 10, Section 3)

a. Let f be a lower semicontinuous convex function from U to $]-\infty, +\infty]$ with a nonempty domain Dom f. Then the set $D(\partial f)$ of points $x \in \text{Dom} f$ where f has a subdifferential is *not empty and is dense in* Dom f.
b. If $J \in \mathscr{L}(U, U^*)$ is the duality operator, the correspondence $x \in D(\partial f) \mapsto Jx + \partial f(x) \subset U^*$ is *surjective*.

8.3. Further Properties of Continuous Convex Functions (Chapter 10, Section 3)

Let f be a *lower semicontinuous convex function* from U to $]-\infty, +\infty]$. Assume that the *interior* Int Dom f of its domain is not empty.

a. $\forall x \in \text{Int Dom } f, \partial f(x)$ is a nonempty *bounded* closed convex subset.
b. $\forall x \in \text{Int Dom } f, \forall v \in U$, the support function of $\partial f(x)$:

$$\sigma(\partial f(x), v) = Df(x)(v)$$

is equal to the derivative from the right of f at x.

c. The mapping $\{x, v\} \in \text{Int Dom } f \times U \to Df(x)(v)$ is *upper semicontinuous*. (Thus the correspondence $\partial f(.)$ is upper hemicontinuous from Int Dom f to U^*.)

d. $\forall x \in \text{Int Dom } f, \forall v \in U$, we have

$$f(x) - f(x - v) \leq Df(x)(v) \leq f(x + v) - f(x).$$

8.4. Properties of the Subdifferential (Chapter 10, Section 3)

Let $f : U \to]-\infty, +\infty]$ be a lower semicontinuous convex function with nonempty domain.

a. $p \in \partial f(x)$ if and only if $x \in \partial f^*(p)$.
b. $\partial f^*(0)$ is the *minimal set* of f (i.e., $\bar{x}$ minimizes f if and only if $\bar{x}$ is a critical point of $\partial f(.)$).
c. f is *Gâteaux-differentiable* at x if f is continuous at x and $\partial f(x)$ contains a *unique element* (which is the *gradient* $Df(x)$).

8.5. Variational Principle (Chapter 10, Section 2)

Let X be a *convex set*, $g : X \to]-\infty, +\infty]$ be *convex*, and $h : U \to]-\infty, +\infty]$ be *convex* and *Gâteaux-differentiable* on X. Then $\bar{x} \in X$ minimizes $x \to g(x) + h(x)$ on X if and only if $\bar{x} \in X$ is a solution to the *variational inequalities*

(5) $\quad \forall x \in X, \quad \langle Dh(\bar{x}), \bar{x} - x \rangle + g(\bar{x}) - g(x) \leq 0.$

In particular, if $g = 0$ and $\bar{x} \in \text{Int } X$, this condition can be written $Dh(\bar{x}) = 0$.

8.6. The Fenchel Theorem (Chapter 10, Section 4)

Let U and V be Hilbert spaces, $L \in \mathcal{L}(U, V)$, and $f : U \to]-\infty, +\infty]$ and $g : V \to]-\infty, +\infty]$ be lower semicontinuous convex functions with nonempty domains. We set

$$v = \inf_{x \in U}(f(x) + g(Lx)); \quad v^* = \inf_{p \in V^*}(f^*(-L^*p) + g^*(p)) \geq -v.$$

a. The following conditions are equivalent.

   i. $(f(\bar{x}) + g(L\bar{x})) + (f^*(-L^*\bar{p}) + g^*(\bar{p})) = 0.$

ii. $L^*\bar{p} \in -\partial f(\bar{x})$ and $\bar{p} \in \partial g(L\bar{x})$ (Lagrangian form).
iii. $L\bar{x} \in \partial g^*(\bar{p})$ and $L^*\bar{p} \in -\partial f(\bar{x})$ (Hamiltonian form).
iv. $\bar{x} \in \partial f^*(-L^*\bar{p}) \cap L^{-1} \partial g^*(\bar{p})$ where $\bar{p}$ is a solution to $0 \in \partial g^*(\bar{p}) - L\partial f^*(-L^*\bar{p}))$ (i.e., a *critical point of* $\partial g^*(.) - L\partial f^*(-L^*.)$).

b. If we assume that $0 \in \text{Int}(L(\text{Dom} f) - \text{Dom } g)$, then there exists $\bar{p} \in V^*$ such that $-v = v^* = f^*(-L^*\bar{p}) + g^*(\bar{p})$.

8.7. Quadratic Programming (Chapter 4, Section 7)

Let $J \in \mathcal{L}(U, U^*)$ be the duality operator of U and $A \in \mathcal{L}(U, V)$ be *surjective*. There exists a *unique* solution $\bar{x}$ to the problem $\alpha(v) = \min_{Ax=v} \frac{1}{2} \|x - u\|^2$ that can be written

$$\bar{x} = u - J^{-1} A^* \bar{p} \quad \text{where} \quad \bar{p} = (AJ^{-1}A^*)^{-1}(Au - v) = -D\alpha(v).$$

8.8. Duality Theorem in Convex Optimization (Chapter 2, Section 6)

Let X be a *convex* subset of a Hilbert space U, $f: X \to \mathbb{R}$ be *convex*, $A: X \to \mathbb{R}^n$ be a mapping whose n components A_i are *convex*, $B \in \mathcal{L}(U, \mathbb{R}^m)$ and $v \in \mathbb{R}^m$.

Let $K = \{x \in X$ such that $Ax \leq 0$ and $Bx = v\}$ and $\alpha = \inf_{x \in K} f(x)$. If we posit the *constraint qualification* condition

$$\begin{cases} \text{i.} & \exists \tilde{x} \in X \quad \text{such that} \quad A\tilde{x} \ll 0 \quad \text{and} \quad B\tilde{x} = v. \\ \text{ii.} & v \in \text{Int } B(X). \end{cases}$$

then there exist $\bar{p} \in \mathbb{R}_+^n$ and $\bar{q} \in \mathbb{R}^m$ such that

$$\alpha = \inf_{x \in X} [f(x) + \langle \bar{p}, A(x) \rangle + \langle \bar{q}, Bx - v \rangle].$$

In this case $\bar{x} \in K$ minimizes f on K if and only if

$$\begin{cases} \text{i.} & f(\bar{x}) + \langle \bar{p}, A\bar{x} \rangle + \langle \bar{q}, B\bar{x} - v \rangle \\ & = \min_{x \in X} [f(x) + \langle \bar{p}, Ax \rangle + \langle \bar{q}, Bx - v \rangle]. \\ \text{ii.} & \langle \bar{p}, A\bar{x} \rangle = 0. \end{cases}$$

9. NONLINEAR ANALYSIS

9.1. The Browder-Ky Fan Theorem (Chapter 15, Section 2)

Let U and V be Hilbert spaces, $L \in \mathcal{L}(U, V)$, $X \subset U$ be a *compact convex* subset and $S: X \to V$ be an *upper hemicontinuous* correspondence with nonempty *closed convex* images. We posit the following *boundary condition*

(1) $\qquad \forall x \in X, \quad -S(x) \cap cl(LT_X(x)) \neq \emptyset.$

It ensues that

(2) $\begin{cases} \text{i.} & \exists \bar{x} \in X \quad \text{such that} \quad 0 \in S(\bar{x}). \\ \text{ii.} & \forall y \in L(X), \quad \exists x \in X \quad \text{such that} \quad y \in Lx + S(x). \end{cases}$

We point out that if S_1 and S_2 satisfy the preceding assumptions, so does $\lambda_1 S_1 + \lambda_2 S_2$ when $\lambda_1, \lambda_2 > 0$.

9.2. The Browder Theorem (Chapter 15, Section 2)

If $U = V = \mathbb{R}^n$ and if X is the unit ball, then any upper hemicontinuous correspondence $S: X \to U$ with nonempty closed convex values satisfying

(3) $\qquad\qquad \text{If} \quad \|x\| = 1, \quad \sigma(S(x), x) \geq 0.$

has a critical point and satisfies $X \subset (1 + S)(X)$.

9.3. Surjectivity of Coercive Correspondences (Chapter 15, Section 3)

Any *coercive upper hemicontinuous* correspondence from $\mathbb{R}^n$ to $\mathbb{R}^n$ *with nonempty closed convex values* is *surjective*.

9.4. The Kakutani Fixed Point Theorem (Chapter 15, Section 3)

Let $X \subset U$ be a *convex compact* subset and $S: X \to U$ be an *upper hemicontinuous* correspondence with *closed convex* values.
a. Either one of the following assumptions

$\begin{cases} \text{i.} & S \text{ maps } X \text{ into } X. \\ \text{ii.} & S \text{ is } inward \ (\forall x \in X, S(x) \cap (x + T_X(x)) \neq \emptyset). \\ \text{iii.} & S \text{ is } outward \ (\forall x \in X, S(x) \cap (x - T_X(x)) \neq \emptyset). \end{cases}$

implies the existence of a *fixed point* $\bar{x} \in X$ of S.
b. If S is *outward*, then S is *surjective* in the sense that $\forall y \in X, \exists x \in X$ such that $y \in S(x)$.

9.5. The Leray-Schauder Theorem (Chapter 15, Section 2)

Let $U = \mathbb{R}^n, X \subset \mathbb{R}^n$ be *convex compact with a nonempty interior*, V be a Hilbert space and $S: X \times [0, 1] \to V$ be an upper hemicontinuous correspondence with closed convex values.

We assume that

(4) $\begin{cases} \text{i.} & \forall x \in X, \quad -S(x,0) \cap cl(LT_X(x)) \neq \emptyset. \\ \text{ii.} & \forall x \in \partial X, \quad \forall \lambda \in [0,1[, \quad 0 \notin S(x,\lambda). \end{cases}$

Then there exists a critical point $\bar{x} \in X$ of the correspondence $x \to S(x,1)$.

9.6. The Debreu-Gale-Nikaïdo Theorem (Chapter 15, Section 8)

Let $M^n = \{x \in \mathbb{R}^n_+ \text{ such that } \sum x_i = 1\}$ and $S: M^n \to \mathbb{R}^n$ be a correspondence with nonempty values satisfying

$\begin{cases} \text{i.} & S \text{ is upper hemicontinuous.} \\ \text{ii.} & \forall x \in M^n, \quad S(x) - \mathbb{R}^n_+ \text{ is closed and convex.} \\ \text{iii.} & \forall x \in M^n, \quad \sigma(S(x), x) \geq 0 \quad \text{(Walras law).} \end{cases}$

Then there exists $\bar{x} \in M^n$ such that $S(\bar{x}) \cap \mathbb{R}^n_+ \neq \emptyset$.

9.7. The Perron-Frobenius Property of Positive Correspondences (Chapter 15, Section 9)

Let S be an upper *hemicontinuous* correspondence with *compact convex* values from M^n to $\mathbb{R}^n_+$ satisfying

$\begin{cases} \text{i.} & \forall x \in M^n, \quad \sigma(S(x), p_0) > 0. \\ \text{ii.} & x \to \sigma(S(x), p_0) \text{ is continuous (where } p_0 = (1,1,\ldots,1)). \end{cases}$

Then
a. $\exists \bar{x} \in M^n, \exists \bar{\lambda} > 0$ such that $\bar{\lambda} \bar{x} \in S(\bar{x})$ (Perron's property).
b. $\forall \mu > \sup_{x \in M^n} \sigma(G(x), p_0), \forall y \in M^n, \exists x \in M^n, \exists \beta > 0$ such that $\mu x \in S(x) + \beta y$ (Frobenius's property).

9.8. Variational Inequalities (Chapter 15, Section 5)

Let $X \subset U$ be *convex compact* and S be an *upper semicontinuous* correspondence from X to U^* with nonempty *compact convex values*. There exists a critical point $\bar{x} \in X$ of the correspondence $x \to S(x) + N_X(x)$ (which is equal to S on Int X).

9.9. Quasi-Variational Inequalities (Chapter 15, Section 6)

Let $X \subset U$ be *compact convex*, S an *upper hemicontinuous* correspondence from X to X with *nonempty closed convex values* and $f: X \times X \to \mathbb{R}$ a function satisfying

$$\begin{cases} \text{i.} & \forall y \in X, \quad x \to f(x,y) \text{ is } \textit{lower semicontinuous}. \\ \text{ii.} & \forall x \in X, \quad y \to f(x,y) \text{ is } \textit{concave}. \\ \text{iii.} & \sup_{y \in X} f(y,y) \leq 0. \end{cases}$$

We posit the following consistency condition: The subset $\{x \in X$ such that $\sup_{y \in S(x)} f(x,y) \leq 0\}$ is closed. Then there exists $\bar{x} \in X$ satisfying

$$\begin{cases} \text{i.} & \bar{x} \in S(\bar{x}). \\ \text{ii.} & \sup_{y \in S(\bar{x})} f(\bar{x},y) \leq 0. \end{cases}$$

9.10. Tangent Cones (Chapter 15, Section 4)

Let $X \subset U$ be a nonempty closed convex subset.
a. $N_X(x) = \partial \Psi_X(x) = \{p \in U^* \text{ such that } \langle p, x \rangle = \sigma_X(p)\}$.
b. $T_X(x) = N_X(x)^- = cl\left\{\bigcup_{\lambda > 0} \lambda(X - x)\right\}$.
c. If $x \in \text{Int } X$, $T_X(x) = U$.
d. If $x \in X \subset Y$, $T_X(x) \subset T_Y(x)$.
f. $T_{\prod_{i=1}^n X_i}(x^1, \ldots, x^n) = \prod_{i=1}^n T_{X_i}(x^i)$.
g. $T_{\overline{A(X)}}(Ax) = \overline{AT_X(x)}$.
h. If $0 \in \text{Int}(L(X) - Y)$, $T_{X \cap L^{-1}(Y)}(x) = T_X(x) \cap L^{-1}T_Y(Lx)$.
i. If P is a closed convex cone, $v \in T_P(x)$ if and only if $\langle p, v \rangle \leq 0$ for all $p \in P^-$ satisfying $\langle p, x \rangle = 0$.

10. MINIMAX INEQUALITIES

10.1. The Lop-Sided Minimax Inequality (Chapter 2, Section 7)

Let X be a *convex compact* subset, Y be a convex subset and $\varphi : X \times Y \to \mathbb{R}$ satisfying

$$\begin{cases} \text{i.} & \forall y \in Y, \quad x \to \varphi(x,y) \text{ is } \textit{convex} \text{ and } \textit{lower semicontinuous}. \\ \text{ii.} & \forall x \in X, \quad y \to \varphi(x,y) \text{ is } \textit{concave}. \end{cases}$$

There exists $\bar{x} \in X$ such that

$$\sup_{y \in Y} \varphi(\bar{x}, y) = \sup_{y \in Y} \inf_{x \in X} \varphi(x,y) = \inf_{x \in X} \sup_{y \in Y} \varphi(x,y).$$

10.2. The Lasry Equality (See [AAA], Chapter 5, Section 6)

Let X be a *convex compact* subset, Y be a convex subset and $\varphi: X \times Y \to \mathbb{R}$ satisfying

$$\begin{cases} \text{i.} & \forall y \in Y, \quad x \to \varphi(x, y) \quad \text{is } lower\ semicontinuous. \\ \text{ii.} & \forall x \in X, \quad y \to \varphi(x, y) \quad \text{is } concave. \end{cases}$$

There exists $\bar{x} \in X$ such that

$$\sup_{y \in Y} \varphi(\bar{x}, y) = \inf_{C \in \mathscr{C}(Y,X)} \sup_{y \in Y} \varphi(C(y), y) = \sup_{D \in \mathscr{C}(X,Y)} \inf_{x \in X} \varphi(x, D(x)).$$

10.3. The Ky Fan Inequality (See [AAA], Chapter 5, Section 6)

Let X be a *convex compact* subset and $\varphi: X \times X \to \mathbb{R}$ satisfy

$$\begin{cases} \text{i.} & \forall y \in X, \quad x \to \varphi(x, y) \quad \text{is } lower\ semicontinuous. \\ \text{ii.} & \forall x \in X, \quad y \to \varphi(x, y) \quad \text{is } concave. \\ \text{iii.} & \forall y \in X, \quad \varphi(y, y) \leq 0. \end{cases}$$

There exists $\bar{x} \in X$ such that $\sup_{y \in X} \varphi(\bar{x}, y) \leq 0$.

11. SOBOLEV SPACES, CONVOLUTION, AND FOURIER TRANSFORM

11.1. Sobolev Spaces (Chapter 7, Sections 1, 2 and 3 and Chapter 9, Section 1.)

a. The following injections are *continuous*, each space being dense in the larger spaces (except $\mathscr{D}(\Omega)^{\circledast}$): if $m > k$,

$$\mathscr{D}(\Omega) \subset H_0^m(\Omega) \subset H_0^k(\Omega) \subset L^2(\Omega) \subset H^{-k}(\Omega) \subset H^{-m}(\Omega) \subset \mathscr{D}(\Omega)^{\circledast}$$

b. The derivative D^k is a continuous linear operator from $H_0^{m+|k|}(\Omega)$ to $H_0^m(\Omega)$ ($m \geq 0$) and from $H^{-m}(\Omega)$ to $H^{-m-|k|}(\Omega)$.
c. The space $H^{-m}(\Omega)$ consists of sums $\sum_{|k| \leq m} D^k f_k$ where f_k belongs to $L^2(\Omega)$ for all $|k| \leq m$.
d. The duality operator of $H_0^m(\Omega)$ is the differential operator $\sum_{|k| \leq m} (-1)^{|k|} D^{2k}$.

11.2. Extension operators (Chapter 7, Sections 4 and 6; and Chapter 9, Sections 1 and 5)

a. $\forall m \geq 0$, $\mathscr{D}(\mathbb{R}^n)$ is *dense* in $H^m(\mathbb{R}^n)$ [i.e., $H^m(\mathbb{R}^n) = H_0^m(\mathbb{R}^n)$].

b. The extension operator π_0 by zero outside of Ω is an isometry from $H_0^m(\Omega)$ to $H^m(\mathbb{R}^n)$.

c. Let $\rho \in \mathscr{L}(H^m(\mathbb{R}^n), H^m(\Omega))$ denote the restriction operator. If Ω is *regular*, there exists $\pi \in \mathscr{L}(H^m(\Omega), H^m(\mathbb{R}^n))$ such that

$$\begin{cases} \text{i.} & \forall |k| \leq m, \quad \rho D^k \pi x = D^k x. \\ \text{ii.} & \forall x \in H_0^m(\Omega), \quad \pi x = \pi_0 x. \end{cases}$$

d. The space $\mathscr{E}(\bar{\Omega})$ is dense in $H^m(\Omega)$.

11.3. Compactness Property (Chapter 7, Section 5 and Chapter 9, Section 6)

Let Ω be *regular* (and thus, *bounded*). If $m > k \geq 0$, the injections from $H^m(\Omega)$ to $H^k(\Omega)$, from $H_0^m(\Omega)$ to $H_0^k(\Omega)$, and from $H^{-k}(\Omega)$ to $H^{-m}(\Omega)$ are *compact*.

11.4. The Trace Theorem (Chapter 7, Section 8 and Chapter 9, Section 5)

Let Ω be *regular* and Γ denote its boundary. The trace operator $\gamma = \gamma_0 \times \ldots \times \gamma_{m-1}$ is a *surjective continuous linear* operator from $H^m(\Omega)$ onto $\prod_{j=0}^{m-1} H^{m-j-1/2}(\Gamma)$. The kernel of γ is $H_0^m(\Omega)$.

11.5. Sobolev and Interpolation Inequalities

Let $\Omega \subset \mathbb{R}^n$ be regular. If $s > n/2$, then $H^s(\mathbb{R}^n) \subset \mathscr{C}_\infty(\mathbb{R}^n)$; the inclusion is continuous. If $r < s < t$, we have

$$\forall x \in H^t(\Omega), \quad \|x\|_s \leq \|x\|_r^\theta \|x\|_t^{1-\theta} \quad \text{where } \theta = \frac{t-s}{t-r}.$$

11.6. Convolution (Chapter 6, Sections 4 to 6)

a. Let $\lambda \in L^1(\mathbb{R}^n)$. The convolution operator $\lambda *$ is a continuous linear operator from $H^s(\mathbb{R}^n)$ to $H^s(\mathbb{R}^n)$ for all $s \in \mathbb{R}$.

b. If $\int \lambda = 1$, if λ has a compact support and if $\lambda_h = (1/h^n) \lambda\left(\frac{\cdot}{h}\right)$, the operator $\lambda_h * \in \mathscr{L}(H^s(\mathbb{R}^n), H^s(\mathbb{R}^n))$ converges pointwise to the identity.

c. If λ is continuously differentiable with compact support and if $x \in H^s(\mathbb{R}^n)$, then $\lambda * x \in H^{s+1}(\mathbb{R}^n)$ and $D_i(\lambda * x) = D_i \lambda * x$.

11.7. Fourier Transform (Chapter 9, Sections 2 and 3)

a. For all $s \in \mathbb{R}$, the Fourier transform F is an *isomorphism* from $H^s(\mathbb{R}^n)$ onto the space $\hat{H}^s(\mathbb{R}^n) = L^2(\mathbb{R}^n, a_s)$ where $a_s(\omega) = (1 + \|\omega\|^2)^s$. Its inverse is $\bar{F}$ and $F^* = \bar{F}$. We also have $(x, y) = (Fx, Fy)$ when $x \in H^s(\mathbb{R}^n)$, $y \in H^{-s}(\mathbb{R}^n)$.

b. If $\lambda \in L^1(\mathbb{R}^n)$ and $x \in H^s(\mathbb{R}^n)$ (where $s \in \mathbb{R}$), then
$$F(\lambda * x) = F(\lambda) F(x).$$

c. We also have the formulas
$$D^k F(x) = F((-2i\pi\omega)^k x).$$
$$(2i\pi\xi)^k F(x)(\xi) = F[D^k(x)](\xi)$$
$$F(\tau_a x) = e^{-2i\pi\langle \cdot, a \rangle} F(x)$$
$$F(x_h) = (Fx)(.h)$$

Exercises

CHAPTER 1

Section 1

1. Let $\{V,((.,.))\}$ be a prehilbert space. Show that the mapping $\{x,y\} \to ((x,y))$ from $V \times V$ to $\mathbb{R}$ is continuous.
2. Let $\{x_n\}$ be a sequence of real numbers satisfying $\sum_{n=1}^{\infty} |x_n|^2 < +\infty$. Show that $|\sum_{n=1}^{\infty} x_n x_{n+1}| \le \sum_{n=1}^{\infty} x_n^2$.
3. Let V be a prehilbert space and x and y points of V.
   a. Show that $((x,y))^2 = ((x,x)) \cdot ((y,y))$ if and only if x and y are linearly dependent.
   b. Deduce from this that $\|x+y\| = \|x\| + \|y\|$ if and only if $x=0$ or there exists $\lambda \ge 0$ such that $y = \lambda x$.
4. Let V be a normed space such that $\|x+y\|^2 + \|x-y\|^2 = 2[\|x\|^2 + \|y\|^2]$ for all x, y in V. Show that the mapping $\{x,y\} \to \frac{1}{4}[\|x+y\|^2 - \|x-y\|^2]$ from $V \times V$ to $\mathbb{R}$ defines a scalar product on V.
5. Show that in an infinite dimensional Hilbert space every compact subset has an empty interior.
6. Let V be a Hilbert space. Show that every vector subspace of V that is finite dimensional is closed.
7. (Pythagorean theorem.) Let V be a prehilbert space and x and y vectors of V. Show that x and y are "orthogonal," that is, $((x,y)) = 0$, if and only if $\|x+y\|^2 = \|x\|^2 + \|y\|^2$.
8. Let $l^2(\mathbb{N})$ be the Hilbert space of real sequences $\{x_n\}_{n \in \mathbb{N}}$ such that $\sum_n |x_n|^2 < +\infty$. We define the subset $Q = \{\{x_n\} \in l^2(\mathbb{N})$ such that $|x_n| \le 1/n$ for all $n \ge 1\}$ (called the Hilbert cube).
   a. Show that Q is compact.
   b. Show that Q is not contained in any finite dimensional vector subspace of $l^2(\mathbb{N})$.
9. Let $\{x_n\}$ be a sequence in a Hilbert space V and x a vector of V. We

say that the sequence $\{x_n\}$ "converges weakly" to x if $\lim_{n\to\infty}((x_n,y)) = ((x,y))$ for all $y \in V$. We say that $\{x_n\}$ "converges strongly" to x if $\lim_{n\to\infty} \|x - x_n\| = 0$ (i.e., if $\{x_n\}$ converges to x for the topology of V).

a. Show that if the sequence $\{x_n\}$ converges strongly to x, it converges weakly to x.
b. Show that the converse of item **a** is in general not true. (Consider, for example, the sequence $\{e^n\} \in l^2(\mathbb{N})$ defined by $e_n^n = 1$ and $e_p^n = 0$ if $p \neq n$.)
c. Show that the sequence $\{x_n\}$ converges strongly to x if and only if $\{x_n\}$ converges weakly to x and $\lim_{n\to\infty} \|x_n\| = \|x\|$.

Section 2

10. Let E and F be two Hilbert spaces. Show that if E is finite dimensional, every linear mapping from E to F is continuous.

11. Let E and F be two Hilbert spaces and A a continuous linear mapping from E to F. We set

$$a_1 = \sup_{\substack{x \neq 0 \\ x \in E}} \frac{\|Ax\|}{\|x\|}, \qquad a_2 = \sup_{\substack{x \in E \\ \|x\|=1}} \|Ax\|$$

$$a_3 = \sup_{\substack{\|x\| \leq 1 \\ x \in E}} \|Ax\|,$$

$$a_4 = \inf\{c > 0 \text{ such that } \|Ax\| \leq c\|x\| \text{ for all } x \in E\}.$$

Show that $\|A\| = a_1 = a_2 = a_3 = a_4$.

12. Let E and F be two Hilbert spaces and f a mapping from E to F such that
    a. $f(x+y) = f(x) + f(y)$ for all $x, y \in E$.
    b. $\exists K > 0$ such that $\|f(x)\| \leq K$ for all $x \in E$, $\|x\| \leq +1$. Show that f is continuous and linear.

13. Let E and F be two Hilbert space and f a continuous mapping from E to F such that $f(x+y) = f(x) + f(y)$ for all x, y in E. Show that f is a linear mapping.

Section 4

14. Let V be a *finite dimensional* Hilbert space and A a nonempty closed subset of V.
    a. Show that for all $x \in V$ there exists $a \in A$ such that $\|x - a\| = \min_{y \in A} \|x - y\|$.
    b. Show by means of a counterexample that if A is not convex, the solution $a \in A$ is not in general unique.

15. Let H be a Hilbert space and $P: H \to H$ a mapping such that $((x - Px, Px - Py)) \geq 0, \forall x, y \in H$.
    a. Let $C = P(H)$ be the image of P. Show that $((x - Px, Px - z)) \geq 0$ for all $x \in H$ and all $z \in \text{Conv } C$ where
    $$\text{Conv } C = \left\{ z = \sum_{i=1}^{n} \alpha_i z_i \text{ such that } \sum_{i=1}^{n} \alpha_i = 1, \alpha_i \geq 0 \text{ and } z_i \in C \, i = 1, \ldots, n \right\}.$$
    b. Show that $((x - Px, Px - z)) \geq 0$ for all $z \in \overline{\text{Conv } C}$.
    c. Show that P is the best approximation projector onto $\overline{\text{Conv } C}$. Conclude from this that C is closed and convex.
    d. Show that a mapping $P: H \to H$ is a best approximation projector onto a closed convex set C if and only if $((x - Px, Px - Py)) \geq 0$, $\forall x, y \in H$.

Section 5

16. Let E be a prehilbert space, F a complete vector subspace of E, and X a nonempty complete convex subset of F.
    a. For all $x \in E$ we denote by $t_F(x)$ [respectively, $t_X(x)$] the best approximation projector of x onto F (respectively, on X). Show that $t_X(x)$ is the projection from $t_F(x)$ on X.
    b. Show that the preceding result is not true in general if we assume simply that F is nonempty complete convex subset of E.
17. Let V be a prehilbert space. We say that a mapping A from V to V is "monotone" if $((Au - Av, u - v)) \geq 0$ for all $u, v \in V$.
    a. Show that A is monotone if and only if we have $\alpha > 0$, $\|v - u + \alpha(Av - Au)\| \geq \|v - u\|$ for all $u, v \in V$, and $\alpha > 0$.
    b. Conclude from this that if A is monotone, the mappings $I + \alpha A$ are injective for all $\alpha > 0$ (where I denotes the identity mapping from V to V).
18. Let X be a nonempty convex subset of a Hilbert space V and $\{x_n\}$ a sequence of points of X. Show that if $x \in V$ satisfies $\lim_{n \to \infty} ((x_n, y)) = ((x, y))$ for all $y \in V$, then x belongs to X.
19. Let X be a nonempty closed convex subset of a Hilbert space V. We define for all $x \in X$ the sets
    $$T(X, x) = \overline{\{\lambda(y - x) | y \in X, \lambda > 0\}}$$
    and
    $$N(X, x) = \{u \in V | ((u, x)) \geq ((u, y)) \text{ for all } y \in X\},$$

which are called, respectively, the tangent cone and the normal cone to X at x.

    a. Show that for all $x \in X$ the sets $T(X,x)$ and $N(X,x)$ are closed convex cones with vertex 0.

    b. Show that for all $x \in X$,

$$T(X,x) = N^\ominus(X,x) \quad \text{and} \quad N(X,x) = T^\ominus(X,x),$$

20. Let X and Y be nonempty closed convex subsets of a Hilbert space V. We consider the cones $T(X,x)$ and $N(X,x)$ for $x \in X$, as defined in exercise 19.

    a. Show that if $x \in \overset{\circ}{X}$, then $T(X,x) = V$ and $N(X,x) = \{0\}$.

    b. Suppose that $X \subset Y$ and show that for all $x \in X$, $T(X,x) \subset T(Y,x)$ and $N(Y,x) \subset N(X,x)$.

    c. Suppose that $X + Y$ is a closed subset of V and show that for all $x \in X, y \in Y$:

$$T(X + Y, x + y) = \overline{T(X,x) + T(Y,y)}$$

and

$$N(X + Y, x + y) = N(X,x) \cap N(Y,Y).$$

(Use the result of exercise 19b.)

21. Let V be a Hilbert space and consider the cones $T(X,x)$ and $N(X,x)$ to a nonempty closed convex subset X of V, as defined in exercise 19.

    a. Show that $T(V,x) = V$ and $N(V,x) = \{0\}$ for all $x \in V$.

    b. Let $X = \{x \in V \text{ such that } \|x\| \leq +1\}$. Show that for all $x \in X$ such that $\|x\| = +1$, we have $N(X,x) = \{\lambda x \text{ such that } \lambda \geq 0\}$ and $T(X,x) = \{u \in V \text{ such that } ((u,x)) \leq 0\}$.

    c. Let x_0 be a point of V. Show that

$$T(\{x_0\},x_0) = \{0\} \quad \text{and} \quad N(\{x_0\},x_0) = V.$$

(Use the result of exercise 19b.)

22. Let V be a Hilbert space and P a continuous linear mapping from V to V. Show that P is an orthogonal projector onto a closed vector subspace of V if and only if $PP = P$ and $((Px,y)) = ((x,Py))$ for all $x, y \in V$.

Section 7

23. Let $l^2(\mathbb{N})$ be the Hilbert space of real sequences $\{x_n\}_{n \geq 1}$ such that $\sum_{n=1}^\infty |x_n|^2 < +\infty$ and let H be the subset of $l^2(\mathbb{N})$ consisting of those

sequences that have at most a finite number of nonzero terms. We define the sequence $\{e^i\}$, $i \in \mathbb{N}$, of H by $e_i^i = +1$ and $e_j^i = 0$ if $i \neq j$.
  a. Show that H is a vector subspace distinct from $l^2(\mathbb{N})$ and that the sequence $\{e^i\}$ is an "algebraic base" for H.
  b. Show that the sequence $\{e^i\}$ is an "orthonormal base" of $l^2(\mathbb{N})$.
24. Let V be a Hilbert space.
  a. Show that every compact subset of V is closed and bounded.
  b. Let $\mathcal{H} = \{e_n\}_{n \in \mathbb{N}}$ be an orthonormal family of a infinite dimensional Hilbert space V. Show that $\mathcal{H}$ is closed and bounded but not compact.
25. Let V be a prehilbert space and $x_1, \ldots, x_n$ vectors of V. We call the Gram determinant of these vectors, denoted by $G(x_1, \ldots, x_n)$, the determinant of the matrix of coefficients $((x_i, x_j))$, $i = 1, \ldots, n$, $j = 1, \ldots, n$.
  a. Show that $G(x_1, \ldots, x_n) \geq 0$ and that $G(x_1, \ldots, x_n) > 0$ if and only if the vectors are linearly independent.
  b. Suppose that the x_i's are linearly independent and denote by H the vector subspace they generate. Show that $d(x, H)^2 = G(x, x_1, \ldots, x_n)/G(x_1, \ldots, x_n)$ for all $x \in V$.

CHAPTER 2

Section 1

1. Let M be a closed vector subspace of a Hilbert space V and x_0 a point of V such that $x_0 \notin M$. Show that there exists a continuous linear form f defined on V such that $f(x_0) = +1$ and $f(x) = 0$ for all $x \in M$.
2. Let V be a Hilbert space and $x_0 \neq 0$ a point of V. Show that there exists a continuous linear form f defined on V such that

$$f(x_0) = \|x_0\| \quad \text{and} \quad \sup_{\substack{x \neq 0 \\ x \in V}} \frac{|f(x)|}{\|x\|} = +1.$$

3. Let M be a vector subspace of a Hilbert space V and x_0 a point of V. Show that $x_0 \in \bar{M}$ if and only if there exists no continuous linear form f defined on V such that $f(x_0) \neq 0$ and $f(x) = 0$ for $x \in M$.
4. Let V be a Hilbert space and $B = \{x \in V \text{ such that } \|x\| < +1\}$ the open unit ball of V. Show that for all $x_0 \in V$ such that $\|x_0\| = +1$ there exists a continuous linear form f, not identically zero, defined on V and satisfying $f(x_0) \geq \sup_{x \in B} f(x)$.

5. Let $\{x_n\}_{n \geq 1}$ be a sequence of a Hilbert space V, $\{\alpha_n\}_{n \geq 1}$ a sequence of real numbers and γ a real number, $\gamma > 0$. Show that there exists a continuous linear form f defined on V such that $f(x_n) = \alpha_n$ for all $n \geq 1$ and $\|f\| \leq \gamma$ if and only if for all k and all real numbers $\beta_1, \ldots, \beta_k$:

$$\left| \sum_{i=1}^{k} \beta_i \alpha_i \right| \leq \gamma \left\| \sum_{i=1}^{k} \beta_i x_i \right\|.$$

(Consider the vector subspace of V generated by the x_n's.)

6. Let V be a Hilbert space and $\{x_n\}_{n \geq 1}$ an orthogonal sequence of V. Show that the sequence $\{x_n\}$ is a base of V if and only if for all $x \in V$ such that $((x, x_n)) = 0$, for all $n \geq 1$, we have $x = 0$.

Sections 4 and 5

7. Let V be a Hilbert space. We call a closed half space every subset $D_{u,\alpha} = \{x \in V \text{ such that } ((u,x)) \leq \alpha\}$ where $\alpha \in \mathbb{R}, u \in V, u \neq 0$. Show that every closed convex subset X of V is an intersection of closed half-affine spaces containing X.

CHAPTER 4

Section 1

1. Let $\{a_n\}_{n \in \mathbb{N}}$ be a sequence of real numbers such that for every sequence $\{u_n\}_{n \in \mathbb{N}}$ of real numbers

$$\sum_{n \in \mathbb{N}} u_n^2 < +\infty \quad \text{implies} \quad \sum_{n \in \mathbb{N}} a_n^2 u_n^2 < +\infty.$$

Show that $\sup_{n \in \mathbb{N}} |a_n| < +\infty$.

[Consider the operators A_n from $l^2(\mathbb{N})$ to itself defined by $A_n u = v$ with $u = \{u_m\}_{m \in \mathbb{N}}$, $v = \{v_m\}_{m \in \mathbb{N}}$ and $v_m = a_n u_m$ for $m \leq n$, $v_m = 0$ for $m > n$ and apply the Banach-Steinhaus theorem. Also give a proof by direct computation and compare.]

2. Let $\{a_n\}_{n \in \mathbb{N}}$ be a sequence of real numbers such that for every sequence $\{b_n\}_{n \in \mathbb{N}}$:

$$\sum_{n \in \mathbb{N}} b_n^2 < +\infty \quad \text{implies} \quad \sum_{n \in \mathbb{N}} |a_n b_n| < +\infty.$$

Show that $\sum_{n \in \mathbb{N}} a_n^2 < +\infty$. (*Hint*: See the preceding exercise.)

CHAPTER 5

1. Let A be the set of sequences $a = \{a_n\}_{n\in\mathbb{N}}$ of *strictly* positive real numbers. For $a \in A$, we denote by $l^2(\mathbb{N}, a)$ the space of real sequences $u = \{u_n\}_{n\in\mathbb{N}}$ such that $\sum_{n=0}^{\infty} a_n u_n^2 < +\infty$. Let D be the space of sequences $\{u_n\}_{n\in\mathbb{N}}$ having at most a finite number of nonzero terms.

  a. Show that $l^2(\mathbb{N}, a)$ is a Hilbert space for the scalar product

$$((u,v)) = \sum_{n=0}^{\infty} a_n u_n v_n$$

and that D is a dense subspace of $l^2(\mathbb{N}, a)$, $\forall a \in A$.

  b. Show that $D = \bigcap_{a \in A} l^2(\mathbb{N}, a)$.

  c. Show that a necessary and sufficient condition for the inclusion $l^2(\mathbb{N}, a) \subset l^2(\mathbb{N}, b)$ to hold is that

  (1) $$\sup_{n\in\mathbb{N}} \frac{b_n}{a_n} < +\infty.$$

  d. Let a and b be two sequences satisfying (1). Show that $l^2(\mathbb{N}, a)$ is a normal subspace of $l^2(\mathbb{N}, b)$. Determine the dual space of $l^2(\mathbb{N}, a)$ when $l^2(\mathbb{N}, b)$ is a pivot space.

  e. Write these results in the special case where $b = \{b_n\}$ with $b_n = 1$, $\forall n \in \mathbb{N}$. Compare this with the results of Section 6 for $L^2(\Omega, a)$.

2. Let $l^2(\mathbb{N})$ be the space of sequences $u = \{u_n\}_{n\in\mathbb{N}}$ such that $\sum_{n=0}^{\infty} u_n^2 < +\infty$ with the scalar product $((u,v)) = \sum_{n=0}^{\infty} u_n v_n$. Let $\lambda = \{\lambda_n\}_{n\in\mathbb{N}}$ be an arbitrary sequence of real numbers. We set

$$D(\Lambda) = \left\{ u \in l^2(\mathbb{N}) \quad \text{such that} \quad \sum_{n=0}^{\infty} \lambda_n^2 u_n^2 < +\infty \right\},$$

and we define the operator Λ from $D(\Lambda)$ to $l^2(\mathbb{N})$ by $\Lambda u = \{\lambda_n u_n\}_{n\in\mathbb{N}}$ for all $u = \{u_n\} \in D(\Lambda)$.

  a. Show that $(D(\Lambda), \Lambda)$ is a closed unbounded operator.

  b. Show that $D(\Lambda)$ is dense of $l^2(\mathbb{N})$. Calculate $(D(\Lambda^*), \Lambda^*)$.

CHAPTER 6

1. Let x and $y \in L^1(\mathbb{R}^n)$. Let $f : \mathbb{R}^{2n} \to \mathbb{R}$ be the function defined by

$$f(\omega, \zeta) = x(\omega - \zeta) y(\zeta) \quad \forall \omega, \zeta \in \mathbb{R}^n.$$

Using the Fubini theorem show that $f \in L^1(\mathbb{R}^{2n})$, since for almost all

$\omega \in \mathbb{R}^n$ the function

$$\zeta \to x(\omega - \zeta)y(\zeta), \quad \zeta \in \mathbb{R}^n$$

belongs to $L^1(\mathbb{R}^n)$ and

$$x * y(\omega) = \int x(\omega - \zeta)y(\zeta)\, d\zeta$$

(with the definition of $x * y$ as given in the text).

2. Let $x, y, z \in \mathscr{C}_0(\mathbb{R}^n)$. Show that

$$x * y = y * x, \quad (x * y) * z = x * (y * z), \quad x * (y + z) = x * y + x * z.$$

3. Let $\lambda \in \mathscr{D}(\Omega)$ such that $\lambda \geq 0$ and $\int_{\mathbb{R}^n} \lambda(\omega)\, d\omega = 1$ and let λ_h for $h > 0$ be defined by

$$\lambda_h(\omega) = \frac{1}{h^n} \lambda\left(\frac{\omega}{h}\right) \quad \forall \omega \in \mathbb{R}^n,$$

Let $x : \mathbb{R}^n \to \mathbb{R}^n$ be a continuous function. Show that $\lambda_h * x$ is an infinitely differentiable function from $\mathbb{R}^n$ to $\mathbb{R}^n$ and that $\lambda_h * x$ converges to x uniformly on every compact subset as $h \to 0$. Show that if x is monotone, $\lambda_h * x$ is monotone. [We say that a function y is monotone if $((y(\omega_1) - y(\omega_2)), \omega_1 - \omega_2)) \geq 0 \quad \forall \omega_1, \omega_2 \in \mathbb{R}^n$.]

CHAPTER 7

1. (Poincaré inequality). Show that there exists a constant $C > 0$ such that

   (1) $$\int_a^b u^2(\omega)\, d\omega \leq C \int_a^b u'^2(\omega)\, d\omega$$

   for all $u \in H_0^1(]a, b[)$. [First establish (1) for $u \in \mathscr{D}(]a, b[)$ using the equality $u^2(\omega) = \int_a^\omega 2u'(\zeta)u(\zeta)\, d\zeta$ and the Cauchy-Schwarz inequality.)

2. Let $\Omega =]a, b[$ be a bounded interval of $\mathbb{R}$. Let $\{\omega_k\}$ be a sequence of points of Ω such that $\omega_k \to b$ as $k \to \infty$. Show that for all $\varphi \in \mathscr{D}(\Omega)$ the sequence $\{D^k\varphi(\omega_k)\}$ has at most a finite number of nonzero terms. Deduce from this that it is possible to define a linear form $T \in \mathscr{D}^*(\Omega)$ by the formula

$$T(\varphi) = \sum_{k \in \mathbb{N}} D^k\varphi(\omega_k) \quad \forall \varphi \in \mathscr{D}(\Omega).$$

Show that $T \notin H^{-m}(\Omega) \quad \forall m \in \mathbb{N}$. Deduce from this that

$$\mathscr{D}^*(\Omega) \neq \bigcup_{m \in \mathbb{N}} H^{-m}(\Omega).$$

(One can show, moreover, that $T \in \mathscr{D}^*(\Omega)$, the topological dual of $\mathscr{D}(\Omega)$ for a suitable topology, which proves that $\mathscr{D}^*(\Omega) \neq \bigcup_{m \in \mathbb{N}} H^{-m}(\Omega)$).

CHAPTER 10

Section 1

1. Show that the conjugate of the function $\varphi : t \to |t|^\alpha/\alpha$ from $\mathbb{R}$ to $\mathbb{R}$, $\alpha > +1$, is the function $t \to |t|^{\alpha^*}/\alpha^*$ from $\mathbb{R}$ to $\mathbb{R}$ where α^* is defined by $(1/\alpha) + (1/\alpha^*) = +1$.
2. Let V be a Hilbert space, φ an even continuous convex function from $\mathbb{R}$ to $\mathbb{R}$, and F the mapping $x \to \varphi(\|x\|)$ from V to $\mathbb{R}$.
   a. Show that F is convex and continuous from V to $\mathbb{R}$.
   b. Show that the conjugate function of F is the mapping $p \to \varphi^*(\|p\|_*)$ from V^* to $]-\infty, +\infty]$.
3. Let V be a Hilbert space. Deduce from exercises 1 and 2 that the conjugate of the function $x \to \|x\|^\alpha/\alpha$ from V to $\mathbb{R}$, $\alpha > +1$, is the function $p \to \|p\|_*^{\alpha^*}/\alpha^*$ from V^* to $\mathbb{R}$ where α^* is defined by $(1/\alpha) + (1/\alpha^*) = +1$.
4. Let $\{f_i\}_{i \in I}$ be a family of functions from a Hilbert space V to $]-\infty, +\infty]$.
   a. Show that $(\inf_{i \in I} f_i)^* = \sup_{i \in I} f_i^*$.
   b. Show that $(\operatorname{Sup}_{i \in I} f_i)^* \leq \operatorname{Inf}_{i \in I} f_i^*$.
5. Let X be a subset of a Hilbert space V, Ψ_X the indicator function of X and $\overline{\operatorname{co}}(X)$ the closed convex hull of X in V. Show that $\Psi_X^{**} = \Psi_{\overline{\operatorname{co}}(X)}$.

Section 2

6. Let f be a mapping from a Hilbert space V to $\mathbb{R}$ that is Fréchet-differentiable at $x_0 \in V$. Show that f is Gâteaux-differentiable at x_0 and that the two derivatives coincide. [Recall that f if Fréchet-differentiable at x_0 if there exists $Df(x_0) \in V^*$, called the *Fréchet derivative*, such that
$$\lim_{\|h\| \to 0} \left[\frac{f(x_0 + h) - f(x_0) - \langle Df(x_0), h \rangle}{\|h\|} \right] = 0.$$
7. Let V be a Hilbert space, x_0 a point of V and f the mapping $x \to \frac{1}{2}\|x - x_0\|^2$ from V to $\mathbb{R}$.
   a. Show that f is continuously differentiable and that the derivative, $f'(x) \in V^*$, of f at x is the mapping $y \to ((x - x_0, y))$ from V to $\mathbb{R}$.
   b. Let X be a nonempty closed convex subset of V. Deduce from

part **a** that the projection $t(x_0)$ of x_0 onto X (i.e., $f(t(x_0)) = \min_{x \in X} f(x)$) satisfies $((t(x_0) - x_0, t(x_0) - y)) \leq 0$ for all $y \in X$. (Compare this with the proof of this proposition in Chapter 1.)

8. Let M be a nonempty closed convex subset of a Hilbert space V and $t(.)$ the best approximation projector from V onto M. Show that the mapping
$$f : x \to \tfrac{1}{2} \inf_{y \in M} \|x - y\|^2$$
from V to $\mathbb{R}$ is continuously differentiable and that the derivative, $f'(x) \in V^*$, of f at x is the mapping $y \to ((x - t(x)), y))$ from V to $\mathbb{R}$.

9. Let X be an **open** convex subset of a Hilbert space V and f a mapping from X to $\mathbb{R}$ that is Gateaux-differentiable at every $x \in X$. Show that f is convex if and only if
$$f(y) \geq f(x) + \langle Df(x), y - x \rangle \quad \text{for all} \quad y, x \in X.$$

10. Let X be an **open** convex subset of a Hilbert space V and f a mapping from X to $\mathbb{R}$ that is Gateaux-differentiable at every $x \in X$. Show that f is convex if and only if
$$\langle Df(x) - Df(y), x - y \rangle \geq 0 \quad \text{for all} \quad x, y \in X.$$

11. Let X be a nonempty closed convex subset of a Hilbert space V. For $x \in X$ we define the set
$$T(X, x) = \overline{\{\lambda(y - x) \text{ such that } y \in X, \lambda \geq 0\}}$$
called the tangent cone to X at x.
  a. Show that the mapping $d_X : x \to d_X(x) = \inf_{y \in X} \|y - x\|$ from V to $\mathbb{R}$ is convex and continuous.
  b. Show that
$$T(X, x) = \{v \in V \quad \text{such that} \quad Dd_X(x)(v) \leq 0\} \quad \text{for all} \quad x \in X.$$

12. Let V be a Hilbert space assumed to be identified with its dual and F the mapping $x \to \|x\|$ from V to $\mathbb{R}$.
  a. Show that F is convex and continuous.
  b. Show that $\partial F(x) = x/\|x\|$ for all $x \in V, x \neq 0$.
  c. Show that $\partial F(0) = \{x \in V \text{ such that } \|x\| \leq +1\}$.

13. Let f and g be two mappings from a Hilbert space V to $]-\infty, +\infty]$ with nonempty domains.
  a. Show that $\partial(\lambda f)(x) = \lambda \partial f(x)$ for all $\lambda > 0, x \in V$.
  b. Show that $\partial f(x) + \partial g(x) \subset \partial(f + g)(x)$ for all $x \in V$.
  c. Let $f : \mathbb{R} \to \mathbb{R}$ be defined by $f(x) = -|x|$. Show that $\partial f(0) = \emptyset$. Let $g = -f$. Show that $\partial f(0) + \partial g(0) \neq \partial(f + g)(0)$.

14. Let f be a convex function from $\mathbb{R}$ to $\mathbb{R}$.
    a. Show that f has a derivative from the left $f'_-(x)$ and a derivative from the right $f'_+(x)$ at every $x \in \mathbb{R}$.
    b. Show that $\partial f(x) = [f'_-(x), f'_+(x)]$ for all $x \in \mathbb{R}$.
15. Let f be a convex function from $\mathbb{R}$ to $]-\infty, +\infty]$ such that $\text{Dom} f = [0, +\infty[$ and having a derivative from the right, $f'_+(0)$, at zero.
    a. Show that $\partial f(0) =]-\infty, f'_+(0)]$.
    b. Find $\partial f(x)$ for $x > 0$. (*Hint*: See exercise 14.)
16. Let f be a function from a Hilbert space V to $]-\infty, +\infty]$ having a nonempty domain.
    a. Show that $\bar{x} \in \text{Dom} f$ is a minimum for the function f on its domain if and only if $0 \in \partial f(\bar{x})$.
    b. Let f be a convex function from $\mathbb{R}$ to $\mathbb{R}$. Show that $\bar{x} \in \mathbb{R}$ is a minimum for f on $\mathbb{R}$ if and only if $f'_-(\bar{x}) \leq 0 \leq f'_+(\bar{x})$. (Use exercise 15.)
17. Let f be a convex lower semicontinuous function from a Hilbert space V to $]-\infty, +\infty]$ with nonempty domain.
    a. Show that for every finite sequence $x_1, \ldots, x_n, x_i \in \text{Dom} f$, we have
    $$\langle p_1, x_1 - x_2 \rangle + \langle p_2, x_2 - x_3 \rangle + \ldots + \langle p_n, x_n - x_1 \rangle \geq 0$$
    for all $p_i \in \partial f(x_i), i = 1, \ldots, n$.
    b. Deduce from this that the "subdifferential is monotone," that is, that $\langle p_1 - p_2, x_1 - x_2 \rangle \geq 0$ for all $x_1, x_2 \in \text{Dom} f, p_1 \in \partial f(x_1), p_2 \in \partial f(x_2)$. (Compare this result with exercise 18.)

Section 10

18. [A converse of exercise 17. For the general case (g multivalued) see R. T. Rockefellar, *Convex Analysis* Princeton University Press, 1970].
    Let H be a Hilbert space, and let $g : H \to H$ be a continuous function such that for every finite family $x_1, \ldots, x_n$ of points of H,
    $$\langle g(x_1), x_2 - x_1 \rangle + \langle g(x_2), x_3 - x_2 \rangle + \ldots + \langle g(x_n), x_1 - x_n \rangle \leq 0.$$
    We define a function $f : H \to \mathbb{R}$ by
    $$f(x) = \int_0^1 \langle g(tx), x \rangle \, dt \quad \forall x \in H.$$
    a. Let $x, y \in H$ and $n \in \mathbb{N}$. We set
    $$\lambda_n = \sum_{k=0}^{n-1} \left\langle g\left(\frac{k}{n} x\right), \frac{x}{n} \right\rangle + \sum_{m=0}^{n-1} \left\langle g\left(x + \frac{m}{n}(y-x)\right), \frac{y-x}{n} \right\rangle$$
    $$+ \sum_{p=0}^{n-1} \left\langle g\left(\left(1 - \frac{p}{n}\right)y\right), -\frac{y}{n} \right\rangle.$$

Show that $\lambda_n \geq 0$. Deduce from this that

$$f(y) - f(x) \leq \int_0^1 \langle g(x + t(y-x)), y - x \rangle \, dt.$$

Interchanging the roles of x and y, deduce that

$$f(y) - f(x) = \int_0^1 \langle g(x + t(y-x)), y - x \rangle \, dt,$$

for all $x, y \in H$.

b. Show that f is Fréchet-differentiable and that $Df = g$.
c. Show that f is convex.

CHAPTER 11

1. Let $\{k_n\}$ be a bounded sequence of real numbers, and let K be the operator from $l^2(\mathbb{N})$ to itself defined by

$$K : \{u_n\} \to \{k_n u_n\}.$$

Show that K is compact if and only if $\lim_{n \to \infty} k_n = 0$. What are the eigenvalues of K?

2. For every sequence $\{a_n\}$ of *strictly* positive real numbers we denote by $l^2(\mathbb{N}, a)$ the Hilbert space of sequences $\{u_n\}$ such that $\sum_{n \in \mathbb{N}} a_n u_n^2 < +\infty$ with the scalar product $((u, v)) = \sum_{n \in \mathbb{N}} a_n u_n v_n$.

   a. Show that the operator J_a defined by $J_a : \{u_n\} \to \{\sqrt{a_n} u_n\}$ is an isomorphism from $l^2(\mathbb{N}, a)$ onto $l^2(\mathbb{N})$.
   b. Let $\{a_n\}$ and $\{b_n\}$ be two sequences of strictly positive real numbers such that $\lim_{n \to \infty}(a_n/b_n) = 0$. Show that $l^2(\mathbb{N}, b) \subset l^2(\mathbb{N}, a)$ and that this embedding is compact. (For instance write $P = J_a^{-1} K J_b$ where P is the embedding and K is a compact operator from $l^2(\mathbb{N})$ to itself).

3. Under the hypothesis and notation of exercise 2, determine the intermediate spaces (see Definition 11.5.1) in the case where $U = l^2(\mathbb{N}, b)$ and $V = l^2(\mathbb{N}, a)$.

4. (Refer to the notation and results of Chapter 8, Section 3). Let $\mathscr{C}_\pi$ be the subspace of $\mathscr{C}(\mathbb{R}, \mathbb{C})$ consisting of periodic functions of period 1. Let H_π^m be the completion of $\mathscr{C}_\pi^\infty$ for the scalar product

$$((x, y)) = \sum_{0 \leq j \leq m} \int_{-1/2}^{+1/2} D^j x(\omega) \overline{D^j y(\omega)} \, d\omega.$$

Let $\{e_k\}$ be the orthonormal base of $L_\mathbb{C}^2(-\tfrac{1}{2}, +\tfrac{1}{2})$ introduced in Chapter 8, Section 3. Let $a^m = \{a_n^m\}_n$ be the sequence defined by $a_n^m = (1 + n^2)^m$.

Show that the linear operator $x \to \{((x_k, e_k))\}_{k \in \mathbb{N}}$ is an isomorphism from H_π^m onto the space

$$l_\mathbb{C}^2(\mathbb{N}, a^m) = \left\{ \{u_n\}_{n \in \mathbb{N}} \text{ such that } u_n \in \mathbb{C} \text{ and } \sum_{n \in \mathbb{N}} a_n^m |u_n|^2 < +\infty \right\}.$$

Deduce from this that the embedding $H_\pi^k \subset H_\pi^m (k > m)$ is compact. (See exercise 2.) Determine the spaces U^s of Definition 11.5.1 in the case where $U = H_\pi^k$, $V = H_\pi^m$.

5. Let H be a pivot space and $A \in \mathscr{L}(H, H)$ be a self-adjoint compact operator. Let $\{e_n\}$ be an orthonormal base and $\{\mu_n\}$ a sequence such that

$$Ax = \sum_{n=1}^{\infty} \mu_n((e_n, x))e_n \qquad \forall x \in H.$$

Let M be the set $M = \{\mu_0, \mu_1, \ldots\} \cup \{0\}$. For every function f from M to $\mathbb{R}$ we define the operator $f(A)$ by

(1) $$f(A)x = \sum_{n=0}^{\infty} f(\mu_n)((e_n, x))e_n \qquad \forall x \in H$$

a. Show that $f(A)$ is self-adjoint.
b. Let $\mathscr{F}$ be the algebra of functions from M to $\mathbb{R}$. Show that the mapping σ defined by $\sigma(f) = f(A)$ is a homomorphism from the algebra $\mathscr{F}$ to $\mathscr{L}(H, H)$. Show that if f is a polynomial function

$$f(t) = a_0 + a_1 t + \ldots + a_k t^k \qquad \forall t \in M,$$

then

$$f(A) = a_0 I + a_1 A + \ldots + a_k A^k.$$

c. Show that if f is continuous at zero, the operator $f(A)$ is compact if and only if $f(0) = 0$.

6. With the same notation as the preceding exercise, we suppose in addition that $A \geq 0$, that is, $((Ax, x)) \geq 0 \quad \forall x \in H$.

a. Show that $\mu_n \geq 0 \quad \forall n \in \mathbb{N}$.
b. Let $f : \mathbb{R}_+ \to \mathbb{R}$ be the function $f(t) = \sqrt{t}$. We define $A^{1/2} = f(A)$. Show that $(A^{1/2})^2 = A$. Show that $A^{1/2}$ is compact and self-transposed.
c. Show that $\|A\| = \sup_{n \in \mathbb{N}} \mu_n$.
d. Let u_n be the sequence of real numbers defined by

$$u_0 = 0, \qquad u_{n+1} = u_n + \frac{\alpha - u_n^2}{k}$$

where α and k are constants such that $4\alpha \leq k^2$. Show that the sequence $\{u_n\}$ is increasing and converges to $\sqrt{\alpha}$. Show that for fixed k the convergence of u_n to $\sqrt{\alpha}$ is uniform with respect to α in the interval $[0, k^2/4]$.

e. We define a sequence of operators A_n by

$$A_0 = 0, \qquad A_{n+1} = A_n + \frac{1}{2\sqrt{\|A\|}}(A - A_n^2).$$

Show that the sequence A_n converges in $\mathscr{L}(H, H)$ to the operator $A^{1/2}$. Show that the A_n's are compact and that the sequence A_n is increasing (i.e. $A_{n+1} - A_n \geq 0$, in the sense that: $(A_{n+1}x, x) \geq (A_n x, x) \quad \forall x \in H$.

7. Let H be a Hilbert space and $\{e_n\}$ an orthonormal base of H. Let $\{\mu_n\}$ be a bounded sequence and $A \in \mathscr{L}(H, H)$ the operator defined by

$$Ax = \sum_{n=0}^{\infty} \mu_n((e_n, x))e_n \qquad \forall x \in H,$$

a. Show that

$$\|A\| = \sup_{\|x\| \leq 1} ((Ax, x)) = \sup_{n \in \mathbb{N}} |\mu_n|.$$

b. Show that $\|A^k\| = \|A\|^k$ and that $Ax \neq 0$ implies $A^k x \neq 0 \quad \forall k \in \mathbb{N}$.

c. Suppose that $\mu_n \to 0$ as $n \to \infty$. Show that for all $x \in H$ there exists an integer $m \in \mathbb{N}$ such that $\{\|A^k x\|^{1/k}\}$ converges to μ_m.

8. Let K be a bounded set in a Hilbert space H such that for every $\varepsilon > 0$ there exists a *finite* dimensional subspace H_ε for which $d(x, H_\varepsilon) < \varepsilon$ for all $x \in K$.

a. Let $P_{1/k}$ be the orthogonal projector onto $H_{1/k}$. Show that $\|P_{1/k}x - x\| \leq 1/k \quad \forall x \in K$.

b. Let P be an orthogonal projector onto a finite dimensional space. Show that for every sequence $\{z_n\}_{n \in \mathbb{N}}$ of points of K there exists a subsequence $\{y_p = z_{n_p}\}_{p \in \mathbb{N}}$ such that the sequence of projectors $\{Py_p\}_{p \in \mathbb{N}}$ is convergent.

c. Let $\{x_n\}_{n \in \mathbb{N}}$ be a sequence of points of K. We define a family of subsequences $\{y_n^k\}_{n \in \mathbb{N}}$ by recursion on k: (i) $y_n^0 = x_n \quad \forall n \in \mathbb{N}$, and (ii) $\{y_n^k\}_{n \in \mathbb{N}}$ is a subsequence of the sequence $\{y_n^{k-1}\}_{n \in \mathbb{N}}$, such that the sequence of projectors $\{Py_n^k\}_{n \in \mathbb{N}}$ is convergent. Show that the diagonal sequence $\{y_n^n\}_{n \in \mathbb{N}}$ is convergent.

d. Show that K is relatively compact.

9. Let K be a compact set in a Hilbert space H. We are going to show that for every $\varepsilon > 0$ there exists a continuous function $f : K \to H$ such that (i) $f(K)$ is contained in a finite dimensional subspace, and (ii) $\|f(x) - x\| \leq \varepsilon$ for all $x \in K$.

    a. Let $\varepsilon > 0$. Show that there exist $x_1, \ldots, x_n$ in K such that K is contained in the union of the balls $B(x_i, \varepsilon)$ with center x_i and radius ε.

    b. Let $\alpha_i : K \to \mathbb{R}$ be the continuous function defined by $\alpha_i(x) = (\varepsilon - \|x - x_i\|)_+$, $i = 1, \ldots, n$ (where $t_+ = \frac{1}{2}(t + |t|)$). Show that $\alpha = \alpha_1 + \ldots + \alpha_n$ is a continuous function with *strictly* positive values, that $\beta_i = \alpha_i / \alpha$ is a continuous function that vanishes outside of the ball $B(x_i, \varepsilon)$ and that $\sum_{i=1}^{n} \beta_i(x) = 1$.

    c. We set $f(x) = \sum_{i=1}^{n} \beta_i(x) x_i$, $\forall x \in K$. Show that f is a continuous function that satisfies (1) and (2).

10. Show with the help of exercises 8 and 9 that a set K in a Hilbert space H is compact if and only if both (i) K is closed and bounded, and (ii) $\forall \varepsilon > 0$ there exists a finite dimensional subspace F_ε such that $d(x, F_\varepsilon) \leq \varepsilon$ for all $x \in K$.

11. Let K be a compact set in a Hilbert space H and let F_ε be subspaces satisfying condition (ii) of exercise 10. Let $\rho = \sup_{x \in K} \|x\|$.

    a. Show that the sets

$$A_\varepsilon = \{x \in H \text{ such that } \|x\| \leq \rho \text{ and } d(x, F_\varepsilon) \leq \varepsilon\}$$

are closed and convex for all $\varepsilon > 0$ and contain K.

    b. Show that $A = \bigcap_{0 < \varepsilon \leq 1} A_\varepsilon$ is a compact convex set. (Use exercise 10.)

    c. Show that the closed convex hull of K is compact.

CHAPTER 13

1. Let Ω be an open set of $\mathbb{R}^n$. Let $f : \Omega \times \mathbb{R} \to \mathbb{R}$ be a continuous function such that

$$|f(\omega, t)| \leq a(t) + b|t| \qquad \forall t \in \mathbb{R}, \qquad \forall \omega \in \Omega$$

where $a \in L^2(\Omega)$ and $b > 0$. For every function $x : \Omega \to \mathbb{R}$ let $F_x : \Omega \to \mathbb{R}$ be the function defined by

$$F_x(\omega) = f(\omega, x(\omega)) \qquad \forall \omega \in \Omega.$$

    a. Show that $x \in L^2(\Omega)$ implies that $F_x \in L^2(\Omega)$.

    d. Show that the mapping $x \to F_x$ from $L^2(\Omega)$ to itself is continuous. (Show that $x_n \to x$ implies $F_{x_n} \to F_x$.)

    c. We denote by $\mathscr{C}^k$ the space of k-times continuously differentiable functions.

Suppose that the function f is $\mathscr{C}^1$ and that

$$|f'_t(\omega, t)| \leq c \quad \forall \omega \in \Omega, \quad \forall t \in \mathbb{R}.$$

Let $x \in L^2(\Omega)$, and let L_x be the linear mapping from $L^2(\Omega)$ to itself defined by

$$z = L_x y \quad \text{if and only if} \quad z(\omega) = f'_t(\omega, x(\omega)) y(\omega) \quad \forall \omega \in \Omega.$$

Show that L_x is continuous. Show that the mapping $x \to F_x$ is Gâteaux-differentiable and that its derivative at the point x is L_x.

2. Let $\{f_n\}_{n \in \mathbb{N}}$ be a sequence of continuous functions from $\mathbb{R}$ to $\mathbb{R}$ such that

$$|f_n(t)| \leq a_n + b|t| \quad \forall n \in \mathbb{N}, \quad \forall t \in \mathbb{R}$$

with $a = \{a_n\} \in l^2(\mathbb{N})$ and $b > 0$. For every sequence $x = \{x_n\}$ we denote by $y = \{y_n\} = F(x)$ the sequence defined by

$$y_n = f_n(x_n).$$

a. Show that for every $x \in l^2(\mathbb{N})$ we have $F(x) \in l^2(\mathbb{N})$.
b. Show that the mapping F is continuous from $l^2(\mathbb{N})$ to itself.
c. Suppose that the functions $f_n, n \in \mathbb{N}$ are $\mathscr{C}^2$ and that there exists a constant c such that

$$|f'_n(t)| \leq c \quad \text{and} \quad |f''_n(t)| \leq c \quad \forall n \in \mathbb{N}, \quad \forall t \in \mathbb{R}.$$

Show that $F : l^2(\mathbb{N}) \to l^2(\mathbb{N})$ is Fréchet-differentiable.

CHAPTER 14

1. (On semigroups of nonlinear contraction; for the general theory see H. Brezis, *Opérateurs Maximaux Monotones*, Mathematics Studies No. 5, North-Holland).

   Let $f : \mathbb{R}^n \to \mathbb{R}^n$ be a locally Lipschitz function. Suppose that there exists $R > 0$ such that

   (1) $\qquad \|x\| \leq R \quad \text{implies} \quad ((f(x), x)) \geq 0.$

   a. Let $x_0 \in \mathbb{R}^n$. Show that the differential equation

   (2) $\qquad x(0) = x_0, \quad \dfrac{dx}{dt}(t) = -f(x(t))$

   has a unique solution $x : [0, +\infty[\to \mathbb{R}^n$ that is $\mathscr{C}^1$ and such that $\|x(t)\| \leq \sup(R, \|x_0\|)$. (Calculate $d/dt \|x\|^2$.)

b. For all $t \geq 0$, we define a mapping $S(t)$ from $\mathbb{R}^n$ to itself by setting $S(t)x_0 = x(t)$ where x is the solution of equation (2). Show that $S(0) = I$ and
$$S(t_1 + t_2) = S(t_1)S(t_2) \qquad \forall t_1, t_2 \geq 0$$
Show that the mapping $t \to S(t)x_0$ is $\mathscr{C}^1$ and that

(3) $$\lim_{t > 0, t \to 0} \frac{S(t)x_0 - x_0}{t} = -f(x_0).$$

c. Show that $S(t)$ is a contraction for every $t \geq 0$, that is,

(4) $$\|S(t)x_0 - S(t)y_0\| \leq \|x_0 - y_0\| \qquad \forall t \geq 0, \quad \forall x_0, y_0 \in \mathbb{R}^n$$

if and only if f is monotone, that is,

(5) $$((f(x_0) - f(y_0), x_0 - y_0)) \geq 0 \qquad \forall x_0, y_0 \in \mathbb{R}^n.$$

[To show (4) implies (5), use (3). To show (5) implies (4) calculate
$$\frac{d(x-y)^2}{dt}$$
where y is the solution of $y(0) = y_0$, $dy/dt = -f(y)$.]

2. Let C be a nonempty closed convex set of $\mathbb{R}^n$. Show that the function $g : \mathbb{R}^n \to \mathbb{R}$ defined by $g(x) = d(x, C)^2$ is convex and differentiable and that the derivative $g' : \mathbb{R}^n \to \mathbb{R}^n$ is monotone and Lipschitz. Calculate the semigroup obtained by taking $f = g'$ in exercise 1.

3. Let $f : \mathbb{R}^n \to \mathbb{R}^n$ be a $\mathscr{C}^1$ function.

a. Show that f is monotone (see exercise 17, Chapter 1) if and only if
$$((f'(x)z, z)) \geq 0 \qquad \forall x, z \in \mathbb{R}^n.$$
(Calculate $d\alpha/dt$ with $\alpha(t) = ((f(x + t(y - x)) - f(x), y - x))$.)

b. Suppose now that there exists $c > 0$ such that
$$((f'(x)z, z)) \geq c \|z\|^2 \qquad \forall x, z \in \mathbb{R}^n.$$
Using the notation of exercise 1, show that
$$\left\| \frac{dx}{dt}(t) \right\|^2 \leq e^{-2ct} \left\| \frac{dx}{dt}(0) \right\|^2 = -e^{-2ct} \|f(x_0)\|^2.$$

Deduce from this that $x(t)$ converges, as $t \to +\infty$, to a point x_∞ such that $f(x_\infty) = 0$.

4. Let $g : \mathbb{R} \to \mathbb{R}^n$ be a $\mathscr{C}^1$ function such that
$$((g'(x)z, z)) \geq c \|z\|^2 \qquad \forall x, z \in \mathbb{R}^n.$$

a. Show that

$$((g(x), x)) \geq -\|g(0)\| \cdot \|x\| + c\|x\|^2 \qquad \forall x \in \mathbb{R}^n.$$

(Calculate $d\alpha/dt$ with $\alpha(t) = ((g(tx) - g(0), x))$.)
Conclude from this that g is *coercive*, that is,

$$\lim_{\|x\| \to \infty} \frac{((g(x), x))}{\|x\|} = +\infty.$$

b. Show that

$$((g(x) - g(y), x - y)) \geq c\|x - y\|^2 \qquad \forall x, y \in \mathbb{R}^n.$$

Show that g is injective.

c. Show that g is a bijection from $\mathbb{R}^n$ onto $\mathbb{R}^n$. (In order to show that $y \in g(\mathbb{R}^n)$ apply the result of exercise 3(b) to the function $f(x) = g(x) - y$.)

5. Let $g: \mathbb{R}^n \to \mathbb{R}^n$ be a continuous monotone function (see exercise 4) that is coercive (see exercise 4a). For every $h > 0$, let $\lambda_n \in \mathcal{D}(\Omega)$ be defined as in exercise 3, Chapter 6, and let $g_h: \mathbb{R}^n \to \mathbb{R}$ be defined by

$$g_h(x) = \lambda_h * g + hx, \qquad \forall x \in \mathbb{R}^n.$$

a. Show that g_h converges to g as h approaches zero ($h > 0$), uniformly on every compact set.

b. Show that g_h is $\mathscr{C}^1$ and that

$$((g_h(x) - g_h(y), x - y)) \geq h\|x - y\|^2.$$

Conclude from this that

$$((g_h'(x)z, z)) \geq h\|z\|^2 \qquad \forall x, z \in \mathbb{R}^n.$$

Show that g_h is a bijection from $\mathbb{R}^n$ onto $\mathbb{R}^n$. (Use exercise 4.)

c. Show that for every $b \geq 0$ there exists $a \geq 0$ such that

$$((g_h(x), x)) \geq -a + b\|x\|, \qquad \forall x \in \mathbb{R}^n, \quad \forall h \in]0, 1].$$

d. Let $\bar{x} \in \mathbb{R}^n$, and let $\{x_m\}$ be a sequence such that $g_{1/m}(x_m) = \bar{x}$ (see part b).
Show with the help of (c) that $\{x_m\}$ is bounded. Conclude from this that there exists $y \in \mathbb{R}^n$ such that $g(y) = \bar{x}$ (use a) and that g is surjective.

e. Show that if $h: \mathbb{R}^n \to \mathbb{R}^n$ is monotone and continuous, the mappings $I + \alpha h$ are bijective for every $\alpha > 0$ (where I is the identity mapping from $\mathbb{R}^n$ to $\mathbb{R}^n$).

CHAPTER 15

1. Let H be a Hilbert space, S an upper hemicontinuous correspondence from H to itself, and $f:H \to H$ a continuous function. Show that Sf is an upper hemicontinuous correspondence.

2. Let X be a nonempty closed convex set in a Hilbert space H and S an upper hemicontinuous correspondence from X to H with nonempty compact convex values such that
$$S(X) = \{y \text{ such that } \exists x \in X, y \in S(x)\}$$
is contained in a compact set K of X. Show that S has a fixed point. (Consider the restriction of S to the closed convex hull of K that is a compact convex set contained in X according to exercise 11, Chapter 11.)

3. (Theorem of H. Schaefer.) Let H be a Hilbert space and S an upper hemicontinuous correspondence from H to itself with nonempty compact convex values and such that (i) for every $\rho > 0$, the image of the ball $B_\rho : S(B_\rho) = \{y \text{ such that } \exists x \text{ such that } \|x\| \leq \rho \text{ and } y \in S(x)\}$ is contained in a compact set, and (ii) the set $E = \{x \in H \text{ such that } \exists \lambda \in [0,1] \text{ such that } x \in \lambda S(x)\}$ is bounded.
    a. Let $\rho > 0$. Show that there exists ρ_0 such that B_{ρ_0} contains $S(B_\rho)$.
    b. Let ρ be such that B_ρ contains E. Let $\rho_1 > \rho$ and let ρ_2 be such that $B\rho_2$ contains $S(B\rho_1)$. Finally let f be the projection onto $B\rho_1$. Show that the correspondence $T:B\rho_2 \to B\rho_2$ defined by $T(x) = S(f(x))$ has a fixed point. (See exercise 2.)
    c. Show that every fixed point of T is a fixed point of S. (Remark that $f(x) = \lambda x$ with $\lambda \in [0,1]$, and use condition (ii)). Conclude that S has a fixed point.

4. Let f be a continuous function from $\mathbb{R}$ to $\mathbb{R}$, and let $\Omega = [a,b]$ be a bounded interval of $\mathbb{R}$.
    a. Show that for every $u \in H^1(\Omega)$ the mapping $f \circ u$ is continuous. (See Proposition 7.5.1.)
    b. Show that the mapping $F:u \to f \circ u$ is continuous from $H^1(\Omega)$ to the space $\mathscr{C}(\bar\Omega)$ of continuous functions on $\bar\Omega$ with the norm $\|v\| = \sup_{\omega \in \Omega} |v(\omega)|$.
    c. Show that for every function $v \in L^2(\Omega)$ there exists one and only one function u such that
    $$(1) \qquad u \in H^2_0(\Omega), \quad \frac{d^2 u(\omega)}{d\omega^2} = v(\omega) \quad \text{in } \Omega$$
    and that the operator $L:v \to u$ is continuous and linear from $L^2(\Omega)$ to $H^2_0(\Omega)$.

d. Let i and j be the continuous injections
$$i: C(\bar{\Omega}) \to L^2(\Omega), \qquad j: H_0^2(\Omega) \to H^1(\Omega),$$
and let $G = j.L.i.F : H^1(\Omega) \to H^1(\Omega)$. Show that G is a continuous mapping and that for every ball
$$B(0, \rho) = \{u \in H^1(\Omega) \mid \|u\|_{H^1} \leq \rho\}|$$
the image $G(B(0, \rho))$ is relatively compact in $H^1(\Omega)$.

e. Show that $u \in H^1(\Omega)$ satisfies $u = G(u)$ if and only if

(2) $\quad \begin{cases} u: \bar{\Omega} \to \mathbb{R} \text{ is } \mathscr{C}^2 \\ u(a) = 0, \quad u(b) = 0 \\ u''(\omega) = f(u(\omega)) \quad \forall \omega \in \Omega. \end{cases}$

f. Suppose that f satisfies
$$-f(t)t \leq c_1 + c_2|t| \qquad \forall t \in \mathbb{R}$$
where c_1 and c_2 are constants. Show that there exist constants c_3 and c_4 such that for every $\lambda \in [0, 1]$ and for every u satisfying the equation

(3) $\quad \begin{cases} u: \bar{\Omega} \to \mathbb{R} \text{ is } \mathscr{C}^2 \\ u(a) = 0, \quad u(b) = 0 \\ u''(\omega) = \lambda f(u(\omega)) \end{cases}$

we have

(4) $\quad \int_a^b u'^2(\omega) \, d\omega \leq c_3 + c_4 \left(\int_a^b u^2(\omega) \, d\omega \right)^{1/2}.$

Conclude from (4) and from the Poincaré inequality (exercise 1, Chapter 7) that there exists a constant c_5 such that $\|u\|_{H^1(\Omega)} \leq c_5$ for all u satisfying (3).

g. Show that $u \in H^1(\Omega)$ satisfies $u = \lambda G(u)$ if and only if (3) is satisfied. Deduce from this that the set
$$\{u \in H^1(\Omega) \text{ such that } \exists \lambda \in [0, 1] \text{ such that } u = \lambda G(u)\}$$
is contained in the ball $B(0, c_5)$ of $H^1(\Omega)$.

h. Show with the help of Schaefer's theorem (exercise 3) that there exists $u \in H^1(\Omega)$ such that $u = G(u)$. Conclude that equation (2) has at least one solution.

INDEX

Adjoint, formal, 283
Adjoint equation, in optimal control, 327
Adjoint of an unbounded operator, 107
Altman theorem, 361
Appell polynomials, 139
Approximation, by convolution, 133
   by Fourier series, 172
   by orthogonal polynomials, 167
   processes, 247
   of solutions to Neumann problems, 312
   by spline functions, 176-181
   by step functions, 175
Arrow-Debreu-Nash theorem, 376

Baire theorem, 74
Banach-Steinhauss theorem, 76
Banach theorem, 80-81
Bernouilli polynomials, 140
Bessel inequality, 24
Best approximation theorem, 15
   processes, 247
Bipolar lemma, 61
Boundary condition, 359
Boundary value problems, abstract, 299
   elliptic, 308, 324
   parabolic, 346

Cauchy-Schwarz inequality, 4
Closed family of operators, 99-103, 146
Closed graph theorem, 83
Closed range theorem, 85
Closed unbounded operator, 107
Coercive correspondences, 361, 371
Collectively injective operators, 99
Compactness property, 157, 207
Compact operators, 233, 239

Completely observable and/or accessible systems, 351
Completion, 53, 109, 110
Complex vector spaces, 5
Conjugate functions, 210
Continuous bilinear operators, 9, 77
Continuous partition of unity, 357
Convolution operator, 2, 13
Convolution product, of distributions, 163
   of functions, 128

Debreu-Gale-Nikaido theorem, 377
Demand function, 94, 378
Density criterion, 29, 59
Derivative, of distributions, 148, 277
Derivative from the right, 214
Differential operational equations, 342
Dirichlet problem, abstract, 298
   for elliptic operators, 308, 325
   parabolic, 347
Distributions, 147, 186
Domain, of a formal adjoint, 284
   of a formal operator, 292
   of $f: U \to]-\infty, +\infty]$, 210
   maximal, 105
   minimal, 104
   tensor product by a, 271
   of an unbounded operator, 106, 272, 303
Dual, closed subspaces, 62
   dense subspaces, 63
   of finite products, 61
   of $H^m(\Omega)$, 160
   of $H_0^m(\Omega)$, 146, 186
   of $\hat{H}^s(\mathbf{R}^n)$, 126
   of $L^2(\Omega, a)$, 124
   of a minimal domain, 104

421

quotient spaces, 62
  realization of the, 55
  of a space of Hilbert-Schmidt operators, 263
  of a tensor product, 267
Dual base, 54
  norm, 10
  scalar product, 52
Duality operator, 52, 151, 161
  pairing, 55
Duality theorem, in convex optimization, 35

Elliptic operators, 66, 67, 108, 242, 301, 321, 339
Embedding, 60
Euler-Lagrange equations, 225, 234
Extension of operators, by density, 12, 14
  by duality, 317
  by linearity, 28
  by zero, 152, 207
Extension operator, 158, 206, 207
External representation of a system, 350

Farkas lemma, 61
Fenchel theorem, 227
Formal adjoint, 283
Formal operator, 292
Fourier series, 172
Fourier transform, 188
Fredholm alternative, 242
Fubini theorem, 123

Gateaux derivative, 214
Gradient, 214
Gram-Schmidt orthonormalization process, 23
Granas, 362
Green formula, 284, 292

Hamiltonian, 228, 323, 326
Hamiltonian equations, 229, 324
Hankel operator, 351
Hausdorff completion, 110
Hermite polynomials, 142, 171
Hilbertian polar cone, 17
Hilbert-Schmidt operators, 257
  system, 348
Hilbert space, 4

orthogonal complement, 17
  sum, 111
  tensor product, 266
Hille-Philips theorem, 338

Indicator, 212
Infinitesimal generator, 335
Interface problem, 308
Internal representation of a system, 351
Interpolation spaces, 114, 195, 244
Inversion theorem, 192
Inward correspondences, 365

Kakutani theorem, 365
Kernel theorem, 277, 279
Ky Fan inequality, 357

Lagrange multipliers, 224, 322
Lagrangian, 36, 230
Laguerre polynomials, 170
Lax-Milgram theorem, 66
Least squares, 88
Lebesgue theorem, 123
Left inverse, 86
Legendre polynomials, 169
Leray-Schauder theorem, 362
Lions-Stampacchia theorem, 67
List of vector spaces of functions and distributions:
  $\mathscr{C}_0\,(\Omega)$ = space of continuous functions with compact support in Ω, 121
  $\mathscr{D}\,(\Omega)$ = space of infinitely differentiable functions with compact support in Ω, 134, 145
  $\mathscr{S}\,(\mathbf{R}^n)$ = space of infinitely differentiable rapidly decreasing functions, 190
  $l^2\,(\mathbb{N})$ = space of square-summable sequences, 7
  $l^2\,(\mathbb{N}, F)$ = vector sequences, 274
  $L^p\,(\Omega)$ = Lebesgue spaces of classes of p, integrable functions, 122
  $L^2\,(\Omega, F)$ = Lebesgue space of square integrable vector functions, 275
  $L^2\,(\Omega, a)$ = space of square integrable functions for the weight $a(\cdot)$, 124
  $\hat{H}^s\,(\mathbf{R}^n)$ = Fourier transform of Sobolev spaces, 126
  $H_0^m\,(\Omega)$ = minimal Sobolev space, 146, 186
  $H^m\,(\Omega)$ = maximal Sobolev space, 153, 187

$H^m(\Omega, F)$ = Sobolev space of vector functions, 278
$H^{-m}(\Omega)$ = Sobolev space of distributions, 146, 186
$H^s(\Gamma)$ = Sobolev spaces on the boundary, 206
$H^1(\Omega, \text{div})$ = Domain of div., 288

Mean ergodic theorem, 79
Minimal domain, 104
Minisup theorem, 40

Nash or noncooperative equilibrium, 69, 376
Neumann problem, abstract, 297
   for elliptic operators, 308
   for parabolic operators, 347
Normal cone, 218, 358, 367-370

Orthogonal base, 22
   left inverse, 87
   polynomial, 167
   projector, 20
   right inverse, 89
Oscillation, of a function, 133, 175
Outward correspondence, 365

Pareto minimum, 46
Parseval-Plancherel theorem, 192
Perron-Frobenius theorem, 379
Perturbation by compact operators, 235, 251-254
Pivot space, 57
Poisson formula, 193
Pontryagin maximum principle, 234
Projector of best approximation, 16
Pseudo inverse, 91

Quadratic programming, 92
Quasivariational inequalities, 376

Quotient space, 21, 62, 110

Realization of the dual, 55
Reproducing kernel, 115
Riesz-Fredholm theorem, 236
Riesz theorem, 7
Right inverse, 88

Saddle point, 41
Scalar product, 4, 5, 6
   dual, 52
   final, 108
   initial, 99
Semigroup of operators, 333
Semiscalar product, 3
Separation theorems, 30, 31, 32
Sobolev inequalities, 195
Spline functions, 177
Sturm-Liouville problem, 305
Subdifferential, 216
Support function, 32
System, input-output, 348

Tangent cone, 358, 367-370
Tensor product, of continuous linear operators, 270
   of Hilbert spaces, 266
Trace theorem, 162, 207
Transpose, 58
Transversality condition, 324, 328

Uniform boundedness theorem, 73
Unilateral boundary value problems, 319
Upper semicontinuous correspondences, 356

Variational inequalities, 67, 370
von Neumann mean ergodic theorem, 80
von Neumann minimax theorem, 41

Walras equilibrium, 95, 377

APPLIED ABSTRACT ANALYSIS

CONTENTS

Chapter 1 Metric Spaces: Definitions and Examples

    Introduction

    1. Preliminaries: The Field of Real Numbers
    2. Metric Spaces
    3. The Extended Real Numbers $\bar{\mathbb{R}} = [-\infty, +\infty]$
    4. Fields with an Absolute Value
    5. Banach Spaces
    6. The Normed Space $\mathbb{R}^n$
    7. The Spaces l^1 and l^∞
    8. The Space $\mathscr{U}(X)$ of Bounded Functions
    9. The Space $\mathscr{C}_\infty(X)$ of Continuous Bounded Functions
    10. The Space $\mathscr{L}(E, F)$ of Continuous Linear Mappings
    11. Hilbert Spaces
    12. The Hölder and the Minkowski Inequalities
    13. Fréchet Spaces

Chapter 2 Topological Properties of Metric Spaces

    Introduction

    1. Balls and Diameters
    2. Closure and Closed Sets
    3. Interior and Open Sets
    4. Neighborhoods
    5. Cluster Points of a Sequence
    6. Compact Sets
    7. Convex Sets

Chapter 3 Continuous Functions

    Introduction

    1. Continuous and Uniformly Continuous Functions
    2. Examples of Continuous and Uniformly Continuous Functions
    3. Linear and Multilinear Mappings